INTRODUCTION TO OPERATIONS RESEARCH

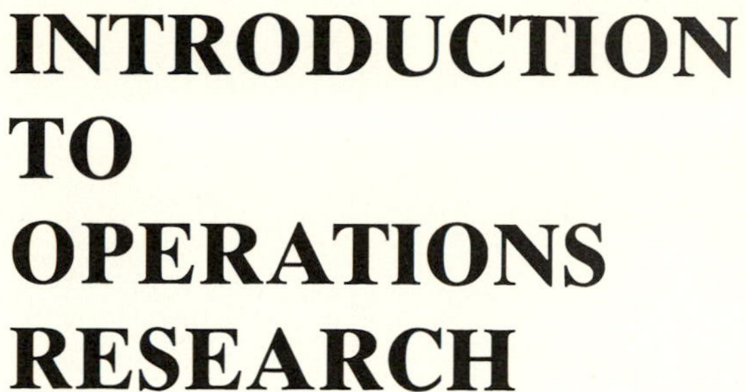

INTRODUCTION TO OPERATIONS RESEARCH

JOSEPH G. ECKER

MICHAEL KUPFERSCHMID

Rensselaer Polytechnic Institute

JOHN WILEY & SONS

NEW YORK CHICHESTER
BRISBANE TORONTO SINGAPORE

Library of Congress Cataloging in Publication Data
Ecker, Joseph G.
Introduction to operations research.
Includes bibliographies and indexes.
1. Operations research. I. Kupferschmid, Michael.
II. Title.
T57.6.E33 1988 658.4'03'4 87-23147
ISBN 0-471-88445-6

Printed in the United States of America

10 9 8 7 6 5 4 3 2 1

To Juanita and to
Kelly and Steve

To Gail

ABOUT THE AUTHORS

JOSEPH G. ECKER is Professor and Chairman of the Department of Mathematical Sciences at Rensselaer Polytechnic Institute in Troy, New York. All of his degrees were received from the University of Michigan—a B.A. in mathematics in 1964, an M.S. in mathematics in 1966, and a Ph.D. in mathematics in 1968. Dr. Ecker has been a faculty member at RPI since 1968 and teaches a variety of courses in operations research and mathematics. He is also an affiliated faculty member in RPI's Department of Decision Sciences and Engineering Systems. In 1975 he was awarded a Fulbright–Hays Research Award and a NATO Postdoctoral Fellowship in Science, and spent the year as a visiting professor in the Center for Operations Research and Econometrics at the Catholic University of Louvain in Belgium. In 1983 he was a visiting professor in the mathematics department of the École Polytechnique Fédérale de Lausanne in Switzerland. His research interests are in operations research and include the theory and applications of linear and nonlinear programming, multiple objective programming, geometric programming, and algorithm development. He is the author or coauthor of more than 45 research publications and is a consultant for several major corporations. Dr. Ecker has been an associate editor of *Operations Research* and is currently an editor of *SIAM Review*.

MICHAEL KUPFERSCHMID is a computing consultant in the Information Technology Services Department at Rensselaer Polytechnic Institute and an adjunct faculty member in RPI's Department of Decision Sciences and Engineering Systems. He received a B.S. in electrical engineering from RPI in 1968 and worked for the next three years at Sikorsky Aircraft designing and flight testing helicopter autopilots. After returning to RPI for a Master of Engineering degree, which was awarded in 1972, he studied theatre engineering at the Yale School of Drama and worked for six more years in industry as a control systems engineer. In 1978 he resigned his position as design supervisor for the controls division of the J. R. Clancy Company and returned to RPI for graduate work in operations research and statistics, leading to a M.S. in 1980 and a Ph.D. in 1981. Dr. Kupferschmid has taught courses in operations research and in computing at RPI and is an author of seven research papers. His research interests are in the experimental evaluation of algorithm performance, the development of nonlinear optimization methods, and the applications of mathematical programming. Dr. Kupferschmid is a registered Professional Engineer.

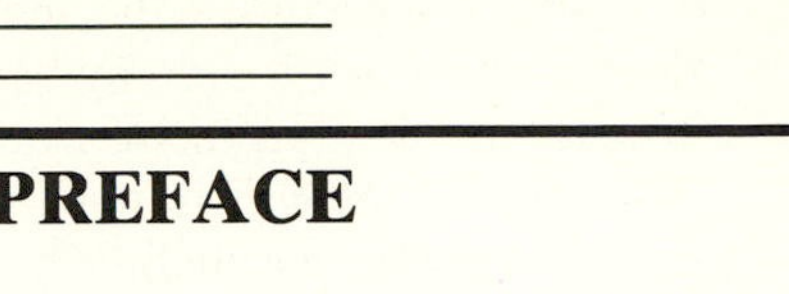

PREFACE

This textbook is intended for use in a two-semester sequence of courses introducing the mathematical methods of operations research. Part I can also be used alone for a one-semester course on linear programming. We have chosen to provide deep and thorough coverage of the most important methods in operations research rather than a superficial treatment of a larger number of topics. The level of exposition is appropriate for juniors and seniors who are majoring in engineering, computer science, mathematics, and quantitative methods in management.

The basic techniques of operations research are simple and straightforward, and only a small amount of advanced mathematics is needed for a technically accurate introduction to the subject. This textbook assumes a knowledge of high school elementary algebra and a familiarity with simple matrix notation such as would be introduced in the first class of an undergraduate linear algebra course. In addition, Chapters 9 and 10 assume a knowledge of elementary differential calculus, and Part III assumes a knowledge of elementary probability and statistics. A concise appendix on matrix notation makes the book accessible to students who have not previously had any linear algebra.

Both the style of exposition and the mathematical notation have been chosen to reflect the simplicity of the subject, and the readability of the text is considered more important than rigor. Examples are used extensively to introduce and motivate the topics. In this way, the presentation reflects the inductive process of scientific discovery rather than imitating the retrospective deduction that is typical in research papers. This approach gives the student the opportunity to rediscover the important results personally instead of merely reading about them in theorems. Proofs are given for some results, but only after the result has been illustrated by example, and only when the proof provides a constructive method for solving problems. Thus the book is not a treatise on mathematical theory.

The simplicity of the methods used in the book means that they can be deeply understood even by beginning students of the subject, and the treatment is mathematically precise even though the results are often stated informally. Thus the discussion is not a cookbook tabulation of trite formulae, and the student should reasonably be expected to understand the mathematical basis for the techniques in addition to being able to apply them.

Answers to selected exercises are given at the end of the book, and a separate answer book is available that contains complete solutions to all of the exercises.

The development of this text benefited greatly from the comments and suggestions made by our colleagues during its use in preliminary form over the past four years. In particular, we express our gratitude to Professors Carlton E. Lemke (Rensselaer Polytechnic Institute), Richard T. Wong (Purdue University), and Thomas M. Liebling (École Polytechnique Fédérale de Lausanne). We also thank the many RPI students who used the preliminary versions, proofread the text, and tested out the exercises. Special thanks are due to Richard Sych, Lori Grieb, Laura Ripans, Carla Bryan, and Robert Bosch.

Joseph G. Ecker
Michael Kupferschmid

CONTENTS

CHAPTER 1

INTRODUCTION

During World War II, the U.S. Army Air Corps suffered many casualties in the course of flying strategic bombing missions over Nazi Germany. An anecdote about that era concerns a study conducted to determine how to reduce those horrible losses. Army B-17 aircraft were examined for holes made by bullets and flak during bombing missions, and the location of each hole was marked on an outline drawing of a B-17. When all of the holes had been marked, it was clear that some parts of a B-17 were much more likely to suffer battle damage than others. The general in charge of the study convened a briefing of military planners to propose that the most heavily damaged areas be provided with additional armor plating. Finding the best places to put the extra armor was very important because only a small amount of weight could be added to the airplanes. At the conclusion of the briefing, after many of those in attendance had agreed to the soundness of the plan, a junior officer in the back of the room timidly raised his hand and was recognized to speak. Clearing his throat nervously, the lieutenant asked in a small voice if it might not be better to armor the parts of the airplane showing the fewest holes rather than the parts with the most severe damage. "After all," he pointed out, "the airplanes that you measured the holes in are the ones that came back."

1.1 OPERATIONS RESEARCH

The preceding story is probably apocryphal, but the question of where to put the armor has many features that are typical of problems in **operations**

research. The most obvious thing about the question, particularly in light of the observation at the end of the story, is that the problem seems rather messy and ill-posed. The statistics about the holes in the airplanes ought to be useful somehow, but it is not precisely clear how they should be used or how to get other information that might help. The hardest thing about the problem seems to be constructing a useful conceptual **model** which embodies the significant information and can be used confidently to reason about the situation. Second, the emphasis is on making a **decision** about where to put the extra armor, and what is sought is scientific assistance in making that decision. Third, in addition to airplanes, people and organizations are involved, so where to put the armor is not an abstract theoretical question. If such a study was actually undertaken, the decision about where to put the armor must have been of paramount importance to the aircrews flying the B-17s and to the military commanders who were trying to control casualties and win the war.

1.2 NATURE AND SCOPE OF OPERATIONS RESEARCH

The term **operations research** means different things to different people, and it is therefore difficult to give a concise definition that everyone would applaud. However, a few of its distinguishing characteristics are so conspicuous that many practitioners would probably find the following definition satisfactory, at least as far as it goes.

> Operations research is the use of quantitative models to analyze and predict the behavior of systems that are influenced by human decisions.

Operations research employs numerical data obtained by measuring and counting, rather than qualitative descriptions of reality, and proceeds by systematic reasoning and the objective analysis of data. It works with simplified, approximate representations of complex systems, constructed by deliberately ignoring details that are thought to be unimportant. Although these attributes give operations research some similarity to the physical sciences, it differs from them in its objects of study; rather than being governed entirely by laws of nature, the systems that are usually studied using operations research are also governed, at least partly, by rules, organizational structures, and other manifestations of human intention and design. The object of most operations research studies is to provide a rational basis for decisions about what organizations should do.

In the anecdote, holes in the airplanes (or rather, in the airplanes that came back) were counted, their positions were measured, and the locations of the holes were marked on a picture of a B-17. The problem of battle damage to bombers was analyzed by using this quantitative description rather than, say, eyewitness accounts of incidents in which flak was encountered. The general's analysis ignored the details of what happened to the bullets and flak pellets after they made the holes, including (unintentionally) the effects of the projectiles that made holes in the airplanes that never returned. The addition of armor was limited by the carrying capacity of the airplanes, but where to put the armor was under human control. Whether to send the B-17s over Germany at all was, of course, also ultimately under human control.

Operations research differs somewhat in philosophical outlook from most physical sciences because of the nature of the systems that it studies. The analytical approach to problem solving is typically **reductionistic** in that it approaches complicated problems by decomposing them into simpler parts each of which can be understood more easily in isolation. Unfortunately, the behavior of organizations and other complex systems is often determined more by the interconnections between parts than it is by the inner mechanisms of the parts themselves, and in many cases the phenomena of interest in an operations research study would be rendered invisible by conceptually dissecting the system. The challenge to the operations research analyst is thus to construct models that give a simplified description of **mechanisms,** but which are also **holistic** in that they do not assume away important aspects of system **structure.** Operations research is often described as a kind of **systems analysis** because of its preoccupation with the connections rather than with the parts.

1.3 HISTORY AND DEVELOPMENT OF OPERATIONS RESEARCH

Operations research uses results from many disciplines, including mathematics, probability and statistics, and economics, so its origins can be traced far back into the history of science. However, it did not really come into existence as a recognizable discipline until World War II, when scientific methods were first successfully used for the analysis of military operations. During the war British and American planners used quantitative methods in allocating spare parts and other scarce materiel, in searching for enemy submarines, in coordinating aircraft detection by multiple radars, and in other tactical problems. The widespread use of mathematical techniques for solving problems of this sort was new and experimental at that time, so the work was naturally viewed as research into ways of conducting military operations, as distinguished from the straightforward application of traditional methods. The analytical approach proved to be of such great value that it soon lost its experimental status and became a permanent part of routine military planning, but the name operations research persisted. Operations research still plays an important role in both strategic and tactical defense studies, but today the majority of its applications are in business, government, industry, engineering, and science.

Many improvements in the methods of operations research were made soon after the war, partly because of the continuing interest of scientists who had worked on the military applications. These improvements in technique, together with the success of operations research during the war, the increased competitive pressures on industry during the postwar period, the growing size and complexity of business organizations and government, and the increasing availability of electronic computers, led to the gradual adoption of operations research as a tool for planning and decision making in all kinds of nonmilitary applications.

Today most large industrial firms, financial institutions, and government agencies have permanent operations research staffs, and many smaller organizations employ outside consultants to perform operations research analyses. The findings of operations research groups have a significant and ever-increasing effect on project management, production scheduling, product distribution, the evaluation of alternative investment opportunities, and other critical activities in which executive decisions can benefit from the results of quantitative analysis.

The analytical methods of operations research are also commonly used in engineering design work, in many areas of applied mathematics, and in scientific research applications. There is a large and vigorous community of academic and industrial scientists who work full time on improving operations research methods, and many universities offer curricula leading to advanced degrees in the subject. Continuing refinement of methods, particularly in concert with further advances in computing, will undoubtedly result in future growth in the application and importance of operations research.

1.4 OVERVIEW OF OPERATIONS RESEARCH METHODS

As pointed out in Section 1.1, an operations research study begins by constructing a simplified conceptual model of the real system that is of interest and then analyzes the model as though it were the real system. Because of the need to incorporate numerical data and other objective facts, operations research models are almost invariably **mathematical models,** and the analysis employs mathematical methods.

Optimization and **uncertainty** are dominant themes in operations research. Many real problems have the property that it is easy to discover a naive solution but very difficult to find a solution that is in some sense optimal, or the best that can be achieved. For example, in deciding how often to replenish a factory's inventory of parts, it might be easy to write down a policy that usually prevents running out of parts. It is always much more difficult to specify a policy that makes such stockouts unlikely at the lowest possible cost, and because of random variations in the rates at which the various parts are used, it may never be possible to specify a policy that guarantees no stockouts will ever occur.

The need to make optimal choices in uncertain situations determines the two main categories of mathematical methods used in operations research:

- Optimization
  - Linear programming
  - Integer programming
  - Nonlinear programming
  - Dynamic programming
- Applied probability
  - Queueing theory
  - Inventory models
  - Discrete-event simulation

Section 1.6 describes how the presentation of these mathematical methods is organized in this book. The optimization methods considered here are known collectively as **mathematical programming** methods (in this use the word *programming* does not necessarily have anything to do with computers). The topics in applied probability listed above are all used in predicting the outcome of a sequence of uncertain events.

1.5 PERSPECTIVE OF THIS TEXT

As may have been suggested by the introductory anecdote, real operations research problems do not come neatly packaged and ready for mathematical analysis. Human affairs are dominated not by technical considerations but by culture, personalities, and politics. The situations in which it is potentially of most value to apply operations research are therefore precisely those that are most clouded by factors that are difficult to quantify and measure, and in which the mechanisms that really matter are most complicated and obscure. Building a trustworthy analytical model is the most important part of any operations research study, but it is also the most difficult except in simple textbook examples.

This book discusses some simple examples (in Chapters 2 and 8, and in many of the exercises) in an attempt to suggest the flavor of the model-building process. Mostly, however, it is about mathematical methods and their use in analyzing operations research models after the models have already been con-

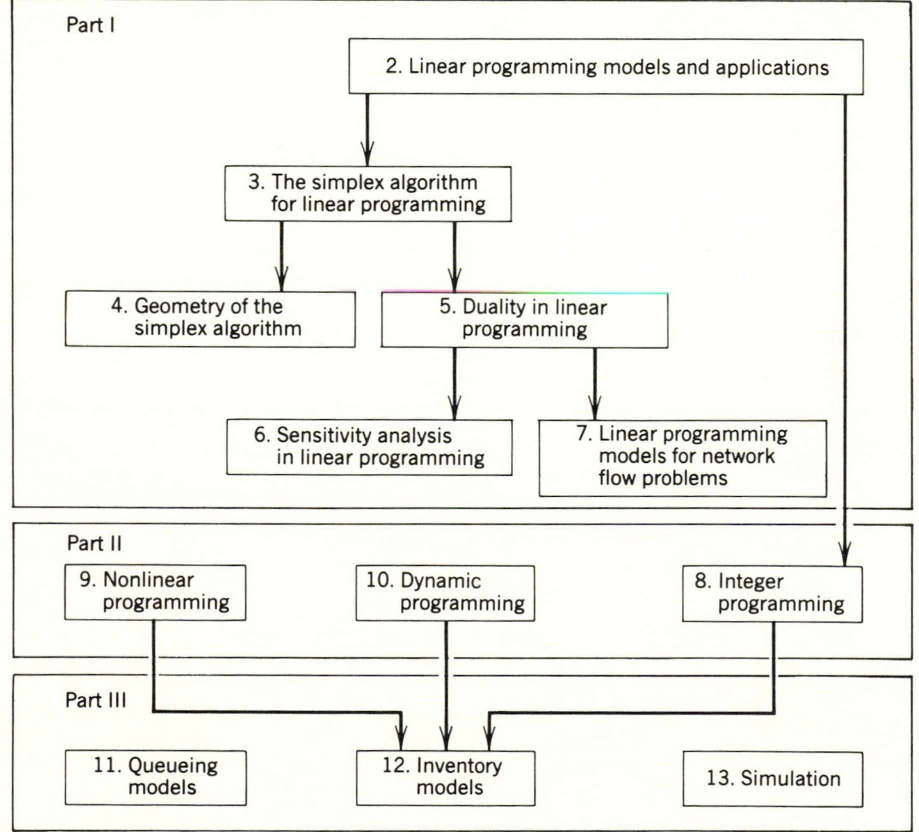

FIGURE 1.1 Organization of the book.

structed. In particular, this book makes no pretense of addressing the psychology and political science that are involved in conducting a real-life operations research study and in implementing policy decisions based on the results. The best way (and very probably the only way) to learn about those things is by apprenticeship and experience, and the best preparation we can hope to provide in a textbook is facility with the analytical methods themselves.

1.6 ORGANIZATION OF THIS TEXT

Figure 1.1 shows how coverage of the methods outlined in Section 1.4 is divided into the parts and chapters of this book. The arrows indicate logical precedence; thus, for example, the exposition of Chapter 5 assumes that the reader already has an understanding of the material covered in Chapters 3 and 2. There are some other connections that are not shown in the diagram, such as references in the exercises of one chapter to those of another, but those connections may be ignored without loss of continuity.

SELECTED REFERENCES

Ackoff, R. L., and Rivett, P., *A Manager's Guide to Operations Research,* Wiley, New York, 1963.

Churchman, C. W., Ackoff, R. L., and Arnoff, E. L., *Operations Research* (Part I), Wiley, New York, 1957.

Churchman, C. W., *The Systems Approach,* Dell Publishing, New York, 1968.

Lanchester, F. W., Mathematics in Warfare, and other articles in Part 4 of J. R. Newman, *The World of Mathematics,* Simon and Schuster, New York, 1956, pp. 2136–2179.

Miser, H. J., Operations Research and Systems Analysis, *Science* **209:**139–146 (1980).

Saaty, T. L., Operations Research: Some Contributions to Mathematics, *Science* **178:**1061–1070 (1972).

EXERCISES

1.1 (This is the only essay question in the book.) Reread Section 1.1 and consider the armor location problem in more detail.

(a) Describe as completely as you can the system that is under study in this problem. What are the parts, and how are they connected?

(b) Describe the conceptual model employed by the general in thinking about the problem. In what respects does this model neglect important interactions between the parts of the system?

(c) Describe the lieutenant's conceptual model, and contrast it to the general's. How could the quantitative description of the casualty problem be changed to reflect the lieutenant's interest in the airplanes that were shot down? What additional information would be useful in improving the analysis, and how would you use it?

(d) Speculate about the political and psychological factors that might affect the conduct of such a project. Should Air Corps officers or enlisted people collect the

necessary data, or should that be done by civilian employees? Should the final decision about where to put the extra armor be made by the scientific advisors, or by the senior officers of the bomber command? How would aircrew and airplane maintenance personnel, including those who were injured by flak on earlier missions, react to the data collection process, and to the installation of the extra armor?
(e) Should anything further be done after the extra armor is installed? What if it is found that the extra armor does not reduce casualties after all?

PART ONE

LINEAR PROGRAMMING

CHAPTER 2

LINEAR PROGRAMMING MODELS AND APPLICATIONS

Many practical problems in commerce and industry involve finding the best way to allocate scarce resources among competing activities. During and just after World War II, a technique called **linear programming** was developed for mathematically solving certain kinds of resource allocation problems. Since then linear programming has found many applications, and it has become one of the most important tools of operations research. This chapter introduces linear programming and uses several examples to illustrate the process of formulating problems for solution by linear programming.

2.1 FORMULATING LINEAR PROGRAMMING MODELS

We begin by considering a simple example.

An Introductory Resource Allocation Problem

The Oakwood Furniture Company has 12.5 units of wood on hand from which to manufacture tables and chairs. Making a table uses two units of wood and making a chair uses one unit. Oakwood's distributor will pay $20 for each table and $15 for each chair, but he will not accept more than eight chairs and he wants at least twice as many chairs as tables. How many tables and chairs should the company produce so as to maximize its revenue? We will formulate this question as a linear programming problem.

Let x_1 be the number of tables produced and x_2 be the number of chairs. The variables x_1 and x_2 are called **decision variables,** and a plan to manufacture x_1 tables and x_2 chairs is called a **production program.** The income z that could be obtained from selling x_1 tables and x_2 chairs would be $z = 20x_1 + 15x_2$, but the distributor's requirements and the available stock of wood impose certain restrictions on the numbers of tables and chairs that can be sold. The limit of eight chairs implies that $x_2 \leq 8$, and the need for at least two chairs with each table implies that $x_2 \geq 2x_1$. Also, x_1 and x_2 must be chosen so that $2x_1 + x_2 \leq 12.5$ in order to keep from using more than the available supply of wood. Finally, it wouldn't make much sense to consider negative numbers of tables or chairs, so $x_1 \geq 0$ and $x_2 \geq 0$. These requirements on x_1 and x_2 are called **constraints,** and a production program that simultaneously satisfies all of the constraints is said to be **feasible.**

Production programs (x_1, x_2) for the Oakwood problem are represented in Figure 2.1. Points representing feasible production programs are called **feasible points,** and the set of all feasible points is called the **feasible set** or **feasible region.** The feasible set is marked X in Figure 2.1, and all of the points in X satisfy the constraints given above. For example, picking $x_1 = 2$ and $x_2 = 7$ would satisfy all of the requirements and would yield a revenue of $z = \$145$.

Many other combinations (x_1, x_2) would also yield a revenue of $z = \$145$, of course; namely, all of the combinations for which $20x_1 + 15x_2 = 145$. Other combinations of x_1 and x_2 can be found that also satisfy the constraints, and that yield larger revenues z, and it will obviously be the objective of Oakwood to find the feasible combination (x_1, x_2) that makes z as big as possible. The function $z = 20x_1 + 15x_2$ is called the **objective function** for the problem.

The line corresponding to the equation $z = 20x_1 + 15x_2 = 145$ is called a **contour** of the objective function, and it is shown in Figure 2.2, passing through the point (2, 7). Several other objective function contours, having other values of z, are also shown in Figure 2.2.

From Figure 2.2 it is clear that the contours of the objective function are parallel lines, and that the value of the objective function z increases as the contour lines move away from the origin of the graph. In order for a production program (x_1, x_2) to be acceptable, it is necessary for it to fall inside the feasible

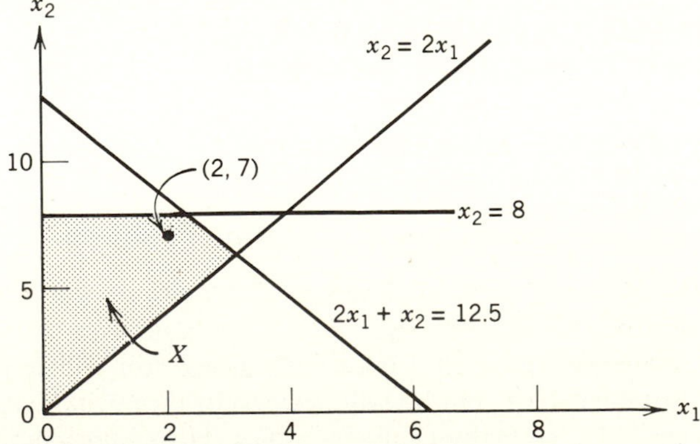

FIGURE 2.1 Feasible Set for the Oakwood Problem.

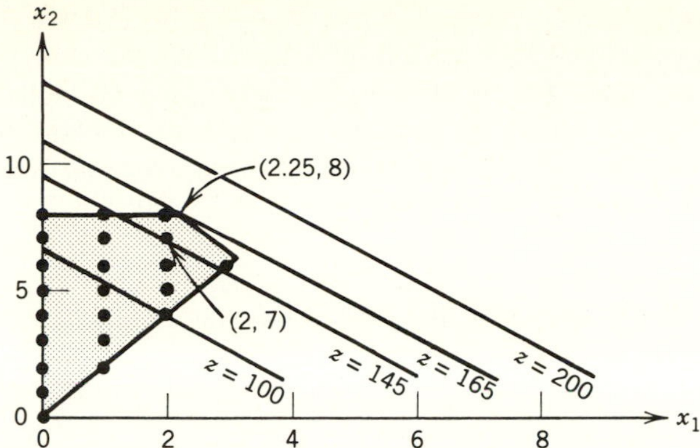

FIGURE 2.2 Objective Function Contours for the Oakwood Problem.

region or be on an edge. Because none of the points on the contour for $z = 200$ are in the feasible set, a revenue of \$200 can never be obtained. In fact, of the objective function contours that intersect the feasible set in at least one point, the one that has the highest value is the line $z = 165$. That line has only one feasible point, namely, $x_1 = 2.25$ and $x_2 = 8$, so the best the Oakwood Company could do would be to use that production program. The set of decision variable values that gives the best objective function value while satisfying all of the constraints is called the **optimal solution** of the linear programming problem, and the corresponding objective function value is called the **optimal value.** The process of finding the maximum or minimum value of an objective function, possibly subject to constraints, is generally referred to as **optimization,** and linear programming is often called **linear optimization.**

Assumption of Continuity

In practice, Oakwood's distributor probably would not be interested in buying a quarter of a table. We shall see later on that if the solution to a linear programming problem turns out to be integers, then it is the best integer solution that can be obtained by any method. If an integer solution is required and the linear programming solution turns out to be noninteger, then another technique called **integer programming** is usually required to find the best integer solution. We will take up integer programming in Chapter 8. Sometimes, as in the simple problem considered above, we can use the graphical method to find the best integer solution. We simply find the objective function contour of highest value passing through a feasible integer point. The feasible integer solutions are marked by a solid circle (●) in Figure 2.2. Oakwood's best integer production program would actually be to make two tables and eight chairs, for a revenue of \$160.

The fact that linear programming can yield noninteger solutions may seem like a defect in the technique, but often finding an integer solution is not required, and even when it is it may not cause serious trouble. In many problems the decision variables measure things such as gallons of oil, and need not have integer values. In other instances the granularity required in the decision variables may be small enough compared to their values so that rounding can safely be used to obtain an optimal or near-optimal integer solution. Also, the units in which the decision variables are measured is important in determining whether or not a noninteger solution is acceptable. For example, if the Oakwood problem had referred to units of chairs and tables rather than individual pieces, then making 2.25 units of tables would probably be acceptable if 1 unit represented 1000 pieces. However, it is important to remember that linear programming always assumes the decision variables can have real values, and to consider whether this assumption is appropriate in the formulation of a problem.

Sensitivity of the Optimal Solution

Often the data in a linear programming problem are not known with as much certainty as one would like. An important question is therefore: How sensitive is the optimal solution to the data? For example, suppose that the distributor in the Oakwood problem is uncertain as to the amount he will be able to get for each chair because of fluctuations in his customers' demand. In particular, suppose that the distributor can pay Oakwood only some amount between \$11 and \$19 for each chair. Will this change the optimal production program? We can answer this question by considering how the objective function contours are affected by the price of chairs.

If chairs bring only \$11 each, the objective function becomes $z_1 = 20x_1 + 11x_2$, whereas if they can be sold for \$19 the objective function becomes $z_2 = 20x_1 + 19x_2$. The contours of these functions passing through the optimal point are shown in Figure 2.3.

It is clear that both of these new objective functions would yield the same optimal program we found before, although the optimal value would change in both cases (to 133 for z_1 and to 197 for z_2). The optimal program also remains unchanged for chair prices between \$11 and \$19. Thus the solution of this linear programming problem is not very sensitive to the price of chairs.

Questions about the sensitivity of linear programming solutions can be answered in ways that are much easier than naively re-solving the problem with new data as we have done here, and the solution can be studied for its sensitivity to problem data other than the coefficients in the objective function. We take up the important matter of sensitivity analysis in Chapter 6.

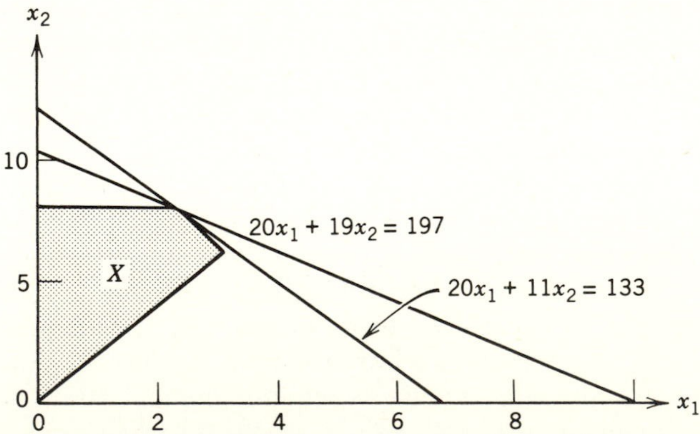

FIGURE 2.3 Objective Function Contours for Varying Chair Prices.

Algebraic Statement of Linear Programming Problems

The geometric interpretation of linear programming problems gives important insights into several aspects of the optimization process, and we will consider it in much greater detail in Chapter 4. However, when the number of decision variables is greater than two or three, it is clumsy or impossible to actually solve a linear programming problem by the graphical method illustrated above. Real linear programming problems commonly contain hundreds or thousands of decision variables and constraints, and for their solution it is necessary to work directly with the algebraic statement of the problem. An algebraic statement of the Oakwood problem is

$$\begin{aligned} &\text{maximize } z = 20x_1 + 15x_2 && \text{objective function} \\ &\text{subject to} \end{aligned}$$

$$\left.\begin{aligned} x_2 &\le 8 \\ 2x_1 - \quad x_2 &\le 0 \\ 2x_1 + \quad x_2 &\le 12.5 \\ x_1, x_2 &\ge 0 \end{aligned}\right\} \text{constraints}$$

Such an algebraic formulation is referred to as a **linear program.** The preceding linear program is also commonly referred to as a linear programming **model** of the word problem stated at the beginning of this section because the constraints and objective of the real problem are represented mathematically in the algebraic formulation. Any linear programming problem can be stated algebraically in several different ways, so there are alternative (mathematically equivalent) models for every problem, but every linear program has the following general form:

maximize
or — a linear function of several variables
minimize

subject to — linear inequality constraints
or
linear equality constraints

In the next section we formulate several more example problems as linear programs to illustrate the process and discuss how decision variables are chosen. With this motivation the reader should be ready for the presentation in Chapter 3 of the simplex algorithm, a powerful analytic method for solving linear programs.

2.2 FURTHER LINEAR PROGRAMMING FORMULATION EXAMPLES

The examples in the remainder of this chapter illustrate some applications of linear programming and demonstrate how to formulate models. There are literally thousands of applications, and the reader should consult some of the references at the end of this chapter to see other application areas. So that the formulation process will not be obscured by unnecessary detail, the examples discussed here are necessarily simplified in that they involve only a few variables and constraints. You should keep in mind that in actual practice linear programs typically have many variables and constraints, and then the sheer volume of numerical data often introduces practical complications that we do not consider here. Nevertheless, these simple problems have forms that are typical of real problems, and the formulation of linear programming models for them requires the same reasoning processes used for larger problems.

The best way (maybe the only way) to learn how to formulate linear programming models is to be involved in the process. It is therefore important for the reader to be an active participant in the study of these examples by stopping after the statement of each problem to ponder what the decision variables should be and to attempt a formulation. Only after you have done this should you go on to study the formulation we provide. Every problem seems obvious once its solution is revealed, and the value of the examples will be greatly reduced if you simply read the solutions first. Remember that linear programming problems typically have several alternative formulations that are mathematically equivalent and therefore equally correct.

Brewery Problem

Microbrewers Incorporated makes four products called Light, Dark, Ale, and Premium. These products are made using the resources of water, malt, hops, and yeast. Microbrewers has a free supply of water, so it is the amount of the other resources that restricts production capacity. The **technology table** (Table 2.1) gives the pounds of each resource required in the production of 1 gal of each product, the pounds of each resource available, and the revenue received for 1 gal of each product. The problem faced by Microbrewers Inc. is to decide how much of each product it should make in order to maximize its revenue. What should be the decision variables in this problem? How is the objective of maximizing revenue then stated mathematically? What constraints on the decision variables are imposed by the statement of the problem and the data in the technology table? Try to formulate the problem as a linear program, algebraically, before reading the formulation discussed below.

Suppose we choose the following decision variables:

x_1 = gallons of Light to be produced
x_2 = gallons of Dark to be produced
x_3 = gallons of Ale to be produced
x_4 = gallons of Premium to be produced

TABLE 2.1 Technology Table for Microbrewers, Inc.

	Light	*Dark*	*Ale*	*Premium*	*Available*
Malt	1	1	0	3	50 lb
Hops	2	1	2	1	150 lb
Yeast	1	1	1	4	80 lb
Revenue	$6	$5	$3	$7	

Then the revenue z (in dollars) that results from a given production program is

$$z = 6x_1 + 5x_2 + 3x_3 + 7x_4$$

and it is clearly Microbrewers' objective to maximize this revenue.

Of course, we cannot use more of each resource than is available. For example, from the technology table we find that

x_1 gallons of Light	uses	$1x_1$ pounds of malt
x_2 gallons of Dark	uses	$1x_2$ pounds of malt
x_3 gallons of Ale	uses	$0x_3$ pounds of malt
x_4 gallons of Premium	uses	$3x_4$ pounds of malt

but only 50 lb of malt is available. Thus the decision variables x_1, x_2, x_3, and x_4 must satisfy the constraint

$$1x_1 + 1x_2 + 0x_3 + 3x_4 \leq 50$$

Similarly, we cannot use more hops or yeast than is available, and this gives rise to the two constraints

$$2x_1 + 1x_2 + 2x_3 + 1x_4 \leq 150$$

and

$$1x_1 + 1x_2 + 1x_3 + 4x_4 \leq 80$$

Finally, we note that it would not make sense to plan on brewing negative amounts of any of the products, so we impose the additional **nonnegativity constraints**

$$x_1 \geq 0 \qquad x_2 \geq 0 \qquad x_3 \geq 0 \qquad x_4 \geq 0$$

An algebraic formulation of the Microbrewers problem is thus to choose values for x_1, x_2, x_3, and x_4 so as to

$$\begin{aligned} &\text{maximize } z = 6x_1 + 5x_2 + 3x_3 + 7x_4 && \text{\} objective} \\ &\text{subject to} \\ &\qquad 1x_1 + 1x_2 + 0x_3 + 3x_4 \leq 50 \\ &\qquad 2x_1 + 1x_2 + 2x_3 + 1x_4 \leq 150 && \text{constraints} \\ &\qquad 1x_1 + 1x_2 + 1x_3 + 4x_4 \leq 80 \\ &\qquad x_1, x_2, x_3, x_4 \geq 0 \end{aligned}$$

The brewery problem has a form typical of many resource allocation problems where m resources are used in making n products, namely,

$$\begin{aligned} &\text{maximize } \sum_{j=1}^{n} c_j x_j \\ &\text{subject to} \\ &\qquad \sum_{j=1}^{n} a_{ij} x_j \leq b_i \qquad i = 1, \ldots, m \\ &\qquad x_j \geq 0 \qquad j = 1, \ldots, n \end{aligned}$$

where

$$b_i = \text{units of resource } i \text{ available}$$
$$a_{ij} = \text{units of resource } i \text{ needed for one unit of product } j$$
$$x_j = \text{units of product } j \text{ to be made}$$
$$\textstyle\sum_{j=1}^{n} a_{ij}x_j = \text{units of resource } i \text{ used in making all } n \text{ products}$$
$$c_j = \text{revenue received for one unit of product } j$$

Oil Refinery Problem

An oil refinery can blend three grades of crude oil to produce regular and super gasoline. Two possible blending processes are available. For each production run the older process uses 5 units of crude A, 7 units of crude B, and 2 units of crude C to produce 9 units of regular and 7 units of super gasoline. The newer process uses 3 units of crude A, 9 units of crude B, and 4 units of crude C to produce 5 units of regular and 9 units of super gasoline for each production run.

Because of prior contract commitments, the refinery must produce at least 500 units of regular gasoline and at least 300 units of super for the next month. It has available 1500 units of crude A, 1900 units of crude B, and 1000 units of crude C. For each unit of regular gasoline produced the refinery receives \$6, and for each unit of super it receives \$9. The problem is to determine how to use the resources of crude oil and the two blending processes to meet the contract commitments and, at the same time, to maximize revenue.

Even though we eventually want to know how many units of each gasoline we should produce, those would not be appropriate decision variables for this problem because what we control is the number of production runs of each process. Thus, for decision variables we choose

$$x_1 = \text{number of production runs of the older process}$$

$$x_2 = \text{number of production runs of the newer process}$$

In x_1 runs of the older process and x_2 runs of the newer process, we produce $9x_1 + 5x_2$ units of regular and $7x_1 + 9x_2$ units of super. Our contract commitments thus require that

$$9x_1 + 5x_2 \geq 500$$

$$7x_1 + 9x_2 \geq 300$$

The available quantities of the three grades of crude oil give rise to the following constraints:

$$5x_1 + 3x_2 \leq 1500$$

$$7x_1 + 9x_2 \leq 1900$$

$$2x_1 + 4x_2 \leq 1000$$

Since the objective is to maximize revenue obtained from the units of gasoline produced, our final model becomes

$$
\begin{aligned}
\text{maximize} \quad & 6(9x_1 + 5x_2) + 9(7x_1 + 9x_2) \\
\text{subject to} \quad & \\
& 9x_1 + 5x_2 \geq 500 \\
& 7x_1 + 9x_2 \geq 300 \\
& \\
& 5x_1 + 3x_2 \leq 1500 \\
& 7x_1 + 9x_2 \leq 1900 \\
& 2x_1 + 4x_2 \leq 1000 \\
& \\
& x_1, x_2 \geq 0
\end{aligned}
$$

It might appear that the decision variables here should be constrained to have only integer values, but for blending processes it is usually permissible to have a fractional production run that uses inputs in the same proportion as a complete run to produce proportional units of output. Thus it is reasonable to assume that the variables in this problem need not be integer.

This model is different from the previous ones in that if the quantities of available resources were changed, the resulting linear program could turn out to be **infeasible.** That is, there might not be any production program that satisfied all of the constraints because if we had too little of some variety of crude oil available it might not be possible to meet the contract commitments. The Oakwood problem and the brewery problem clearly have feasible solutions no matter what quantities of resources are available in those models. We shall see, in Chapter 3, how it is easily discovered whether there are any feasible solutions to a linear program. If a real-life problem is sure to have feasible solutions, as is the case of the Oakwood and brewery problems, then one elementary check on the correctness of the linear programming model is that it must not be infeasible.

Warehouse Problem

A large peanut warehouse has the capacity for indefinitely storing 200 tons of peanuts. There are currently 80 tons of peanuts in the warehouse and the forecasting department has come up with what it thinks are going to be the per-ton prices for peanuts at the next five monthly peanut sales. These projected prices are listed in Table 2.2. There is also a storage cost of $5 per ton assessed on peanuts left in the warehouse after each sale. Because the price of peanuts

TABLE 2.2 Prices at the Next Five Peanut Sales

Month	*Price*
1	120
2	100
3	150
4	180
5	130

fluctuates, it is possible to make a profit by "buying low and selling high." For example, the warehouse could buy peanuts in month 2 at $100 per ton, store them until month 4, and then sell them at $180 per ton. The question is: What should the buying and selling program be in order to maximize the total profit made at the next five peanut sales?

This problem is different in structure from the previous ones in a very important way; namely, it is **dynamic** in the sense that what we do in one period affects what we can do in later periods.

As usual, the first step in formulating this problem as a linear program is to decide what the decision variables should be. Because the buying price and selling price at any given sale is the same, one choice would be to let

$$x_t = \text{amount by which inventory changes after sale } t$$

The decision variables x_t would not be constrained to be nonnegative because negative values would simply reflect the fact that we sold some peanuts. However, with this choice of decision variables, the formulation of the problem as a linear program is not as transparent as it is with other choices of variables. It is usually worthwhile to consider different choices of decision variables and see how the formulation changes.

Suppose instead that we approach the problem as follows. At each sale t, our actions determine three quantities:

$$B_t = \text{amount bought}$$
$$S_t = \text{amount sold}$$
$$K_t = \text{amount stored}$$

Clearly, these three quantities are not independent, but in formulating linear programming models it is often convenient to let the constraints reflect how variables depend on one another, rather than trying to find a set of decision variables that are independent. The relationship between these quantities is that, for each sale,

$$(\text{amount on hand}) + (\text{amount bought}) = (\text{amount sold}) + (\text{amount stored})$$

This relationship between the decision variables is easily stated in the following **equality constraints:**

$$80 + B_1 = S_1 + K_1$$

$$K_{t-1} + B_t = S_t + K_t \quad \text{for } t = 2, 3, 4, 5$$

An additional **inequality constraint** (of the sort we have seen in the earlier models) arises from the fact that we cannot store more than 200 tons of peanuts at any given time:

$$K_t \leq 200 \quad \text{for } t = 1, 2, 3, 4, 5$$

The net profit from sale 1 is given by

$$120S_1 - 120B_1 - 5K_1 = 120(S_1 - B_1) - 5K_1$$

and the profits realized at each subsequent sale are similarly just

$$(\text{current price}) \cdot (\text{net amount sold}) - (\text{storage cost})$$

If we assume that we sell all of the inventory at sale 5, then the objective function becomes

$$P = 120(S_1 - B_1) + 100(S_2 - B_2) + 150(S_3 - B_3) + 180(S_4 - B_4) + 130K_4 - 5(K_1 + K_2 + K_3 + K_4)$$

Of course, all of these decision variables are nonnegative, so our final linear programming model is

$$\begin{aligned}
&\text{maximize} \quad P \\
&\text{subject to} \\
&\qquad 80 + B_1 = S_1 + K_1 \\
&\qquad K_{t-1} + B_t = S_t + K_t \quad \text{for } t = 2, 3, 4, 5 \\
&\qquad K_t \leq 200 \quad \text{for } t = 1, 2, 3, 4, 5 \\
&\qquad S_t, B_t, K_t \geq 0 \quad \text{for } t = 1, 2, 3, 4, 5
\end{aligned}$$

At any given peanut sale we might either buy or sell, but it would not be profitable to both buy and sell. For example, we could buy 100 tons and sell 80 tons but, since the buying and selling price are equal, this is the same as only buying 20 tons. An optimal solution of the above model would indicate for each sale whether we should buy or sell. In particular, if $S_t > B_t$ in an optimal solution, then our net action at sale t would be to sell $S_t - B_t$ units of current inventory. If $S_t \leq B_t$, then we end up adding $B_t - S_t$ units to the current inventory.

There is nothing in the linear programming model that forces one of the variables S_t or B_t to be zero. However, if an optimal solution has $S_t > 0$ and $B_t > 0$ for some t, then a new optimal vector can easily be constructed that has either $S_t = 0$ or $B_t = 0$. For example, if $S_2 = 100$ and $B_2 = 60$, then another feasible point with the same (optimal) objective function value can be obtained by simply letting $S_2 = 40$ and $B_2 = 0$ and leaving all other variables unchanged. The new point is feasible (and optimal) because it is only the difference $S_2 - B_2$ that appears in the constraints (and the objective function).

Having formulated the problem using the preceding decision variables, it is interesting to complete the alternative formulation suggested first, using the variables x_t (see Exercise 2.5).

Chicken and the Egg Problem

The owner of a small chicken farm must determine a laying and hatching program for 100 hens. There are currently 100 eggs in the henhouse, and the hens can be used either to hatch existing eggs or to lay new ones. In each 10-day period, a hen can either hatch 4 eggs or lay 12 new eggs. Chicks that are hatched can be sold for 60 cents each, and every 30 days an egg dealer gives 10 cents each for the eggs accumulated to date. Eggs not being hatched in one period can be kept in a special incubator room for hatching in a later period. The problem is to determine how many hens should be hatching and how many should be laying in each of the next three 10-day periods so that total revenue is maximized.

It is natural in this problem to let the decision variables be

H_t = number of hens hatching in period t

L_t = number of hens laying in period t

for $t = 1, 2$, and 3. Since only 100 eggs are candidates to be hatched in the first period, we have the following constraint:

$$4H_1 \leq 100$$

For the second period the eggs available to be hatched come from two sources: those not hatched in the first period and the new eggs laid in the first period. It will be helpful to identify the eggs not being hatched in each period, so we let

$$I_t = \text{eggs put in the incubator room during period } t$$

Using the variable I_1, the preceding constraint can be reformulated as an equality:

$$4H_1 + I_1 = 100$$

For periods $t = 2$ and $t = 3$, the following holds:

$$\begin{array}{c} \text{eggs hatched in period } t \\ + \\ \text{eggs put in the incubator room} \\ \text{in period } t \end{array} = \begin{array}{c} \text{eggs laid in period } t - 1 \\ + \\ \text{eggs put in the incubator} \\ \text{room in period } t - 1. \end{array}$$

This requirement gives rise to two constraints:

$$4H_2 + I_2 = 12L_1 + I_1$$

$$4H_3 + I_3 = 12L_2 + I_2$$

Also, since only 100 hens are available for hatching and laying in each period, we have the constraints

$$H_t + L_t = 100 \quad \text{for } t = 1, 2, 3$$

Actually, we could formulate these last three constraints as

$$H_t + L_t \leq 100 \quad \text{for } t = 1, 2, 3$$

but any optimal program will surely not have hens sitting idle because it is always more profitable to at least have them laying eggs. Thus, we will use the equality constraints in our model. (Of course, it is sometimes difficult to determine whether an inequality constraint will be satisfied as an equality at the optimal solution, and in that case the constraint should be left as an inequality.)

The objective function to be maximized is the revenue obtained from selling chicks and the eggs ($I_3 + 12L_3$ of them) that have accumulated by the end of period 3. Thus our final linear programming model becomes

$$\begin{array}{ll} \text{maximize} & 60(4H_1 + 4H_2 + 4H_3) + 10(I_3 + 12L_3) \\ \text{subject to} & \\ & 4H_1 + I_1 = 100 \\ & 4H_t + I_t = 12L_{t-1} + I_{t-1} \quad \text{for } t = 2, 3 \\ & H_t + L_t = 100 \qquad\qquad \text{for } t = 1, 2, 3 \\ & H_t, L_t, I_t \geq 0 \qquad\qquad \text{for } t = 1, 2, 3 \end{array}$$

Because this problem deals with whole hens and eggs, one must question the appropriateness of the linear programming model and its underlying assumption that variables take on continuous values. What the chicken farmer really needs is the best integer production program, and there is no reason why an optimal solution to the linear program above will necessarily have that property. However, the numbers of hens and eggs are large enough so that in this case rounding off the optimal linear programming solution to nearest integers will probably yield an integer solution that is both feasible and not too far from optimal.

Nurse Scheduling Problem

A hospital administrator is in charge of scheduling nurses to work each of the six shifts that start every 4 hr. The first shift starts at 8:00 in the morning and each shift lasts 8 hr. The number of nurses required during the day is given in Table 2.3. The problem is to schedule the nurses to meet the requirements and to do so with the minimum number of nurses.

The main difficulty in formulating a linear programming model for this problem is in deciding what to choose as decision variables. One might at first think that the decision variables should be the numbers of nurses working in each of the six periods of the day. However, if we use these variables, then we cannot enforce the restriction that nurses work shifts of eight consecutive hours.

If we focus on how the scheduling actually works, other choices of decision variables come to mind. In particular, what we have control of is how many nurses start working at the beginning of each shift. Thus, suppose we let

$$x_t = \text{number of nurses starting to work on shift } t$$

With these decision variables, constraints arise from the fact that for each of the above periods of the day

$$\begin{bmatrix}\text{number of nurses}\\ \text{starting at the beginning}\\ \text{of the period}\end{bmatrix} + \begin{bmatrix}\text{number of nurses}\\ \text{who started at the}\\ \text{beginning of the}\\ \text{previous period}\end{bmatrix} \geq r_t$$

where r_t is the required number of nurses for period t. Thus the linear programming model for this problem is now easily given as

$$\begin{array}{ll}
\text{minimize} & \displaystyle\sum_{t=1}^{6} x_t \\
\text{subject to} & \\
& x_6 + x_1 \geq 140 \\
& x_1 + x_2 \geq 120 \\
& x_2 + x_3 \geq 160 \\
& x_3 + x_4 \geq 90 \\
& x_4 + x_5 \geq 30 \\
& x_5 + x_6 \geq 60 \\
\text{and} & x_t \geq 0 \qquad t = 1, \ldots, 6
\end{array}$$

TABLE 2.3 Nursing Staff Requirements

Period of the Day	*Number of Nurses Required*
8:00–12:00	140
12:00–16:00	120
16:00–20:00	160
20:00–24:00	90
24:00– 4:00	30
4:00– 8:00	60

2.3 SOME SCIENTIFIC APPLICATIONS OF LINEAR PROGRAMMING

The examples of Section 2.2 are simple versions of practical problems that arise in the work of business people, managers of industry, military planners, and other executives whose main concern is with routine operational matters. In addition to helping with the solution of problems from that sphere of activity, linear programming also finds numerous more abstract applications in engineering design and scientific research. This section presents some examples from those areas of application.

Curve Fitting

Values R_i are obtained from laboratory measurements of a certain physical quantity $R(t)$ at times t_i, $i = 1, \ldots, N$. Random experimental errors introduce noise in the measurements, but it is known theoretically that $R(t)$ depends on time t according to a quadratic relationship of the form

$$R(t) = at^2 + bt + c$$

The **parameters** a, b, and c are unknown and are to be estimated from the experimental measurements R_i. This kind of curve-fitting problem is often referred to as a **polynomial regression.** One approach to the problem is to find parameter values a, b, and c so as to minimize the absolute value of the largest discrepancy between the measured values R_i and corresponding theoretical values $R(t_i)$. For a given set of parameter values a, b, and c, the **deviation** of observation i from the value predicted by the **model function** $R(t)$ is defined as

$$D_i = R_i - [a(t_i)^2 + b(t_i) + c]$$

Using this definition, one way to formally state the mathematical problem of finding the best parameter values by minimizing the largest deviation is

$$\underset{a,b,c}{\text{minimize}} \; [\underset{i}{\text{maximum}} \; |D_i|]$$

The list of parameters beneath the word *minimize* means that the minimization is to be performed by varying the values of those parameters; similarly, the maximum is to be taken over the **absolute deviations** $|D_i|$, with a, b, and c held constant. Because we are minimizing the maximum absolute deviation, such a formulation is commonly called a **minimax** problem.

Finding values for the parameters a, b, and c to minimize the largest absolute deviation might not at first seem like a linear programming problem. Also, deciding on what decision variables to use does not appear to be as important an issue as it always was in the problems of Section 2.2. After all, the only things we can control are the values of the parameters, so they are natural choices as decision variables. There is, however, a dependent variable in this problem, namely, the largest deviation, which depends on the values of the parameters. Suppose we let

$$w = \underset{i}{\text{maximum}} \{|D_i|\}$$

and try to write a linear programming formulation involving that quantity.

Because w is the largest absolute deviation, it must satisfy each of the inequalities

$$w \geq |R_i - [a(t_i)^2 + b(t_i) + c]| \qquad i = 1, \ldots, N$$

Of course, for any given set of parameter values, one (or more) of the deviations D_i will be larger than the others, and so the corresponding inequality will actually hold with equality. The problem, then, is to find values for a, b, c, and w so that each of the inequalities is satisfied and so that w is as small as possible. In other words, we need to

$$\begin{aligned} &\text{minimize } w \\ &\text{subject to} \\ &\qquad w \geq |R_i - [a(t_i)^2 + b(t_i) + c]| \qquad i = 1, \ldots, N \end{aligned}$$

Unfortunately, this problem is not a linear program as it stands because the absolute value of a linear function is not a linear function. However, we can convert it into a linear program by making use of the following elementary fact:

> For any w and y,
> $w \geq |y|$ if and only if $w \geq y$ and $w \geq -y$.

Using this fact, we can replace each of the nonlinear inequality constraints above with two linear inequality constraints. Then the preceding optimization problem is equivalent to the following linear program in the variables w, a, b, and c:

$$\begin{aligned} &\text{minimize } w \\ &\text{subject to} \\ &\qquad w \geq (R_i - [a(t_i)^2 + b(t_i) + c]) \qquad i = 1, \ldots, N \\ &\qquad w \geq -(R_i - [a(t_i)^2 + b(t_i) + c]) \qquad i = 1, \ldots, N \end{aligned}$$

Nothing in the statement of the problem requires the parameters to be nonnegative, so a, b, and c are **free variables,** that is, variables unconstrained in sign. It would be harmless to include a nonnegativity constraint on the variable w, but it would also be superfluous because the above constraints already require w to be nonnegative.

Solving this linear program yields values for the parameters a, b, and c that minimize the largest absolute deviation, and an optimal value of w that is equal to that largest absolute deviation.

Inconsistent Systems of Equations

It is possible for a system of linear equations to be inconsistent, in which case it has no solution. Even though there is no combination of variable values that exactly satisfies such a system, it is always possible to find a "best" point that "comes closest," in some sense, to satisfying the equations. The problem of finding a best point is important because inconsistent systems of linear equations arise in many contexts.

Consider the following general form for a linear system of m equations in n variables:

$$\sum_{j=1}^{n} a_{ij}x_j = b_i \qquad i = 1, \ldots, m$$

Given a set of values $(x_1, x_2, \ldots, x_n)$ for the variables, we define the deviation D_i associated with the ith equation by

$$D_i = \sum_{j=1}^{n} a_{ij}x_j - b_i \qquad i = 1, \ldots, m$$

If the linear system has a solution, and if that solution is used in computing the deviations, then each of the D_i will be zero. However, if no solution exists, then one or more of the deviations will be nonzero no matter what values are given to the variables x_j, and in this case there are many different ways of defining what one means by a point that most nearly satisfies the inconsistent equations. One approach is to define the best set of x_j as the one that minimizes the largest absolute deviation $|D_i|$, just as we defined the best set of parameters in the curve-fitting problem as those that minimize the largest absolute deviation. Following the approach that we used for that problem, we introduce a new variable w and consider the optimization problem

$$\begin{aligned}
&\text{minimize } w \\
&\text{subject to} \\
&\qquad w \geq |D_i| \qquad i = 1, \ldots, m
\end{aligned}$$

Just as we did for the curve-fitting problem, we can reformulate this minimization as the linear program

$$\begin{aligned}
&\text{minimize } w \\
&\text{subject to} \\
&\qquad w \geq D_i \qquad i = 1, \ldots, m \\
&\qquad w \geq -D_i \qquad i = 1, \ldots, m \\
&\qquad D_i = \sum_{j=1}^{n} a_{ij}x_j - b_i \qquad i = 1, \ldots, m
\end{aligned}$$

Note that rather than eliminate the dependent variables D_i, as we did in the curve-fitting problem, we have simply let the equality constraints reflect the relationships between the variables x_j and the D_i. Solution of this model gives the point that best satisfies the inconsistent linear system in the sense of minimizing the largest absolute deviation.

Another reasonable way of picking a point that best satisfies the system is to minimize the sum of the absolute deviations,

$$\sum_{i=1}^{m} |D_i|$$

The optimization problem that results is given by

$$\begin{aligned} &\text{minimize} \quad \sum_{i=1}^{m} |D_i| \\ &\text{subject to} \\ &\qquad D_i = \sum_{j=1}^{n} a_{ij}x_j - b_i \qquad i = 1, \ldots, m \end{aligned}$$

This problem has variables D_i, $i = 1, \ldots, m$ and x_j, $j = 1, \ldots, n$. Unfortunately, because of the nonlinear objective function, it is not a linear program. However, as in the case of the curve-fitting problem, an elementary observation about real numbers permits us to reformulate the problem in linear form:

> For any y, we can find $u \geq 0$ and $v \geq 0$ such that
> $y = u - v$ and either $u = 0$ or $v = 0$ or both,
> and then $|y| = u + v$.

Using this fact, we can write

$$|D_i| = D_i^+ + D_i^- \quad \text{for } i = 1, \ldots, m$$

where $D_i^+ \geq 0$, $D_i^- \geq 0$ and either $D_i^+ = 0$ or $D_i^- = 0$. The preceding optimization can then be recast as the linear program

$$\begin{aligned} &\text{minimize} \quad \sum_{i=1}^{m} (D_i^+ + D_i^-) \\ &\text{subject to} \quad D_i^+ - D_i^- = \sum_{j=1}^{n} a_{ij}x_j - b_i \qquad i = 1, \ldots, m \\ &\qquad\qquad D_i^+ \geq 0 \qquad D_i^- \geq 0 \qquad i = 1, \ldots, m \end{aligned}$$

The optimal solution to this linear program will automatically satisfy the additional condition that either $D_i^+ = 0$ or $D_i^- = 0$ because we are minimizing the sum of those quantities. To see that this must be true, suppose that a feasible solution to the above linear program had $D_i^+ = 7$ and $D_i^- = 2$, corresponding to $D_i = 5$. Then we could find a point that is feasible and has a lower objective function value simply by choosing $D_i^+ = 5$ and $D_i^- = 0$. (This point is feasible because it is only the difference $D_i^+ - D_i^-$ that is important in the constraints.) Thus the minimization process forces the optimal solution to satisfy the condition that either D_i^+ or D_i^- is zero. Note that if this condition had to be enforced explicitly (such as by requiring $D_i^+ \cdot D_i^- = 0$), then the problem would not be a linear program.

Another way to define the best solution to a system of inconsistent equations is as the one that minimizes the sum of the squares of the deviations rather than the sum of their absolute values, but that problem cannot be reformulated as a linear program.

Feasibility Problems

Another important problem, arising in many contexts, is that of determining whether or not a linear system containing both equations and inequalities has a solution. Any such linear system could serve as the constraints for a linear program, and the linear system has a solution if and only if the linear program has at least one feasible point. As we mentioned above, Chapter 3 will show how it can easily be determined whether or not a linear program is feasible. The existence of a feasible point for a linear program is not affected by the objective function, so we can choose an objective function arbitrarily if only feasibility is to be determined.

Another feasibility-type question involves homogeneous linear systems. A **homogeneous** equation has zero for its constant term, and a homogeneous system is thus of the following form:

$$\sum_{j=1}^{n} a_{ij}x_j = 0 \qquad i = 1, \ldots, m$$

A **nonnegative solution** to this system is one that has $x_j \geq 0$ for $j = 1, \ldots, n$. One point that is always a nonnegative solution to the system is the **trivial solution** $(x_1, \ldots, x_n) = (0, \ldots, 0)$. Are there any nonnegative solutions in addition to the trivial one? If there is a nonnegative solution with some components nonzero, then those components are positive and the sum of all the components is positive. This suggests that we can answer the question by considering the following linear program:

$$\begin{aligned} \text{maximize } & \sum_{j=1}^{n} x_j \\ \text{subject to } & \sum_{j=1}^{n} a_{ij}x_j = 0 \qquad i = 1, \ldots, m \\ & x_j \geq 0 \qquad j = 1, \ldots, n \end{aligned}$$

This linear program is always feasible because the trivial solution to the homogeneous system satisfies the constraints. Also, the optimal value of the linear program cannot be negative because the sum of nonnegative variables is always nonnegative. If the optimal value is zero, then there is no nonnegative solution to the homogeneous system other than the trivial solution. However, if the optimal value is positive, then the optimal solution to the linear program is a solution to the homogeneous system that has at least one nonzero component.

Actually, if there is a nonzero nonnegative solution to the homogeneous system, then there is no finite maximum value for the linear program. This is because multiplying each variable in such a solution by the same positive constant gives another feasible solution with an objective function value that is larger by that factor. Thus, if the homogeneous system has a nonnegative solution that is not the trivial solution, the optimal value of the linear program will be infinite. A linear program whose objective function value is infinite is said to be **unbounded,** and Chapter 3 shows how such linear programs are easily recognized.

As a final example, suppose that we are given the system

$$\sum_{j=1}^{n} a_{ij}x_j = b_i \qquad i = 1, \ldots, m$$

$$x_j > 0 \qquad j = 1, \ldots, n$$

Note that this system requires all of the variables to be **strictly positive.** There is an important difference between using inequalities and using strict inequalities in defining a feasible set. In the Oakwood problem, for example, if all of the constraints had been strict inequalities, then points on the boundary of the set X in Figure 2.1 would not have been feasible. Points on the boundary of the feasible set for a linear programming problem must themselves be feasible points. Thus, because the inequalities $x_j > 0$ are strict, they are not in the proper form to be used as constraints in a linear program.

It is possible, however, despite the strict inequalities, to formulate a linear program whose solution determines whether or not there is a point that satisfies the linear system given above. Suppose we introduce a new variable v, and replace the strict inequalities $x_j > 0$ by $x_j \geq v$. The original system has a solution if and only if the new system (obtained by this replacement) has a solution with v strictly positive. To determine whether or not the new system has a solution with $v > 0$, we need only solve the following linear program:

maximize $\quad v$

subject to

$$\sum_{j=1}^{n} a_{ij}x_j = b_i \qquad i = 1, \ldots, m$$

$$x_j \geq v \qquad j = 1, \ldots, n$$

If the maximum value of this linear program is negative or zero, then the original system does not have a solution. Otherwise, this linear program has a feasible point with $v > 0$, and the original system does have a solution.

SELECTED REFERENCES

Arnold, L. R., and Botkin, D., Portfolios to Satisfy Damage Judgements: A Linear Programming Approach, *Interfaces* **8**:38–42 (1978).

Bradley, S. P., Hax, A. C., and Magnanti, T. L., *Applied Mathematical Programming,* Addison-Wesley, Reading, MA, 1977, Chapters 5–7.

Daellenbach, H. G., and Bell, E. J., *User's Guide to Linear Programming,* Prentice-Hall, Englewood Cliffs, NJ, 1970.

Dantzig, G. B., *Linear Programming and Extensions,* Princeton University Press, Princeton, 1963, Chapters 2 and 3.

Driebeek, N. J., *Applied Linear Programming,* Addison-Wesley, Reading, MA, 1977.
Glassey, C. R., and Gupta, V. K., A Linear Programming Analysis of Paper Recycling, *Management Science* **20**:392–408 (1974).
Heckman, L. B., and Taylor, H. M., School Rezoning to Achieve Racial Balance: A Linear Programming Approach, *Journal of Socio-Economic Planning Science* **3**:127–133 (1969).
Jackson, B. L., and Brown, J. M., Using LP for Crude Oil Sales at Elk Hills: A Case Study, *Interfaces* **10**:65–70 (1980).
Salkin, H. M., and Saha, J., *Studies in Linear Programming,* North-Holland—American Elsevier, New York, 1975.
Wagner, H. M., *Principles of Operations Research,* Prentice-Hall, Englewood Cliffs, NJ, 1969, Chapter 2.
Wagner, H. M., Linear Programming and Regression Analysis, *Journal of the American Statistical Association* **54**:206–212 (1959).

EXERCISES

2.1 (a) In the Oakwood problem of Section 2.1, below what value must the price of chairs fall before the optimal solution $(x_1, x_2) = (2.25, 8)$ changes? What would the new optimal solution be?
(b) Suppose the price of chairs rises to \$2000 per chair. Will the optimal solution change? Is there a value that the price of chairs can rise above and change the optimal solution? Explain.
(c) Are there positive values for the prices of tables and chairs so that only tables and no chairs will be produced in an optimal production program?

2.2 Consider the linear program

maximize $x_1 + x_2$

subject to

$$\begin{aligned} -2x_1 + 2x_2 &\leq 1 \\ 16x_1 - 14x_2 &\leq 7 \\ x_1 \geq 0 \qquad x_2 &\geq 0 \end{aligned}$$

(a) Use the graphical method to find an optimal solution (x_1, x_2).
(b) Assume in addition that each variable is restricted to be integer valued. Is a feasible integer point obtained by rounding each component of the optimal point up or down to the nearest integer?
(c) Use the graphical method to find an optimal integer solution.

2.3 (a) Use the graphical method to solve the linear program

minimize $x_1 - x_2$

subject to

$$\begin{aligned} x_1 + 2x_2 &\geq 4 \\ x_2 &\leq 4 \\ 3x_1 - 2x_2 &\leq 0 \\ x_2 &\geq 0 \end{aligned}$$

(b) What would the optimal solution be if the constraint $x_1 \geq 0$ were added to the above model?

2.4 (a) Use the graphical method to solve the linear program

maximize $z = x_1 + 2x_2 + x_3$

subject to

$$x_1 + x_2 + x_3 \leq 4$$

$$x_1 \geq 0 \qquad x_2 \geq 0 \qquad x_3 \geq 0$$

(b) Suppose the objective function in part (a) is replaced by

$$z = x_1 + x_3$$

Use the graphical method to find the set of all optimal solutions.

2.5 (a) For the peanut warehouse problem, state algebraically how the decision variables we used in the formulation are related to the variables

$$x_t = \text{amount by which the inventory changes after sale } t$$

(b) Reformulate the problem using the variables x_t as the only decision variables.

2.6 Show that $w \geq |y|$ if and only if $w \geq y$ and $w \geq -y$.

2.7 In the formulation of the curve-fitting problem in Section 2.3, explain why the constraints in the linear program automatically guarantee $w \geq 0$.

2.8 Formulate a linear programming model for solving the curve-fitting problem of Section 2.3 when the sum of the absolute values of the deviations is to be minimized.

2.9 In the inconsistent equations problem, why can't the minimization of the sum of squares of the deviations be formulated as a linear program?

2.10 A factory has two different machines, each of which can be used to manufacture the same three different products. The following table gives the number of units of product k that can be produced by machine i if machine i is used the entire working day to make product k.

		product		
		1	2	3
machine	1	8	2	9
	2	3	5	6

If a machine is used for a fraction of the day to produce some product, that fraction of an entire day's output of that product is produced, and no production is lost in switching from one product to another. Thus, for example, if machine 1 is used for half a day to produce product 1 and then makes product 2 for the remainder of the day, it will produce 4 units of product 1 and 1 unit of product 2.

All three products can be produced in noninteger amounts. However, products 1, 2, and 3 must be produced in the preassigned proportions $\frac{1}{2}$, $\frac{1}{4}$, $\frac{1}{4}$, respectively; that is, the quantity of product 1 must be $\frac{1}{2}$ the total quantity of all products produced, and equal quantities of products 2 and 3 must be made.

Formulate a linear programming problem that will determine what fraction of the day each machine should be used to produce each product so as to maximize the total quantity of products produced.

2.11 Consider the following linear program:

$$\begin{aligned} \text{minimize } z = -x_1 - x_2 & \\ \text{subject to } -x_1 + x_2 &\leq 3 \\ x_1 - 2x_2 &\leq 2 \\ x_1 \geq 0 \quad x_2 &\geq 0 \end{aligned}$$

(a) Use the graphical method to show that this linear program is feasible but has no optimal solution.
(b) Suppose the constraint

$$2x_1 - 8x_2 \geq 4$$

is added. Either find an optimal point or explain why none exist.

2.12 The owner of a large restaurant is considering how to obtain tablecloths for each of the 7 days that the restaurant is open each week. New tablecloths can be bought for \$10 each, and after being used one day they can be cleaned at either Bud's 1-day service or Mac's 2-day service. Thus, for example, tablecloths used on Monday will be ready for use on Wednesday if sent to Bud or on Thursday if sent to Mac. Bud charges \$3 per tablecloth and Mac's slower service costs \$1 per tablecloth. For the 7 days Monday through Sunday, the restaurant needs 110, 100, 160, 120, 180, 200, and 120 tablecloths, respectively. After each week is over (late Sunday night) all tablecloths are sold for \$2 each because they are unsuitable for another week's use.

Formulate a linear programming model to determine how to obtain the required supply of tablecloths at minimum cost.

2.13 The Apple Company has made a contract with the government to supply 1200 microcomputers this year and 2500 next year. The company has the production capacity to make 1400 microcomputers each year, and it has already committed its production line for this level. Labor and management have agreed that the production line can be used for at most 80 overtime shifts each year, each shift costing the company an additional \$20,000. In each overtime shift, 50 microcomputers can be manufactured. Units made this year but used to meet next year's demand must be stored at a cost of \$100 per unit.
(a) Formulate a linear programming model for finding a production schedule that minimizes cost.
(b) Explain how an optimal production schedule can be obtained by just thinking carefully about the problem, without even considering a linear programming formulation.

2.14 The Wood Products Company uses three machines to produce two products. The following technology table gives the hours on each machine required to produce one unit of each product, and the total time available on each machine during the production period.

	Time Used in Making		*Total Machine Time Available in*
Machine	*Product 1*	*Product 2*	*Production Period*
Lathe	1.1	2.0	1000
Sander	3.0	4.5	2000
Polisher	2.5	1.3	1500

The problem faced by the company is that it wants to maximize the total number of products made, but it wants to be sure that the amount of product 1 is at least one third the total number produced.

(a) Formulate a linear programming model for this problem.
(b) Suppose that instead of 3 machines and 2 products, 300 machines are used to produce 200 products. Let

r_{ij} = number of hours on machine j required to produce one unit of product i
c_j = available machine time on machine j

Formulate a linear programming model for this enlarged version of the problem.

2.15 A coal-fired electric plant burns three types of coal to drive steam turbines in order to produce electricity. Federal standards require that emissions from the furnace stack contain no more than 2500 parts per million (ppm) of sulfur oxide and that no more than 15 kilograms per hour (kg/hr) of particulate matter (smoke) be emitted from the stack. The following table gives the amounts of both pollutants that result from burning the three types of coal.

Coal Type	*Sulfur Oxide in Stack Emissions (ppm)*	*Particulates Emitted per Ton of Coal Burned (kg/hr)*
A	1200	1
B	2500	3
C	3700	2

Burning coal A results in 22,000 lb of steam per hour, whereas burning coal B or C, respectively, produces 26,000 or 32,000 lb per hour. The furnace has a capacity for burning 25 tons per hour of any mixture of the three coals. Also, the sulfur oxide emissions that result from burning a mixture of coals is equal to a weighted average of the parts-per-million emissions of the individual coals, where each weight is equal to the proportion of that coal used in the mixture.

Formulate a linear programming model for operating the electric plant so as to maximize the amount of steam generated per hour.

2.16 A company has decided to hire additional workers to serve as temporary help during the prime vacation months of June, July, August, and September. Union work rules permit temporary workers only during those months and require that they be hired only for 1, 2, or 3 months. The number of temporary workers required for the months of June through September are 90, 140, 130, and 80, respectively. The company gets its temporary workers through a placement service, and the following table gives the amount charged by the placement service in each of the 4 months for providing a temporary worker for 1, 2, or 3 months.

Placement Service Charges per Worker

Month Hired	*Hired for k Months* $k = 1$	$k = 2$	$k = 3$
June	\$ 80	\$ 70	\$ 90
July	\$ 50	\$ 90	\$ 60
August	\$100	\$ 80	*
September	\$ 60	*	*

*Indicates that temporary work for this many months is not possible

Formulate a model to find a hiring schedule that minimizes the amount the company must pay the placement service.

2.17 (a) Find an optimal solution to the oil refinery problem of Section 2.2.
(b) Which constraints are **redundant,** in that removing them does not change the feasible set?

2.18 The following pipeline network gives the possible routes for shipping oil from city 1 to city 5.

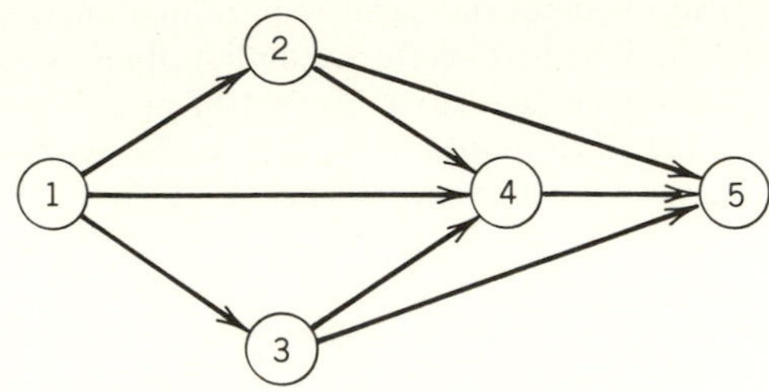

Suppose that

p_{ij} = per-unit cost of shipping oil from city i to city j
c_{ij} = capacity of the route from city i to city j

Formulate a linear programming model for determining the best way to ship 100 units of oil from city 1 to city 5.

2.19 Formulate a linear programming model whose solution will determine whether or not the following system of strict inequalities has a solution:

$$\sum_{j=1}^{n} a_{ij}x_j > 0 \qquad i = 1, \ldots, m$$

2.20 Formulate a linear program that will determine whether or not the system

$$\sum_{j=1}^{n} a_{ij}x_j \le 0 \qquad i = 1, \ldots, m$$

$$\sum_{j=1}^{n} b_j x_j > 0$$

has a solution.

2.21 A company is planning its work force for the next 3 years, year 1, year 2, and year 3. At the beginning of year 1, its work force consists of 200 workers who were hired earlier and have exactly one year left on their contract to work. The company needs exactly 500 workers in year 1, 700 in year 2, and 300 in year 3. At the beginning of each year, workers can be hired for exactly 1 or 2 years. At the end of year 3, the company wants at least 100 workers with exactly 1 year left on their contract. To hire a worker the company must use an employment service, which charges the company for finding workers. If the service provides a worker in year i for exactly j years, then it charges the company a finder's fee of c_{ij} given as the entry in the ith row and jth column of the matrix C:

$$C = \begin{bmatrix} 8 & 3 \\ 4 & 10 \\ 2 & 9 \end{bmatrix}$$

Formulate a linear programming model for hiring required workers that minimizes the total amount the company must pay the employment service.

2.22 Consider the following system of linear equations.

$$x_1 + x_2 = 1$$
$$x_1 + x_2 = 2$$

Clearly, this system has no solution. Formulate a linear program for finding an x such that

$$|x_1 + x_2 - 1| + |x_1 + x_2 - 2|$$

is as small as possible.

CHAPTER 3

THE SIMPLEX ALGORITHM FOR LINEAR PROGRAMMING

In the preceding chapter we were primarily concerned with formulating linear programming problems. In this chapter we show how the simplex algorithm can be used to solve any linear program. The algorithm either finds an optimal vector or shows that no optimal vectors exist.

3.1 STANDARD FORM AND PIVOTING

The models developed in Chapter 2 are all linear programming problems, but they have slightly different forms. In this section we introduce a standard form into which any linear programming problem can be reformulated and a standard bookkeeping scheme for representing the data of a linear program in standard form. Then we show how the standard form can be manipulated to obtain different, but equivalent, representations of the problem. Finding such alternative representations is an essential part of solving linear programs by the simplex algorithm.

Standard Form

A linear program is in **standard form** if it is

- a minimization problem,
- with equality constraints, and
- all variables are nonnegative.

Thus a linear program in standard form looks like:

$$\textit{minimize} \quad z(\mathbf{x}) = d + \mathbf{c}^T\mathbf{x} \quad \textit{subject to} \quad \mathbf{A}\mathbf{x} = \mathbf{b} \quad \textit{and} \quad \mathbf{x} \geq \mathbf{0}$$

where d is a constant,

$$\mathbf{c}^T\mathbf{x} = c_1x_1 + c_2x_2 + \cdots + c_nx_n$$

and $\mathbf{A}\mathbf{x} = \mathbf{b}$ is the matrix form of a linear system of equality constraints. Throughout this book, vectors and matrices appear in boldface type. The ith constraint equation is

$$\mathbf{A}_i\mathbf{x} = b_i, \qquad i = 1, \ldots, m$$

where $\mathbf{A}_i$ is the ith row of the $m \times n$ matrix $\mathbf{A}$, and b_i is the ith component of the vector $\mathbf{b}$. For example, the system of equations

$$\begin{aligned} 2x_1 + 3x_2 - x_3 &= 6 \\ x_1 + 5x_3 &= -2 \end{aligned}$$

has $\mathbf{A}_1 = [2, 3, -1]$ so that $\mathbf{A}_1\mathbf{x}$ is just the dot product of the row $\mathbf{A}_1$ with the vector $\mathbf{x}$. Thus

$$\mathbf{A}_1\mathbf{x} = [2, 3, -1]\begin{bmatrix} x_1 \\ x_2 \\ x_3 \end{bmatrix} = 2x_1 + 3x_2 - x_3$$

and similarly

$$\mathbf{A}_2\mathbf{x} = [1, 0, 5]\begin{bmatrix} x_1 \\ x_2 \\ x_3 \end{bmatrix} = x_1 + 5x_3$$

Two vectors are equal if and only if their corresponding components are equal, so the vector equation

$$\mathbf{A}\mathbf{x} = \begin{bmatrix} \mathbf{A}_1\mathbf{x} \\ \mathbf{A}_2\mathbf{x} \end{bmatrix} = \begin{bmatrix} b_1 \\ b_2 \end{bmatrix} = \mathbf{b}$$

means

$$\mathbf{A}_1\mathbf{x} = b_1 \quad \text{and} \quad \mathbf{A}_2\mathbf{x} = b_2$$

For a brief review of matrix notation, see the appendix.

To illustrate how standard form can be obtained, we show how the linear programming model for the brewery problem of Section 2.2 can be reformulated into standard form. The model for this problem was given as

$$\begin{aligned} &\text{maximize } z = 6x_1 + 5x_2 + 3x_3 + 7x_4 \\ &\text{subject to} \\ &\qquad 1x_1 + 1x_2 + 0x_3 + 3x_4 \leq 50 \\ &\qquad 2x_1 + 1x_2 + 2x_3 + 1x_4 \leq 150 \\ &\qquad 1x_1 + 1x_2 + 1x_3 + 4x_4 \leq 80 \\ &\qquad x_i \geq 0, \qquad i = 1, \ldots, 4 \end{aligned}$$

This linear program is not in standard form because it is not a minimization problem and it does not have equality constraints. However, we can reformulate the problem as a minimization problem by using the fact that

> maximizing $\mathbf{c}^T\mathbf{x}$ subject to $\mathbf{x}$ being feasible is equivalent to minimizing $-\mathbf{c}^T\mathbf{x}$ subject to $\mathbf{x}$ being feasible.

A vector $\mathbf{x}^*$ that is optimal for the minimization problem is also optimal for the maximization problem, and the associated optimal values are negatives of one another.

It is also easy to rewrite the inequality constraints in this problem as equality constraints by introducing the variables

x_5 = amount of malt not used
x_6 = amount of hops not used
x_7 = amount of yeast not used

Then, for example, the first constraint can be rewritten as the equality

$$\underset{\text{(amount of malt used)}}{(x_1 + x_2 + 3x_4)} + \underset{\text{(amount of malt not used)}}{(x_5)} = 50$$

and the other constraints can be rewritten as equalities in a similar way. The new variables x_5, x_6, and x_7 are called **slack variables.**

Thus, by replacing the maximization by an equivalent minimization and introducing the slack variables x_5, x_6, and x_7, the brewery problem can be reformulated in standard form as

$$\begin{array}{lrcl}
\textit{minimize } z(\mathbf{x}) = & -6x_1 - 5x_2 - 3x_3 - 7x_4 & & \\
\textit{subject to} & & & \\
 & 1x_1 + 1x_2 \phantom{{}+ 2x_3} + 3x_4 + 1x_5 \phantom{{}+1x_6+1x_7} & = & 50 \\
 & 2x_1 + 1x_2 + 2x_3 + 1x_4 \phantom{{}+1x_5} + 1x_6 \phantom{{}+1x_7} & = & 150 \\
 & 1x_1 + 1x_2 + 1x_3 + 4x_4 \phantom{{}+1x_5+1x_6} + 1x_7 & = & 80 \\
\textit{and} & \mathbf{x} \geq \mathbf{0} & &
\end{array}$$

Here the matrix $\mathbf{A}$ is given by

$$\mathbf{A} = \begin{bmatrix} 1 & 1 & 0 & 3 & 1 & 0 & 0 \\ 2 & 1 & 2 & 1 & 0 & 1 & 0 \\ 1 & 1 & 1 & 4 & 0 & 0 & 1 \end{bmatrix}$$

$$\mathbf{c}^T = [-6 \quad -5 \quad -3 \quad -7 \quad 0 \quad 0 \quad 0], \quad \text{and} \quad \mathbf{b} = \begin{bmatrix} 50 \\ 150 \\ 80 \end{bmatrix}$$

Section 3.6 shows in detail how to reformulate any linear program into standard form. The above linear program has $d = 0$ in the objective function $z = d + \mathbf{c}^T\mathbf{x}$, but, as we will soon see, nonzero constant terms often arise naturally in the process of solving linear programs.

The Simplex Tableau

Associated with each linear program in standard form is a tableau used as a bookkeeping scheme to record the data that defines the problem. This tableau is called a **simplex tableau** and it has the following form:

		$x_1 \; x_2 \; \cdots \; x_n$
z	$-d$	$\mathbf{c}^T$
	$\mathbf{b}$	$\mathbf{A}$

The first row of this tableau represents the equation

$$z - d = \mathbf{c}^T\mathbf{x}$$

or equivalently, the equation

$$z = d + \mathbf{c}^T\mathbf{x}$$

The last m rows are the constraint rows, and they represent the constraint equations

$$b_i = \mathbf{A}_i\mathbf{x} \qquad i = 1, \ldots, m$$

The simplex tableau associated with the above resource allocation problem is

		x_1	x_2	x_3	x_4	x_5	x_6	x_7	
z	0	−6	−5	−3	−7	0	0	0	
	50	1	1	0	3	1	0	0	Tableau
	150	2	1	2	1	0	1	0	T_1
	80	1	1	1	4	0	0	1	

It is important to realize that the same linear program can be represented by many different, but equivalent, simplex tableaus. For example, given tableau T_1, we might decide to eliminate the variable x_1 from the second constraint equation. (We will see later why we would want to do this.) To accomplish this we could solve the first constraint equation for x_1 in terms of the other variables, obtaining

$$x_1 = 50 - x_2 - 3x_4 - x_5$$

and then substitute this expression for x_1 into the second constraint equation. The second equation then becomes

$$150 = 2(50 - x_2 - 3x_4 - x_5) + x_2 + 2x_3 + x_4 + x_6$$

or

$$50 = -x_2 + 2x_3 - 5x_4 - 2x_5 + x_6$$

Another way to obtain this same equation is to simply add −2 times the first constraint equation to the second constraint equation; that is, in tableau T_1 we add -2 times each entry in the first constraint row to the corresponding entry in the second constraint row.

Similarly, we could eliminate x_1 from the third equation by adding -1 times the first equation to the third. The resulting simplex tableau is

		x_1	x_2	x_3	x_4	x_5	x_6	x_7
z	0	-6	-5	-3	-7	0	0	0
	50	1	1	0	3	1	0	0
	50	0	-1	2	-5	-2	1	0
	30	0	0	1	1	-1	0	1

This tableau represents the same linear programming problem; we have simply rewritten the constraints in a slightly different, but equivalent, form.

The objective function represented by this tableau is

$$z(\mathbf{x}) = -6x_1 - 5x_2 - 3x_3 - 7x_4$$

Suppose that we also wanted to eliminate the variable x_1 from the objective function (we will see later that this type of elimination is required by the simplex method in solving linear programs). To achieve this, we could substitute

$$x_1 = 50 - x_2 - 3x_4 - x_5$$

into the objective function and rewrite $z(\mathbf{x})$ as

$$z(\mathbf{x}) = -6(50 - x_2 - 3x_4 - x_5) - 5x_2 - 3x_3 - 7x_4$$

or

$$z(\mathbf{x}) = -300 + x_2 - 3x_3 + 11x_4 + 6x_5$$

This algebraic substitution process can be achieved by simply adding 6 times the first constraint row to the objective function row. Doing so yields the tableau

		x_1	x_2	x_3	x_4	x_5	x_6	x_7
z	300	0	1	-3	11	6	0	0
	50	1	1	0	3	1	0	0
	50	0	-1	2	-5	-2	1	0
	30	0	0	1	1	-1	0	1

Tableau T_2

Of course, tableau T_2 represents the same linear programming problem represented by tableau T_1. For example, note that $\mathbf{x}^0 = [50, 0, 0, 0, 0, 50, 30]^T$ is a nonnegative vector that satisfies the constraint equations as represented in tableau T_2 and $z(\mathbf{x}^0) = -300$. It is easy to check that $\mathbf{x}^0$ also satisfies the constraint equations as given in tableau T_1, and substituting $\mathbf{x}^0$ into the objective function of tableau T_1 also yields $z(\mathbf{x}^0) = -300$.

Pivoting on a Simplex Tableau

The sequence of elementary row operations performed on tableau T_1 to obtain tableau T_2 is called a **pivot.** The steps in performing a pivot on a tableau are as follows:

To perform a pivot:
• Select a nonzero entry a_{hk} in row h and column k of the matrix $\mathbf{A}$ (row h is called the **pivot row,** column k is called the **pivot column,** and position (h, k) is called the **pivot position**). • Multiply the pivot row by the constant $1/a_{hk}$ so that the entry in the pivot position equals 1. • Then use elementary row operations to make all other tableau entries in the pivot column equal to zero.

The simplex algorithm for solving linear programming problems, devised by George Dantzig in 1947, is a method for systematically choosing a sequence of pivots on a tableau in standard form. The pivots continue until a tableau in one of four final forms is obtained. In the sections below we will see how, from the form of the final tableau, it is easy to either write down an optimal solution to a linear program or deduce that no optimal solution exists.

3.2 CANONICAL FORM

Consider the following tableau for a linear program in standard form:

		x_1	x_2	x_3	x_4	x_5
z	-8	0	-2	7	0	0
	4	0	2	-4	1	0
	6	1	1	5	0	0
	5	0	-2	1	0	1

It is not important to know the origin of this tableau; imagine, for example, that it is the current tableau in some sequence of pivots. Note that from the form of this tableau we can easily write down a feasible solution. Just let $x_2 = 0$ and $x_3 = 0$, and then from the constraint equations it follows that $x_4 = 4$, $x_1 = 6$, and $x_5 = 5$. This yields the feasible vector

$$\mathbf{x}^0 = \begin{bmatrix} 6 \\ 0 \\ 0 \\ 4 \\ 5 \end{bmatrix} \quad \text{with } z(\mathbf{x}^0) = 8 - 2x_2 + 7x_3 = 8$$

The vector $\mathbf{x}^0$ is feasible because it has nonnegative components and (by its construction) satisfies all of the constraint equations.

Canonical Form

It is the special form of the preceding tableau that allows us to so easily write down a feasible solution, and this form is called **canonical form.** A simplex tableau

	$x_1 \cdots x_n$
$-d$	$\mathbf{c}^T$
b_1 $\vdots$ b_m	$\mathbf{A}$

for a linear program in standard form is in canonical form if

- $b_i \geq 0, i = 1, \ldots, m,$
- the matrix $\mathbf{A}$ contains the m identity columns of the $m \times m$ identity matrix $\mathbf{I}_m$, and
- the objective function coefficients corresponding to those m identity columns are zero.

The example tableau above is in canonical form because

- b_1, b_2, and b_3 are nonnegative,
- the fourth, first, and fifth columns of $\mathbf{A}$ form the 3×3 identity matrix

$$\mathbf{I}_3 = \begin{bmatrix} 1 & 0 & 0 \\ 0 & 1 & 0 \\ 0 & 0 & 1 \end{bmatrix}$$

- and the corresponding objective function coefficients c_4, c_1, and c_5 are zero.

The variables corresponding to the m identity columns are called **basic variables** and the remaining variables are called **nonbasic variables.** The feasible solution obtained by setting the nonbasic variables equal to zero and using the constraint equations to solve for the basic variables is called the **basic feasible solution** associated with the tableau. The example tableau is in canonical form with basic variables x_4, x_1, and x_5 and nonbasic variables x_2 and x_3, and the vector

$$\mathbf{x}^0 = \begin{bmatrix} 6 \\ 0 \\ 0 \\ 4 \\ 5 \end{bmatrix}$$

is the basic feasible solution associated with the tableau.

In this example the indices of the basic variables are 1, 4, and 5. Sometimes it will be convenient to order the indices of the basic variables in a sequence so that the ith index in the sequence is the index of the basic variable corresponding to the ith identity column. In our example this gives rise to the ordered sequence

$$S = (4, 1, 5)$$

This ordered sequence of basic variable indices is called the **basic sequence** associated with the canonical form tableau, and we say that the tableau is in canonical form *with respect to* the basic sequence S.

Some standard form tableaus (those corresponding to infeasible linear programs) cannot be placed in canonical form. In Section 3.5 we show how to perform a sequence of pivots on a standard form tableau to obtain either canonical form or one of two infeasible forms. If canonical form is obtained, then the linear program certainly has at least one feasible solution, namely, the basic feasible solution associated with that canonical form.

Finding a Better Basic Feasible Solution

The objective function in the preceding example is given by

$$z(\mathbf{x}) = 8 - 2x_2 + 7x_3$$

Because we are interested in decreasing $z(\mathbf{x})$, and because the cost coefficient of the nonbasic variable x_2 is -2, it may be possible to find a feasible solution with an objective function value less than $z(\mathbf{x}^0) = 8$ by letting the variable x_2 take on some positive value instead of the zero value it has in $\mathbf{x}^0$. To this end, we adjust the nonbasic variables by letting $x_2 = t > 0$ and $x_3 = 0$. Substituting these values in the constraint equations in the above tableau forces the basic variables to assume the values $x_1 = 6 - t$, $x_4 = 4 - 2t$, and $x_5 = 5 + 2t$. (The basic variables can be thought of as dependent variables because their values are completely determined once the nonbasic variables are assigned values.) These substitutions yield the vector

$$\mathbf{x}(t) = \begin{bmatrix} 6 - t \\ 0 + t \\ 0 \\ 4 - 2t \\ 5 + 2t \end{bmatrix} \quad \text{with } z[\mathbf{x}(t)] = 8 - 2t$$

We would like to let t be as large as possible in order to decrease $z[\mathbf{x}(t)]$ as much as possible. By its construction, $\mathbf{x}(t)$ satisfies the constraint equations for any value of t, so it is feasible provided it remains nonnegative, that is, provided

$$6 - t \geq 0 \qquad 4 - 2t \geq 0 \qquad 5 + 2t \geq 0$$

Thus we can let t (that is, x_2) increase up to $t = 2$ and $\mathbf{x}(t)$ remains nonnegative. Letting $t = 2$ yields the improved feasible vector

$$\hat{\mathbf{x}} = \begin{bmatrix} 4 \\ 2 \\ 0 \\ 0 \\ 9 \end{bmatrix} \quad \text{with } z(\hat{\mathbf{x}}) = 4$$

A natural question to ask is whether $\hat{\mathbf{x}}$ is in fact the basic feasible solution associated with some other canonical form tableau for the linear program. If this were the case, x_2 would have to be a basic variable and the corresponding constant column entry would have to be 2. By examining the original tableau

		x_1	x_2	x_3	x_4	x_5
z	-8	0	-2	7	0	0
	4	0	②	-4	1	0
	6	1	1	5	0	0
	5	0	-2	1	0	1

we see that x_2 can indeed be made a basic variable, with a corresponding constant column entry of 2, by performing a pivot on the circled entry. Performing this pivot yields the tableau

		x_1	x_2	x_3	x_4	x_5
z	-4	0	0	3	1	0
	2	0	1	-2	1/2	0
	4	1	0	7	$-1/2$	0
	9	0	0	-3	1	1

This tableau is in canonical form with respect to the basic sequence (2, 1, 5), and the associated basic feasible solution is precisely the vector $\hat{\mathbf{x}}$.

Thus the parametric analysis we first performed using $x_2 = t$ to obtain the vector $\hat{\mathbf{x}}$ can be replaced by simply performing a pivot in a special position of the tableau. It is important to see precisely how this special pivot position is chosen. First, note that for $t > 0$.

$$\mathbf{x}(t) = \begin{bmatrix} 6 - t \\ 0 + t \\ 0 \\ 4 - 2t \\ 5 + 2t \end{bmatrix} \geq \mathbf{0} \quad \begin{array}{l} \text{if and only if} \\ t \leq \frac{6}{1} \text{ and } t \leq \frac{4}{2} \end{array}$$

Thus $\mathbf{x}(t) \geq \mathbf{0}$ if and only if $t \leq \min \{\frac{6}{1}, \frac{4}{2}\}$. This upper bound on t can easily be determined from the original tableau because it is the minimum of the ratios

$$\frac{b_i}{a_{ik}} \quad \text{for all rows } i \text{ having } a_{ik} > 0$$

Note that a row i having $a_{ik} \leq 0$ never limits how large t can be. For example, $x_5(t) = 5 + 2t$ is always nonnegative no matter how large t becomes. The pivot row is the row corresponding to the upper limit on t, namely, that row where the minimum ratio occurs.

In summary, the pivot row h selected by the above analysis is precisely the one for which

$$\frac{b_h}{a_{hk}} = \min \left\{ \frac{b_i}{a_{ik}} \middle| a_{ik} > 0 \right\}$$

Given k, a row selected by this ratio test is called a **minimum ratio row** associated with column k. For a given k, there may be more than one minimum ratio row because of ties in the minimum ratio, and the consequences of this are considered in Section 3.4. Also note that, even if a column k does not have $c_k < 0$, a pivot in a minimum ratio row yields a tableau that is in canonical form because such a pivot maintains $b_i \geq 0$ for each i.

The Simplex Rule for Pivoting

If $c_k < 0$, we have seen that pivoting in column k and in a minimum ratio row associated with that column yields a new canonical form tableau with a basic feasible solution having an objective value less than or equal to the previous objective value. (The objective function would not change if the pivot row h has $b_h = 0$.) We call this rule for selecting a pivot position the **simplex rule.**

SIMPLEX RULE

- Select the pivot column as any column k with $c_k < 0$.
- Given k, select the pivot row h as a minimum ratio row associated with column k.

We show in Section 3.4 that pivoting by the simplex rule is all that is needed to solve most linear programs that are in canonical form.

The Geometry of a Pivot

It is important to realize that performing a pivot on a canonical form tableau corresponds to moving along the boundary of the feasible set from one "corner point" to a neighboring "corner point." We will be more precise about the geometry of pivoting in Chapter 4, but for now a simple example will illustrate this important geometric interpretation of a pivot.

Consider the linear program

$$\begin{aligned} \text{minimize} \quad & -2x_1 - x_2 \\ \text{subject to} \quad & x_1 + x_2 \le 10 \\ & 2x_1 - x_2 \le 5 \\ & x_i \ge 0, \qquad i = 1, 2 \end{aligned}$$

The feasible set for this linear program is given as the shaded region in Figure 3.1 with the four corner points A, B, C, and D. In Chapter 4 we precisely define a corner point as an "extreme point" of the feasible set. Using contours of the objective function, as discussed in Section 2.1, it is not hard to see that point C with $x_1 = 5$ and $x_2 = 5$ is the optimal point. Note in Figure 3.1 how the feasible set is bounded by the lines

$$x_1 + x_2 = 10, \qquad 2x_1 - x_2 = 5, \qquad x_1 = 0, \qquad x_2 = 0$$

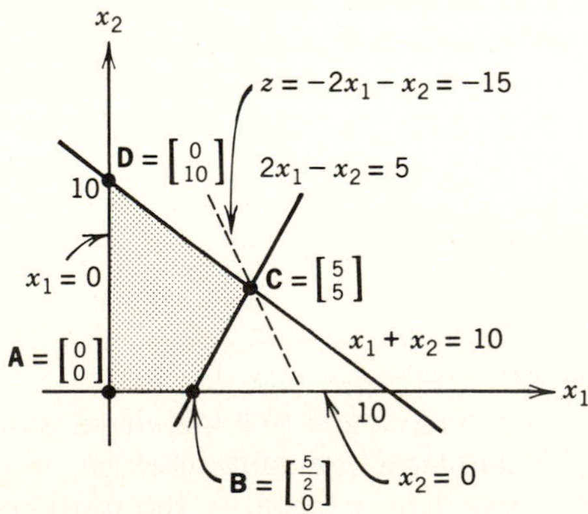

FIGURE 3.1 A feasible set with four corner points.

The objective function contour is given by the dotted line

$$z = -2x_1 - x_2 = -15$$

Introducing slack variables, s_1 and s_2, we obtain the following canonical form tableau T_A for this linear program,

$T_A =$

	x_1	x_2	s_1	s_2
0	−2	−1	0	0
10	1	1	1	0
5	2	−1	0	1

with associated basic feasible solution having

$x_1 = 0$ and $x_2 = 0$

Because the basic feasible solution for tableau T_A has $x_1 = 0$ and $x_2 = 0$, we say that this tableau corresponds to point **A** in the feasible region.

Pivoting in the x_1 column of the above tableau by the simplex rule gives the tableau, T_B, namely,

$T_B =$

	x_1	x_2	s_1	s_2
0	0	−2	0	1
15/2	0	3/2	1	−1/2
5/2	1	−1/2	0	1/2

with associated basic feasible solution having

$x_2 = \frac{5}{2}$ and $x_2 = 0$

This tableau corresponds to the corner point **B** in Figure 3.1 and the pivot corresponded to moving along the boundary from the corner point **A** to the corner point **B**. Note that this pivot corresponds to increasing the variable x_1 (that is nonbasic in T_A) from 0 up to the minimum ratio of $\frac{5}{2}$ for the x_1 column in T_A. As Figure 3.1 shows, increasing x_1 above $\frac{5}{2}$ while keeping $x_2 = 0$ leads to points outside of the feasible set.

Performing the circled pivot in tableau T_B leads to the optimal tableau

$T_C =$

	x_1	x_2	s_1	s_2
15	0	0	4/3	1/3
5	0	1	2/3	−1/3
5	1	0	1/3	1/3

with associated basic feasible solution having

$x_1 = 5$ and $x_2 = 5$

This tableau corresponds to the optimal point **C** in Figure 3.1 and the pivot corresponded to moving from the corner point **B** along the boundary of the feasible set to the corner point **C**.

3.3 OPTIMAL, UNBOUNDED, AND INFEASIBLE FORMS

There are four possible final forms for a simplex tableau. In this section we describe those forms and discuss how to recognize and interpret them.

Optimal Form

A vector $\mathbf{x}^*$ is called a **minimizing vector** if $z(\mathbf{x}) \geq z(\mathbf{x}^*)$ for every feasible solution $\mathbf{x}$. For the example in Section 3.2 we found a feasible vector $\hat{\mathbf{x}}$ with

$z(\hat{\mathbf{x}}) = 4$, so the minimum value for the linear program must be less than or equal to 4. To see if there is a feasible vector with an objective value smaller than 4, we need only inspect the tableau associated with the basic feasible solution $\hat{\mathbf{x}}$:

		x_1	x_2	x_3	x_4	x_5
z	-4	0	0	3	1	0
	2	0	1	-2	1/2	0
	4	1	0	7	$-1/2$	0
	9	0	0	-4	1	1

In this tableau the objective function is given by

$$z(\mathbf{x}) = 4 + 3x_3 + 1x_4$$

Because any feasible vector must have $x_3 \geq 0$ and $x_4 \geq 0$, it follows that

$$z(\mathbf{x}) = 4 + 3x_3 + 1x_4 \geq 4 \quad \text{for any feasible vector } \mathbf{x}$$

Given $z(\hat{\mathbf{x}}) = 4$, we conclude that $z(\mathbf{x}) \geq z(\hat{\mathbf{x}})$ for any feasible vector, and therefore that $\hat{\mathbf{x}}$ is a minimizing vector for the linear program.

This reasoning depends only on the fact that the tableau is in canonical form and has a cost vector $\mathbf{c} \geq \mathbf{0}$. In general, if $\mathbf{x}^*$ is the basic feasible solution for such a tableau, then $z(\mathbf{x}^*) = d$. Also, $z(\mathbf{x}) = d + \mathbf{c}^T\mathbf{x} \geq d$ for any feasible vector $\mathbf{x}$ because $\mathbf{c}^T\mathbf{x} \geq 0$. This implies that $z(\mathbf{x}) \geq z(\mathbf{x}^*)$ for any feasible $\mathbf{x}$. Thus a tableau in canonical form is said to be in

optimal form if $c_j \geq 0$ for each j

A tableau must be in canonical form in order to be in optimal form. If a tableau is not in canonical form, then the condition $\mathbf{c} \geq \mathbf{0}$ does not permit us to find an optimal vector easily, or even to conclude that optimal vectors exist.

Unbounded Form

Not all tableaus in canonical form can be pivoted into optimal form. For example, consider the tableau

		x_1	x_2	x_3	x_4	x_5
z	-9	0	0	-2	-1	0
	3	0	0	-1	2	1
	1	1	0	0	1	0
	5	0	1	-4	1	0

which is in canonical form with respect to the basic sequence (5, 1, 2). Here, the basic feasible solution is $\mathbf{x}^0 = [1, 5, 0, 0, 3]^T$. Suppose we increase the nonbasic variable x_3 in order to obtain a decrease in z. Letting $x_3 = t > 0$ and $x_4 = 0$ yields

$$\mathbf{x}(t) = \begin{bmatrix} 1 \\ 5 + 4t \\ 0 + t \\ 0 \\ 3 + t \end{bmatrix} \quad \text{with} \quad z[\mathbf{x}(t)] = 9 - 2t$$

By construction, $\mathbf{x}(t)$ satisfies the constraint equations, and no matter how large t becomes, each component of $\mathbf{x}(t)$ remains nonnegative. As $t \to +\infty$, $z[\mathbf{x}(t)] \to -\infty$, and thus the linear program has no finite minimum value and no optimal vector exists. This reasoning depends only on the fact that the above tableau is in canonical form and has cost coefficient $c_3 < 0$ with $a_{i3} \leq 0$ for each i, $i = 1, \ldots, m$. This argument is valid for all such tableaus, so a tableau in canonical form is said to be in

unbounded form if $c_k < 0$ for some k and $a_{ik} \leq 0$ for each i

Two Infeasible Forms

If a tableau is in canonical form, then we know that the linear program has at least one feasible solution, namely, the basic feasible solution associated with that canonical form. However, it is not always possible to obtain a canonical form tableau for every linear program in standard form. One reason for this is that not every system of equations has a solution, and the first infeasible form for a linear program is a consequence of this fact. For example, consider the standard form tableau

		x_1	x_2	x_3	x_4
z	5	2	−3	1	−1
	2	①	0	−1	−1
	1	−1	1	2	0
	−4	0	1	1	−1

In attempting to obtain an identity matrix necessary for canonical form, we might perform the sequence of pivots that results in the following sequence of tableaus.

		x_1	x_2	x_3	x_4
z	1	0	−3	3	1
	2	1	0	−1	−1
	3	0	①	1	−1
	−4	0	1	1	−1

		x_1	x_2	x_3	x_4
z	10	0	0	6	−2
	2	1	0	−1	−1
	3	0	1	1	−1
	−7	0	0	0	0

The last tableau shows that the linear program is infeasible because no vector (nonnegative or otherwise) satisfies the third constraint equation. We say that a linear program in standard form is in

infeasible form 1 if $b_i \neq 0$ for some i and $a_{ij} = 0$ for each j

If the system of equations $\mathbf{Ax} = \mathbf{b}$ has solutions, then no such row could ever be obtained, but even if solutions exist, there may not be any nonnegative solutions. For example, the linear program in standard form with the tableau

		x_1	x_2	x_3	x_4
z	-6	-1	1	-1	0
	8	2	-2	-6	0
	-2	0	5	4	1

has the last constraint equation

$$-2 = 5x_2 + 4x_3 + x_4$$

and clearly no nonnegative vector could satisfy this equation. Thus, we say that a linear program in standard form is in

infeasible form 2 if $b_i < 0$ for some i and $a_{ij} \geq 0$ for each j

One of the major goals of this chapter is to show how to perform a sequence of pivots systematically on a standard form tableau to obtain either optimal form, unbounded form, infeasible form 1, or infeasible form 2. The development is organized as follows:

We first show how to start with canonical form and obtain either optimal form or unbounded form.

Then we show how to start with standard form and obtain either canonical form, infeasible form 1, or infeasible form 2.

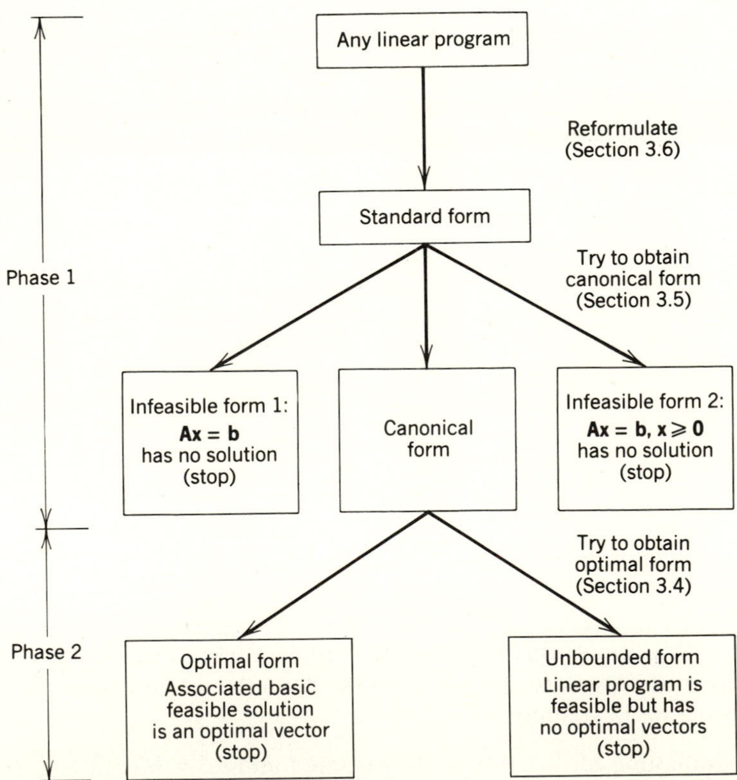

FIGURE 3.2 Solving any linear program.

It might at first seem more natural to start with standard form and do the second task first. However, in our development, knowing how to pivot from canonical form to optimal or unbounded form will allow us to show how to do the second task rather easily.

Figure 3.2 summarizes our approach for solving linear programming problems and lists the sections of this chapter that discuss the various steps in the solution process.

3.4 SOLVING LINEAR PROGRAMS IN CANONICAL FORM

Pivoting to Optimal Form

It is important to realize that a linear program in canonical form is either in

- optimal form, or
- unbounded form, or
- c_k is negative for some k and at least one $a_{ik} > 0$.

In the last case, pivoting by the simplex rule usually leads to optimal or unbounded form. The following illustrates a typical sequence of simplex rule pivots that leads to optimal form. The pivot position in each tableau is circled.

		x_1	x_2	x_3	x_4	x_5	x_6
z	−6	0	−1	−4	0	0	0
	2	1	−3	1	0	0	1
	20	1	4	−1	1	0	0
	1	1	−1	(1)	0	1	0

		x_1	x_2	x_3	x_4	x_5	x_6
z	−2	4	−5	0	0	4	0
	1	0	−2	0	0	−1	1
	21	2	(3)	0	1	1	0
	1	1	−1	1	0	1	0

		x_1	x_2	x_3	x_4	x_5	x_6
z	33	22/3	0	0	5/3	17/3	0
	15	4/3	0	0	2/3	−1/3	1
	7	2/3	1	0	1/3	1/3	0
	8	5/3	0	1	1/3	4/3	0

In the first tableau there are two choices for the pivot column using the simplex rule, and a reasonable strategy is to select the column with the most negative c_j. Increasing the nonbasic variable in such a column gives the greatest initial decrease in the objective function value. However, this strategy does not always give the pivot column that leads to the largest decrease in $z(\mathbf{x})$ after the pivot. For example, a pivot in the x_3 column of the first tableau above gives a decrease of 4 in $z(\mathbf{x})$, but a little extra work shows that a pivot in the x_2 column results in $z(\mathbf{x})$ decreasing by 5. The question of how various column selection strategies affect the overall performance of the simplex algorithm has been studied, and the strategy of selecting the column with the most negative c_j seems to work about as well as any.

What Can Go Wrong: Degeneracy and Cycling

We remarked above that pivoting on canonical form tableaus by the simplex rule usually leads to optimal form or unbounded form. To see what might go wrong, consider the following sequence of tableaus obtained by using the simplex rule and pivoting in the most negative c_j column.

		x_1	x_2	x_3	x_4	x_5	x_6	x_7
z	3	−3/4	20	−1/2	6	0	0	0
	0	(1/4)	−8	−1	9	1	0	0
	0	1/2	−12	−1/2	3	0	1	0
	1	0	0	1	0	0	0	1

		x_1	x_2	x_3	x_4	x_5	x_6	x_7
z	3	0	−4	−7/2	33	3	0	0
	0	1	−32	−4	36	4	0	0
	0	0	(4)	3/2	−15	−2	1	0
	1	0	0	1	0	0	0	1

		x_1	x_2	x_3	x_4	x_5	x_6	x_7
z	3	0	0	−2	18	1	1	0
	0	1	0	(8)	−84	−12	8	0
	0	0	1	3/8	−15/4	−1/2	1/4	0
	1	0	0	1	0	0	0	1

		x_1	x_2	x_3	x_4	x_5	x_6	x_7
z	3	1/4	0	0	−3	−2	3	0
	0	1/8	0	1	−21/2	−3/2	1	0
	0	−3/64	1	0	(3/16)	1/16	−1/8	0
	1	−1/8	0	0	21/2	3/2	−1	1

		x_1	x_2	x_3	x_4	x_5	x_6	x_7
z	3	−1/2	16	0	0	−1	1	0
	0	−5/2	56	1	0	(2)	−6	0
	0	−1/4	16/3	0	1	1/3	−2/3	0
	1	5/2	−56	0	0	−2	6	1

		x_1	x_2	x_3	x_4	x_5	x_6	x_7
z	3	−7/4	44	1/2	0	0	−2	0
	0	−5/4	28	1/2	0	1	−3	0
	0	−1/6	−4	−1/6	1	0	(1/3)	0
	1	0	0	1	0	0	0	1

		x_1	x_2	x_3	x_4	x_5	x_6	x_7
z	3	−3/4	20	−1/2	6	0	0	0
	0	(1/4)	−8	−1	9	1	0	0
	0	1/2	−12	−1/2	3	0	1	0
	1	0	0	1	0	0	0	1

The last and first tableaus in this sequence are identical, so continued pivoting in the most negative column will just repeat the sequence. This phenomenon is called **cycling.** In the above sequence every pivot row h had $b_h = 0$, so the objective function value never decreased. A linear program is called **degenerate** if some $b_i = 0$ in at least one of its canonical form tableaus. A linear program that is not degenerate is called **nondegenerate.**

If a linear program is nondegenerate, each canonical form tableau has every $b_i > 0$, and the objective function value therefore strictly decreases at every simplex rule pivot. Because of this strict decrease in the objective function, no sequence of simplex rule pivots can ever result in two tableaus that are identical if the linear program is nondegenerate. Thus, if cycling occurs, the linear program must be degenerate. Of course, not all degenerate linear programs cycle, and even if some canonical form tableaus have constant terms $b_i = 0$, such tableaus might not be encountered while pivoting to obtain optimal form.

It is actually extremely rare to find a degenerate problem that cycles. When cycling does occur, it can be avoided by changing the pivot rule. In the above example, always pivoting in the column having the least negative c_j does not result in cycling, and gives the following sequence of tableaus.

		x_1	x_2	x_3	x_4	x_5	x_6	x_7
z	3	−3/4	20	−1/2	6	0	0	0
	0	1/4	−8	−1	9	1	0	0
	0	1/2	−12	−1/2	3	0	1	0
	1	0	0	(1)	0	0	0	1

		x_1	x_2	x_3	x_4	x_5	x_6	x_7
z	7/2	−3/4	20	0	6	0	0	1/2
	1	1/4	−8	0	9	1	0	1
	1/2	(1/2)	−12	0	3	0	1	1/2
	1	0	0	1	0	0	0	1

		x_1	x_2	x_3	x_4	x_5	x_6	x_7
z	17/4	0	2	0	21/2	0	3/2	5/4
	3/4	0	−2	0	15/2	1	−1/2	3/4
	1	1	−24	0	6	0	2	1
	1	0	0	1	0	0	0	1

Two pivots have resulted in optimal form.

Ways to Prevent Cycling

Pivoting on the least negative c_j does not always prevent cycling, but there are pivot rules that do. Consider the tableau

		x_1	x_2	x_3	x_4	x_5	x_6
z	-5	0	0	-2	-8	0	0
	4	0	0	2	-1	1	0
	6	0	0	3	5	0	1
	0	0	1	-4	1	0	0
	4	1	0	1	-1	0	0

If the x_3 column is selected as the pivot column, then both the first and the second constraint rows are minimum ratio rows. For each of these two rows, that is, for $i = 1$ and $i = 2$, consider the successive ratios

$$\left(\frac{b_i}{a_{i3}}, \frac{a_{i1}}{a_{i3}}, \frac{a_{i2}}{a_{i3}}, \ldots, \frac{a_{in}}{a_{i3}}\right)$$

namely,

$$\left(\tfrac{4}{2}, \tfrac{0}{2}, \tfrac{0}{2}, \tfrac{2}{2}, -\tfrac{1}{2}, \tfrac{1}{2}, \tfrac{0}{2}\right) \quad \text{for } i = 1$$

and

$$\left(\tfrac{6}{3}, \tfrac{0}{3}, \tfrac{0}{3}, \tfrac{3}{3}, \tfrac{5}{3}, \tfrac{0}{3}, \tfrac{1}{3}\right) \quad \text{for } i = 2$$

Compare the successive ratio rows having ties, one column at a time, from left to right, until a column is found where the minimum entry in the column is unique. (In the preceding tableau this happens in column 5.) The row in which this unique minimum occurs is called the **minimum successive ratio row** associated with the selected pivot column k. There will always be a column where the minimum entry is unique because otherwise two or more rows of the tableau would be identical, and this cannot happen because the tableau is in canonical form.

We can now state one pivot rule that guarantees that cycling will not occur, namely, the **successive ratio rule:**

SUCCESSIVE RATIO RULE

- Select the pivot column as any column k with $c_k < 0$.
- Given k, select the pivot row h as the minimum successive ratio row associated with column k.

In the preceding example there are two possible pivot columns. If the x_3 column is chosen, then, as we saw, the first constraint row is the pivot row selected by the successive ratio rule. If the x_4 column is selected as the pivot column, then the usual minimum ratio is unique, and the minimum successive ratio row is this unique minimum ratio row. In other words, for pivot columns where there is no tie in the usual minimum ratio, the successive ratio rule reduces to the simplex rule.

Another pivot rule that also prevents cycling is the recently discovered **smallest index rule:**

SMALLEST INDEX RULE

- Select the pivot column k as that column having the negative cost entry c_k with the smallest index k.
- Given k, select the pivot row h as the minimum ratio row having the smallest row index h.

The smallest index rule always determines a unique pivot position. For the preceding example the pivot position determined by the smallest index rule is in the x_3 column and the first constraint row.

It is not difficult to show that the smallest index rule prevents cycling, but the argument is not particularly instructive and will not be given here. However, the argument that the successive ratio rule prevents cycling is a constructive one, and it provides some additional insight into the process of pivoting. After we consider the nondegenerate case, we conclude this section by showing that the successive ratio rule prevents cycling.

Convergence in the Nondegenerate Case

Although the simplex rule for pivoting may lead to cycling, it does not do so if the linear program is nondegenerate. To see why this is true, we first note that there are at most

$$p = \binom{n}{m} m! = \frac{n!}{(n - m)!}$$

basic sequences when A is $m \times n$. We will show later that if two tableaus are in canonical form with respect to the same basic sequence, and if one tableau is obtained from the other by pivoting, then they must be identical tableaus. Thus, in any sequence of canonical form tableaus obtained by pivoting, at most p can be distinct. This is true even if the problem is degenerate. For a nondegenerate problem each simplex rule pivot gives a strict decrease in the objective function value, so in any sequence of simplex rule pivots starting from a tableau in canonical form, all the tableaus produced must be different. Therefore, either optimal form or unbounded form must be obtained in at most p pivots because otherwise we could generate a sequence of more than p different canonical form tableaus.

When $m = 10$ and $n = 20$, p is about 10^{11}, so even for small linear programs there may be a very large number of canonical form tableaus. One might expect that the simplex algorithm would have to find a great many of these canonical forms before it finds an optimal one. Remarkably, however, empirical evidence accumulated over years of using the simplex algorithm shows that typically only about $1.5m$ to $2m$ canonical forms are examined before an optimal one is found.

Convergence in the Degenerate Case

In the degenerate case, pivoting by the simplex rule does not always give a strict decrease in the objective function value because the pivot row h may have $b_h = 0$. Therefore, as shown by the example of cycling given above, tableaus may repeat if the simplex rule is used on a degenerate problem. However, if we pivot by the successive ratio rule instead of the simplex rule, then we can show that all of the tableaus produced must be different even if the problem is degenerate. To see why this is true we need to look more closely at a canonical form tableau.

A row in a tableau is called **lexicographically positive** if the first nonzero entry in the row is positive. For example, consider the tableau

	x_1	x_2	x_3	x_4	x_5	x_6	x_7
−3	2	2	0	0	−4	0	0
5	−1	3	0	0	1	1	0
0	0	−1	0	1	1	0	0
0	0	1	0	0	1	0	1
2	1	−2	1	0	2	0	0

Here the first, third, and fourth constraint rows are lexicographically positive, but the second constraint row is not. Given any canonical form tableau we can always rearrange the variables so that the identity columns come first, and then each constraint row is lexicographically positive. One such rearrangement in the preceding example gives the tableau

	x_3	x_4	x_6	x_7	x_1	x_2	x_5
−3	0	0	0	0	2	2	−4
5	0	0	1	0	−1	3	1
0	0	1	0	0	0	−1	2
0	0	0	0	1	0	1	1
2	1	0	0	0	1	−2	2

If we start with a tableau where each constraint row is lexicographically positive and pivot by the successive ratio rule, then each constraint row in the next tableau will be lexicographically positive. To see why this is true, we need to examine how each constraint row

$$(b_i, a_{i1}, \ldots, a_{in})$$

in the tableau changes with a pivot on the element $a_{hk} > 0$ in row h and column k. Let

$$(b_i^*, a_{i1}^*, \ldots, a_{in}^*)$$

denote the ith row after the pivot. Then for the pivot row h,

$$(b_h^*, a_{h1}^*, \ldots, a_{hn}^*) = \left(\frac{b_h}{a_{hk}}, \frac{a_{h1}}{a_{hk}}, \ldots, \frac{a_{hn}}{a_{hk}}\right)$$

that is, the new pivot row is just the successive ratios for row h. Because $a_{hk} > 0$ and because the first nonzero element in the old row h is positive, this new row h must be lexicographically positive. For all other constraint rows i with

$i \neq h$, the result of a pivot on element a_{hk} gives

$$(b_i^*, a_{i1}^*, \ldots, a_{in}^*) = (b_i, a_{i1}, \ldots, a_{in}) + (-a_{ik})\left(\frac{b_h}{a_{hk}}, \frac{a_{h1}}{a_{hk}}, \ldots, \frac{a_{hn}}{a_{hk}}\right)$$

That is, after we divide the pivot row by a_{hk}, we multiply it by $(-a_{ik})$ and add it to row i. If $a_{ik} \leq 0$, then this equation shows that the new ith row is lexicographically positive because a nonnegative multiple of a lexicographically positive row is lexicographically positive and the sum of two lexicographically positive rows is also lexicographically positive. We can rewrite the last equation as

$$(b_i^*, a_{i1}^*, \ldots, a_{in}^*) = (a_{ik})\left[\left(\frac{b_i}{a_{ik}}, \frac{a_{i1}}{a_{ik}}, \ldots, \frac{a_{in}}{a_{ik}}\right) - \left(\frac{b_h}{a_{hk}}, \frac{a_{h1}}{a_{hk}}, \ldots, \frac{a_{hn}}{a_{hk}}\right)\right]$$

Note that the term in brackets subtracts the successive ratios of row h from those of row i. Choosing the pivot row h by the successive ratio rule guarantees that the first nonzero difference in this subtraction is positive. So if $a_{ik} > 0$, the last equation shows that the new ith row is lexicographically positive because a positive multiple of a lexicographically positive row is lexicographically positive.

When we pivot on the entry a_{hk}, the new objective function row becomes

$$(-d^*, c_1^*, \ldots, c_n^*) = (-d, c_1, \ldots, c_n) - c_k\left(\frac{b_h}{a_{hk}}, \frac{a_{h1}}{a_{hk}}, \ldots, \frac{a_{hn}}{a_{hk}}\right)$$

that is, we add $(-c_k)$ times the new pivot row to the objective function row. We can rewrite this last equation as

$$(-d^*, c_1^*, \ldots, c_n^*) - (-d, c_1, \ldots, c_n) = (-c_k)\left(\frac{b_h}{a_{hk}}, \frac{a_{h1}}{a_{hk}}, \ldots, \frac{a_{hn}}{a_{hk}}\right)$$

Because $-c_k > 0$, and because the new pivot row is lexicographically positive, we know that the right-hand side of the last equation is a lexicographically positive row.

We will say that a vector $\mathbf{u}$ is **lexicographically greater** than a vector $\mathbf{v}$ if $\mathbf{u} - \mathbf{v}$ is lexicographically positive. The last equation then shows that the new objective function row is lexicographically greater than the old objective function row. Note that if $\mathbf{w}$ is lexicographically greater than $\mathbf{u}$ and if $\mathbf{u}$ is lexicographically greater than $\mathbf{v}$, then $\mathbf{w}$ is lexicographically greater than $\mathbf{v}$; that is, lexicographically greater than is a **transitive relation** (see Exercise 3.13).

In summary, if we start with a tableau in canonical form having all constraint rows lexicographically positive, then performing a sequence of pivots by the successive ratio rule gives a sequence of canonical-form tableaus having all constraint rows lexicographically positive, and for each tableau the objective function row is lexicographically greater than all the previous ones. Thus all the tableaus in the sequence must be different, and we can conclude, just as we did for simplex rule pivots in the nondegenerate case, that no two of these tableaus can be in canonical form with respect to the same basic sequence. As we saw earlier, there are at most

$$p = \frac{n!}{(n-m)!}$$

different canonical form tableaus. Therefore, we must obtain either optimal form or unbounded form in at most p successive ratio rule pivots because otherwise we could generate a sequence of more than p different canonical form tableaus.

3.5 OBTAINING CANONICAL FORM FROM STANDARD FORM

Now that we have shown how to pivot any canonical form tableau to either optimal or unbounded form, we need to consider the problem of getting an initial canonical form for a linear program that is in standard form (see Figure 3.2). A linear program that is in standard form may be infeasible, and thus have no canonical forms, so the search for a canonical form may result in infeasible form 1 or infeasible form 2.

This section shows how pivoting can be used to either obtain canonical form or discover that a linear program is infeasible. The process of trying to get canonical form from standard form is usually referred to as **phase 1** of the simplex algorithm, and trying to obtain optimal form from canonical form is called **phase 2**. Later, in Section 3.8, we describe an alternative phase 1 procedure.

Getting an Identity

Recall that for a tableau to be in canonical form the matrix **A** must contain the columns of the $m \times m$ identity matrix, and the c_j's associated with those columns must be zero. We therefore begin the process of getting canonical form by pivoting to obtain an identity matrix in **A**. Consider the following tableau:

	x_1	x_2	x_3	x_4	x_5
0	1	1	0	0	−2
1	(1)	0	−1	2	−2
4	−1	−1	2	−1	1
5	0	−1	1	1	−1

If we pivot on the circled element, the x_1 column becomes the first identity column, and c_1 becomes zero.

	x_1	x_2	x_3	x_4	x_5
−1	0	1	1	−2	0
1	1	0	−1	2	−2
5	0	(−1)	1	1	−1
5	0	−1	1	1	−1

Continuing this strategy makes the x_2 column the second identity column.

	x_1	x_2	x_3	x_4	x_5
4	0	0	2	−1	−1
1	1	0	−1	2	−2
−5	0	1	−1	−1	1
0	0	0	0	0	0

The row of all zeros in the last tableau shows that the last constraint equation is **redundant**; that is, it depends on the other constraint equations. (In the first tableau the last constraint equation is just the sum of the first two.) Thus this last equation can be eliminated from the last tableau, which results in an equivalent linear program with the tableau

	x_1	x_2	x_3	x_4	x_5
4	0	0	2	−1	−1
1	1	0	−1	2	−2
−5	0	1	−1	−1	1

The only thing that would prevent the above process from obtaining a tableau with an appropriate identity is that a row with each $a_{ij} = 0$ and $b_i \neq 0$ may be obtained, but then the tableau would be in infeasible form 1.

The process of obtaining the identity is called **phase 0** and is summarized as follows.

Phase 0
set $i = 1$;
1 look for an entry $a_{ij} \neq 0$ in row i;
if each $a_{ij} = 0$ and $b_i = 0$, delete row i and go to 2;
if each $a_{ij} = 0$ and $b_i \neq 0$, stop in infeasible form 1;
otherwise, pivot on the first nonzero entry in row i;
2 if $i < n$ replace i by $i + 1$ and go to 1; otherwise stop.

By its construction, this process yields either a tableau that has the m identity columns appearing as columns in $\mathbf{A}$ with the corresponding cost coefficients equal to zero, or a tableau in infeasible form 1.

The Subproblem Technique

As described, the phase 0 technique unfortunately can result in a tableau having some b_i's negative. In that case, since canonical form requires $\mathbf{b} \geq \mathbf{0}$, some additional work must be done to get $\mathbf{b} \geq \mathbf{0}$ while always keeping a set of identity columns with zero costs. To see how this can be accomplished, consider the following example:

	x_1	x_2	x_3	x_4	x_5	x_6	x_7	x_8	x_9
81	0	0	0	2	14	−42	13	0	−12
−55	0	0	0	0	−5	25	−6	1	5
−4	1	0	0	−1	0	10	0	0	1
2	0	1	0	1	−1	1	−1	0	0
1	0	0	1	−1	2	−11	2	0	−1

This tableau is not in canonical form because b_1 and b_2 are negative.

If we pivot in this tableau, the new tableau will surely also have the required identity columns in it, and their associated costs will be zero. Moreover, if the pivot position is properly chosen, b_2 can be increased. If the second constraint row were the objective function row for a linear programming problem, then b_2 would be the negative of the objective function value for that problem, and we could increase b_2 by simply pivoting that linear program toward optimality. This line of reasoning leads us to consider the following **subproblem:**

		x_1	x_2	x_3	x_4	x_5	x_6	x_7	x_8	x_9	
	81	0	0	0	2	14	−42	13	0	−12	
	−55	0	0	0	0	−5	25	−6	1	5	
**	−4	1	0	0	−1	0	10	0	0	1	subproblem
	2	0	1	0	(1)	−1	1	−1	0	0	subproblem
	1	0	0	1	−1	2	−11	2	0	−1	subproblem

The objective function row for the subproblem is marked **, and the constraint rows are all of the original rows having $b_i \geq 0$. Here the objective function entry of the subproblem is -4. The choice of the b_2 row as the subproblem objective function row is arbitrary; we could just as well have selected the b_1 row of the original problem instead because b_1 is also negative. In general, we construct such a subproblem as follows:

> **Constructing the subproblem:**
>
> Let the constraint rows of the subproblem be those constraint rows of the original tableau having nonnegative constant column entries.
>
> Let the subproblem objective function row be a constraint row of the original tableau having a negative constant column entry.

Note that, by its construction, a subproblem is in canonical form, so we can pivot to obtain optimal or unbounded form for the subproblem. The pivot positions are determined by using the simplex rule on the subproblem, but the pivots are performed on the entire tableau.

In our example, pivoting the subproblem to optimality requires two pivots.

		x_1	x_2	x_3	x_4	x_5	x_6	x_7	x_8	x_9	
	77	0	−2	0	0	16	−44	15	0	−12	
	−55	0	0	0	0	−5	25	−6	1	5	
**	−2	1	1	0	0	−1	11	−1	0	1	subproblem
	2	0	1	0	1	−1	1	−1	0	0	subproblem
	3	0	1	1	0	(1)	−10	1	0	−1	subproblem

		x_1	x_2	x_3	x_4	x_5	x_6	x_7	x_8	x_9	
	29	0	−18	−16	0	0	116	−1	0	4	
	−40	0	5	5	0	0	−25	−1	1	0	
**	1	1	2	1	0	0	1	0	0	0	subproblem in optimal form
	5	0	2	1	1	0	−9	0	0	−1	subproblem in optimal form
	3	0	1	1	0	1	−10	1	0	−1	subproblem in optimal form

In this case the subproblem strategy was effective in making b_2 positive. Of course, it can turn out that the optimal value of a subproblem is positive, in which case the corresponding b_i remains negative. Then we would conclude that the original linear program was in infeasible form 2.

To make b_1 positive in the example, we now consider the larger subproblem shown below:

		x_1	x_2	x_3	x_4	x_5	x_6	x_7	x_8	x_9	
	29	0	−18	−16	0	0	116	−1	0	4	
**	−40	0	5	5	0	0	−25	−1	1	0	new larger subproblem
	1	1	2	1	0	0	(1)	0	0	0	
	5	0	2	1	1	0	−9	0	0	−1	
	3	0	1	1	0	1	−10	1	0	−1	

Pivoting the enlarged subproblem toward optimality yields the following tableaus:

		x_1	x_2	x_3	x_4	x_5	x_6	x_7	x_8	x_9	
	−87	−116	−250	−132	0	0	0	−1	0	4	
**	−15	25	55	30	0	0	0	−1	1	0	subproblem
	1	1	2	1	0	0	1	0	0	0	
	14	9	20	10	1	0	0	0	0	−1	
	13	10	21	11	0	1	0	(1)	0	−1	

		x_1	x_2	x_3	x_4	x_5	x_6	x_7	x_8	x_9	
	−74	−106	−229	−121	0	1	0	0	0	3	
**	−2	35	76	41	0	1	0	0	1	(−1)	subproblem in unbounded form
	1	1	2	1	0	0	1	0	0	0	
	14	9	20	10	1	0	0	0	0	−1	
	13	10	21	11	0	1	0	1	0	−1	

Examining the x_9 column in the subproblem shows that the last tableau is in unbounded form and that its objective function entry is still negative (that is, $b_1 = -2$). But now a pivot on the circled entry will make b_1 positive and add nonnegative multiples of that positive value to the other constant column entries. Therefore, such a pivot will always maintain nonnegativity of the b_i's in the current subproblem. Performing the pivot yields the following initial canonical form tableau.

	x_1	x_2	x_3	x_4	x_5	x_6	x_7	x_8	x_9	
−80	−1	−1	2	0	4	0	0	3	0	
2	−35	−76	−41	0	−1	0	0	−1	1	
1	(1)	2	1	0	0	1	0	0	0	canonical form
16	−26	−56	−31	1	−1	0	0	−1	0	
15	−25	−55	−30	0	0	0	1	−1	0	

One further pivot yields optimal form for the original problem.

Pivoting To Form a Subproblem

Before summarizing the entire process of obtaining canonical form, we consider another example starting with the tableau

	x_1	x_2	x_3	x_4	x_5	x_6
75	0	0	5	−1	2	0
−10	1	0	(−1)	0	0	0
−5	0	1	−1	2	−2	0
−15	0	0	−1	1	−1	1

Here it is not possible to form a subproblem immediately because there are no nonnegative b_i's. However, pivoting on a negative a_{ij} will make the b_i in the pivot row positive. Pivoting on the preceding circled entry yields the following tableau:

		x_1	x_2	x_3	x_4	x_5	x_6
	25	5	0	0	−1	2	0
	10	−1	0	1	0	0	0
	5	−1	1	0	2	−2	0
**	−5	(−1)	0	0	1	−1	1

If for any i with $b_i < 0$ it also happened that $a_{ij} \geq 0$ for each j, then we would stop with infeasible form 2. Thus, if the linear program is feasible, it will always be possible to perform such a pivot and get at least one $b_i > 0$. In this example the pivot made both b_1 and b_2 nonnegative, so we can form a subproblem using those rows as the constraints and the row with b_3 as the objective.

This subproblem happens already to be in unbounded form, so we pivot in an unbounded column of the objective function row and obtain the following initial canonical form tableau:

	x_1	x_2	x_3	x_4	x_5	x_6
0	0	0	0	4	−3	5
15	0	0	1	−1	1	−1
10	0	1	0	1	−1	−1
5	1	0	0	−1	1	−1

One further pivot yields optimal form for the original problem.

Summary of the Subproblem Technique

As illustrated by the preceding examples, if a tableau is in standard form with the identity columns appearing as columns in **A** and the corresponding c_j's zero, then the subproblem technique can be applied to determine a sequence of pivots that yields either canonical form or infeasible form 2. In summary, the subproblem technique works as follows.

THE SUBPROBLEM TECHNIQUE

If $\mathbf{b} \geq \mathbf{0}$, then the tableau is in canonical form; stop.
If $b_i < 0$ for each i, then let $i = 1$;

if $a_{ij} \geq 0$ for each j, then the process stops with infeasible form 2;

otherwise, at least one $a_{ij} < 0$; pivot on that entry to get a new tableau with $b_i > 0$.

If at least one $b_i \geq 0$, and at least one $b_i < 0$, then form a subproblem and pivot to obtain optimal form or unbounded form for the subproblem.

If optimal form for the subproblem is obtained and its objective function entry remains negative then the process stops with infeasible form 2 for the original problem.

If unbounded form for the subproblem is obtained and its objective function entry remains negative, then a single pivot will make it positive and allow a larger subproblem to be formed.

Whenever a constant column entry becomes nonnegative in the process of pivoting on a subproblem, its row can be included as a constraint row in a larger subproblem. In particular, as soon as the objective function entry (that is, the b_i) of a subproblem becomes nonnegative, its row can be used in forming a larger subproblem; it is not necessary to retain the old subproblem until optimal or unbounded form is obtained.

Now that we have shown how to obtain canonical form, we can summarize the results needed to solve any linear program in standard form. We do this in the following section by giving a formal statement of the simplex algorithm.

3.6 THE SIMPLEX ALGORITHM

The simplex algorithm for solving any linear program in standard form consists of a sequence of pivots that leads to one of the four final forms.

The Simplex Algorithm

Given a linear program in standard form with simplex tableau

		$x_1 \; x_2 \; \cdots \; x_n$
z	$-d$	$\mathbf{c}^T$
	$\mathbf{b}$	$\mathbf{A}$

where **A** is $m \times n$:

PHASE 1

Pivot using phase 0 and the subproblem technique to obtain an initial canonical form or infeasible form 1 or infeasible form 2.

If canonical form is obtained, go to phase 2; otherwise stop.

PHASE 2

1 If the current canonical form tableau has $c_j \geq 0$, for $j = 1, \ldots, n$, then the current basic feasible solution is optimal. Stop.

Otherwise, select some column k with $c_k < 0$.

If $a_{ik} \leq 0$ for $i = 1, \ldots, m$, then the linear program is unbounded. Stop.

If $a_{ik} > 0$ for some i, pivot by the simplex rule in column k to get a new canonical form. Go to 1.

3.7 REFORMULATING ANY LINEAR PROGRAM INTO STANDARD FORM

Sections 3.4 and 3.5 show how to solve any linear program that is in standard form. In this section we show how to reformulate any linear program so that it is in standard form, completing our description of the procedure outlined in Figure 3.2.

Maximization Problems

Some linear programs are naturally formulated as maximization problems of the form

$$\max \mathbf{c}^T\mathbf{x} \quad \text{subject to} \quad \mathbf{Ax} = \mathbf{b}, \mathbf{x} \geq \mathbf{0}$$

In Section 3.1 we pointed out that since

$$\max\{\mathbf{c}^T\mathbf{x} \mid \mathbf{Ax} = \mathbf{b}, \mathbf{x} \geq \mathbf{0}\} = -\min\{-\mathbf{c}^T\mathbf{x} \mid \mathbf{Ax} = \mathbf{b}, \mathbf{x} \geq \mathbf{0}\}$$

we can instead solve the problem

$$\min -\mathbf{c}^T\mathbf{x} \quad \text{subject to} \quad \mathbf{Ax} = \mathbf{b}, \mathbf{x} \geq \mathbf{0}$$

and remember that the optimal value for the original maximization problem is the negative of the optimal value for the minimization problem. The two problems have the same optimal vectors; it is only the optimal values that are negatives of one another. As an illustration, if the problem were to

$$\max \quad 2x_1 - x_2 - 5x_3 \quad \text{subject to constraints}$$

we would instead

$$\min \ -2x_1 + x_2 + 5x_3 \quad \text{subject to the same constraints}$$

Inequality Constraints

We also mentioned in Section 3.1 that an inequality constraint can be replaced by an equality constraint if an additional nonnegative variable is introduced to represent the amount by which the two sides of the inequality differ. As a further illustration of this process, consider the following example:

$$\begin{array}{ll} \min & 3x_1 - x_2 + 4x_3 \\ \text{subject to} & \\ & x_1 + x_2 + 2x_3 \le 10 \\ & x_1 - x_3 = 5 \\ & 2x_1 - x_2 - x_3 \ge 8 \\ & x_i \ge 0 \quad i = 1, 2, 3 \end{array}$$

Given any vector $x = [x_1, x_2, x_3]^T$, define a slack variable

$$x_4 = 10 - (x_1 + x_2 + 2x_3)$$

The first inequality constraint can now be expressed as $x_4 \ge 0$. Thus the single inequality constraint

$$x_1 + x_2 + 2x_3 \le 10$$

can be replaced by the two constraints

$$\begin{aligned} x_1 + x_2 + 2x_3 + x_4 &= 10 \\ x_4 &\ge 0 \end{aligned}$$

We handle the second inequality constraint in the example in a similar way. First, we multiply it by -1 so that it has the form

$$-2x_1 + x_2 + x_3 \le -8$$

We then introduce the nonnegative slack variable

$$x_5 = -8 - (-2x_1 + x_2 + x_3)$$

and replace the single inequality constraint by the two constraints

$$\begin{aligned} -2x_1 + x_2 + x_3 + x_5 &= -8 \\ x_5 &\ge 0 \end{aligned}$$

Thus, by introducing two slack variables, the original linear program can be equivalently reformulated as

$$\begin{array}{ll} \min & 3x_1 - x_2 + 4x_3 \\ \text{subject to} & \\ & x_1 + x_2 + 2x_3 + x_4 = 10 \\ & x_1 - x_3 = 5 \\ & -2x_1 + x_2 + x_3 + x_5 = -8 \\ & x_i \ge 0 \quad i = 1, \ldots, 5 \end{array}$$

It is clear that this procedure can be used to deal with any number of inequality constraints.

Free Variables

In the linear programming models of Chapter 2, we saw some formulations involving free variables. Recall from that discussion that a free variable is one that is not constrained in sign. The logic of the simplex algorithm depends on having the linear program in standard form, and in particular on the fact that all of the variables are constrained to be nonnegative. Fortunately, it is easy to reformulate any linear program so that all of the variables are nonnegative, and we illustrate two methods of doing so with an example.

Consider the following minimization problem in which some of the variables are free.

$$\begin{array}{ll} \min & 2x_1 + 3x_2 - x_3 + 7x_4 \\ \text{subject to} & \\ & x_1 + 2x_2 - x_3 - 5x_4 = 10 \\ & -2x_1 + 3x_3 - 2x_4 = 5 \\ & x_1 \geq 0 \\ & x_2,\ x_3,\ \text{and } x_4 \text{ free} \end{array}$$

Here only x_1 is constrained to be nonnegative, and the remaining variables are free. The first method for handling free variables uses the following fact.

> Any vector can be written as the difference of two nonnegative vectors.

It is easy to see that this is true. For example, note that

$$\begin{bmatrix} -9 \\ 1 \\ -5 \end{bmatrix} = \begin{bmatrix} 0 \\ 1 \\ 0 \end{bmatrix} - \begin{bmatrix} 9 \\ 0 \\ 5 \end{bmatrix}$$

is one way of writing the vector $[-9, 1, -5]^T$ as the difference of two nonnegative vectors. Thus any free variable x_j can be replaced by the difference of two nonnegative variables, say,

$$x_j = x_j' - x_j'' \quad \text{with } x_j' \geq 0 \text{ and } x_j'' \geq 0$$

As x_j' and x_j'' vary over all nonnegative values, x_j varies over all real values. In our example, replacing each free variable with the difference of two nonnegative variables results in the following reformulated problem that is in standard form:

$$\begin{array}{ll} \min & 2x_1 + 3(x_2' - x_2'') - (x_3' - x_3'') + 7(x_4' - x_4'') \\ \text{subject to} & \\ & x_1 + 2(x_2' - x_2'') - (x_3' - x_3'') - 5(x_4' - x_4'') = 10 \\ & -2x_1 + 3(x_3' - x_3'') - 2(x_4' - x_4'') = 5 \\ & x_1 \geq 0 \quad \begin{bmatrix} x_2' \\ x_3' \\ x_4' \end{bmatrix} \geq \mathbf{0} \quad \begin{bmatrix} x_2'' \\ x_3'' \\ x_4'' \end{bmatrix} \geq \mathbf{0} \end{array}$$

The problem now has seven nonnegative variables, which is three more than in the original problem.

Given a feasible vector for the original problem, it is easy to construct a feasible vector for this reformulated problem. For example, the vector $[4, -9, 1, -5]^T$ is feasible for the original problem and yields an objective value of -55. Letting

$$x_1 = 4 \qquad \begin{bmatrix} x_2' \\ x_3' \\ x_4' \end{bmatrix} = \begin{bmatrix} 0 \\ 1 \\ 0 \end{bmatrix} \qquad \begin{bmatrix} x_2'' \\ x_3'' \\ x_4'' \end{bmatrix} = \begin{bmatrix} 9 \\ 0 \\ 5 \end{bmatrix}$$

gives the components of a vector feasible for the reformulated problem, which also has an objective value of -55. Moreover, given a feasible vector for the reformulated problem with a certain objective value, one can simply use the relations $x_j = x_j' - x_j''$ to obtain a feasible vector for the original problem with the same objective value.

Thus the two problems are **equivalent** in the following sense: Given a feasible vector for one, we can construct a feasible vector for the other yielding the same objective value; if one problem is in optimal form, so is the other; if one is infeasible, so is the other; and if one is unbounded, so is the other.

In general, given a problem of the form

$$\begin{aligned} \min \quad & \mathbf{c}^T\mathbf{x} + \mathbf{a}^T\mathbf{y} \\ \text{subject to} \quad & \\ & \mathbf{A}\mathbf{x} + \mathbf{B}\mathbf{y} = \mathbf{b} \\ & \mathbf{x} \geq \mathbf{0} \qquad \mathbf{y} \text{ free} \end{aligned}$$

we let $\mathbf{y} = \mathbf{u} - \mathbf{v}$, where $\mathbf{u} \geq \mathbf{0}$ and $\mathbf{v} \geq \mathbf{0}$, and solve the problem

$$\begin{aligned} \min \quad & \mathbf{c}^T\mathbf{x} + \mathbf{a}^T(\mathbf{u} - \mathbf{v}) \\ \text{subject to} \quad & \\ & \mathbf{A}\mathbf{x} + \mathbf{B}(\mathbf{u} - \mathbf{v}) = \mathbf{b} \\ & \mathbf{x} \geq \mathbf{0}, \mathbf{u} \geq \mathbf{0}, \mathbf{v} \geq \mathbf{0} \end{aligned}$$

If there are k free variables, this method gives an equivalent problem in standard form having k more variables than the original problem. There is a way of obtaining an equivalent linear program in standard form that only increases the total number of variables by 1. This alternative method uses the following fact:

> Any vector can be written as the difference of two nonnegative vectors with one of the vectors having all components the same.

For example,

$$\begin{bmatrix} -9 \\ 1 \\ -5 \end{bmatrix} = \begin{bmatrix} 0 \\ 10 \\ 4 \end{bmatrix} - \begin{bmatrix} 9 \\ 9 \\ 9 \end{bmatrix}$$

This can always be done in such a way that the vector being subtracted has all components equal to the largest absolute value of the negative components in the original vector. Thus any vector $\mathbf{y} \in \mathrm{R}^k$ can be written

$$\mathbf{y} = \mathbf{u} - \begin{bmatrix} w \\ \vdots \\ w \\ w \\ w \\ \vdots \\ w \end{bmatrix} \quad \text{with} \quad \mathbf{u} \geq \mathbf{0} \text{ and } w \geq 0$$

Note that $\mathbf{u} \in \mathrm{R}^k$ but $w \in \mathrm{R}^1$. It will be convenient to let $\mathbf{e}$ denote a vector having all components equal to 1; that is,

$$\mathbf{e} = [1, 1, \ldots, 1]^T$$

Then

$$\mathbf{y} = \mathbf{u} - \begin{bmatrix} w \\ \vdots \\ w \\ w \\ w \\ \vdots \\ w \end{bmatrix} = \mathbf{u} - w \begin{bmatrix} 1 \\ \vdots \\ 1 \\ 1 \\ 1 \\ \vdots \\ 1 \end{bmatrix} = \mathbf{u} - w\mathbf{e}$$

Suppose now that we are given a linear program of the form

$$\begin{aligned} \min \quad & \mathbf{c}^T\mathbf{x} + \mathbf{a}^T\mathbf{y} \\ \text{subject to} \quad & \\ & \mathbf{A}\mathbf{x} + \mathbf{B}\mathbf{y} = \mathbf{b} \\ & \mathbf{x} \geq \mathbf{0} \qquad \mathbf{y} \text{ free} \end{aligned}$$

Letting $\mathbf{y} = (\mathbf{u} - w\mathbf{e})$ gives the equivalent program in standard form

$$\begin{aligned} \min \quad & \mathbf{c}^T\mathbf{x} + \mathbf{a}^T(\mathbf{u} - w\mathbf{e}) \\ \text{subject to} \quad & \\ & \mathbf{A}\mathbf{x} + \mathbf{B}(\mathbf{u} - w\mathbf{e}) = \mathbf{b} \\ & \mathbf{x} \geq \mathbf{0},\ \mathbf{u} \geq \mathbf{0},\ w \geq 0 \end{aligned}$$

Using only the rules for matrix multiplication, we see that

$$\mathbf{c}^T\mathbf{x} + \mathbf{a}^T(\mathbf{u} - w\mathbf{e}) = \mathbf{c}^T\mathbf{x} + \mathbf{a}^T\mathbf{u} - (\mathbf{a}^T\mathbf{e})w$$

and

$$\mathbf{A}\mathbf{x} + \mathbf{B}(\mathbf{u} - w\mathbf{e}) = \mathbf{A}\mathbf{x} + \mathbf{B}\mathbf{u} - (\mathbf{B}\mathbf{e})w$$

Note that $\mathbf{a}^T\mathbf{e}$ is simply the sum of the components of the vector $\mathbf{a}$ and that $\mathbf{Be}$ is a vector whose components are the sums of the rows of $\mathbf{B}$. Thus, given a linear program with free variables, it is easy to write down a tableau that represents an equivalent standard form for the problem.

We illustrate how this is done using our original problem

$$\text{minimize} \quad 2x_1 + 3x_2 - x_3 + 7x_4$$

subject to

$$\begin{aligned} x_1 + 2x_2 - x_3 - 5x_4 &= 10 \\ -2x_1 + 3x_3 - 2x_4 &= 5 \\ x_1 \geq 0 \quad \text{with} \quad x_2, x_3, x_4 &\ \text{free} \end{aligned}$$

Here, we introduce a single new nonnegative variable w and write down the tableau

		x_1	x_2'	x_3'	x_4'	w
z	0	2	3	−1	7	−9
	10	1	2	−1	−5	4
	5	−2	0	3	−2	−1

The nonnegative variables x_j' are defined by

$$\begin{bmatrix} x_2 \\ x_3 \\ x_4 \end{bmatrix} = \begin{bmatrix} x_2' \\ x_3' \\ x_4' \end{bmatrix} - \begin{bmatrix} w \\ w \\ w \end{bmatrix} = \begin{bmatrix} x_2' \\ x_3' \\ x_4' \end{bmatrix} - w \begin{bmatrix} 1 \\ 1 \\ 1 \end{bmatrix}$$

Note that for the objective function row, the coefficient of the variable w is simply the negative of the sum of the coefficients for the primed variables x_j', $j = 2, 3, 4$. This is also true for each constraint row.

Finally, if $\bar{x}_1, \bar{x}_2', \bar{x}_3', \bar{x}_4'$, and $\bar{w}$ are optimal for the reformulated problem, then

$$\begin{bmatrix} \bar{x}_1 \\ \bar{x}_2' - \bar{w} \\ \bar{x}_3' - \bar{w} \\ \bar{x}_4' - \bar{w} \end{bmatrix}$$

is an optimal vector for the original problem.

3.8 THE METHOD OF ARTIFICIAL VARIABLES

We saw in Section 3.5 how phase 0 and the subproblem technique can be used to obtain an initial canonical form for a linear program that is in standard form. In this section we describe an alternative way of obtaining canonical form from standard form.

Getting $\mathbf{b} \geq \mathbf{0}$

Consider the following tableau for a linear program in standard form:

	x_1	x_2	x_3	x_4	x_5
0	−2	3	2	−1	5
1	1	0	−1	2	−2
−4	1	1	−2	1	−1
5	0	−1	1	1	−1

This tableau is clearly not in canonical form. It does not have the columns of the 3 × 3 identity in **A**, with the corresponding c_j's zero, and it does not have $\mathbf{b} \geq \mathbf{0}$. In Section 3.5 we first used phase 0 to get the identity columns, and then solved subproblems to get $\mathbf{b} \geq \mathbf{0}$. Suppose now that we begin instead by making $\mathbf{b} \geq \mathbf{0}$. This can be accomplished by multiplying any rows having $b_i < 0$ by -1. In our example we would rewrite the tableau as follows.

	x_1	x_2	x_3	x_4	x_5
0	−2	3	2	−1	5
1	1	0	−1	2	−2
4	−1	−1	2	−1	1
5	0	−1	1	1	−1

Because we can always make $\mathbf{b} \geq \mathbf{0}$ in this way, it is without loss of generality for us to assume that we have a linear program of the form

$$\min z = \mathbf{c}^T\mathbf{x} \quad \text{subject to } \mathbf{Ax} = \mathbf{b}, \mathbf{x} \geq \mathbf{0} \text{ with } \mathbf{b} \geq \mathbf{0}$$

Throughout this section we refer to this problem as the **original problem.**

The Artificial Problem

Associated with the original problem is the following **artificial problem:**

$$\min \quad y_1 + y_2 + \cdots + y_p$$

subject to

$$\mathbf{Ax} + \mathbf{Iy} = \mathbf{b}, \mathbf{x} \geq \mathbf{0}, \mathbf{y} \geq \mathbf{0},$$

where **A** is $p \times n$. By its construction, the artificial problem contains the columns of the $p \times p$ identity matrix in the term **Iy**. The variables $y_1, \ldots, y_p$ are called **artificial variables** because they are introduced only for the purpose of providing the identity columns. Note that the objective function for the original problem plays no role in forming the artificial problem.

The objective function for the artificial problem is always nonnegative because it is the sum of nonnegative variables, and so the artificial problem cannot have an unbounded minimum value. Also, $\mathbf{b} \geq \mathbf{0}$ implies that the vector $\mathbf{x} = \mathbf{0}$ and $\mathbf{y} = \mathbf{b}$ is feasible for the artificial problem, and so it cannot end up in one of the infeasible forms. This means that the artificial problem must have an optimal vector and that its minimum value must be nonnegative.

Because we can always make $\mathbf{b} \geq \mathbf{0}$ before forming the artificial problem, and because the artificial problem provides the columns of the $m \times m$ identity matrix in **A**, the artificial problem is almost in canonical form. For our example the artificial problem has the following tableau:

	x_1	x_2	x_3	x_4	x_5	y_1	y_2	y_3
0	0	0	0	0	0	1	1	1
1	1	0	−1	2	−2	1	0	0
4	−1	−1	2	−1	1	0	1	0
5	0	−1	1	1	−1	0	0	1

To obtain canonical form, all we need to do is zero out the costs associated with the identity columns. This can be done by simply adding -1 times each constraint row to the objective row, which produces the following canonical form tableau for the artificial problem:

	x_1	x_2	x_3	x_4	x_5	y_1	y_2	y_3
−10	0	2	−2	−2	2	0	0	0
1	1	0	−1	2	−2	1	0	0
4	−1	−1	(2)	−1	1	0	1	0
5	0	−1	1	1	−1	0	0	1

The method of artificial variables proceeds by first solving the artificial problem.

Feasibility of the Original Problem

Before we consider the use of the artificial problem for obtaining an initial canonical form for the original problem, it is important to understand how solving the artificial problem shows whether or not the original problem is feasible.

> The original problem is feasible if and only if the artificial problem has a minimum value of 0.

To see why this is true, suppose first that $\bar{\mathbf{x}}$ is feasible for the original problem. Then with $\bar{\mathbf{y}} = \mathbf{0}$, the pair $(\bar{\mathbf{x}}, \bar{\mathbf{y}})$ is feasible for the artificial problem and $\bar{y}_1 + \cdots + \bar{y}_p = 0$. Because all other feasible points $(\mathbf{x}, \mathbf{y})$ for the artificial problem have $y_1 + \cdots + y_p \geq 0$, we can conclude that its minimum value is zero. Now suppose that the artificial problem has a minimum value of zero attained at the point $(\mathbf{x}^*, \mathbf{y}^*)$. Given $\mathbf{y}^* \geq \mathbf{0}$ and $y_1^* + \cdots + y_p^* = 0$, it follows that $\mathbf{y}^* = \mathbf{0}$. Then $\mathbf{Ax}^* + \mathbf{Iy}^* = \mathbf{b}$ reduces to $\mathbf{Ax}^* = \mathbf{b}$ and so $\mathbf{x}^*$ is feasible for the original problem.

Thus, if the minimum value of the artificial problem is positive, then the original problem is infeasible and we stop. Otherwise, the minimum value is zero (because it cannot be negative).

Canonical Form for the Original Problem

From the optimal tableau for the artificial problem it is easy to obtain an initial canonical form for the original problem. We illustrate how this is done by continuing with our example. Pivoting according to the simplex algorithm on the above canonical form tableau for the artificial problem yields the following sequence of tableaus.

	x_1	x_2	x_3	x_4	x_5	y_1	y_2	y_3
−6	−1	1	0	−3	3	0	1	0
3	1/2	−1/2	0	(3/2)	−3/2	1	1/2	0
2	−1/2	−1/2	1	−1/2	1/2	0	1/2	0
3	1/2	−1/2	0	3/2	−3/2	0	−1/2	1

	x_1	x_2	x_3	x_4	x_5	y_1	y_2	y_3
0	0	0	0	0	0	2	2	0
2	1/3	−1/3	0	1	−1	2/3	1/3	0
3	−1/3	−2/3	1	0	0	1/3	2/3	0
0	0	0	0	0	0	−1	−1	1

The last tableau is in optimal form for the artificial problem and its minimum value is zero. Thus, in view of the above result, the original problem must have at least one feasible point.

To obtain an initial canonical form for the original problem from the optimal tableau for an artificial problem, we consider two cases:

(i) At least one artificial variable is a basic variable in the optimal tableau.

(ii) All artificial variables are nonbasic.

Note that when the optimal value for the artificial problem is zero, any artificial variable that is basic in the optimal tableau must have the corresponding constant column entry equal to zero.

The preceding optimal tableau for the artificial problem is in case (i) and

> All the coefficients of the variables x_j in the row corresponding to the basic artificial variable are zero.

Because the constant column entry b_3 is also zero, this means that the third constraint in the original problem is redundant. Therefore, we can delete the third row in the artificial problem along with the column corresponding to the basic artificial variable y_3. This results in the tableau

	x_1	x_2	x_3	x_4	x_5	y_1	y_2
0	0	0	0	0	0	2	2
2	1/3	−1/3	0	1	−1	2/3	1/3
3	−1/3	−2/3	1	0	0	1/3	2/3

which has one less artificial variable basic. In fact, here this results in case (ii).

Of course, it is possible for the optimal form of an artificial problem to be in case (i) and

> some coefficients of the variables x_j in the row corresponding to the basic artificial variable are nonzero.

For example, suppose we ended up with the following optimal tableau for an artificial problem.

	x_1	x_2	x_3	y_1	y_2
0	2	1	0	2	0
1	1	2	1	1	0
0	−2	(−1)	0	−1	1

Here the artificial variable y_2 is basic and not all coefficients of the variables x_j in the first constraint row are zero. In this case we select one of these nonzero coefficients and pivot on it to obtain a tableau for the artificial problem having one less basic artificial variable. For example, performing the preceding marked pivot yields the following tableau having all artificial variables nonbasic.

	x_1	x_2	x_3	y_1	y_2
0	0	0	0	1	1
1	−3	0	1	−1	2
0	2	1	0	1	−1

This tableau is in case (ii) for the artificial problem.

Thus, given an optimal tableau for the artificial problem in case (i), the above examples illustrate how we can obtain case (ii) by either deleting rows or performing pivots to get all artificial variables nonbasic.

Suppose then we have an optimal tableau in case (ii) for the artificial problem. An initial canonical form for the original problem is obtained as follows:

- Delete the artificial objective function row.
- Delete the columns corresponding to the artificial variables.

In our first example this yields the partial tableau

	x_1	x_2	x_3	x_4	x_5
2	1/3	−1/3	0	1	−1
3	−1/3	−2/3	1	0	0

corresponding to the constraint equations in the original problem. Then, to obtain the desired canonical form, we

- add the original objective function to the partial tableau and pivot to get zeros above the identity columns.

This process is illustrated in the following tableaus:

	x_1	x_2	x_3	x_4	x_5	
0	−2	3	2	−1	5	add the original objective
2	1/3	−1/3	0	(1)	−1	
3	−1/3	−2/3	1	0	0	

	x_1	x_2	x_3	x_4	x_5	
2	−5/3	8/3	2	0	4	
2	1/3	−1/3	0	1	−1	pivot makes $c_4 = 0$
3	−1/3	−2/3	(1)	0	0	

	x_1	x_2	x_3	x_4	x_5	
−4	−1	4	0	0	4	
2	1/3	−1/3	0	1	−1	pivot makes $c_3 = 0$
3	−1/3	−2/3	1	0	0	

Of course, this final tableau is in canonical form.

Actually, it is not necessary to add an artificial variable for each constraint row in the original problem. One need only introduce enough artificial variables so that the artificial problem has an appropriate identity matrix (see Exercise 3.25).

The method of artificial variables is somewhat simpler to implement on a computer than is the method discussed in Section 3.5 of using phase 0 and subproblems, so practical computer codes for solving linear programs use the method of artificial variables for phase 1 of the simplex algorithm.

3.9 PIVOT MATRICES AND THE REVISED SIMPLEX METHOD

Most practical computer codes for solving linear programs are not straightforward implementations of the simplex algorithm as it is presented in Section 3.6. One important refinement is based on the observation that it is not really necessary to compute all of the elements of each tableau in solving a linear program; we need to compute only enough of the elements to determine the next pivot position. In selecting a pivot position by the simplex rule, we could examine cost coefficients only until finding the first negative c_k, then compute the column of coefficients a_{ik} below c_k, and finally find the constant column elements b_i for which $a_{ik} > 0$. These tableau entries are the only ones needed to determine a next pivot position, and the other elements of the tableau need not even be found. Of course, we would have to keep track of the pivots we performed and remember the current basic sequence at each step of the process, so that a final tableau could be constructed once optimal or unbounded form was attained. This typically requires considerably less work than updating the whole tableau at each iteration, particularly for large problems.

In this section we present a **revised simplex algorithm** that uses the idea just outlined to organize the calculations so that they are faster, numerically more stable, and require less storage than the basic algorithm of Section 3.6

Pivot Matrices

The revised simplex method keeps track of what pivots have been performed by means of pivot matrices. A **pivot matrix** is a square matrix $\mathbf{Q}$ such that premultiplying a tableau

$M =$

$-d$	$\mathbf{c}^T$
$\mathbf{b}$	$\mathbf{A}$

by $\mathbf{Q}$ has the same effect as performing a sequence of pivots on M. Such matrices are an important tool in the revised simplex method, and understanding how they are constructed also provides further insight into pivoting.

If M_1 is obtained from a tableau M by a single pivot, then it is actually quite easy to obtain a pivot matrix $\mathbf{Q}_1$ such that $M_1 = \mathbf{Q}_1 M$. We simply use the following slogan:

Do unto the identity as you would do unto M.

To illustrate this slogan and the construction of a pivot matrix $\mathbf{Q}_1$ so that $M_1 = \mathbf{Q}_1 M$, consider the following tableau:

$M =$

	x_1	x_2	x_3	x_4
-3	0	1	0	-2
3	1	1	0	1
2	0	-4	1	②

Using the slogan, the pivot matrix $\mathbf{Q}_1$ is obtained by successively "doing unto the identity matrix $\mathbf{I}$ as one would do unto M" while performing the pivot. Because tableau M has three rows, the identity we use here is

$$\mathbf{I}_3 = \begin{bmatrix} 1 & 0 & 0 \\ 0 & 1 & 0 \\ 0 & 0 & 1 \end{bmatrix}$$

The last row of this 3×3 identity is multiplied by $1/2$ to obtain the last row of $\mathbf{Q}_1$, then -1 times that row is added to the second row of the identity, and finally 2 times that row is added to the first row of the identity. Those are the row operations that we would perform on M to do the pivot. This gives the matrix

$$\mathbf{Q}_1 = \begin{bmatrix} 1 & 0 & 1 \\ 0 & 1 & -1/2 \\ 0 & 0 & 1/2 \end{bmatrix}$$

Note that the first column of a pivot matrix is always the first identity column because the objective function row of a tableau is never a pivot row. From the rules for matrix multiplication, $\mathbf{Q}_1 M$ is just the tableau that results from performing the pivot on M. Thus

$M_1 = \mathbf{Q}_1 M =$

	x_1	x_2	x_3	x_4
-1	0	-3	1	0
2	1	3	$-1/2$	0
1	0	-2	$-1/2$	1

Actually, matrix $\mathbf{Q}_1$ can be obtained directly from tableau M_1. Because M is in canonical form, it contains all the columns of the 3×3 identity except the first. But the first column of $\mathbf{Q}_1$ is always the first identity column, so we need only know what the other identity columns are transformed into by the pivot operations. These identity columns are in M, however, so we know that they are transformed into the corresponding columns of M_1.

If a sequence of pivots (that is, a sequence of row operations) is performed on M to obtain M^*, then to obtain a matrix $\mathbf{Q}$ such that $M^* = \mathbf{Q}M$ we simply use the slogan and successively perform the same sequence of row operations on the identity.

The Revised Simplex Method

To apply the revised simplex method we need to have first obtained canonical form. Thus, suppose that for a certain linear program we have the following canonical form tableau:

$M_0 =$

	x_1	x_2	x_3	x_4	x_5	x_6	x_7
−4	0	1	−3	2	−5	1	0
2	1	−1	⓵	1	2	−1	0
1	0	4	−2	1	1	3	1

Pivoting by the simplex rule, we might select the circled entry as the pivot position. We would then perform the pivot to obtain the next tableau, M_1, and continue pivoting if M_1 is not in optimal or unbounded form. In the revised simplex method not all of tableau M_1 is computed. Only enough of M_1 is found to determine the next pivot. We illustrate this in our example where

$$M_1 = \mathbf{Q}_1 M_0 \quad \text{and} \quad \mathbf{Q}_1 = \begin{bmatrix} 1 & 3 & 0 \\ 0 & 1 & 0 \\ 0 & 2 & 1 \end{bmatrix}$$

Note that to form the pivot matrix $\mathbf{Q}_1$ we use only the entries in the pivot column of M_0. We now find just enough of M_1 to determine the next pivot. Using matrix multiplication, we calculate entries in the objective function row until the first negative c_j is discovered. Then we calculate the entries in that column and the entries that are needed in the constant column. That is, using

$$M_1 = \mathbf{Q}_1 M_0 = \begin{bmatrix} 1 & 3 & 0 \\ 0 & 1 & 0 \\ 0 & 2 & 1 \end{bmatrix}$$

	x_1	x_2	x_3	x_4	x_5	x_6	x_7
−4	0	1	−3	2	−5	1	0
2	1	−1	1	1	2	−1	0
1	0	4	−2	1	1	3	1

we calculate the following entries of M:

$M_1 =$

	x_1	x_2	x_3	x_4	x_5	x_6	x_7
*	3	−2	*	*	*	*	*
*	*	−1	*	*	*	*	*
5	*	⓶	*	*	*	*	*

The next pivot matrix $\mathbf{Q}_2$ does not depend on the $*$ entries in M_1, and they are not calculated. In fact, from the preceding partial information about M_1, the next tableau M_2 is given by

$$M_2 = \mathbf{Q}_2 M_1 \quad \text{where} \quad \mathbf{Q}_2 = \begin{bmatrix} 1 & 0 & 1 \\ 0 & 1 & 1/2 \\ 0 & 0 & 1/2 \end{bmatrix}$$

Because we do not know all of M_1, we use

$$M_2 = \mathbf{Q}_2 M_1 = \mathbf{Q}_2(\mathbf{Q}_1 M_0) = (\mathbf{Q}_2\mathbf{Q}_1)M_0$$

Thus, $M_2 = \mathbf{P}_2 M_0$ where $\mathbf{P}_2 = \mathbf{Q}_2\mathbf{Q}_1$. Matrix $\mathbf{P}_2$ is a pivot matrix that upon premultiplying M_0 has the effect of performing the pivot of $\mathbf{Q}_1$ followed by the pivot of $\mathbf{Q}_2$. Actually, calculating $\mathbf{Q}_2\mathbf{Q}_1$ does not require as much work as a usual matrix multiplication because we know exactly what $\mathbf{Q}_2$ does to a matrix upon premultiplication. It adds the third row to the first row; it adds half of the third row to the second row; and it multiplies the third row by $\frac{1}{2}$. Thus, given $\mathbf{Q}_1$, we easily calculate $\mathbf{P}_2 = \mathbf{Q}_2\mathbf{Q}_1$ as

$$\mathbf{P}_2 = \begin{bmatrix} 1 & 5 & 1 \\ 0 & 2 & 1/2 \\ 0 & 1 & 1/2 \end{bmatrix}$$

Now, with this **updated pivot matrix** $\mathbf{P}_2$, we continue the process and calculate just enough of M_2 to determine the next pivot. Using

$$M_2 = \mathbf{P}_2 M_0 = \begin{bmatrix} 1 & 5 & 1 \\ 0 & 2 & 1/2 \\ 0 & 1 & 1/2 \end{bmatrix}$$

	x_1	x_2	x_3	x_4	x_5	x_6	x_7
−4	0	1	−3	2	−5	1	0
2	1	−1	1	1	2	−1	0
1	0	4	−2	1	1	3	1

we calculate the following entries of M_2:

$M_2 =$

	x_1	x_2	x_3	x_4	x_5	x_6	x_7
*	5	0	0	8	6	−1	*
*	*	*	*	*	*	−1/2	*
5/2	*	*	*	*	*	(1/2)	*

Here the next pivot matrix $\mathbf{Q}_3$ is given by

$$\mathbf{Q}_3 = \begin{bmatrix} 1 & 0 & 2 \\ 0 & 1 & 1 \\ 0 & 0 & 2 \end{bmatrix}$$

Now with $\mathbf{P}_3 = \mathbf{Q}_3\mathbf{P}_2$ we calculate enough of $M_3 = \mathbf{P}_3 M_0$ to determine the next pivot if M_3 is not in optimal or unbounded form. As indicated by $\mathbf{Q}_3$, $\mathbf{P}_3$ is obtained from $\mathbf{P}_2$ by adding twice $\mathbf{P}_2$'s third row to its first row; adding its third row to its second row; and multiplying its third row by 2. Thus

$$\mathbf{P}_3 = \begin{bmatrix} 1 & 7 & 2 \\ 0 & 3 & 1 \\ 0 & 2 & 1 \end{bmatrix}$$

In general, suppose k pivots have been determined and $\mathbf{P}_k$ denotes the updated pivot matrix. We then calculate enough of the next tableau

$$M_k = \mathbf{P}_k M_0$$

to determine the $(k + 1)$st pivot if M_k is not in optimal or unbounded form. Let $\mathbf{Q}_{k+1}$ be the pivot matrix for the $(k + 1)$st pivot. With

$$\mathbf{P}_{k+1} = \mathbf{Q}_{k+1}\mathbf{P}_k,$$

we continue the process until optimal or unbounded form is attained.

Of course, optimal form is attained if no negative c_j is found when the objective function row is calculated. In our example using $\mathbf{P}_3 M_0$ to calculate the objective function row and the constant column of M_3 yields

$$M_3 = \mathbf{P}_3 M_0 = \begin{bmatrix} 1 & 7 & 2 \\ 0 & 3 & 1 \\ 0 & 2 & 1 \end{bmatrix} \begin{array}{|c|ccccccc|} \hline & x_1 & x_2 & x_3 & x_4 & x_5 & x_6 & x_7 \\ -4 & 0 & 1 & -3 & 2 & -5 & 1 & 0 \\ \hline 2 & 1 & -1 & 1 & 1 & 2 & -1 & 0 \\ 1 & 0 & 4 & -2 & 1 & 1 & 3 & 1 \\ \hline \end{array}$$

$$= \begin{array}{|c|ccccccc|} \hline & x_1 & x_2 & x_3 & x_4 & x_5 & x_6 & x_7 \\ 12 & 7 & 2 & 0 & 11 & 11 & 0 & 2 \\ \hline 7 & * & * & * & * & * & * & * \\ 5 & * & * & * & * & * & * & * \\ \hline \end{array}$$

Thus M_3 is in optimal form, and if we knew its basic sequence, we could then read off an optimal basic feasible solution.

Of course, it is easy to keep track of the current basic sequence while performing iterations of the revised simplex method. In our example the initial tableau M_0 has the basic sequence

$$S_0 = (1, 7)$$

Because we know the pivot position at each pivot, we can keep track of how the basic sequence changes. At each pivot the index of the basic variable that becomes nonbasic is replaced by the index of the nonbasic variable that becomes basic. Thus the basic sequences for M_1, M_2, and M_3 above are, respectively,

$$S_1 = (3, 7)$$
$$S_2 = (3, 2)$$
$$S_3 = (3, 6)$$

Given the final basic sequence S_3 and the partial information about M_3, we know that an optimal basic feasible solution is given by

$$x_3^* = 7, x_6^* = 5 \quad \text{with} \quad x_j^* = 0 \text{ otherwise; and } z(\mathbf{x}^*) = -12$$

Tableaus with the Same Basic Sequence

By using pivot matrices, it is easy to obtain the important result that we used earlier to show that the simplex algorithm converges. The result we used was the following.

> If a tableau M^* is obtained from a tableau M by a sequence of pivots, and if M and M^* are both in canonical form with respect to the same basic sequence, then $M = M^*$.

To see why this result is true, we use the fact that $M^* = \mathbf{Q}M$ where $\mathbf{Q}$ is a pivot matrix. To show that $M^* = M$, it suffices to show that $\mathbf{Q} = \mathbf{I}$. We know that the first column of $\mathbf{Q}$ is the first column of $\mathbf{I}$. If $\mathbf{I}^k$ denotes the kth identity column, then the column of M^* containing $\mathbf{I}^k$ is precisely the column of M that contains $\mathbf{I}^k$. But by the definition of matrix multiplication, a given column of M^* is obtained by multiplying the corresponding column of M by $\mathbf{Q}$. Thus

$$\mathbf{I}^k = \mathbf{Q}\mathbf{I}^k \quad \text{for } k = 2, \ldots, m$$

but, again by the definition of matrix multiplication, $\mathbf{Q}\mathbf{I}^k$ is simply the kth column of $\mathbf{Q}$, namely, $\mathbf{Q}^k$. Therefore,

$$\mathbf{I}^k = \mathbf{Q}^k \quad \text{for } k = 2, \ldots, m$$

and so $\mathbf{Q} = \mathbf{I}$ and $M^* = M$.

3.10 COMPUTER SOLUTION OF LINEAR PROGRAMS

We now know how to solve any linear programming problem, but for large problems we must, of course, use a computer to perform the pivot calculations. Computers, however, do not use exact rational arithmetic, and roundoff errors are introduced into the calculations. Because of this we must be careful to ensure that an implementation of the simplex algorithm on a computer performs in the way it should theoretically. In this section we discuss some consequences of not using rational arithmetic in performing a sequence of pivots. We also discuss some techniques that have been successful in reducing the time required to solve linear programs.

Computer Cycling

Cycling means that, while performing a sequence of pivots, two tableaus are obtained that are in canonical form with respect to the same basic sequence. If rational arithmetic is used, then, as we noted in Section 3.4, the two tableaus must be identical. Thus the same sequence of pivots could be repeated indefinitely without obtaining optimal or unbounded form. Cycling that occurs when rational arithmetic is used is called **classical cycling.** The example of cycling given in Section 3.4 was, of course, an example of classical cycling. Cycling that occurs because the arithmetic is done using the finite precision of a computer is called **computer cycling.**

To illustrate the difference between classical and computer cycling, consider the following canonical form tableau:

		x_1	x_2	x_3	x_4	x_5	x_6	x_7
z	3	$-3/4$	20	$-1/2$	6	0	0	0
	10^{-9}	$1/4$	-8	-1	9	1	0	0
	0	(1/2)	-12	$-1/2$	3	0	1	0
	1	0	0	1	0	0	0	1

This tableau is identical to the one used to exhibit classical cycling in the example of Section 3.4, except that the b_1 entry is 10^{-9} instead of 0.

If rational arithmetic is used, pivoting by the simplex rule on this tableau does not lead to cycling. In fact, if the first simplex rule pivot is the one circled in the preceding tableau, then the following sequence of tableaus is obtained.

		x_1	x_2	x_3	x_4	x_5	x_6	x_7
z	3	0	2	$-5/4$	21/2	0	3/2	0
	10^{-9}	0	-2	$-3/4$	15/2	1	$-1/2$	0
	0	1	-24	-1	6	0	2	0
	1	0	0	(1)	0	0	0	1

		x_1	x_2	x_3	x_4	x_5	x_6	x_7
z	17/4	0	2	0	21/2	0	3/2	5/4
	$3/4 + 10^{-9}$	0	-2	0	15/2	1	$-1/2$	3/4
	1	1	-24	0	6	0	2	1
	1	0	0	1	0	0	0	1

Thus optimal form is obtained after two pivots. Even though this example does not exhibit classical cycling, it could exhibit computer cycling. Suppose, for example, that a particular implementation set $\epsilon = 10^{-8}$ as the tolerance for zero; that is, numbers less than 10^{-8} are considered to be zero. Such an implementation would permit the same sequence of pivots that lead to classical cycling in Section 3.4. Thus this example would exhibit computer cycling because the seventh tableau would have the same basic sequence as the first. Of course, when computer cycling occurs, it is extremely unlikely because of roundoff errors that the tableaus associated with a repeating basic sequence will be numerically identical.

Computer cycling can often be prevented by the same methods discussed in Section 3.4 for preventing classical cycling. However, because of the computational cost of schemes like the successive ratio rule for pivoting, and because of the inflexibility in choosing pivots by the smallest index rule, practical computer codes instead either ignore the problem altogether or manage numerical degeneracy by a variety of other means, including the temporary introduction of small perturbations into the problem data when cycling is detected.

Controlling Roundoff Errors

The roundoff errors that occur using finite precision arithmetic can sometimes lead to obtaining a tableau that is in optimal form, but the associated basic feasible solution $\mathbf{x}^*$ is not feasible and even if it is feasible, it may not be optimal. That is, if we substitute $\mathbf{x}^*$ into the constraints of the original tableau, it might not satisfy the equations to within an acceptable tolerance. Also, if we replace the objective function row in the final tableau with the original objective function row and then perform the calculations to make the cost coefficients of the basic variables equal to zero, it might happen that some nonbasic variable has a negative cost coefficient. These problems occur because of the accumulation of roundoff errors in the sequence of pivots used to solve the problem.

One way of controlling the accumulation of roundoff errors is called **reinversion.** Given a tableau M, suppose after a certain number of pivots a tableau M^* is obtained that is in canonical form with respect to the basic sequence $S = (s_1, \ldots, s_m)$. From our discussion of pivot matrices in Section 3.9 it is easy to obtain a pivot matrix $\mathbf{Q}$ such that

$$\mathbf{Q}M^* = M$$

We simply let the first column of $\mathbf{Q}$ be the first identity column; and for $i = 1, \ldots, m$, column $i + 1$ of $\mathbf{Q}$ is the column in M corresponding to the basic variable x_{s_i} in M^*. For example, if

$M =$

*	* · · · *	−2	3
*	* · · · *	1	−2
*	* · · · *	4	5

and

$M^* =$

*	* · · · *	0	0
*	* · · · *	0	1
*	* · · · *	1	0

$$\mathbf{Q} = \begin{bmatrix} 1 & 3 & -2 \\ 0 & -2 & 1 \\ 0 & 5 & 4 \end{bmatrix}$$

The important thing here is that $\mathbf{Q}$ is formed using the columns of the original tableau M and thus $\mathbf{Q}$ does not have any accumulated roundoff errors. Since $\mathbf{Q}M^* = M$, we could in principle use the inverse $\mathbf{Q}^{-1}$ of $\mathbf{Q}$ to recompute M^* as

$$M^* = \mathbf{Q}^{-1}M$$

and this accounts for the fact that the technique is referred to as reinversion. Given $\mathbf{Q}$, at most m pivots are required to find its inverse $\mathbf{Q}^{-1}$, even though many more than m pivots may have been used in finding M^* from M originally. In practice, rather than computing $\mathbf{Q}^{-1}$ explicitly, one might simply pivot in M to obtain the final basic sequence associated with M^*, and this also requires no more than m pivots. Because less arithmetic is performed in reconstructing M^* from the original data in the most direct way, the roundoff errors introduced are usually less than those in the version of M^* that was originally obtained by the larger number of pivots. The tableau found by reinversion is the one used for subsequent pivots, and after a certain number of pivots on that tableau another reinversion might be done.

Tolerances and Errors

Because of roundoff errors, it would be remarkable if the basic feasible solution $\bar{\mathbf{x}}$ of a current tableau precisely satisfied the original equality constraints. For this reason feasibility tolerances must be used in practical computer implementations. If

$$\mathbf{A}_i\mathbf{x} = b_i$$

is the ith constraint equation in the original problem, then the ***i*th row error** associated with the current solution $\mathbf{x}$ is given by

$$|\mathbf{A}_i\bar{\mathbf{x}} - b_i| \qquad i = 1, \ldots, m$$

Some implementations require these row errors to be computed every so often, and if these errors exceed some prespecified tolerance, then a reinversion is required.

Another kind of tolerance that is usually set in most implementations is the **pivot rejection tolerance.** If a coefficient a_{hk} is very near zero, then it is rejected as a pivot element. A tolerance of 10^{-6} is typical.

Other Practical Considerations

Practical computer programs for solving real linear programming problems typically use the revised simplex method and incorporate various refinements designed to increase execution speed and reduce storage requirements. A few commonly used refinements will be described below to suggest the sort of algorithmic considerations that are important in the design of a practical code.

In searching for a pivot column, production codes typically construct a **candidate list** of 5 to 10 columns, and pick the best column from among those few. At each iteration only enough columns are computed to yield the required number of candidate columns, and the remaining columns of the tableau are not even calculated.

In most large linear programming problems, the coefficient matrix is **sparse**, containing mostly zero entries. When this is true, considerable memory can be saved by representing the pivot matrix used in the revised simplex method as a product of elementary matrices, and computing time can be saved by avoiding unnecessary arithmetic operations. Because an elementary matrix is a square matrix differing from the identity in only one row or column, it can be stored very compactly by recording only the elements of the nonunit vector column and its position in the matrix. Periodically, to limit the number of elementary matrices in the representation, a new pivot matrix is computed from the original problem data and a new compact representation is found for it in terms of elementary matrices. If the problem is sparse enough, then the work required to do this is more than compensated for by the efficiencies that the representation permits. In addition to saving storage and time, representation of the pivot matrix as a product of elementary matrices also helps to control the propagation of roundoff errors.

Many real linear programming problems include constraints that are simple bounds on the variables. Unfortunately, the size of such a problem may be greatly increased if slack and surplus variables are used to convert the bound constraints into equalities. Because of the special structure of bound constraints, it is possible to avoid such an increase in problem size by instead enforcing the bounds through appropriate changes to the rules for pivoting and optimality testing. The new rules are more complicated than those we have presented here, but the use of such a bounding method instead of slack and surplus variables can result in significant savings of both time and storage for problems having many bound constraints.

It is often possible to save time and storage by capitalizing on special structure in the coefficient matrix of a linear programming problem. For example, when the rows and columns of a tableau can be arranged in such a way that the nonzero elements form a staircase pattern of blocks coupled together by only a few rows, then it may be possible to decompose the problem into several relatively easy subproblems and one master program. The subproblems have objective functions containing variable parameters that are adjusted by the master program, and an iterative technique is used to repeatedly solve the subproblems with parameter values adjusted to account for the coupling between them. There is an extensive literature on decomposition and other techniques for handling problems with special structure, and such methods have occasionally been used in practical codes for very large problems.

Some linear programming problems have structures that are so special that they lead naturally to very special algorithms; an example is the transportation problem discussed in Chapter 7. Although the method we will use for the transportation problem is really just a variant of the simplex algorithm, it may be hard to recognize it as such because it is so highly specialized to take advantage of problem structure.

SELECTED REFERENCES

Beale, E. M. L., Cycling in the Dual Simplex Algorithm, *Naval Research Logistics Quarterly* **2:**269–275 (1955).

Bland, R. G., New Finite Pivoting Rules for the Simplex Method, *Mathematics of Operations Research* **2:**103–107 (1977).

Dantzig, G. B., *Linear Programming and Extensions,* Princeton University Press, Princeton, 1963.

Gass, S. I., *Linear Programming: Methods and Applications* (4th ed.), McGraw-Hill, New York, 1969.

Gass, S. I., Comments on the Possibility of Cycling with the Simplex Method, *Operations Research* **27:**848–852 (1979).

Kotiah, T. C. T., and Steinberg, D. I., On the Possibility of Cycling with the Simplex Method, *Operations Research* **26:**374–376 (1978).

Lasdon, L. S., *Optimization Theory for Large Systems,* Macmillan, New York, 1970.

Murtagh, B. A., *Advanced Linear Programming,* McGraw-Hill, New York, 1981.

Murty, K. G., *Linear Programming,* Wiley, New York, 1983.

Orchard-Hays, W., *Advanced Linear Programming Computing Techniques,* McGraw-Hill, New York, 1968.

Spivey, A. W., and Thrall, R. M., *Linear Optimization,* Holt, Rinehart, and Winston, New York, 1970.

EXERCISES

3.1 Consider the following linear program:

$$\begin{aligned} \text{maximize} \quad & 2x_1 + 3x_2 \\ \text{subject to} \quad & x_1 + x_2 \le 4 \\ & -x_1 + 2x_2 \ge -1 \\ & x_1 \ge 0, x_2 \ge 0 \end{aligned}$$

(a) Reformulate an equivalent problem that is in standard form.

(b) Put the data for the reformulated problem into a simplex tableau and pivot, if necessary, to obtain an initial canonical form.
(c) Perform simplex rule pivots on the tableau of part (b) to obtain an optimal vector.
(d) Draw the feasible set for the original problem in x_1x_2 space and label the points in this set that correspond to the basic feasible solutions obtained while performing the pivots in part (c).

3.2 (a) Reformulate the following linear program into standard form:

$$\begin{aligned} \text{minimize} \quad & -2x_1 + x_2 \\ \text{subject to} \quad & x_1 + 2x_2 \le 10 \\ & x_1 - x_2 \le -5 \\ & x_1 \ge 0, x_2 \ge 0 \end{aligned}$$

(b) Show that the reformulated problem is equivalent to the original problem in the following sense: Given a feasible point for one of the problems, one can construct a feasible point for the other problem that has the same objective function value. Give a precise rule for constructing such vectors.
(c) If two linear programs are equivalent in the sense discussed in part (b), and if one problem has a finite optimal value, then why must the other problem have the same optimal value?
(d) If two linear programs are equivalent in the sense discussed in part (b), and if one problem is feasible but has no finite optimal value, then why must this also be true of the other problem?
(e) If two linear programs are equivalent in the sense discussed in part (b), and if one problem has no feasible vectors, then why must this be true of the other problem as well?

3.3 The following tableau represents a linear program in standard form:

	x_1	x_2	x_3	x_4	x_5
-9	0	b	e	0	0
a	1	c	1	0	0
2	0	d	-1	1	0
4	0	-1	1	0	1

Give conditions on the parameters a, b, c, d, and e so that

(i) the tableau is in optimal form,
(ii) the tableau is in unbounded form,
(iii) the tableau is in infeasible form 2,
(iv) the tableau is in optimal form and the feasible region is unbounded.

3.4 (a) Reformulate the following linear program into standard form:

$$\begin{aligned} \text{maximize} \quad & 5x_1 + 7x_2 - 2x_3 + 3x_4 - 6x_5 \\ \text{subject to} \quad & x_1 + x_2 + x_3 + x_4 - x_5 = 1 \\ & x_i \ge 0 \quad i = 1, \ldots, 5 \end{aligned}$$

(b) Perform a single pivot that will result in an initial canonical form that is also in optimal form.
(c) Give a rule for simply writing down an optimal solution to any linear program of the form

$$\text{maximize} \quad \mathbf{c}^T\mathbf{x} \quad \text{subject to} \quad x_1 + \cdots + x_n = 1 \quad \text{and} \quad \mathbf{x} \ge \mathbf{0}$$

3.5 (a) Solve the following linear program:

$$\begin{aligned} \text{minimize} \quad & 2x_1 - 7x_2 - 3x_3 \\ \text{subject to} \quad & x_1 + 2x_2 + x_3 \le 5 \\ & 2x_1 \qquad + x_3 \le 10 \\ & x_i \ge 0 \qquad i = 1, 2, 3 \end{aligned}$$

(b) Is the linear program degenerate? Explain.

3.6 Solve the following linear program:

$$\begin{aligned} \text{maximize} \quad & -2x_1 - x_2 + 4x_3 \\ \text{subject to} \quad & 3x_1 - x_2 + 2x_3 \le 25 \\ & -x_1 - x_2 + 2x_3 \le 20 \\ & -x_1 - x_2 + x_3 \le 5 \\ & x_i \ge 0 \qquad i = 1, 2, 3 \end{aligned}$$

3.7 Consider the following linear program in canonical form:

$$\begin{aligned} \text{minimize} \quad & 2x_1 - x_2 + 3x_4 \\ \text{subject to} \quad & x_1 + x_2 + x_4 + x_5 = 10 \\ & 3x_1 - x_2 + x_3 - 2x_4 = 6 \\ & x_i \ge 0 \qquad i = 1, \ldots, 6 \end{aligned}$$

(a) Write down a canonical form simplex tableau M that represents this linear program.
(b) What is the basic sequence S associated with this canonical form tableau and which variables are basic and nonbasic?
(c) Write down the basic feasible solution associated with the tableau M in part (a).
(d) Show that solving the first constraint equation for x_2 in terms of the other variables and algebraically substituting this expression for x_2 into the objective function and the second constraint equation gives the same representation of the linear program that is obtained by performing a simplex rule pivot on the x_2 column of tableau M in part (a).

3.8 (a) Use the subproblem technique to find an initial canonical form for the linear program

$$\begin{aligned} \text{minimize} \quad & 2x_1 - x_2 + x_3 + x_4 \\ \text{subject to} \quad & x_1 + 2x_2 \qquad - x_4 = 5 \\ & -x_1 + x_2 - 2x_3 + x_4 = 10 \\ & x_i \ge 0 \qquad i = 1, \ldots, 4 \end{aligned}$$

(b) Use the method of artificial variables to find an initial canonical form for the linear program in (a).

3.9 The following tableau is the initial canonical form that we obtained in Section 3.1 for the brewery problem:

	x_1	x_2	x_3	x_4	x_5	x_6	x_7
0	−6	−5	−3	−7	0	0	0
50	1	1	0	3	1	0	0
180	2	1	2	1	0	1	0
80	1	1	1	4	0	0	1

Obtain an optimal form for this linear program.

3.10 Consider the following canonical form tableau:

	x_1	x_2	x_3	x_4	x_5	x_6	x_7
−9	0	−3	−7	0	0	4	0
2	0	2	1	1	0	−1	0
6	−1	−1	3	0	0	3	1
4	0	4	2	0	1	−2	0

(a) If the x_2 column is selected as the pivot column, what is the pivot row selected by the successive ratio rule?
(b) If the x_3 column is selected as the pivot column, what is the pivot row selected by the successive ratio rule?
(c) What is the pivot column and pivot row selected by the smallest index rule?
(d) Suppose the element $b_2 = 6$ were replaced by $b_2 = 0$. How could we rewrite the linear program so that all the constraint rows would be lexicographically positive?

3.11 (a) For a tableau in canonical form, if $c_k < 0$ and there is a tie for the minimum ratio row, as is the case for the tableau in Exercise 3.10, why can we conclude that the linear program is degenerate?
(b) Could we draw the same conclusion if the pivot column k had $c_k \geq 0$? Explain.

3.12 Use the method of artificial variables to solve the following linear programming problems:

(a) minimize $2x_1 + x_2$

subject to

$$x_1 - x_2 \geq -1$$
$$x_1 - x_2 \leq 1$$
$$x_1 \geq 0, x_2 \geq 0$$

(b) minimize $2x_1 + x_2$

subject to

$$x_1 - x_2 \leq -1$$
$$x_1 - x_2 \geq 1$$
$$x_1 \geq 0, x_2 \geq 0$$

3.13 Show that if **w**, **u**, and **v** are all vectors of the same size and if **w** is lexicographically greater than **u** and if **u** is lexicographically greater than **v**, then **w** is lexicographically greater than **v**.

3.14 Solve the chicken and the egg problem of Section 2.2.

3.15 At times $t_1 = 1$, $t_2 = 2$, $t_3 = 4$, and $t_4 = 5$, measurements of a quantity y are found to equal 5, 13, 30, and 45, respectively. Find parameters a, b, and c for which the curve

$$y(t) = at^2 + bt + c$$

best fits these measured values in the sense that the following sum is minimized:

$$\sum_{i=1}^{4} |D_i|$$

where $D_i = (at_i^2 + bt_i + c) -$ (observed value of y at time t_i). See the curve-fitting problem in Section 2.3 for a linear programming formulation of this problem.

3.16 Solve the standard form linear programs with the following initial simplex tableaus:

(a)

	x_1	x_2	x_3	x_4	x_5
3	1	0	0	1	0
−1	1	1	0	−1	0
−4	−1	0	1	−1	0
1	1	0	0	0	1

(b)

	x_1	x_2	x_3	x_4	x_5
−1	0	0	−1	0	1
−1	1	0	0	2	−1
−1	0	0	1	1	−1
−2	−3	1	5	0	−2

(c)

	x_1	x_2	x_3	x_4	x_5	x_6
0	2	5	3	0	0	0
5	1	1	0	1	0	0
15	2	1	2	0	1	0
8	1	1	1	0	0	1

(d)

	x_1	x_2	x_3	x_4	x_5	x_6	x_7	x_8
0	−1	0	2	1	−3	4	0	4
10	1	0	−1	2	1	−3	4	1
−5	0	1	1	−3	−1	2	1	−1
0	1	−1	2	−1	−1	0	0	−2

3.17 Find an optimal vector for the linear program

$$\begin{aligned} \text{minimize} \quad & x_1 + 2x_2 - x_3 \\ \text{subject to} \quad & x_1 - x_2 + x_3 \leq 1 \\ & x_1 + x_2 - 2x_3 \leq 4 \\ & x_1 \geq 0,\ x_2 \text{ and } x_3 \text{ free} \end{aligned}$$

3.18 (a) Reformulate the following linear program into standard form:

$$\begin{aligned} \text{maximize} \quad & x_1 - x_2 + 2x_3 \\ \text{subject to} \quad & 2x_1 + 3x_2 \leq 4 \\ & x_1 - x_3 \geq 2 \\ & x_1 + 2x_2 = 1 \\ & x_3 \geq 0,\ x_1 \text{ and } x_2 \text{ free} \end{aligned}$$

(b) Solve the preceding linear program.

3.19 Find a nonnegative solution to the following system of equations or show that no such solution exists.

$$\begin{aligned} x_1 + x_2 - x_4 + x_5 &= -1 \\ -x_1 + x_3 + x_4 + x_6 &= 1 \\ x_2 - x_3 + x_7 &= -1 \end{aligned}$$

3.20 Consider the following linear program:

$$\begin{array}{ll} \text{minimize} & 4x_1 \qquad - x_3 \\ \text{subject to} & x_1 + x_2 - x_3 = -1 \\ & -x_1 + x_2 \qquad = 1 \\ & x_i \geq 0 \quad i = 1, 2, 3 \end{array}$$

(a) Use the subproblem technique to find an initial canonical form.
(b) Use the method of artificial variables to find an initial canonical form.

3.21 (a) Formulate a linear programming problem for finding a vector satisfying

$$4x_1 + x_2 \leq 5 \quad \text{and} \quad x_1 \geq 0, x_2 \geq 0$$

and having the maximum of

$$2x_1 - x_2 \quad \text{and} \quad -3x_1 + 2x_2$$

as small as possible.
(b) Find a solution of the linear program.

3.22 If in the linear program

$$\text{minimize } \mathbf{c}^T\mathbf{x} \quad \text{subject to} \quad \mathbf{Ax} \leq \mathbf{b} \text{ and } \mathbf{x} \geq \mathbf{0},$$

the vectors **c** and **b** are nonnegative, show that the vector $\mathbf{x} = \mathbf{0}$ is optimal.

3.23 True or False
(a) Given a linear program in canonical form, if we pivot using the simplex rule, then a strict decrease in the objective function value is obtained.
(b) If the feasible region for a linear program is unbounded, then no finite optimal value exists.
(c) If a linear program is in canonical form, then a pivot by the successive ratio rule yields a tableau with each constraint row being lexicographically positive.
(d) A linear program in canonical form always has a feasible solution.
(e) If **A** is a 3×6 matrix, a linear program with the feasible set $\{\mathbf{x} \mid \mathbf{Ax} = \mathbf{b}, \mathbf{x} \geq \mathbf{0}\}$ may have 125 basic feasible solutions.
(f) The tableau below for a linear program in standard form shows that the linear program has no finite minimum value.

−2	−1	0	1	0
1	−1	1	0	1
0	−2	0	1	−1

(g) Every tableau in standard form can be put into canonical form by an appropriate sequence of pivots.
(h) If a linear program is feasible and if an artificial variable remains as a basic variable after the artificial problem has been solved, then the corresponding constant column entry could be positive.
(i) If a tableau is in canonical form and if the cost coefficient of a nonbasic variable is negative, then the associated basic feasible solution could not be optimal.

3.24 For a linear program in standard form with the tableau

	x_1	x_2	x_3	x_4	x_5	x_6
−9	0	4	5	0	2	0
4	1	2	4	0	1	0
2	0	−1	−1	0	1	1
5	0	3	1	1	−1	0

the optimal basic feasible solution is $\mathbf{x} = [4, 0, 0, 5, 0, 2]^T$. Find the next best basic feasible solution.

3.25 (a) Use a variation of the method of artificial variables that uses only one artificial variable to find an initial canonical form for the linear program in standard form with the tableau

	x_1	x_2	x_3	x_4	x_5
0	0	1	2	−1	4
6	1	−1	0	3	−3
5	0	1	−1	−1	1

(b) Solve the linear program.

3.26 Suppose that tableau M^* is obtained from tableau M by a sequence of pivots, where

$M =$

	x_1	x_2	x_3	x_4	x_5
−9	0	−8	10	−1	0
5	0	−1	−5	3	1
2	1	1	1	−1	0

$M^* =$

	x_1	x_2	x_3	x_4	x_5
38.5	12.5	0	0	0	4.5
3.5	0.5	0	−2	1	0.5
5.5	1.5	1	−1	0	0.5

Find a pivot matrix $\mathbf{Q}$ such that $M^* = \mathbf{Q}M$.

3.27 (a) Pivot on the cycling example of Section 3.4 using the pivot column as the one with the most negative cost coefficient but with the pivot row selected by the successive ratio rule. Show that optimal form can be obtained in five such pivots.
(b) Pivot on the cycling example using the smallest index rule and show that optimal form is obtained after six pivots. Note that the first five pivots are identical to the ones that lead to cycling.

3.28 (a) Write down a linear program for finding a nonnegative vector $\mathbf{x}$ satisfying

$$2x_1 + x_2 \leq 10 \text{ and } \mathbf{x} \geq \mathbf{0}$$

and having $|x_1 - 2x_2| + |-3x_1 - x_2|$ as small as possible.
(b) Solve the resulting linear program.

3.29 Suppose that the following tableau M is the initial canonical form tableau for a linear program that is being solved by the revised simplex method.

$M =$

	x_1	x_2	x_3	x_4	x_5	x_6	x_7
0	2	0	−2	7	0	5	0
20	−4	0	2	3	1	2	0
110	3	1	−1	1	0	1	0
80	1	0	4	4	0	−1	1

(a) Write down a pivot matrix **Q** such that **Q**M gives the next tableau M_1
(b) Using the following empty tableau M_1, fill in the entries that would be calculated by the revised simplex method in determining the next pivot. Do not fill in more entries than are necessary to determine the next pivot (only six entries need to be filled in).

M_1 =

	x_1	x_2	x_3	x_4	x_5	x_6	x_7

(c) Use the entries in M_1 to determine the next pivot matrix and update **Q** so that **Q**M gives the next tableau M_2.

3.30 Find two points, $\bar{\mathbf{x}}$ and $\hat{\mathbf{x}}$, in the region defined by

$$\begin{aligned} x_1 - 2x_2 + 5x_3 &\leq 25 \\ x_1 + x_2 + x_3 &\leq 10 \\ -2x_1 + 3x_2 - 7x_3 &\leq 20 \\ x_i \geq 0 \qquad i &= 1, 2, 3 \end{aligned}$$

so that $\mathbf{c}^T\bar{\mathbf{x}} - \mathbf{c}^T\hat{\mathbf{x}}$ is as large as possible where

$$\mathbf{c}^T\mathbf{x} = x_1 - x_2 + 2x_3$$

3.31 Consider the following canonical form tableau and its associated basic feasible solution, $\bar{\mathbf{x}}$.

	x_1	x_2	x_3	x_4	x_5
−8	0	−5	4	0	0
4	0	1	−1	1	0
2	1	−1	4	0	0
6	0	2	8	0	1

$$\bar{\mathbf{x}} = \begin{bmatrix} 2 \\ 0 \\ 0 \\ 4 \\ 6 \end{bmatrix}$$

Find a vector **v** such that the basic feasible solution, $\hat{\mathbf{x}}$, obtained by performing the simplex rule pivot on the tableau can be written as

$$\hat{\mathbf{x}} = \bar{\mathbf{x}} + \lambda\mathbf{v}$$

where λ is the minimum ratio associated with the pivot, that is, $\lambda = 3$.

CHAPTER 4

GEOMETRY OF THE SIMPLEX ALGORITHM

In Chapter 3 we were mainly concerned with algebraic properties of linear programs and with the various forms of simplex tableaus. In this chapter we discuss important geometric interpretations of many of those algebraic properties and give a geometric interpretation of the simplex algorithm.

4.1 GEOMETRY OF PIVOTING

We will use the following linear program to illustrate most of the new ideas in this section.

$$\begin{aligned}
&\text{minimize} \quad -2x_1 - x_2 \\
&\text{subject to} \\
&\qquad \begin{array}{rcl} x_1 + x_2 & \le & 6 \\ x_1 - x_2 & \le & 2 \\ x_1 \qquad & \le & 3 \\ x_2 & \le & 6 \end{array} \quad \text{and} \quad \begin{bmatrix} x_1 \\ x_2 \end{bmatrix} \ge \mathbf{0}
\end{aligned}$$

Graphical Representation of the Feasible Set

Because our example problem has only two variables, we can graphically represent the feasible set X as the shaded region in Figure 4.1. This sort of

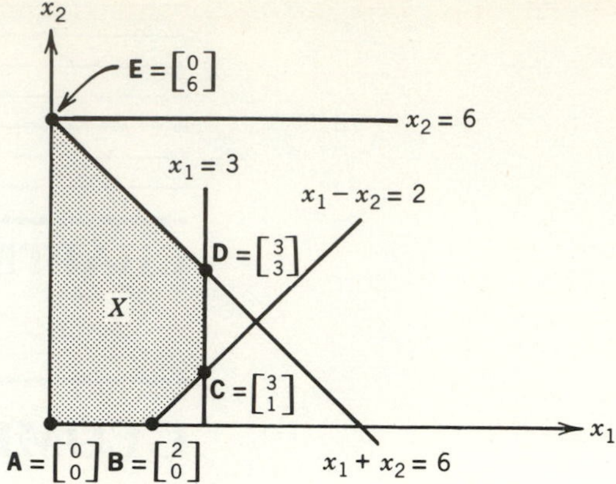

FIGURE 4.1 Graphical Representation of the Feasible Set.

picture should be familiar from Chapter 2, where we used a graphical representation to introduce the idea of linear programming.

The set of points satisfying a linear inequality constraint is called the **half-space** associated with the inequality. For example, in the preceding linear program, the set

$$\{\mathbf{x} \mid x_1 + x_2 \leq 6\}$$

is the half-space associated with the constraint $x_1 + x_2 \leq 6$. The feasible set for any linear program is the intersection of a finite number of half-spaces. In our example the feasible set X is the intersection of the six half-spaces associated with the four inequality constraints and the two nonnegativity constraints.

In general, the half-space associated with an inequality constraint $\mathbf{a}^T\mathbf{x} \leq \text{b}$ is

$$\{\mathbf{x} \mid \mathbf{a}^T\mathbf{x} \leq \text{b}\}$$

and the set

$$\{\mathbf{x} \mid \mathbf{a}^T\mathbf{x} = \text{b}\}$$

is called the **hyperplane** associated with the half-space. In two-dimensional space, R^2, a hyperplane is a one-dimensional line; in R^3 a hyperplane is a two-dimensional plane. In the above example there are six hyperplanes (lines) associated with the constraints, namely, the lines bounding the feasible region X in Figure 4.1.

Points **A, B, C, D,** and **E** in Figure 4.1 are special feasible points. For one thing each of these points is the intersection of two or more of the hyperplanes. Note that for each of the points **A, B, C,** and **D,** there are exactly two hyperplanes that intersect at the point. Point **E,** however, is at the intersection of three hyperplanes, namely, $x_1 + x_2 = 6$, $x_1 = 0$, and $x_2 = 6$. We shall see that this means the linear program is degenerate; that is, some canonical form for this problem will have at least one $b_i = 0$.

Points **A** through **E** are also special because they are at vertices of the feasible set. A feasible point is a **vertex** or **extreme point** of a feasible set if it is not the midpoint of some line segment contained in the feasible set. For example, point **D** in Figure 4.1 is an extreme point, but the point $[1, 0]^T$ is not an extreme point because it is the midpoint of the line segment from **A** to **B,** which is contained in X. For convenience in referring to line segments, we let

$$[\mathbf{P}, \mathbf{Q}]$$

denote the line segment between points **P** and **Q.** By examining Figure 4.1, it is easy to see that points **A, B, C, D,** and **E** are the only extreme points of the feasible set X.

An **edge** of a feasible region is a line segment between two extreme points such that no point on the line segment is the midpoint of two distinct feasible points not on the line segment. For example, in Figure 4.1, [**A, B**] is an edge but [**A, D**] is not.

Extreme Points and Basic Feasible Solutions

The following fact may at first seem remarkable, but we shall see that it is a natural consequence of the structure of linear programming problems.

> The extreme points of the feasible set correspond to the basic feasible solutions associated with the canonical form tableaus for the linear program.

Before showing why this is true in general, we first illustrate the correspondence by using the preceding example. In that example adding four slack variables S_i, $i = 1, \ldots, 4$, yields a linear program in canonical form with the following tableau:

$T_{\mathbf{A}} =$

	x_1	x_2	s_1	s_2	s_3	s_4
0	−2	−1	0	0	0	0
6	1	1	1	0	0	0
2	1	−1	0	1	0	0
3	1	0	0	0	1	0
6	0	1	0	0	0	1

Suppose we now apply the simplex algorithm and pivot this tableau to optimal form. This results in the following sequence of tableaus:

$T_{\mathbf{B}} =$

	x_1	x_2	s_1	s_2	s_3	s_4
4	0	−3	0	2	0	0
4	0	2	1	−1	0	0
2	1	−1	0	1	0	0
1	0	(1)	0	−1	1	0
6	0	1	0	0	0	1

$T_{\mathbf{C}} =$

	x_1	x_2	s_1	s_2	s_3	s_4
7	0	0	0	−1	3	0
2	0	0	1	(1)	−2	0
3	1	0	0	0	1	0
1	0	1	0	−1	1	0
5	0	0	0	1	−1	1

$T_D =$

	x_1	x_2	s_1	s_2	s_3	s_4
9	0	0	1	0	1	0
2	0	0	1	1	−2	0
3	1	0	0	0	1	0
3	0	1	1	0	−1	0
3	0	0	−1	0	1	1

The important thing to note is that the basic feasible solution associated with the initial tableau T_A has x_1 and x_2 components that are precisely those of point **A.** The same correspondence holds between the tableaus T_B, T_C, and T_D and the extreme points **B, C,** and **D,** respectively.

Thus, pivoting by the simplex algorithm gives a sequence of basic feasible solutions that correspond to extreme points of the feasible region. In the above example the extreme points (basic feasible solutions) visited by the simplex algorithm are on a path of consecutive edges, and this is always true whenever the simplex algorithm is applied to a linear program in canonical form.

Two extreme points connected by an edge are called **adjacent,** and in the example each pivot results in moving from one extreme point to an adjacent extreme point. This is always true if the linear program is nondegenerate but, as we shall see, it is not necessarily true if the linear program is degenerate because a pivot may yield the same extreme point.

Of course, there may be more than one path of adjacent extreme points that the simplex algorithm can follow to reach the optimal extreme point. From the feasible region for our example in Figure 4.1, it is clear that there are only two different paths from **A** to the optimal point **D,** but in higher dimensions there are usually a large number of such paths. In R^3, for example, it is not hard to imagine a feasible set for which there are hundreds of possible paths that the simplex algorithm could follow (see Exercise 4.9). The path that is followed depends on the pivot columns chosen in the successive tableaus. For example, in tableau T_A above, choosing the x_1 column as the pivot column results in increasing the nonbasic variable x_1 and going along the edge [**A, B**] until x_1 reaches the value of the minimum ratio for that column, that is, until $x_1 = 2$, which occurs at the extreme point **B.** The subsequent pivots lead from **B** to **C** and finally to the optimal extreme point **D.** If the x_2 column is selected for the pivot column in tableau T_A, then the following alternative sequence of tableaus leads to optimal form:

$T_A =$

	x_1	x_2	s_1	s_2	s_3	s_4
0	−2	−1	0	0	0	0
6	1	1	1	0	0	0
2	1	−1	0	1	0	0
3	1	0	0	0	1	0
6	0	(1)	0	0	0	1

$T_E =$

	x_1	x_2	s_1	s_2	s_3	s_4
6	−2	0	0	0	0	1
0	(1)	0	1	0	0	−1
8	1	0	0	1	0	1
3	1	0	0	0	1	0
6	0	1	0	0	0	1

$T_{E'}$ =

	x_1	x_2	s_1	s_2	s_3	s_4
6	0	0	2	0	0	−1
0	1	0	1	0	0	−1
8	0	0	−1	1	0	2
3	0	0	−1	0	1	(1)
6	0	1	0	0	0	1

T_D =

	x_1	x_2	s_1	s_2	s_3	s_4
9	0	0	1	0	1	0
3	1	0	0	0	1	0
2	0	0	1	1	−2	0
3	0	0	−1	0	1	1
3	0	1	1	0	−1	0

This time the sequence of basic feasible solutions corresponds to the path from **A** to **E** and then to the optimal point **D**, and three pivots are required. It is important to note, however, that not each pivot in the above sequence results in a new basic feasible solution. The tableau T_E shows that the linear program is degenerate, and although the pivot results in tableau $T_{E'}$ with a new basic sequence, the new basic feasible solution is the same as the one associated with T_E.

If in T_A we select a_{12} as the pivot position, the simplex algorithm follows the path **A** to **E** to **D** in exactly two pivots. The following sequence of tableaus is generated on this path:

T_A =

	x_1	x_2	s_1	s_2	s_3	s_4
0	−2	−1	0	0	0	0
6	1	(1)	1	0	0	0
2	1	−1	0	1	0	0
3	1	0	0	0	1	0
6	0	1	0	0	0	1

$T_{E''}$ =

	x_1	x_2	s_1	s_2	s_3	s_4
6	−1	0	1	0	0	0
6	1	1	1	0	0	0
8	2	0	1	1	0	0
3	1	0	0	0	1	0
0	−1	0	−1	0	0	1

T_D =

	x_1	x_2	s_1	s_2	s_3	s_4
9	0	0	1	0	1	0
3	0	1	1	0	−1	0
2	0	0	1	1	−2	0
3	1	0	0	0	1	0
3	0	0	−1	0	1	1

The final tableau is the same as the tableaus T_D obtained earlier, except that the constraint rows are permuted. The sequence of pivots from T_A to T_E to $T_{E'}$ to T_D illustrated how a basic feasible solution can be repeated when the simplex algorithm is used to solve a linear program that is degenerate. However, as we saw in Section 3.4, such repetition does not necessarily happen, and the sequence of pivots from $T_{\mathbf{A}}$ to $T_{\mathbf{E}''}$ to $T_{\mathbf{D}}$ above illustrates that fact. This example also shows that pivoting in the column with the most negative c_j does not always produce optimal form in the smallest number of pivots.

The General Case

Above we used an example to illustrate that basic feasible solutions correspond to extreme points of the feasible set for a linear program. We will now show in general that a basic feasible solution for a linear program in canonical form is an extreme point of the feasible set defined by

$$\{\mathbf{x} \mid \mathbf{Ax} = \mathbf{b}, \mathbf{x} \geq \mathbf{0}\}$$

Suppose that the linear program

$$\min \mathbf{c}^T\mathbf{x} \qquad \text{subject to} \qquad \mathbf{Ax} = \mathbf{b}, \mathbf{x} \geq \mathbf{0}$$

is in canonical form and that $\bar{\mathbf{x}}$ is the associated basic feasible solution. To show that $\bar{\mathbf{x}}$ is an extreme point, we need to show that it cannot be the midpoint of a line segment between two distinct feasible points, say $\mathbf{x}^0$ and $\mathbf{x}^1$. We will do this by showing that if $\bar{\mathbf{x}}$ is assumed to be the midpoint of the line segment $[\mathbf{x}^0, \mathbf{x}^1]$, then $\mathbf{x}^0$ and $\mathbf{x}^1$ must be the same point.

If $\bar{\mathbf{x}}$ is the midpoint of the line segment $[\mathbf{x}^0, \mathbf{x}^1]$, then we write it algebraically as

$$\bar{\mathbf{x}} = \tfrac{1}{2}\mathbf{x}^0 + \tfrac{1}{2}\mathbf{x}^1$$

This vector equation means that the components of the vector $\bar{\mathbf{x}}$ are equal to the corresponding components of the vector $\frac{1}{2}\mathbf{x}^0 + \frac{1}{2}\mathbf{x}^1$. In particular,

$$\bar{x}_j = \tfrac{1}{2}x_j^0 + \tfrac{1}{2}x_j^1$$

for each index j corresponding to a nonbasic variable. But $\bar{x}_j = 0$ for each such index and, since $x_j^0/2 \geq 0$ and $x_j^1/2 \geq 0$, it follows that $x_j^0 = 0$ and $x_j^1 = 0$. So the components of $\mathbf{x}^0$ and $\mathbf{x}^1$ corresponding to the nonbasic components of $\bar{\mathbf{x}}$ must be zero.

Now suppose that the basic sequence of the tableau is $S = (k_1, \ldots, k_m)$ so that $\bar{x}_{k_i} = b_i$. Because $\mathbf{Ax}^0 = \mathbf{b}$ and $x_j^0 = 0$ for all nonbasic indices, it must be that $x_{k_i}^0 = b_i$. For example, if the canonical form tableau looked like

	x_1	x_2	x_3	x_4	x_5
$-d$	0	0	c_3	c_4	0
b_1	0	1	a_{13}	a_{14}	0
b_2	0	0	a_{23}	a_{24}	1
b_3	1	0	a_{33}	a_{34}	0

with basic sequence $S = (2, 5, 1)$

then the fact that $\mathbf{x}^0$ satisfies the constraint equations and that $x_3^0 = 0 = x_4^0$ means that $x_1^0 = b_3$, $x_2^0 = b_1$, and $x_5^0 = b_2$. Similarly, $x_{k_i}^1 = b_i$; thus, $\bar{\mathbf{x}} = \mathbf{x}^0 = \mathbf{x}^1$ and so the basic feasible solution $\bar{\mathbf{x}}$ cannot be the midpoint of a line segment between two distinct feasible points. Therefore $\bar{\mathbf{x}}$ is an extreme point of the feasible set.

Alternative Representation of the Feasible Set

The feasible region for the example in Figure 4.1 is defined in terms of the inequalities on the variables x_1 and x_2, and the slack variables s_1 through s_4 that are used in the reformulation to obtain equality constraints do not appear in Figure 4.1. However, for each point $\bar{\mathbf{x}} = (\bar{x}_1, \bar{x}_2)$ in Figure 4.1, we can readily determine the values of the slack variables $\bar{\mathbf{s}} = (\bar{s}_1, \bar{s}_2, \bar{s}_3, \bar{s}_4)$ so that

$$\begin{bmatrix} \bar{\mathbf{x}} \\ \bar{\mathbf{s}} \end{bmatrix}$$

satisfies the equality constraints of the reformulated problem. The value of the slack variable $\bar{s}_i$ is the amount by which the point $\bar{\mathbf{x}}$ fails to satisfy the ith inequality constraint, $\mathbf{A}_i\mathbf{x} \leq b_i$, with equality. An inequality constraint that is satisfied with equality is said to be **active** or **tight,** and an inequality constraint that is satisfied as a strict inequality is said to be **inactive.** For example, at point **C** in Figure 4.1, the second and third inequality constraints are tight because point **C** lies on the hyperplanes defined by those inequalities, but the first and fourth inequality constraints are not active because $\mathbf{A}_1\mathbf{x} < b_1$ and $\mathbf{A}_4\mathbf{x} < b_4$. Thus, at point **C,** s_2 and s_3 must be zero, while s_1 and s_4 must be positive. Tableau T_C, which represents point **C,** shows that this is indeed true, and it gives the actual values $s_1 = 2$ and $s_4 = 5$.

Each hyperplane used in defining the feasible region in Figure 4.1 can be identified as the set of points where one of the variables of the problem equals zero. For example, the hyperplane

$$\{(x_1, x_2) \mid x_1 + x_2 = 6\}$$

can be identified as the hyperplane where $s_1 = 0$, because $s_1 = 6 - (x_1 + x_2)$. Thus the hyperplane associated with the ith inequality constraint is identified as the one where $s_i = 0$. Similarly, the hyperplanes associated with the nonnegativity constraints are identified as the ones where $x_1 = 0$ and $x_2 = 0$. Figure 4.2 below uses these identifications for the constraint hyperplanes to provide an alternative description of the feasible set X.

Examining Figure 4.2 and keeping in mind the preceding interpretation of the slack variables, we can see from the geometry of the feasible set that a pivot from **A** to **B** results in increasing x_1 and decreasing the slack variables s_1, s_2,

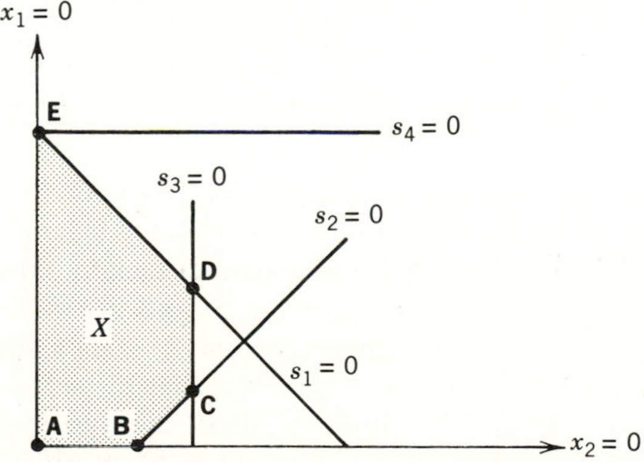

FIGURE 4.2 Alternative Representation of the Feasible Set.

and s_3. A pivot from **B** to **C** results in increasing both x_1 and x_2 and decreasing the slack variables s_1, s_3, and s_4. Finally, the pivot from **C** to the optimal extreme point **D** results in further increasing x_2 while increasing the slack variable s_2 and decreasing the slack variables s_1 and s_4. Of course, the numerical values of all these changes are easily found from the variable values in the tableaus that we computed earlier.

As we remarked earlier, the extreme points in Figure 4.2 correspond to intersection points of two or more of the hyperplanes. For example, extreme point **C** is where $s_2 = 0$ and $s_3 = 0$. Of course, in R^2 only two intersecting hyperplanes (lines) are required to uniquely determine a point. Three hyperplanes intersect at the degenerate extreme point **E**, and thus **E** can be determined as the intersection of the hyperplanes

$$x_1 = 0 \quad \text{and} \quad s_4 = 0$$

or

$$s_1 = 0 \quad \text{and} \quad s_4 = 0$$

or

$$x_1 = 0 \quad \text{and} \quad s_1 = 0$$

We earlier found three tableaus that represent the extreme point **E**, namely, $T_{\mathbf{E}}$, $T_{\mathbf{E}'}$, and $T_{\mathbf{E}''}$. For these tableaus

$T_{\mathbf{E}}$ has x_1 and s_4 nonbasic
$T_{\mathbf{E}'}$ has s_1 and s_4 nonbasic
$T_{\mathbf{E}''}$ has x_1 and s_1 nonbasic

In general, given a tableau in canonical form, the dimension of the feasible set is equal to the number of nonbasic variables. In the preceding example the canonical form tableau $T_{\mathbf{A}}$ has six variables, but the dimension of the feasible set is equal to 2 even if it is considered a subset of R^6.

Graphical Interpretation of Canonical Form Tableaus

As a final example, suppose the following canonical form tableau is given for some linear program.

	x_1	x_2	x_3	x_4	x_5
0	0	1	−1	1	0
10	1	1	1	1	0
5	0	1	1	1/6	1

This linear program can be interpreted graphically, just as the preceding example was, even though none of the variables are identified as slacks. We simply treat the basic variables in the same way we treated the slack variables in the earlier example. Here the feasible set has dimension 3, and it can be represented as the intersection of five hyperplanes where, for $j = 1, \ldots, 5$, the jth hyperplane is the set of points where $x_j = 0$. A picture of the feasible set and the associated hyperplanes is given in Figure 4.3. In R^3, three of the hyperplanes are needed to determine an extreme point. For example, $\mathbf{C} = [0, 4, 6]^T$ is the intersection of the hyperplanes $x_1 = 0$, $x_2 = 0$, and $x_5 = 0$.

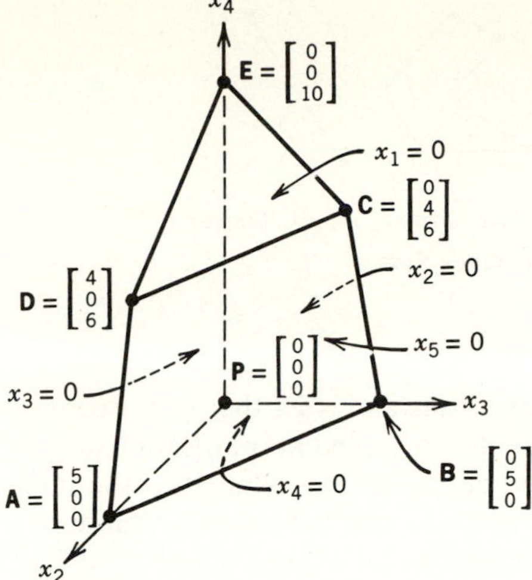

FIGURE 4.3 A Feasible Set in R³.

With this representation of the feasible set we can determine all of the feasible points of the original problem. For example, the basic feasible solution associated with the preceding tableau corresponds to point **P** in Figure 4.3, which has $x_2 = x_3 = x_4 = 0$. Because there are five variables in the problem, a feasible point must have five components. The other two components are implicit in Figure 4.3 as the amounts by which the constraints $x_1 \geq 0$ and $x_5 \geq 0$ are inactive. Since

$$x_1 = 10 - (x_2 + x_3 + x_4)$$

and

$$x_5 = 5 - (x_2 + x_3 + x_4/6)$$

we see that, at point **P,** $x_1 = 10$ and $x_5 = 5$. As another example, point **C** corresponds to the feasible $\mathbf{x} = [0, 0, 4, 6, 0]^T$.

4.2 CONVEX SETS

The feasible regions represented graphically in Figures 4.1 and 4.2 have the property that the line segment between any two feasible points is contained in the feasible set. This property holds for the feasible set of any linear program.

Definition of a Convex Set

In general, a set having the property that it contains the line segment between any two of its points is called a convex set, that is:

A set S is a **convex set** if and only if $\mathbf{x} \in S$ and $\mathbf{y} \in S$ implies $[\mathbf{x}, \mathbf{y}] \subseteq S$.

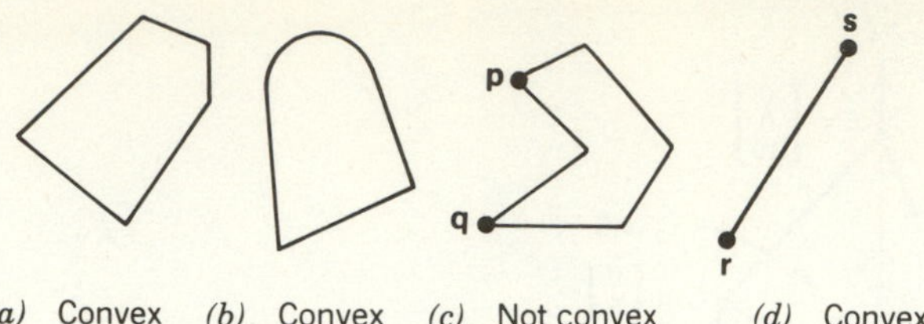

FIGURE 4.4 Examples of Convex and Nonconvex Sets.

Figure 4.4 contains examples in R^2 of convex sets and a set that is not convex. The set in Figure 4.4*c* is not convex because the line segment $[\mathbf{p}, \mathbf{q}]$, for example, is not contained in the set. Note that this set cannot be obtained as the intersection of half-spaces. Although the set in 4.4*b* with the curved boundary is convex, it cannot be obtained as the intersection of a finite number of half-spaces and so it cannot be the feasible region for a linear program. The sets in Figures 4.4*a* and 4.4*b* are two-dimensional convex sets, but the set in Figure 4.4*d* is a one-dimensional convex set, namely, the line segment $[\mathbf{r}, \mathbf{s}]$ defined by points $\mathbf{r}$ and $\mathbf{s}$ in R^2.

A convex set that is the intersection of half-spaces is called a **convex polyhedron.** Thus the feasible sets of linear programs are always convex polyhedra.

To show that a set S is convex, we would select two arbitrary points $\mathbf{x}$ and $\mathbf{y}$ in S and show that each point of the line segment $[\mathbf{x}, \mathbf{y}]$ is contained in S. Any point $\mathbf{w} \in [\mathbf{x}, \mathbf{y}]$ can be expressed as

$$\mathbf{w} = \lambda\mathbf{x} + (1 - \lambda)\mathbf{y} \quad \text{for some } 0 \leq \lambda \leq 1$$

We say that $\mathbf{w}$ is a **convex combination** of $\mathbf{x}$ and $\mathbf{y}$. Note that if $\lambda = 0$, then $\mathbf{w} = \mathbf{y}$; if $\lambda = 1$, then $\mathbf{w} = \mathbf{x}$; and if $\lambda = \frac{1}{2}$, then $\mathbf{w}$ is the midpoint of the line segment $[\mathbf{x}, \mathbf{y}]$.

Convexity of the Feasible Set

We mentioned earlier that the feasible set X for a linear program is always a convex set. We now show algebraically that this is true, using the definition of convexity. Consider the following linear program in standard form.

$$\min \mathbf{c}^T\mathbf{x} \qquad \text{subject to} \quad \mathbf{x} \in X = \{\mathbf{x} \in R^n \mid \mathbf{A}\mathbf{x} = \mathbf{b}, \mathbf{x} \geq \mathbf{0}\}$$

To show that X is convex, we suppose that $\mathbf{x}^0 \in X$ and $\mathbf{x}^1 \in X$, and we show that $[\mathbf{x}^0, \mathbf{x}^1]$ is contained in X; that is, we show that $\mathbf{w} = \lambda\mathbf{x}^0 + (1 - \lambda)\mathbf{x}^1$ belongs to X for each λ, $0 \leq \lambda \leq 1$. Note that

$$\mathbf{x}^0 \in X \quad \text{implies } \mathbf{A}\mathbf{x}^0 = \mathbf{b} \text{ and } \mathbf{x}^0 \geq \mathbf{0}$$

and

$$\mathbf{x}^1 \in X \quad \text{implies } \mathbf{A}\mathbf{x}^1 = \mathbf{b} \text{ and } \mathbf{x}^1 \geq \mathbf{0}$$

Since $\lambda \geq 0$ and $(1 - \lambda) \geq 0$, we see that $\lambda\mathbf{x}^0 + (1 - \lambda)\mathbf{x}^1 \geq \mathbf{0}$. Also, using basic properties of matrix algebra, we calculate

$$\begin{aligned}\mathbf{Aw} &= \mathbf{A}(\lambda\mathbf{x}^0 + (1 - \lambda)\mathbf{x}^1) \\ &= \mathbf{A}(\lambda\mathbf{x}^0) + \mathbf{A}((1 - \lambda)\mathbf{x}^1) \\ &= \lambda(\mathbf{Ax}^0) + (1 - \lambda)(\mathbf{Ax}^1) \\ &= \lambda\mathbf{b} + (1 - \lambda)\mathbf{b} \\ &= \mathbf{b}\end{aligned}$$

Thus $\mathbf{Aw} = \mathbf{b}$ and $\mathbf{w} \geq \mathbf{0}$, and so $\mathbf{w} \in X$ and hence X is a convex set. No matter what system of linear inequalities or linear equations defines the feasible set for a linear program, an argument similar to the one above can be given to show that the feasible set is convex.

4.3 MULTIPLE OPTIMAL SOLUTIONS

Graphically, it is easy to see how a linear program can have more than one optimal extreme point (or basic feasible solution). Consider, for example, the linear program

$$\min z = -x_1 - x_2 \qquad \text{subject to } \mathbf{x} \in X$$

where the feasible region X is shown in Figure 4.5.

It is clear from Figure 4.5 that the extreme point **C** is optimal because there is no feasible point with a smaller objective value than $z = -4$. However, because the contour line $z = -4$ is parallel to the edge [**C**, **D**], each point on that edge is optimal.

Finding All Optimal Solutions

The existence of multiple optimal vectors can also be recognized from the form of the optimal tableau. For example, the linear program represented in

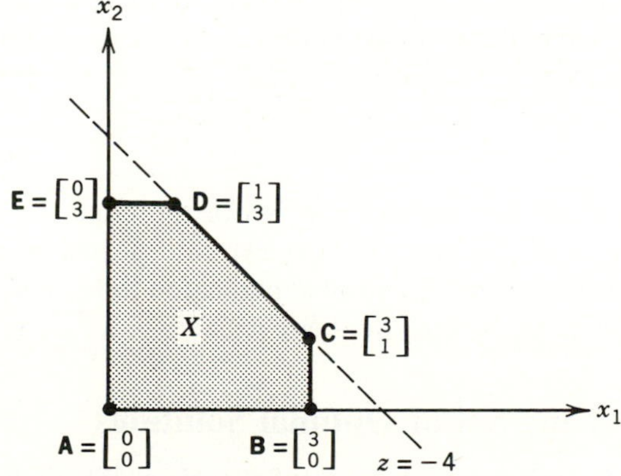

FIGURE 4.5 A Linear Program with Multiple Optimal Vectors.

Figure 4.5 has the following canonical form tableau:

	x_1	x_2	x_3	x_4	x_5
0	−1	−1	0	0	0
4	1	1	1	0	0
3	1	0	0	1	0
3	0	1	0	0	1

Two pivots, leading from **A** to **B** to **C** in Figure 4.5, yield the following optimal tableau:

$T_C =$

	x_1	x_2	x_3	x_4	x_5
4	0	0	1	0	0
1	0	1	1	−1	0
3	1	0	0	1	0
2	0	0	−1	(1)	1

From tableau T_C we see that an increase in the nonbasic variable x_4 is possible without affecting the objective function value of $z = -4$. In fact, since the minimum ratio for the x_4 column is 2, we can increase x_4 up to 2 and remain feasible. Performing the circled pivot yields the optimal tableau

$T_D =$

	x_1	x_2	x_3	x_4	x_5
4	0	0	1	0	0
3	0	1	0	0	1
1	1	0	1	0	−1
2	0	0	−1	1	1

which corresponds to extreme point **D** in Figure 4.5. Of course, the objective function row is not affected by such a pivot. Tableau T_D has a nonbasic variable, x_5, with a zero cost coefficient, but a pivot in the x_5 column only leads back to tableau T_C. In general:

> A linear program has multiple optimal solutions if a nonbasic variable in an optimal tableau has a cost coefficient of zero and if pivoting in that column changes the optimal vector.

In this example the optimal set consists of the edge [**C**, **D**], the set of all convex combinations of the two optimal extreme points **C** and **D**. In higher dimensions the optimal set can be more complicated than an edge of the feasible set but, as we show next, it must be a convex set.

Convexity of the Set of Optimal Solutions

If the set of optimal vectors consists of a single point, then the set is trivially convex. If $\mathbf{x}^0$ and $\mathbf{x}^1$ are distinct optimal vectors, we can show that the optimal set is convex by showing that the convex combination $\mathbf{w} = \lambda\mathbf{x}^0 + (1 - \lambda)\mathbf{x}^1$ is optimal for any λ, $0 \leq \lambda \leq 1$. We have already seen that such a vector $\mathbf{w}$ is feasible, so to show that $\mathbf{w}$ is optimal we need only show that $\mathbf{c}^T\mathbf{w}$ equals the optimal objective function value d^*. But

$$\begin{aligned}\mathbf{c}^T\mathbf{w} &= \mathbf{c}^T(\lambda\mathbf{x}^0 + (1 - \lambda)\mathbf{x}^1)\\ &= \mathbf{c}^T(\lambda\mathbf{x}^0) + \mathbf{c}^T((1 - \lambda)\mathbf{x}^1)\\ &= \lambda(\mathbf{c}^T\mathbf{x}^0) + (1 - \lambda)(\mathbf{c}^T\mathbf{x}^1)\\ &= \lambda d^* + (1 - \lambda)d^*\\ &= d^*\end{aligned}$$

and so any convex combination of optimal points is optimal. Therefore, the optimal set for any linear program is a convex set.

Optimal Rays

As another example of multiple optimal solutions, consider the linear program with the following optimal tableau:

$T_0 =$

	x_1	x_2	x_3	x_4	x_5
−9	0	1	0	0	0
2	1	1	(1)	−1	0
6	0	1	2	−1	1

with optimal basic feasible solution $\mathbf{x}^0 = \begin{bmatrix}2\\0\\0\\0\\6\end{bmatrix}$

Here the nonbasic variables x_3 and x_4 both have zero cost coefficients. A pivot can be performed in the x_3 column yielding another optimal tableau

$T_1 =$

	x_1	x_2	x_3	x_4	x_5
−9	0	1	0	0	0
2	1	1	1	−1	0
2	−2	−1	0	(1)	1

with optimal basic feasible solution $\mathbf{x}^1 = \begin{bmatrix}0\\0\\2\\0\\2\end{bmatrix}$

A third optimal extreme point is obtained by performing the circled pivot on T_1, which yields the tableau T_2.

$T_2 =$

	x_1	x_2	x_3	x_4	x_5
−9	0	1	0	0	0
4	−1	0	1	0	1
2	−2	−1	0	1	1

with optimal basic feasible solution $\mathbf{x}^2 = \begin{bmatrix}0\\0\\4\\2\\0\end{bmatrix}$

Examining tableaus T_0, T_1, and T_2 shows that the basic feasible solutions $\mathbf{x}^0$, $\mathbf{x}^1$, and $\mathbf{x}^2$ are the only optimal extreme points, and all possible convex combinations of $\mathbf{x}^0$, $\mathbf{x}^1$, and $\mathbf{x}^2$ are also optimal points because of the convexity of the optimal set. However, this problem has other optimal points as well, not corresponding to basic feasible solutions.

First consider tableau T_0, which is repeated:

T_0 =

	x_1	x_2	x_3	x_4	x_5
−9	0	1	0	0	0
2	1	1	1	−1	0
6	0	1	2	−1	1

In tableau T_0 it is not possible to pivot in the x_4 column, but because the cost coefficient c_4 is zero, increasing the variable x_4 results in other optimal vectors. More specifically, in tableau T_0, letting $x_4 = t_1 > 0$ with $x_2 = x_3 = 0$ yields $x_1 = 2 + t_1$ and $x_5 = 6 + t_1$. Thus we obtain the vector $\mathbf{x}(t_1)$, which satisfies

$$\mathbf{x}(t_1) = \begin{bmatrix} 2 + t_1 \\ 0 \\ 0 \\ t_1 \\ 6 + t_1 \end{bmatrix} = \mathbf{x}^0 + t_1 \begin{bmatrix} 1 \\ 0 \\ 0 \\ 1 \\ 1 \end{bmatrix} = \mathbf{x}^0 + t_1\mathbf{u}$$

where $\mathbf{u} = [1, 0, 0, 1, 1]^T$. Thus, increasing x_4 leads to the **ray** of optimal vectors starting at $\mathbf{x}^0$ and going forever in the direction of the vector $\mathbf{u}$.

Next consider tableau T_2, which is repeated:

T_2 =

	x_1	x_2	x_3	x_4	x_5
−9	0	1	0	0	0
4	−1	0	1	0	1
2	−2	−1	0	1	1

The x_1 column of T_2 yields another optimal ray by letting $x_1 = t_2 > 0$ with $x_2 = x_5 = 0$ so that $x_3 = 4 + t_2$ and $x_4 = 2 + 2t_2$. This gives the vector:

$$\mathbf{x}(t_2) = \begin{bmatrix} t_2 \\ 0 \\ 4 + t_2 \\ 2 + 2t_2 \\ 0 \end{bmatrix} = \mathbf{x}^2 + t_2 \begin{bmatrix} 1 \\ 0 \\ 1 \\ 2 \\ 0 \end{bmatrix} = \mathbf{x}^2 + t_2\mathbf{v}$$

where $\mathbf{v} = [1, 0, 1, 2, 0]^T$.

Thus the set of optimal vectors for this problem consists of all possible convex combinations of pairs of points chosen from $\mathbf{x}^0$, $\mathbf{x}^1$, $\mathbf{x}^2$, $\mathbf{x}^0 + t_1\mathbf{u}$, $\mathbf{x}^2 + t_2\mathbf{v}$. Figure 4.6 shows the feasible set as drawn from the representation of tableau T_0. The optimal set is that unbounded two-dimensional face of the constraint polyhedron that lies in the $x_2 = 0$ hyperplane.

The techniques we have used to find the set of all optimal vectors can be applied to discover properties of the feasible set itself. For example, from tableau T_1 above we can see that there are three edges of the feasible set incident to the extreme point $\mathbf{x}^1$ because in each of the three nonbasic columns in T_1 a pivot is possible. Two of the pivots lead, as we have seen, to the optimal extreme points $\mathbf{x}^0$ and $\mathbf{x}^2$. The minimum ratio pivot in the x_2 column of T_1 also leads to a new extreme point, but, of course, it is not optimal. Performing this pivot

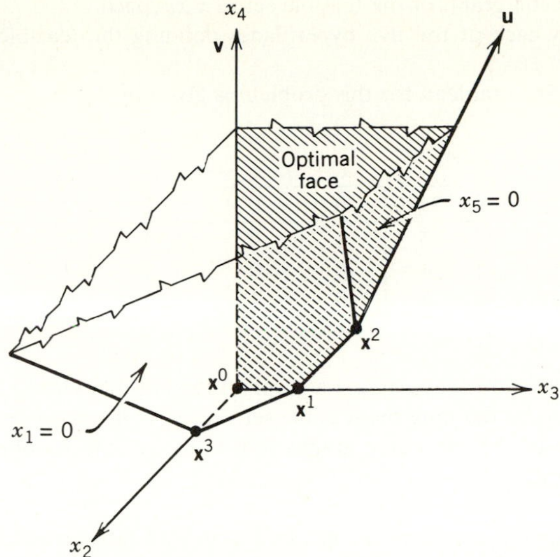

FIGURE 4.6 An Unbounded Set of Optimal Vectors.

yields the nonoptimal tableau:

$$T_3 = \begin{array}{|c|ccccc|} \multicolumn{1}{c}{} & x_1 & x_2 & x_3 & x_4 & \multicolumn{1}{c}{x_5} \\ \hline -11 & -1 & 0 & -1 & 1 & 0 \\ \hline 2 & 1 & 1 & 1 & -1 & 0 \\ 4 & -1 & 0 & 1 & 0 & 1 \\ \hline \end{array} \quad \text{with the basic feasible solution} \quad \mathbf{x}^3 = \begin{bmatrix} 0 \\ 2 \\ 0 \\ 0 \\ 4 \end{bmatrix}$$

The point $\mathbf{x}^3$ is the only other extreme point in this linear program.

SELECTED REFERENCES

Dantzig, G. B., *Linear Programming and Extensions,* Princeton University Press, Princeton, 1963, Chapter 7.

Murty, K. G., *Linear Programming,* Wiley, New York, 1983, Chapter 3.

Spivey, A. W., and Thrall, R. M., *Linear Optimization,* Holt, Rinehart, and Winston, New York, 1970, Chapter 3.

EXERCISES

4.1 Consider a linear program in standard form with the following simplex tableau:

$$M = \begin{array}{|c|ccccc|} \multicolumn{1}{c}{} & x_1 & x_2 & x_3 & x_4 & \multicolumn{1}{c}{x_5} \\ \hline 0 & -5 & -3 & 0 & 0 & 0 \\ \hline 5 & 1 & 0 & 1 & 0 & 0 \\ 15 & 1 & 2 & 0 & 1 & 0 \\ 8 & 1 & 1 & 0 & 0 & 1 \\ \hline \end{array}$$

(a) Construct a graph of the feasible set in x_1x_2 space.
(b) Identify each of the five hyperplanes defining the feasible set as $x_i = 0$ for some i, $1 \leq i \leq 5$.
(c) The optimal tableau for this problem is given by

$M^* =$

	x_1	x_2	x_3	x_4	x_5
34	0	0	2	0	3
5	1	0	1	0	0
4	0	0	1	1	−2
3	0	1	−1	0	1

Using the tableau M^*, construct a graph of the feasible set in x_3x_5 space and label each hyperplane defining the feasible set with $x_i = 0$, for some $i \in \{1, 2, 3, 4, 5\}$.
(d) In each of the preceding graphs that represent the feasible set, identify the optimal point.

4.2 Consider the linear program in standard form with tableau M and associated basic feasible solution $\mathbf{x}^0$:

$M =$

	x_1	x_2	x_3	x_4	x_5	x_6
−5	0	0	1	0	0	0
15	1	1	1	0	0	−1
5	0	−1	0	0	1	0
10	0	0	−1	1	0	−1

$$\mathbf{x}^0 = \begin{bmatrix} 15 \\ 0 \\ 0 \\ 10 \\ 5 \\ 0 \end{bmatrix}$$

(a) Find all optimal extreme points.
(b) Find a nonzero vector $\mathbf{v}$ such that the point $\mathbf{x}^0 + t\mathbf{v}$ is optimal for any $t \geq 0$.
(c) Describe the set of all optimal solutions.

4.3 Consider the linear program in standard form with the following tableau M:

$M =$

	x_1	x_2	x_3	x_4	x_5
0	0	−2	4	−1	0
5	0	−1	−2	1	1
2	1	1	1	1	0

(a) How many edges of the feasible set are incident to the basic feasible solution given by tableau M?
(b) Obtain a tableau in optimal form for this linear program.
(c) In the optimal tableau, circle the pivot position that leads to the smallest increase in the objective function value.
(d) Performing the pivot in part (c) gives the second best extreme point. What is the third best extreme point?
(e) Describe a general method for finding all the extreme points for a feasible set and illustrate your method by finding all extreme points in the feasible set defined by tableau M.

4.4 Show that the feasible set for the following linear program is a convex set.

$$\begin{aligned} \text{minimize} \quad & \mathbf{c}^T\mathbf{x} \\ \text{subject to} \quad & \mathbf{Ax} \geq \mathbf{b} \\ & \mathbf{x} \geq \mathbf{0} \end{aligned}$$

4.5 Consider the linear program

$$\begin{aligned} \text{minimize} \quad & -x_1 - x_2 \\ \text{subject to} \quad & -x_1 + x_2 \leq 2 \\ & x_1 - x_2 \leq 2 \\ & 0 \leq x_1 \leq 4 \quad \text{and} \quad 0 \leq x_2 \leq 4 \end{aligned}$$

(a) Draw a picture of the feasible set in R^2. Which extreme point is optimal?
(b) If the origin is the initial extreme point used by the simplex algorithm, how many simplex rule pivots are required to reach the optimal point?
(c) Using the initial canonical form obtained by adding slack variables, x_i, $i = 3, 4, 5, 6$, pivot using the simplex rule to obtain the optimal point and indicate on the graph of the feasible set which edges were traversed in reaching the optimal point.
(d) Identify the six hyperplanes that define the feasible set as $x_i = 0$ for $i = 1, 2, \ldots, 6$.
(e) Identify each extreme point visited by the simplex rule pivots in part (c) as the intersection of exactly two of the hyperplanes in part (d).
(f) Redraw the graph of the feasible set so that all of the intersection points of the 6 defining hyperplanes are given. You should find 12 such points.
(g) One of the intersection points in part (f) is $(x_1, x_2) = (0, 4)$. Show how to perform one pivot on the initial canonical form tableau of part (c) to obtain this intersection point.
(h) Using the graph in part (f), show how to perform exactly two pivots (not necessarily by the simplex rule) to obtain the optimal point.

4.6 Find all optimal solutions to the linear program in standard form with the following simplex tableau.

	x_1	x_2	x_3	x_4
0	1	0	0	0
0	1	0	1	−2
1	0	1	−2	0

4.7 This exercise shows how to construct a linear program in two variables so that given any positive integer M, the simplex algorithm will visit exactly M extreme points in reaching the optimal point.
(a) Use the following rule to construct a set of integer points (x, y) in xy space.

$x^1 = 0$ and $y^1 = 0$.

Let $i = 1$

1 $x^{i+1} = x^i + 1$ and $y^{i+1} = y^i + M - i$

If $M - i = 1$, stop; otherwise, replace i by $i + 1$ and go to 1.

(b) Show that the equation of the line passing through (x^{i+1}, y^{i+1}) and (x^i, y^i) is given by

$$y - [iM - \tfrac{1}{2}i(i + 1)] = (M - i)(x - i) \quad \text{for } i = 1, 2, \ldots, M$$

and show that this simplifies to

$$(i - M)x + y = \tfrac{1}{2}i(i - 1)$$

(c) Generate another sequence of integer points (x, y) using a rule similar to that in part (a) where the role of x and y are interchanged, namely,

$x^1 = 0$ and $y^1 = 0$.

Let $i = 1$

1 $y^{i+1} = y^i + 1$ and $x^{i+1} = x^i + M - i$

If $M - i = 1$, stop; otherwise, replace i by $i + 1$ and go to 1.

In this case show that the line passing through the $(i + 1)$st and ith point generated is given by

$$x + (i - M)y = \tfrac{1}{2}i(i - 1)$$

(d) Now consider the following linear program.

$$\begin{aligned} \text{maximize} \quad & x_1 + x_2 \\ \text{subject to} \quad & (i - M)x_1 + x_2 \leq \tfrac{1}{2}i(i - 1) \qquad i = 1, 2, \ldots, M \\ & x_1 + (i - M)x_2 \leq \tfrac{1}{2}i(i - 1) \qquad i = 1, 2, \ldots, M \\ & x_1 \geq 0 \quad \text{and} \quad x_2 \geq 0 \end{aligned}$$

Construct the feasible set in the case $M = 4$.

(e) Show that the linear program in part (d) has exactly $2M$ extreme points and that exactly M of them will be visited by the simplex algorithm when it starts from the initial canonical form obtained by adding slack variables. Note that the origin is the only degenerate vertex and so $M + 1$ pivots are required to reach the optimal point.

4.8 (a) If C_1 and C_2 are convex sets, show that the intersection of C_1 and C_2 is a convex set.
(b) Show that the feasible set for any linear program can be written as the intersection of a finite number of half-spaces.
(c) Using part (a) and part (b), why is the feasible set for any linear program a convex set?

4.9 Given a positive integer N, show how to construct a linear program with a feasible set in R^3 so that there are N distinct paths (with each path consisting of two edges) that the simplex algorithm could follow in reaching the optimal extreme point.

4.10 The following tableau represents a linear program in standard form.

		x_1	x_2	x_3	x_4	x_5
	0	0	0	1	0	0
$M =$	5	1	−3	1	2	0
	2	0	−2	−1	1	1

(a) Find all the optimal basic feasible solutions.
(b) Is the set of all optimal solutions bounded? Explain.
(c) Describe the set of all optimal solutions.

CHAPTER 5

DUALITY IN LINEAR PROGRAMMING

Soon after the development of the simplex algorithm, it was discovered that every linear program has a related linear program called its **dual,** and that a solution to either problem can be constructed from a solution to the other. The discovery of dual linear programs has had an important impact on both computational methods and theoretical developments in linear programming and has also influenced the development of other optimization methods. In this chapter we study the relationship between a linear program and its dual, and we show how the dual problem can be used in some cases to provide important economic information about the underlying linear programming model. Also, after showing how to form the dual of any linear program, we present some important applications of linear programming duality.

5.1 THE STANDARD DUAL PAIR

We will refer to the following pair of linear programming problems as the **standard dual pair:**

minimize $\mathbf{c}^T\mathbf{x}$	maximize $\mathbf{b}^T\mathbf{y}$
subject to $\mathbf{A}\mathbf{x} \geq \mathbf{b}$	subject to $\mathbf{A}^T\mathbf{y} \leq \mathbf{c}$
$\mathbf{x} \geq \mathbf{0}$	$\mathbf{y} \geq \mathbf{0}$

Each of these problems is the dual of the other. In some settings it is conventional to refer to one of the problems as the **primal** linear program and to the other problem as the **dual;** which of the problems is identified as the primal depends on the context of the discussion.

The following linear programs are in the form of the standard dual pair:

minimize $6x_1 + 2x_2 + 3x_3$
subject to

$$\begin{bmatrix} 1 & 1 & 0 \\ 1 & -1 & 1 \end{bmatrix} \begin{bmatrix} x_1 \\ x_2 \\ x_3 \end{bmatrix} \geq \begin{bmatrix} 2 \\ 1 \end{bmatrix}$$

$$\mathbf{x} \geq \mathbf{0}$$

maximize $2y_1 + y_2$
subject to

$$\begin{bmatrix} 1 & 1 \\ 1 & -1 \\ 0 & 1 \end{bmatrix} \begin{bmatrix} y_1 \\ y_2 \end{bmatrix} \leq \begin{bmatrix} 6 \\ 2 \\ 3 \end{bmatrix}$$

$$\mathbf{y} \geq \mathbf{0}$$

Note the following structural relationships between the two problems:

- One problem is a minimization and the other is a maximization.
- The problems have inequality constraints, and the inequalities for the minimization problem have the opposite sense of those for the maximization problem.
- The coefficient matrix for the constraints in one problem is the transpose of the coefficient matrix for the constraints in the other problem.
- The objective function coefficients in one problem are used to form the constant column in the other problem.

These structural relationships give rise to important algebraic **duality relations** between the linear programs in the standard dual pair.

Duality Relations

Some of the duality relations that exist between the two linear programs in the standard dual pair are as follows:

(i) If $\mathbf{x}$ is feasible for the minimization problem and $\mathbf{y}$ is feasible for the maximization problem, then $\mathbf{c}^T\mathbf{x} \geq \mathbf{b}^T\mathbf{y}$.

(ii) If one of the problems has an optimal vector, then so does the other problem, and the optimal values are equal.

(iii) If both problems have feasible vectors, then both problems have optimal vectors.

(iv) If one problem is feasible but has no finite optimal value, then the other problem is infeasible.

(v) Both problems may be infeasible.

It is easy to see why most of these relations hold. To see why (i) is true, assume that $\mathbf{x}$ is feasible for the minimization problem and that $\mathbf{y}$ is feasible for the maximization problem. Then

$$\mathbf{c} \geq \mathbf{A}^T\mathbf{y} \text{ and } \mathbf{x} \geq \mathbf{0} \quad \text{imply that} \quad \mathbf{x}^T\mathbf{c} \geq \mathbf{x}^T(\mathbf{A}^T\mathbf{y})$$
$$\mathbf{A}\mathbf{x} \geq \mathbf{b} \text{ and } \mathbf{y} \geq \mathbf{0} \quad \text{imply that} \quad \mathbf{y}^T(\mathbf{A}\mathbf{x}) \geq \mathbf{y}^T\mathbf{b}$$

Since $\mathbf{x}^T(\mathbf{A}^T\mathbf{y}) = (\mathbf{A}\mathbf{x})^T\mathbf{y} = \mathbf{y}^T(\mathbf{A}\mathbf{x})$ we see that $\mathbf{x}^T\mathbf{c} \geq \mathbf{y}^T\mathbf{b}$, which means that $\mathbf{c}^T\mathbf{x} \geq \mathbf{b}^T\mathbf{y}$, and this establishes relation (i).

If $\mathbf{x}$ is feasible for the minimization problem, then relation (i) shows that

$\mathbf{c}^T\mathbf{x}$ is an upper bound on $\mathbf{b}^T\mathbf{y}$ for any $\mathbf{y}$ that is feasible for the maximization problem. Similarly, if $\mathbf{y}$ is feasible for the maximization problem, then $\mathbf{b}^T\mathbf{y}$ is a lower bound on $\mathbf{c}^T\mathbf{x}$ for any $\mathbf{x}$ that is feasible for the minimization problem. Thus neither linear program can be unbounded. Therefore, if both problems are feasible, they must both have optimal vectors, and this establishes relation (iii).

Relation (iv) also follows immediately from (i). If the minimization problem is feasible but has no finite minimum value, then the maximization problem could not have a feasible vector. The reason for this is that if $\bar{\mathbf{y}}$ were feasible for the maximization problem, then relation (i) implies that

$$\mathbf{c}^T\mathbf{x} \geq \mathbf{b}^T\bar{\mathbf{y}} \text{ for each } \mathbf{x} \text{ feasible for the minimization problem}$$

Such a lower bound on $\mathbf{c}^T\mathbf{x}$ is impossible if the minimization problem has no finite minimum value. A similar argument establishes that the minimization problem is infeasible if the maximization problem is unbounded.

We establish relation (v) by exhibiting a particular standard dual pair where both problems are infeasible, namely, the following dual problems where x and y are in R^1.

$$\begin{array}{llcll} \text{minimize} & -x & \qquad & \text{maximize} & y \\ \text{subject to} & 0x \geq 1 & & \text{subject to} & 0y \leq -1 \\ & x \geq 0 & & & y \geq 0 \end{array}$$

Clearly, both of these problems are infeasible.

Of the duality relations listed above, only (ii) remains to be established.

(ii) If one of the problems has an optimal vector, then so does the other problem, and the optimal values are equal.

In the next section we prove this relation by showing how to construct an optimal vector for one problem from an optimal tableau for the other.

5.2 GETTING THE DUAL SOLUTION FROM A PRIMAL SOLUTION

Solving one of the problems in the standard dual pair provides information that allows us easily to obtain a solution to the other problem. Before showing why this is true in general, we illustrate the construction of an optimal dual vector by solving the example introduced earlier.

Relationships between Optimal Tableaus

The example we gave earlier of a standard dual pair is repeated here:

minimize $6x_1 + 2x_2 + 3x_3$ subject to

$$\begin{bmatrix} 1 & 1 & 0 \\ 1 & -1 & 1 \end{bmatrix} \begin{bmatrix} x_1 \\ x_2 \\ x_3 \end{bmatrix} \geq \begin{bmatrix} 2 \\ 1 \end{bmatrix} \qquad \mathbf{x} \geq \mathbf{0}$$

maximize $2y_1 + y_2$ subject to

$$\begin{bmatrix} 1 & 1 \\ 1 & -1 \\ 0 & 1 \end{bmatrix} \begin{bmatrix} y_1 \\ y_2 \end{bmatrix} \leq \begin{bmatrix} 6 \\ 2 \\ 3 \end{bmatrix} \qquad \mathbf{y} \geq \mathbf{0}$$

Neither of these problems is in standard form. To reformulate the minimization problem for solution by the simplex method, we change the sense of the constraint inequalities and then add slack variables s_1 and s_2 to obtain the following tableau:

$$M = \begin{array}{c|ccccc} & x_1 & x_2 & x_3 & s_1 & s_2 \\ \hline 0 & 6 & 2 & 3 & 0 & 0 \\ \hline -2 & -1 & -1 & 0 & 1 & 0 \\ -1 & -1 & 1 & -1 & 0 & 1 \end{array}$$

Pivoting yields the following optimal tableau M^*:

$$M^* = \begin{array}{c|ccccc} & x_1 & x_2 & x_3 & s_1 & s_2 \\ \hline -10 & 0 & 0 & 1 & 4 & 2 \\ \hline 3/2 & 1 & 0 & 1/2 & -1/2 & -1/2 \\ 1/2 & 0 & 1 & -1/2 & -1/2 & 1/2 \end{array}$$

This tableau shows that the minimization problem has an optimal value of $\mathbf{c}^T\mathbf{x}^* = 10$ and an optimal vector

$$\mathbf{x}^* = \begin{bmatrix} 3/2 \\ 1/2 \\ 0 \end{bmatrix}$$

To reformulate the maximization problem into standard form, we change the sense of the optimization and add slack variables w_1, w_2, and w_3 to obtain the tableau:

$$T = \begin{array}{c|ccccc} & y_1 & y_2 & w_1 & w_2 & w_3 \\ \hline 0 & -2 & -1 & 0 & 0 & 0 \\ \hline 6 & 1 & 1 & 1 & 0 & 0 \\ 2 & 1 & -1 & 0 & 1 & 0 \\ 3 & 0 & 1 & 0 & 0 & 1 \end{array}$$

Two pivots yield the following optimal tableau:

$$T^* = \begin{array}{c|ccccc} & y_1 & y_2 & w_1 & w_2 & w_3 \\ \hline 10 & 0 & 0 & 3/2 & 1/2 & 0 \\ \hline 2 & 0 & 1 & 1/2 & -1/2 & 0 \\ 4 & 1 & 0 & 1/2 & 1/2 & 0 \\ 1 & 0 & 0 & -1/2 & 1/2 & 1 \end{array}$$

Remembering that we had to change the sense of the optimization in reformulating the original problem into standard form, this tableau shows that the maximization problem has an optimal value of $\mathbf{d}^T\mathbf{y}^* = 10$. The optimal vector is

$$\mathbf{y}^* = \begin{bmatrix} 4 \\ 2 \end{bmatrix}$$

The optimal tableaus M^* and T^* are repeated below for convenience in comparing them to one another.

$M^* =$

	x_1	x_2	x_3	s_1	s_2
−10	0	0	1	4	2
3/2	1	0	1/2	−1/2	−1/2
1/2	0	1	−1/2	−1/2	1/2

$T^* =$

	y_1	y_2	w_1	w_2	w_3
10	0	0	3/2	1/2	0
2	0	1	1/2	−1/2	0
4	1	0	1/2	1/2	0
1	0	0	−1/2	1/2	1

Note that the optimal vector for the minimization problem, $\mathbf{x}^* = [\frac{3}{2}, \frac{1}{2}, 0]^T$, has components that are exactly the cost coefficients of the slack variables w_1, w_2, and w_3 in the optimal tableau T^* for the maximization problem. Similarly, the optimal vector for the maximization problem, namely, $\mathbf{y}^* = [4, 2]^T$, has components that are exactly the cost coefficients of the slack variables s_1 and s_2 in the optimal tableau M^* for the minimization problem. Also, of course, the objective function value entries in the two optimal tableaus have the same magnitude and opposite sign because as we pointed out above the optimal values for the original minimization and maximization problems are equal.

Constructing an Optimal Dual Vector

The relationships illustrated by the above example hold in general for the optimal tableaus of dual linear programs, and from the optimal tableau for either problem in the standard dual pair we can always construct the optimal solution to the other problem by using the following fact.

> The optimal vector for the maximization problem has components equal to the cost coefficients of the slack variables in the optimal tableau for the minimization problem, and conversely.

To show that this is true in general, we will solve the minimization problem of the standard dual pair symbolically, and use the other duality relations to prove that the resulting cost coefficients for the slack variables are indeed the elements of a vector that is optimal for the dual problem. First, recall the standard dual pair:

$$\begin{array}{llll} \text{minimize} & \mathbf{c}^T\mathbf{x} & \text{maximize} & \mathbf{b}^T\mathbf{y} \\ \text{subject to} & \mathbf{A}\mathbf{x} \geq \mathbf{b} & \text{subject to} & \mathbf{A}^T\mathbf{y} \leq \mathbf{c} \\ & \mathbf{x} \geq \mathbf{0} & & \mathbf{y} \geq \mathbf{0} \end{array}$$

As in the preceding example, we introduce slack variables $\mathbf{s} \geq \mathbf{0}$ to reformulate the minimization problem into standard form, obtaining the following simplex tableau:

$$M = \begin{array}{c|c|c} & \mathbf{x} & \mathbf{s} \\ \hline 0 & \mathbf{c}^T & \mathbf{0} \\ \hline -\mathbf{b} & -\mathbf{A} & \mathbf{I} \end{array}$$

If the minimization problem has an optimal vector $\mathbf{x}^*$, then applying the simplex algorithm to M must yield an optimal tableau M^*. Suppose that M^* is given by

$$M^* = \begin{array}{c|c|c} & \mathbf{x} & \mathbf{s} \\ \hline -d & \mathbf{u}^T & \mathbf{v}^T \\ \hline \mathbf{b}^* & \mathbf{D} & \mathbf{B} \end{array}$$

Our task is then to show that $\mathbf{v}$ is optimal for the maximization problem.

We know from Section 3.9 that, because the optimal tableau M^* is obtained from the starting tableau M by pivoting, there must be a pivot matrix $\mathbf{Q}$ such that

$$M^* = \mathbf{Q}M$$

Recall that the first column of a pivot matrix is always the first column of an $(m + 1) \times (m + 1)$ identity matrix, and note that the last m identity columns appear as the rightmost partition of tableau M. Because we can get the required pivot matrix $\mathbf{Q}$ by "doing unto those m identity columns as we would do unto M", and because the pivots that produce M^* do just that, we can simply read off the rightmost m columns of $\mathbf{Q}$ as the columns of M^* corresponding to the variables $\mathbf{s}$. The pivot matrix thus produced is

$$\mathbf{Q} = \left[\begin{array}{c|c} 1 & \mathbf{v}^T \\ \hline \mathbf{0} & \mathbf{B} \end{array}\right]$$

For the minimization problem of our example,

$$M^* = \begin{array}{c|ccc|cc} & x_1 & x_2 & x_3 & s_1 & s_2 \\ \hline -10 & 0 & 0 & 1 & 4 & 2 \\ \hline 3/2 & 1 & 0 & 1/2 & -1/2 & -1/2 \\ 1/2 & 0 & 1 & -1/2 & -1/2 & 1/2 \end{array}$$

$$= \begin{array}{c|c|c} & \mathbf{x} & \mathbf{s} \\ \hline -d & \mathbf{u}^T & \mathbf{v}^T \\ \hline \mathbf{b} & \mathbf{D} & \mathbf{B} \end{array}$$

so

$$\mathbf{Q} = \left[\begin{array}{c|c} 1 & \mathbf{v}^T \\ \hline \mathbf{0} & \mathbf{B} \end{array}\right] = \left[\begin{array}{c|cc} 1 & 4 & 2 \\ \hline 0 & -1/2 & -1/2 \\ 0 & -1/2 & 1/2 \end{array}\right] = \begin{bmatrix} 1 & 4 & 2 \\ 0 & -1/2 & -1/2 \\ 0 & -1/2 & 1/2 \end{bmatrix}$$

Using the general form of the pivot matrix **Q** to compute the elements of M^* in terms of the elements of M, we obtain

$$\underset{\mathbf{Q}}{\left[\begin{array}{c|c} 1 & \mathbf{v}^T \\ \hline \mathbf{0} & \mathbf{B} \end{array}\right]} \underset{M}{\begin{array}{|c|c|c|} \hline 0 & \mathbf{c}^T & \mathbf{0}^T \\ \hline -\mathbf{b} & -\mathbf{A} & \mathbf{I} \\ \hline \end{array}} = \underset{M^*}{\begin{array}{|c|c|c|} \hline -\mathbf{v}^T\mathbf{b} & (\mathbf{c}^T - \mathbf{v}^T\mathbf{A}) & \mathbf{v}^T \\ \hline -\mathbf{B}\mathbf{b} & -\mathbf{B}\mathbf{A} & \mathbf{B} \\ \hline \end{array}}$$

The tableau M^* is in optimal form, so $\mathbf{v}^T \geq \mathbf{0}$ and

$$\mathbf{u}^T = \mathbf{c}^T - \mathbf{v}^T\mathbf{A} \geq \mathbf{0}$$

so

$$\mathbf{A}^T\mathbf{v} \leq \mathbf{c}$$

Thus, $\mathbf{v}$ is feasible for the maximization problem. Also,

$$-d = -\mathbf{v}^T\mathbf{b}$$

or

$$d = \mathbf{b}^T\mathbf{v}$$

However, $d = \mathbf{c}^T\mathbf{x}^*$ is the optimal value for the minimization problem, so, by duality relation (i), no feasible vector for the maximization problem can give an objective function value greater than d. Because $\mathbf{v}$ is feasible and $\mathbf{b}^T\mathbf{v} = d$, the vector $\mathbf{v}$ must be optimal.

This argument shows how an optimal vector for the maximization problem can be constructed from the optimal tableau for the minimization problem, and that the optimal values of the two problems are equal. A similar argument can be used to show that an optimal vector for the maximization problem has its components equal the cost coefficients of the slack variables in the optimal tableau for the minimization problem (see Exercise 5.4). Taken together, these constructive arguments prove duality relation (ii).

5.3 ECONOMIC INTERPRETATION OF DUAL VARIABLES

In many linear programming models, the dual variables have an important economic interpretation. In this section we illustrate this interpretation by considering a simple resource allocation model.

A certain firm has three resources R_1, R_2, and R_3 and makes four products P_1, P_2, P_3, and P_4. Table 5.1 gives the units of each resource required to make

TABLE 5.1 Technology Table

	P_1	P_2	P_3	P_4	*Available*
R_1	2	1	1	1	30
R_2	1	0	2	1	15
R_3	1	1	1	0	5
Revenue	6	1	4	5	

one unit of each product, and it also gives the selling price (in dollars) of each product and the amount of each resource available. For example, product P_3 sells for \$4 and requires 1, 2, and 1 units of the resources R_1, R_2, and R_3, respectively.

Given the available resources, the firm would like to determine the amount x_j of product P_j that it should produce in order to maximize the revenue obtained from selling the products. Thus the firm would like to solve the linear program

$$\begin{aligned}
&\text{maximize } 6x_1 + x_2 + 4x_3 + 5x_4 \\
&\text{subject to} \\
&\qquad 2x_1 + x_2 + x_3 + x_4 \leq 30 \\
&\qquad x_1 + 2x_3 + x_4 \leq 15 \\
&\qquad x_1 + x_2 + x_3 \leq 5 \\
&\qquad \text{and } x_j \geq 0 \quad j = 1, \ldots, 4
\end{aligned}$$

This problem has the form of the maximization problem in our standard dual pair, and when reformulated into standard form, it is represented by the tableau

$$T = \begin{array}{|c|ccccccc|}
\multicolumn{1}{c}{} & x_1 & x_2 & x_3 & x_4 & s_1 & s_2 & \multicolumn{1}{c}{s_3} \\
\hline
0 & -6 & -1 & -4 & -5 & 0 & 0 & 0 \\
\hline
30 & 2 & 1 & 1 & 1 & 1 & 0 & 0 \\
15 & 1 & 0 & 2 & 1 & 0 & 1 & 0 \\
5 & 1 & 1 & 1 & 0 & 0 & 0 & 1 \\
\hline
\end{array}$$

where s_i is the slack variable associated with resource R_i. An optimal tableau for this problem is

$$T^* = \begin{array}{|c|ccccccc|}
\multicolumn{1}{c}{} & x_1 & x_2 & x_3 & x_4 & s_1 & s_2 & \multicolumn{1}{c}{s_3} \\
\hline
80 & 0 & 0 & 7 & 0 & 0 & 5 & 1 \\
\hline
10 & 0 & 0 & -2 & 0 & 1 & -1 & -1 \\
10 & 0 & -1 & 1 & 1 & 0 & 1 & -1 \\
5 & 1 & 1 & 1 & 0 & 0 & 0 & 1 \\
\hline
\end{array}$$

The optimal production program is to make 5 units of product P_1, 10 units of product P_4, and 0 units of products P_2 and P_3. This optimal program yields the maximum revenue of \$80.

Now suppose that, prior to production, someone offered to buy one unit of resource R_2. The firm would realize some revenue from the sale, of course. However, if the firm sells one unit of resource R_2, the optimal production program $\mathbf{x}^*$ may be affected, and the revenue obtained from production may be reduced. It is thus important for the firm to answer the following question.

> What is the minimum price that could be charged for that one unit of resource R_2 if the total revenue is to remain $\geq$ \$80?

The slack variable $s_2 = 0$ in the optimal tableau T^*, so all of resource R_2 is used in the optimal production program $\mathbf{x}^*$. Selling one unit of R_2 prior to production therefore has the same effect as insisting that one unit be left over after production, so that $s_2 = 1$ in the new optimal solution. The effect on the optimal solution of requiring that $s_2 = 1$ can be determined by simply letting $s_2 = 1$ in the optimal tableau T^*. By rewriting the equations represented by T^* so that s_2 appears on the left side along with the constant column, we obtain the following new tableau:

$T' =$

	x_1	x_2	x_3	x_4	s_1	s_3
$80 - 5s_2$	0	0	7	0	0	1
$10 + s_2$	0	0	-2	0	1	-1
$10 - s_2$	0	-1	1	1	0	-1
5	1	1	1	0	0	1

As long as $10 + s_2 \geq 0$ and $10 - s_2 \geq 0$, tableau T' remains in canonical form and in optimal form, and the revenue from production is $80 - 5s_2$. If the total revenue is to remain at least \$80, then

$$\text{(revenue from production)} + \text{(revenue from sale)} \geq 80$$

or

$$(80 - 5s_2) + s_2 \cdot \text{(selling price/unit)} \geq 80.$$

Solving the last inequality answers the question posed earlier:

$$\text{(selling price/unit)} \geq \$5.$$

This minimum price of \$5 per unit for resource R_2 is called the **shadow price** or **imputed price** of the resource. A similar analysis using the s_3 column in T shows that the shadow price for resource R_3 is \$1 per unit, provided $s_3 \leq 5$.

Combining the s_1 column with the constant column would not result in a canonical form tableau because the slack variable s_1 is basic in the optimal tableau T^*. However, since $s_1 = 10$ in the optimal solution $\mathbf{x}^*$, we know that 10 units of resource R_1 are left over after production and are not used in the optimal production program. Thus the firm could sell units of resource R_1 for nothing, provided $s_1 \leq 10$, and R_1 has a shadow price of zero.

The shadow prices determined above for the resources R_1, R_2, and R_3 are as follows:

$$\begin{bmatrix} \text{shadow price for resource } R_1 \\ \text{shadow price for resource } R_2 \\ \text{shadow price for resource } R_3 \end{bmatrix} = \begin{bmatrix} 0 \\ 5 \\ 1 \end{bmatrix}$$

Using the results of Section 5.2, we can read off the optimal dual vector as the cost coefficients of the slack variables in tableau T^*. This vector,

$$\mathbf{y}^* = \begin{bmatrix} 0 \\ 5 \\ 1 \end{bmatrix}$$

has elements that are just the shadow prices of the resources R_1, R_2, and R_3.

This correspondence holds in general and constitutes the economic interpretation of the optimal dual variables.

> The optimal vector for the dual of a resource allocation problem has components that are the shadow prices of the resources.

The Complementary Slackness Conditions

As shown by the preceding example, the shadow price of a resource that is not used up is zero. The shadow prices in a resource allocation problem are just the optimal values of the dual variables associated with the primal resource availability constraints, so dual variables corresponding to slack constraints in the primal problem are zero at optimality.

Our first example of a standard primal–dual pair also shows how the optimal values of dual variables provide information about which constraints in the primal problem are satisfied with equality. The problems and their optimal vectors are repeated:

$$\begin{array}{ll} \text{minimize} & 6x_1 + 2x_2 + 3x_3 \\ \text{subject to} & \\ \multicolumn{2}{l}{\begin{bmatrix} 1 & 1 & 0 \\ 1 & -1 & 1 \end{bmatrix} \begin{bmatrix} x_1 \\ x_2 \\ x_3 \end{bmatrix} \geq \begin{bmatrix} 2 \\ 1 \end{bmatrix}} \\ & \mathbf{x} \geq \mathbf{0} \end{array} \qquad \begin{array}{ll} \text{maximize} & 2y_1 + y_2 \\ \text{subject to} & \\ \multicolumn{2}{l}{\begin{bmatrix} 1 & 1 \\ 1 & -1 \\ 0 & 1 \end{bmatrix} \begin{bmatrix} y_1 \\ y_2 \end{bmatrix} \leq \begin{bmatrix} 6 \\ 2 \\ 3 \end{bmatrix}} \\ & \mathbf{y} \geq \mathbf{0} \end{array}$$

$$\mathbf{x}^* = \begin{bmatrix} 3/2 \\ 1/2 \\ 0 \end{bmatrix} \qquad \mathbf{y}^* = \begin{bmatrix} 4 \\ 2 \end{bmatrix}$$

Here the optimal vector $\mathbf{x}^*$ has its first two components positive and the first two constraints in the maximization problem are active at $\mathbf{y}^*$; also, $\mathbf{y}^*$ has both components positive and both constraints in the minimization problem are active at $\mathbf{x}^*$. (Recall from Section 4.1 that an inequality constraint is said to be active if it holds with equality.)

It is true in general that the optimal vector for one of the problems in a standard dual pair provides the following information about which constraints in the other problem are satisfied with equality.

> If the ith component of an optimal vector for one problem in the standard dual pair is positive, then the ith constraint in the other problem is active.

To see why this is always true, recall that in establishing duality relation (i) we showed

$$\mathbf{c}^T\mathbf{x} \geq \mathbf{y}^T(\mathbf{A}\mathbf{x}) \geq \mathbf{b}^T\mathbf{y}$$

whenever $\mathbf{x}$ is feasible for the minimization problem and $\mathbf{y}$ is feasible for the maximization problem. If $\mathbf{x}$ and $\mathbf{y}$ are optimal for their respective problems, then $\mathbf{c}^T\mathbf{x} = \mathbf{b}^T\mathbf{y}$ and

$$\mathbf{c}^T\mathbf{x} = \mathbf{y}^T(\mathbf{A}\mathbf{x}) = \mathbf{b}^T\mathbf{y}$$

so

$$\mathbf{c}^T\mathbf{x} - (\mathbf{y}^T\mathbf{A})\mathbf{x} = 0 \quad \text{and} \quad \mathbf{y}^T(\mathbf{A}\mathbf{x}) - \mathbf{y}^T\mathbf{b} = 0$$

Rewriting these two equations slightly yields the following **complementary slackness conditions.**

COMPLEMENTARY SLACKNESS CONDITIONS

$$\mathbf{x}^T(\mathbf{c} - \mathbf{A}^T\mathbf{y}) = 0 \quad \text{and} \quad \mathbf{y}^T(\mathbf{A}\mathbf{x} - \mathbf{b}) = 0$$

at optimality

The significance of the complementary slackness conditions lies in the fact that if the dot product of two nonnegative vectors is zero, then whenever a component in one of the vectors is positive the corresponding component in the other vector must be zero. Thus, if $\mathbf{x}$ and $\mathbf{y}$ are optimal vectors, then the complementary slackness conditions have the following implications:

$y_i > 0$ implies that the ith primal constraint $\mathbf{A}_i\mathbf{x} \geq b_i$ is active

$x_j > 0$ implies that the jth dual constraint $\mathbf{A}_j^T\mathbf{y} \leq c_j$ is active

The complementary slackness conditions can also be used to check whether or not two feasible vectors $\mathbf{x}$ and $\mathbf{y}$ are, respectively, optimal for the minimization and maximization problems in the standard dual pair. If $\mathbf{x}$ and $\mathbf{y}$ are feasible for their respective problems and satisfy the complementary slackness conditions, then working backward through the above derivation shows that $\mathbf{c}^T\mathbf{x} = \mathbf{b}^T\mathbf{y}$ (see Exercise 5.9). Therefore, by duality relation (i) we can conclude that $\mathbf{x}$ and $\mathbf{y}$ are optimal.

5.4 FINDING THE DUAL OF ANY LINEAR PROGRAM

If a linear program has the form of one of the problems in the standard dual pair, then we can simply write down the other problem as its dual, and the duality relations of Section 5.1 hold for these two problems. To find the dual of a linear program that is not already in the form of one of the problems in the standard dual pair, we use the following procedure:

- Reformulate the problem algebraically so that it is in the form of one of the problems in the standard dual pair.
- Then simply write down the corresponding dual problem.

This method always works because any linear program can be reformulated so that it has the form of one of the problems in the standard dual pair. In this section we illustrate the method by using it to construct the duals of several example problems that are not originally in the form of either problem in the standard dual pair.

Dual of a Linear Program in Standard Form

Recall from Section 3.1 that a linear program is in standard form when it is written as

$$\begin{aligned} &\text{minimize} \quad \mathbf{c}^T\mathbf{x} \\ &\text{subject to} \quad \mathbf{A}\mathbf{x} = \mathbf{b} \\ &\qquad\qquad\quad \mathbf{x} \geq \mathbf{0} \end{aligned}$$

This is different from the form of either problem in the standard dual pair. Replacing $\mathbf{Ax} = \mathbf{b}$ with the two conditions $\mathbf{Ax} \leq \mathbf{b}$ and $\mathbf{Ax} \geq \mathbf{b}$, we can reformulate this problem as

$$\begin{aligned} &\text{minimize} \quad \mathbf{c}^T\mathbf{x} \\ &\text{subject to} \quad \mathbf{A}\mathbf{x} \geq \mathbf{b} \\ &\qquad\qquad -\mathbf{A}\mathbf{x} \geq -\mathbf{b} \\ &\qquad\qquad\quad \mathbf{x} \geq \mathbf{0} \end{aligned}$$

which can be rewritten as

$$\begin{aligned} &\text{minimize} \quad \mathbf{c}^T\mathbf{x} \\ &\text{subject to} \quad \begin{bmatrix} \mathbf{A} \\ -\mathbf{A} \end{bmatrix} \mathbf{x} \geq \begin{bmatrix} \mathbf{b} \\ -\mathbf{b} \end{bmatrix} \\ &\qquad\qquad\qquad \mathbf{x} \geq \mathbf{0} \end{aligned}$$

This reformulated problem has the form of the minimization problem in the standard dual pair, so we can write down its dual as

$$\begin{aligned} &\text{maximize} \quad [\mathbf{b}^T, \ -\mathbf{b}^T] \begin{bmatrix} \mathbf{u} \\ \mathbf{v} \end{bmatrix} \\ &\text{subject to} \quad [\mathbf{A}^T, \ -\mathbf{A}^T] \begin{bmatrix} \mathbf{u} \\ \mathbf{v} \end{bmatrix} \leq \mathbf{c} \\ &\qquad\qquad\qquad\qquad\qquad \begin{bmatrix} \mathbf{u} \\ \mathbf{v} \end{bmatrix} \geq \mathbf{0} \end{aligned}$$

For notational convenience we have partitioned the dual variables into two vectors $\mathbf{u}$ and $\mathbf{v}$, to correspond with the partitioning of the cost vector $[\mathbf{b}^T, \ -\mathbf{b}^T]$ and the constraint coefficient matrix $[\mathbf{A}^T, \ -\mathbf{A}^T]$. This dual problem can be rewritten as

$$\begin{aligned} &\text{maximize} \quad \mathbf{b}^T(\mathbf{u} - \mathbf{v}) \\ &\text{subject to} \quad \mathbf{A}^T(\mathbf{u} - \mathbf{v}) \leq \mathbf{c} \\ &\qquad\qquad\quad \mathbf{u} \geq \mathbf{0}, \ \mathbf{v} \geq \mathbf{0} \end{aligned}$$

As shown in Section 3.7, a free variable can be replaced by the difference of two nonnegative variables. Thus the previous problem is equivalent to

$$\begin{aligned} \text{maximize } & \mathbf{b}^T\mathbf{y} \\ \text{subject to } & \mathbf{A}^T\mathbf{y} \leq \mathbf{c} \\ & \mathbf{y} \text{ free} \end{aligned}$$

The three previous problems are, of course, all equivalent, and each of them is a dual of the original problem. However, because of the simple form of the last version, we will usually refer to it as "the" dual of the original problem. In addition to the standard dual pair, we can now write down and use the dual pair consisting of a standard form linear program and its simple dual.

$$\begin{aligned} \text{minimize } & \mathbf{c}^T\mathbf{x} \\ \text{subject to } & \mathbf{A}\mathbf{x} = \mathbf{b} \\ & \mathbf{x} \geq \mathbf{0} \end{aligned} \qquad \begin{aligned} \text{maximize } & \mathbf{b}^T\mathbf{y} \\ \text{subject to } & \mathbf{A}^T\mathbf{y} \leq \mathbf{c} \\ & \mathbf{y} \text{ free} \end{aligned}$$

Given the maximization problem in this dual pair, we could go through the above construction in reverse order and find that the minimization problem is its dual. Because we constructed the dual above by first writing the original problem in the form of one of the problems in the standard dual pair, it is easy to see that the duality relations of Section 5.1 hold for this dual pair just as they do for the standard dual pair.

A More Complicated Example

It may be necessary to perform several algebraic operations on a primal problem in order to get it into the form of one of the problems in a dual pair that you know. As an example of this, consider the following linear program:

$$\begin{aligned} \text{maximize } & \mathbf{a}^T\mathbf{y} + \mathbf{b}^T\mathbf{w} \\ \text{subject to } & \mathbf{A}\mathbf{y} + \mathbf{B}\mathbf{w} \geq \mathbf{c} \\ & \mathbf{D}\mathbf{w} \leq \mathbf{d} \\ & \mathbf{w} \geq \mathbf{0},\ \mathbf{y} \text{ free} \end{aligned}$$

Letting $\mathbf{y} = (\mathbf{u} - \mathbf{v})$, where $\mathbf{u} \geq \mathbf{0}$ and $\mathbf{v} \geq \mathbf{0}$, and multiplying the first set of inequality constraints by (-1) gives the equivalent problem

$$\begin{aligned} \text{maximize } & \mathbf{a}^T(\mathbf{u} - \mathbf{v}) + \mathbf{b}^T\mathbf{w} \\ \text{subject to } & -\mathbf{A}(\mathbf{u} - \mathbf{v}) - \mathbf{B}\mathbf{w} \leq -\mathbf{c} \\ & \mathbf{D}\mathbf{w} \leq \mathbf{d} \\ & \mathbf{u} \geq \mathbf{0},\ \mathbf{v} \geq \mathbf{0},\ \mathbf{w} \geq \mathbf{0} \end{aligned}$$

Using the rules of matrix multiplication, this problem can be rewritten as

$$\text{maximize } [\mathbf{a}^T, -\mathbf{a}^T, \mathbf{b}^T]\begin{bmatrix} \mathbf{u} \\ \mathbf{v} \\ \mathbf{w} \end{bmatrix}$$

$$\text{subject to} \quad \begin{bmatrix} -\mathbf{A} & \mathbf{A} & -\mathbf{B} \\ \mathbf{0} & \mathbf{0} & \mathbf{D} \end{bmatrix} \begin{bmatrix} \mathbf{u} \\ \mathbf{v} \\ \mathbf{w} \end{bmatrix} \le \begin{bmatrix} -\mathbf{c} \\ \mathbf{d} \end{bmatrix}$$

$$\text{and} \quad \begin{bmatrix} \mathbf{u} \\ \mathbf{v} \\ \mathbf{w} \end{bmatrix} \ge \mathbf{0}$$

where $\mathbf{0}$ indicates a matrix of the appropriate size with all entries zero. This problem has the form of the maximization problem in the standard dual pair and so its dual is

$$\text{minimize} \quad [-\mathbf{c}^T, \mathbf{d}^T] \begin{bmatrix} \mathbf{x} \\ \mathbf{y} \end{bmatrix}$$

$$\text{subject to} \quad \begin{bmatrix} -\mathbf{A}^T & \mathbf{0} \\ \mathbf{A}^T & \mathbf{0} \\ -\mathbf{B}^T & \mathbf{D}^T \end{bmatrix} \begin{bmatrix} \mathbf{x} \\ \mathbf{y} \end{bmatrix} \ge \begin{bmatrix} \mathbf{a} \\ -\mathbf{a} \\ \mathbf{b} \end{bmatrix}$$

$$\begin{bmatrix} \mathbf{x} \\ \mathbf{y} \end{bmatrix} \ge \mathbf{0}$$

The above problem can be rewritten as

$$\begin{aligned} \text{minimize} \quad & -\mathbf{c}^T\mathbf{x} + \mathbf{d}^T\mathbf{y} \\ \text{subject to} \quad & -\mathbf{A}^T\mathbf{x} \ge \mathbf{a} \\ & \mathbf{A}^T\mathbf{x} \ge -\mathbf{a} \\ & -\mathbf{B}^T\mathbf{x} + \mathbf{D}^T\mathbf{y} \ge \mathbf{b} \\ & \mathbf{x} \ge \mathbf{0}, \quad \mathbf{y} \ge \mathbf{0} \end{aligned}$$

or in the simpler form

$$\begin{aligned} \text{minimize} \quad & -\mathbf{c}^T\mathbf{x} + \mathbf{d}^T\mathbf{y} \\ \text{subject to} \quad & \mathbf{A}^T\mathbf{x} = -\mathbf{a} \\ & -\mathbf{B}^T\mathbf{x} + \mathbf{D}^T\mathbf{y} \ge \mathbf{b} \\ & \mathbf{x} \ge \mathbf{0}, \quad \mathbf{y} \ge \mathbf{0} \end{aligned}$$

Dual of the Transportation Problem

An important application area of linear programming concerns the problem of how to ship units of a commodity from several supply points to several destinations so that the total shipping cost is minimized. This problem is called a **transportation problem,** and in Chapter 7 we formulate a linear programming model for this problem that has the following form:

$$\text{minimize} \quad \sum_{i=1}^{m} \sum_{j=1}^{n} c_{ij}x_{ij}$$

$$\text{subject to } \sum_{j=1}^{n} x_{ij} = a_i \qquad i = 1, \ldots, m$$

$$\sum_{i=1}^{m} x_{ij} = b_j \qquad j = 1, \ldots, n$$

$$x_{ij} \geq 0 \quad \text{for each } i \text{ and } j$$

We study the transportation problem in detail in Chapter 7 and develop an efficient method of solution that uses the dual of the above linear program. Of course, the transportation problem is in standard form and we already have seen how to find the dual of such a problem. However, before we can write down the dual, we need to know the structure of the matrix of constraint equation coefficients. It is easiest to illustrate this structure and the construction of the dual by using an example. We consider a transportation problem having $m = 2$ and $n = 3$. In this case there are six variables and letting

$$\mathbf{x} = [x_{11}, x_{12}, x_{13}, x_{21}, x_{22}, x_{23}]^T$$

the (primal) transportation model becomes

$$\text{minimize} \quad [c_{11}, c_{12}, c_{13}, c_{21}, c_{22}, c_{23}]^T\mathbf{x}$$

$$\text{subject to} \quad \left[\begin{array}{ccc|ccc} 1 & 1 & 1 & 0 & 0 & 0 \\ 0 & 0 & 0 & 1 & 1 & 1 \\ \hline 1 & 0 & 0 & 1 & 0 & 0 \\ 0 & 1 & 0 & 0 & 1 & 0 \\ 0 & 0 & 1 & 0 & 0 & 1 \end{array}\right] \mathbf{x} = \begin{bmatrix} a_1 \\ a_2 \\ b_1 \\ b_2 \\ b_3 \end{bmatrix}$$

$$\mathbf{x} \geq \mathbf{0}$$

Given this explicit matrix for the constraint equations, we can now easily write down the dual of this standard from linear program. Introducing dual variables u_i to go with the a_i's and dual variables v_j to go with the b_j's gives the following dual program:

$$\text{maximize} \quad a_1u_1 + a_2u_2 + b_1v_1 + b_2v_2 + b_3v_3$$

$$\text{subject to} \begin{bmatrix} 1 & 0 & 1 & 0 & 0 \\ 1 & 0 & 0 & 1 & 0 \\ 1 & 0 & 0 & 0 & 1 \\ 0 & 1 & 1 & 0 & 0 \\ 0 & 1 & 0 & 1 & 0 \\ 0 & 1 & 0 & 0 & 1 \end{bmatrix} \begin{bmatrix} u_1 \\ u_2 \\ v_1 \\ v_2 \\ v_3 \end{bmatrix} \leq \begin{bmatrix} c_{11} \\ c_{12} \\ c_{13} \\ c_{21} \\ c_{22} \\ c_{23} \end{bmatrix}$$

$$u_i \text{ free} \qquad i = 1, 2$$
$$v_j \text{ free} \qquad j = 1, 2, 3$$

Note in particular that the dual constraint associated with the constant c_{ij} has the form

$$u_i + v_j \leq c_{ij}$$

Generalizing from the result, the dual for an arbitrary transportation problem having the preceding primal is

$$\begin{aligned} &\text{maximize} && \sum_{i=1}^{m} a_i u_i + \sum_{j=1}^{n} b_j v_j \\ &\text{subject to} && u_i + v_j \le c_{ij} \quad \text{for each } i \text{ and } j, \\ &&& \mathbf{u} \text{ and } \mathbf{v} \text{ free} \end{aligned}$$

5.5 THE DUAL SIMPLEX METHOD

In many linear programming applications, including the important area of sensitivity analysis that we consider in Chapter 6, the following situation arises. After solving the model and obtaining a tableau in optimal form, a change in the data or a change in the model itself results in a new tableau that still has nonnegative cost coefficients and a set of basic variables, but the constant column is no longer nonnegative.

To obtain an optimal solution to this new problem, we could apply a phase 1 procedure (such as the subproblem technique or artificial variables) to obtain an initial canonical form, but there is no guarantee that the cost coefficients would remain nonnegative. Thus, after obtaining canonical form, we would then have to apply phase 2 of the simplex algorithm to solve the problem. There is, however, another solution approach that usually requires much less work. We illustrate the approach and the pivot rules that it uses by considering a simple example.

Suppose that the simplex method has been used to solve a linear program, and that the following optimal tableau has been obtained:

	x_1	x_2	x_3	x_4	x_5
70	0	2	1	0	4
50	0	1	−2	1	−2
5	1	1	1	0	−1

Now suppose that the optimal solution represented by this tableau is not acceptable, and that the additional constraint

$$x_2 + x_3 \ge 10$$

is imposed. Introducing a slack variable x_6 for this constraint, the enlarged problem has the following standard form tableau:

	x_1	x_2	x_3	x_4	x_5	x_6
70	0	2	1	0	4	0
50	0	1	−2	1	−2	0
5	1	1	1	0	−1	0
−10	0	−1	−1	0	0	1

This tableau is no longer in optimal form, nor even in canonical form, because it has $b_3 = -10 < 0$. However, if a tableau has nonnegative cost entries and has a set of basic variables, then

> while maintaining nonnegativity of the cost coefficients, it is possible to pivot to obtain a nonnegative constant column or to show that the problem is infeasible.

To see how to derive a set of pivot rules for accomplishing this, consider the following dual pair:

minimize $\mathbf{c}^T\mathbf{x}$	maximize $-\mathbf{b}^T\mathbf{y}$
subject to $\mathbf{Ax} \le \mathbf{b}$	subject to $-\mathbf{A}^T\mathbf{y} \le \mathbf{c}$
$\mathbf{x} \ge \mathbf{0}$	$\mathbf{y} \ge \mathbf{0}$

and we assume that $\mathbf{c} \ge \mathbf{0}$

Introducing vectors **u** and **s** of nonnegative slack variables and reformulating these problems into standard form yields the following simplex tableaus:

$M =$

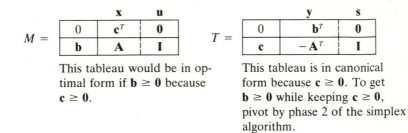

	x	**u**
0	$\mathbf{c}^T$	**0**
b	**A**	**I**

This tableau would be in optimal form if $\mathbf{b} \ge \mathbf{0}$ because $\mathbf{c} \ge \mathbf{0}$.

$T =$

	y	**s**
0	$\mathbf{b}^T$	**0**
c	$-\mathbf{A}^T$	**I**

This tableau is in canonical form because $\mathbf{c} \ge \mathbf{0}$. To get $\mathbf{b} \ge \mathbf{0}$ while keeping $\mathbf{c} \ge \mathbf{0}$, pivot by phase 2 of the simplex algorithm.

In the dual tableau T, we know that pivoting by phase 2 of the simplex algorithm will yield optimal or unbounded form. Below, we explicitly write down the rules for performing phase 2 on the dual tableau T. Because these rules are obtained by applying the simplex algorithm to the dual, we refer to this method of pivoting as the **dual simplex method.** Note, in particular, that the column in T below b_h has entries $-a_{hj}$ for $j = 1, \ldots, n$. Thus the minimum ratio for the b_h column is given by

$$\min\left\{\frac{c_j}{-a_{hj}} \;\middle|\; -a_{hj} > 0\right\}$$

or equivalently, by

$$\max\left\{\frac{c_j}{a_{hj}} \;\middle|\; a_{hj} < 0\right\}$$

Thus, applying phase 2 of the simplex algorithm to tableau T yields the following pivot rules:

THE DUAL SIMPLEX METHOD

> Start with a tableau having $\mathbf{c} \geq \mathbf{0}$
>
> 1 select an h having $b_h < 0$; if $\mathbf{b} \geq \mathbf{0}$, the tableau is in optimal form; stop
> 2 if $-a_{hj} \leq 0$ for $j = 1, \ldots, n$ then the dual is unbounded and the primal is infeasible; stop
> 3 select k so that
>
> $$\frac{c_k}{a_{hk}} = \max\left\{\frac{c_j}{a_{hj}} \,\middle|\, a_{hj} < 0\right\}$$
>
> 4 pivot on a_{hk} and go to step 1

We write down these rules in this particular form because if we perform pivots on tableau M using these rules then the pivots will maintain nonnegativity of the cost coefficients and either obtain nonnegativity of the constant column or give an infeasible tableau. To see why this always happens, we continue with our example and write down the tableaus corresponding to M and T. We have rearranged the columns in our example tableau so that it has the form of the previous tableau, M.

$M =$

	x_2	x_3	x_5	x_4	x_1	x_6
70	2	1	4	0	0	0
50	1	−2	−2	1	0	0
5	1	1	−1	0	1	0
−10	−1	(−1)	0	0	0	1

$T =$

	y_1	y_2	y_3	y_4	y_5	y_6
−70	50	5	−10	0	0	0
2	−1	−1	1	1	0	0
1	2	−1	(1)	0	1	0
4	2	1	0	0	0	1

The dual simplex pivot rules give the circled position in T. More importantly, however, applying these rules to M gives the circled position in M. Performing these pivots in M and T gives the following two tableaus where again we have rearranged the columns so that the identity appears last.

$M =$

	x_2	x_6	x_5	x_4	x_1	x_3
60	1	1	4	0	0	0
70	3	−2	−2	1	0	0
−5	0	1	(−1)	0	1	0
10	1	1	0	0	0	1

$T =$

	y_1	y_2	y_5	y_4	y_3	y_6
−60	70	−5	10	0	0	0
1	−3	0	−1	1	0	0
1	2	−1	1	0	1	0
4	2	(1)	0	0	0	1

Again, the circled pivots are determined by the dual simplex method, and performing the pivots yields the two optimal tableaus.

$M =$

	x_2	x_6	x_1	x_4	x_5	x_3
40	1	5	4	0	0	0
80	3	−4	−2	1	0	0
5	0	−1	−1	0	1	0
10	1	1	0	0	0	1

$T =$

	y_1	y_6	y_5	y_4	y_3	y_2
−40	80	5	10	0	0	0
1	−3	0	−1	1	0	0
5	4	1	1	0	1	0
4	2	1	0	0	0	1

In general, as is illustrated by this sequence of tableaus, pivoting by the simplex method on T is equivalent to pivoting by the dual simplex method on M. For any of these pairs of tableaus, it is easy to see how the data in one of the tableaus can be used to obtain the data in the other tableau. In particular, the data in each pair are related in the same way that the data in the preceding general tableaus M and T are related.

We showed in Chapter 3 that pivoting on T by the simplex method is sure to lead to either optimal or unbounded form (provided cycling does not occur). Because of the duality relations, it follows that pivoting on M by the dual simplex method leads to either optimal or infeasible form (provided cycling does not occur). Note that because the dual represented by tableau T is feasible, the primal represented by tableau M is surely not unbounded.

Another Example of Dual Simplex Pivoting

We conclude this section with another example of pivoting by the dual simplex method on a tableau having nonnegative cost coefficients and a set of basic variables. In this example we do not rearrange the columns so that the identity columns appear last, and in general there is no need to do so.

The following tableau would be in canonical form except that there are two rows with $b_i < 0$. Arbitrarily selecting the first row as the pivot row, we obtain the following sequence of circled pivots by the dual simplex method.

	x_1	x_2	x_3	x_4	x_5	x_6
−10	0	0	3	1	2	0
−5	1	0	−1	0	(−1)	0
2	0	0	2	3	0	1
−7	0	1	2	−1	−1	0

$$\frac{2}{-1} > \frac{3}{-1}$$

	x_1	x_2	x_3	x_4	x_5	x_6
−20	2	0	1	1	0	0
5	−1	0	1	0	1	0
2	0	0	2	3	0	1
−2	−1	1	3	(−1)	0	0

$$\frac{1}{-1} > \frac{2}{-1}$$

	x_1	x_2	x_3	x_4	x_5	x_6
−22	1	1	4	0	0	0
5	−1	0	1	0	1	0
−4	(−3)	3	11	0	0	1
2	1	−1	−3	1	0	0

	x_1	x_2	x_3	x_4	x_5	x_6
70/3	0	2	23/3	0	0	1/3
19/3	0	−1	−8/3	0	1	−1/3
4/3	1	−1	−11/3	0	0	−1/3
2/3	0	0	2/3	1	0	1/3

5.6 USING THE DUAL TO COMPUTATIONAL ADVANTAGE

As mentioned in Chapter 3, experience with the simplex algorithm shows that the number of pivots required to solve a linear program is usually in the range of $1\frac{1}{2}$ to 2 times the number of constraints. Thus, if a linear program has many more constraints than variables, work can sometimes be saved by solving its dual and then using the results of Section 5.2 to construct a solution for the original problem. In this section we present an example to illustrate this fact.

A Difficult Primal Problem Having an Easy Dual

The following linear programming problem is specially constructed (see Exercise 4.7) to illustrate the possible advantages of solving a dual problem instead of the primal. For an arbitrary positive integer M, consider the following linear program:

$$\begin{array}{lrl}
\text{maximize} & x_1 + x_2 & \\
\text{subject to} & (i - M)x_1 + x_2 \leq i(i - 1) & i = 1, \ldots, M \\
& x_1 + (i - M)x_2 \leq i(i - 1) & i = 1, \ldots, M \\
& x_1 \geq 0, \quad x_2 \geq 0 &
\end{array}$$

The feasible set has $2M$ extreme points and is pictured for $M = 4$ in Figure 5.1, along with the optimal point $\mathbf{x}^*$. Introducing slack variables to put the problem into standard form yields an initial canonical form tableau with an associated basic feasible solution corresponding to the extreme point at the origin in Figure 5.1. There are two choices for the first pivot, corresponding to the two paths leading from the origin to $\mathbf{x}^*$. The problem is degenerate because more than two constraint hyperplanes intersect at an extreme point, namely, the origin.

The first simplex rule pivot on the initial canonical form tableau is a degenerate pivot, that is, it does not result in moving to an adjacent extreme point. However, subsequent simplex rule pivots are not degenerate and move the solution along whichever path is being followed, from one extreme point to the next, until $\mathbf{x}^*$ is reached after a total of exactly $(M + 1)$ pivots. Thus, in the

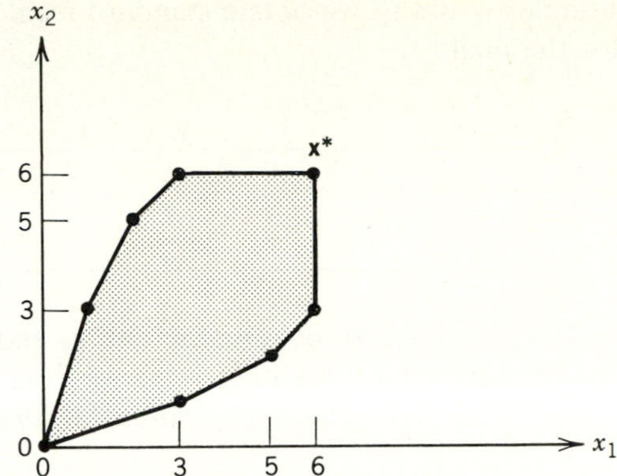

FIGURE 5.1 Feasible Set and Optimal Point for $M = 4$.

example above with $M = 4$, it takes five pivots to obtain optimality. The optimal point $\mathbf{x}^*$ is

$$x_1^* = x_2^* = M(M - 1)/2$$

so in Figure 5.1, $x^* = [6, 6]^T$.

By choosing M, we can produce a linear program for which any given number of pivots will be required to reach optimal form. If $M = 1000$, for example, each of the two paths from the origin to $\mathbf{x}^*$ will have 1000 edges, and the simplex algorithm will take 1001 pivots to obtain optimal form. Thus, because the computational work required to solve a linear program grows with the number of pivots needed, we can make this problem arbitrarily difficult for the simplex algorithm simply by increasing M.

Now consider the dual of the linear program. The dual has $2M$ nonnegative variables but only two constraints. For example, in the case when $M = 4$, the primal problem is

$$\begin{array}{ll} \text{maximize} & x_1 + x_2 \\ \text{subject to} & \begin{bmatrix} -3 & 1 \\ -2 & 1 \\ -1 & 1 \\ 0 & 1 \\ 1 & -3 \\ 1 & -2 \\ 1 & -1 \\ 1 & 0 \end{bmatrix} \begin{bmatrix} x_1 \\ x_2 \end{bmatrix} \le \begin{bmatrix} 0 \\ 1 \\ 3 \\ 6 \\ 0 \\ 1 \\ 3 \\ 6 \end{bmatrix} \end{array}$$

$$x_1 \ge 0 \quad \text{and} \quad x_2 \ge 0$$

so its dual is

$$\begin{array}{ll} \text{minimize} & 0y_1 + 1y_2 + 3y_3 + 6y_4 + 0y_5 + 1y_6 + 3y_7 + 6y_8 \\ \text{subject to} & \\ & -3y_1 - 2y_2 - 1y_3 + 0y_4 + 1y_5 + 1y_6 + 1y_7 + 1y_8 \ge 1 \\ & 1y_1 + 1y_2 + 1y_3 + 1y_4 - 3y_5 - 2y_6 - 1y_7 + 0y_8 \ge 1 \end{array}$$

$$y_i \ge 0 \qquad i = 1, \ldots, 8$$

Introducing slack variables s_1 and s_2, we obtain standard form and the following simplex tableau T for the dual:

$T =$

	y_1	y_2	y_3	y_4	y_5	y_6	y_7	y_8	s_1	s_2
0	0	1	3	6	0	1	3	6	0	0
1	−3	−2	−1	0	1	1	1	1	−1	0
1	1	1	1	1	−3	−2	−1	0	0	−1

This tableau already has two identity columns under y_8 and y_4, so an initial canonical form is easily obtained by pivoting in those columns to make the corresponding costs zero. This yields the initial canonical form tableau

$T^* =$

	y_1	y_2	y_3	y_4	y_5	y_6	y_7	y_8	s_1	s_2
−12	12	7	3	0	12	7	3	0	6	6
1	−3	−2	−1	0	1	1	1	1	−1	0
1	1	1	1	1	−3	−2	−1	0	0	−1

This tableau is in optimal form, so only two pivots were required to solve the dual problem. The optimal vector $\mathbf{x}^*$ for the original problem has components that are the cost coefficients of the slack variables in T, so

$$x_1^* = 6 \text{ and } x_2^* = 6$$

Now matter how large M is, two pivots will always yield optimal form for the dual. Thus, the work required to get optimal form for the dual can be made arbitrarily less than that required to solve the primal:

$$\frac{\text{pivots to solve the dual}}{\text{pivots to solve the primal}} = \frac{2}{M + 1}$$

In general, of course, more than two pivots might be needed to solve a linear program having two constraints. This example was contrived to provide a dramatic illustration of the savings in work that can result from solving the dual of a problem that has many more constraints than variables, and the savings in other cases will usually be less than those obtained in this example. Nevertheless, the computational advantages that sometimes exist in solving the dual are sufficiently great that practical computer implementations of the simplex algorithm usually allow for solving the dual and using the dual solution to construct the solution to the original problem.

5.7 THEOREMS OF THE ALTERNATIVE

In the mathematical theory relating to systems of linear inequalities and equations, many results have the following form:

> Either system 1 has a solution or system 2 has a solution
> but not both.

Such a result is commonly referred to as a **theorem of the alternative** because it describes two alternatives that are mutually exclusive. The most famous theorem of the alternative was established by Farkas:

FARKAS' RESULT

> For any $m \times n$ matrix $\mathbf{A}$ and any vector $\mathbf{b} \in \mathrm{R}^n$,
>
> either (1) $\mathbf{Ax} = \mathbf{b}$, $\mathbf{x} \geq \mathbf{0}$ has a solution
>
> or (2) $\mathbf{A}^T\mathbf{y} \leq \mathbf{0}$, $\mathbf{b}^T\mathbf{y} > 0$ has a solution
>
> but not both.

Note that system 1 is just the constraints for a linear program in standard form, so one implication of Farkas' result is that if a linear program in standard form is infeasible, then the corresponding system 2 must have a solution.

For $n = 2$ it is easy to see geometrically why Farkas' result is true. First, recall the following two elementary facts:

- If two vectors make an acute angle with one another, then their dot product is positive.
- If $\mathbf{Ax} = \mathbf{b}$ for $\mathbf{x} \geq \mathbf{0}$, then $\mathbf{b}$ is a nonnegative linear combination of the columns of $\mathbf{A}$.

Now suppose that $\mathbf{A}$ is 2×2 and has columns $\mathbf{A}_1$ and $\mathbf{A}_2$. In Figure 5.2 the set of all vectors that lie in the cross-hatched region is the set of all nonnegative linear combinations $[x_1\mathbf{A}_1 + x_2\mathbf{A}_2]$ of the columns $\mathbf{A}_1$ and $\mathbf{A}_2$. The vectors that make an obtuse angle to the columns of $\mathbf{A}$ (that is, all the vectors $\mathbf{y}$ for which $\mathbf{A}^T\mathbf{y} \leq \mathbf{0}$) lie in the shaded region.

If the vector $\mathbf{b}$ lies in the cross-hatched region (that is, if system 1 has a solution), then by the construction of Figure 5.2 it is clear that no vector $\mathbf{y}$ in the shaded region could make an acute angle with $\mathbf{b}$; thus $\mathbf{b}^T\mathbf{y} \leq 0$ and system 2 has no solution. On the other hand, if $\mathbf{b}$ does not lie in the cross-hatched region (that is, if $\mathbf{Ax} \neq \mathbf{b}$ so that system 1 has no solution), then from Figure 5.2 it is easy to see that there must be a vector $\mathbf{y}$ in the shaded region that makes an acute angle with $\mathbf{b}$; thus there is a $\mathbf{b}$ such that $\mathbf{b}^T\mathbf{y} > \mathbf{0}$ and system 2 has a solution.

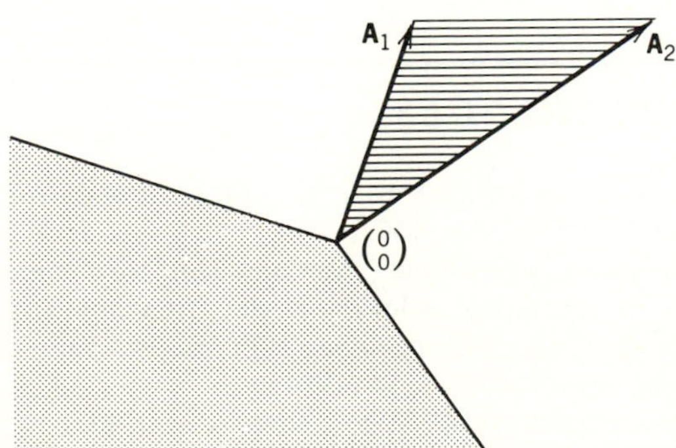

FIGURE 5.2 Geometric Interpretation of Farkas' Result

To see why Farkas' result is true in general, we can use the duality relations. To determine whether or not system 2 has a solution, we need only solve the linear program

$$\begin{aligned} \text{P:} \quad & \text{maximize} && \mathbf{b}^T\mathbf{y} \\ & \text{subject to} && \mathbf{A}^T\mathbf{y} \le \mathbf{0} \\ & && \mathbf{y} \text{ free} \end{aligned}$$

The dual of this problem is

$$\begin{aligned} \text{D:} \quad & \text{minimize} && \mathbf{0}^T\mathbf{x} \\ & \text{subject to} && \mathbf{A}\mathbf{x} = \mathbf{b} \\ & && \mathbf{x} \ge \mathbf{0} \end{aligned}$$

Suppose that system 1 has a solution $\bar{\mathbf{x}}$, so that $\bar{\mathbf{x}}$ is feasible for problem D. Because $\mathbf{0}^T\bar{\mathbf{x}} = 0$ and because no feasible vector can give an objective function value less than 0, $\bar{\mathbf{x}}$ must be optimal for problem D, and the optimal value is zero. Then by the duality relations, the optimal objective function value for problem P is zero. This means that $\mathbf{b}^T\mathbf{y} \le 0$ for each feasible vector $\mathbf{y}$, so system 2 has no solution. Of course, this also shows that if system 2 has a solution, then system 1 does not. Thus not both systems can have solutions.

On the other hand, suppose that system 2 has no solution, so that $\mathbf{b}^T\mathbf{y} \le 0$ for each $\mathbf{y}$ such that $\mathbf{A}^T\mathbf{y} \le \mathbf{0}$. But $\mathbf{y} = \mathbf{0}$ satisfies $\mathbf{A}^T\mathbf{y} \le \mathbf{0}$ and gives $\mathbf{b}^T\mathbf{y} = 0$. Therefore, $\mathbf{y} = \mathbf{0}$ is optimal for problem P. Thus, by the duality relations, problem D must have an optimal vector $\mathbf{x}$. If the vector $\mathbf{x}$ is optimal, then it is certainly feasible for problem D, so system 1 has a solution. This also shows that if system 1 has no solution, then system 2 must have a solution. Thus, one of the systems must have a solution.

Combining the results of the last two paragraphs completes the proof of the Farkas result using duality theory. Linear programming had not yet been invented when Farkas proved his result in 1902, so he used a much different argument. The same kind of argument we used can be used to prove many similar theorems of the alternative involving linear systems. The first step in the approach is always to find a linear program whose solution will determine whether or not one of the systems has a solution. Then using the dual of this linear program and the duality relations, it is usually easy to see why the result holds.

SELECTED REFERENCES

Bradley, S. P., Hax, A. C., and Magnanti, T. L., *Applied Mathematical Programming,* Addison-Wesley, Reading, MA, 1977.

Gass, S., *Linear Programming* (4th ed.), McGraw-Hill, New York, 1975.

Hadley, G., *Linear Programming,* Addison-Wesley, Reading, MA, 1962.

Lemke, C. E., The Dual Method of Solving a Linear Programming Problem, *Naval Research Logistics Quarterly* **1:**48–54 (1954).

Murty, K. G., *Linear Programming,* Wiley, New York, 1983.

Spivey, A. W., and Thrall, R. M., *Linear Optimization,* Holt, Rinehart, and Winston, New York, 1970.

EXERCISES

5.1 (a) Write down the dual of the Oakwood Furniture problem given in Section 2.1.
(b) Obtain an initial standard form tableau for the dual and pivot it to optimal form.
(c) From the optimal tableau for the dual, read off an optimal vector for the Oakwood problem.
(d) Suppose someone wanted to buy one unit of the 12.5 units of oak that are available for production. What is the least amount that we should charge for that unit of oak?
(e) Write down an initial standard form tableau for the Oakwood problem and pivot it to optimal form. Read off the optimal vector for the dual from this tableau and verify that it agrees with the one found in part (b).
(f) What economic interpretation can be given to the optimal value of the dual variable y_1?

5.2 Find the dual of each of the following problems:

(a) maximize $x_1 + x_2$
subject to

$$\begin{aligned} -2x_1 + 2x_2 &\le 1 \\ 16x_1 - 14x_2 &\le 7 \\ x_1 \ge 0,\ x_2 &\ge 0 \end{aligned}$$

(b) minimize $2x_1 - 7x_2 - 3x_3$
subject to

$$\begin{aligned} x_1 + 2x_2 + x_3 &\le 5 \\ 2x_1 \qquad + x_3 &\le 10 \\ x_j \ge 0 \quad j &= 1, 2, 3 \end{aligned}$$

(c) minimize $x_1 - x_2$
subject to

$$\begin{aligned} x_1 + 2x_2 &\ge 4 \\ x_2 &\le 4 \\ 3x_1 - 2x_2 &\le 0 \\ x_1 \text{ free},\ x_2 &\ge 0 \end{aligned}$$

5.3 Consider the following linear program:

maximize $x_1 + 2x_2 + 5x_3 + 4x_4$
subject to

$$\begin{aligned} x_1 + x_2 + x_3 + x_4 &\le 10 \\ x_j \ge 0 \quad j &= 1, \ldots, 4 \end{aligned}$$

(a) Write down the dual of this linear program.
(b) Solve the dual by inspection.

5.4 Consider the maximization problem

maximize $\mathbf{b}^T\mathbf{y}$
subject to $\mathbf{A}^T\mathbf{y} \le \mathbf{c}$

$$\mathbf{y} \ge \mathbf{0}$$

(a) If T is the standard form tableau obtained by adding slack variables $\mathbf{w}$, and

T^* is an optimal tableau in the form

		y	**w**
$T^* =$	d^*	$\mathbf{s}^T$	$\mathbf{r}^T$
	$\mathbf{c}^*$	$\mathbf{S}$	$\mathbf{R}$

then explain why the matrix $\mathbf{Q}$ such that $T^* = \mathbf{Q}T$ is given by

$$\mathbf{Q} = \begin{bmatrix} 1 & \mathbf{r}^T \\ \mathbf{0} & \mathbf{R} \end{bmatrix}$$

(b) Using the fact that $\mathbf{s} \geq \mathbf{0}$ and $\mathbf{r} \geq \mathbf{0}$, show that $\mathbf{x}^* = \mathbf{r}$ is feasible for the dual of the maximization problem.
(c) Using the fact that the optimal value of the maximization problem is d^*, show that $\mathbf{c}^T\mathbf{r} = d^*$ and explain why we can then conclude that $\mathbf{x}^* = \mathbf{r}$ is optimal for the dual of the maximization problem.

5.5 In Section 5.4 we saw the linear program

$$\begin{aligned} \text{minimize} \quad & \mathbf{c}^T\mathbf{x} \\ \text{subject to} \quad & \mathbf{A}\mathbf{x} = \mathbf{b} \\ & \mathbf{x} \geq \mathbf{0} \end{aligned}$$

has a dual given by

$$\begin{aligned} \text{maximize} \quad & \mathbf{b}^T\mathbf{y} \\ \text{subject to} \quad & \mathbf{A}^T\mathbf{y} \leq \mathbf{c} \\ & \mathbf{y} \text{ free} \end{aligned}$$

Let M be the initial standard form tableau and let M^* be an optimal tableau. In this case there are no slack variables in M, and so we cannot directly use the result stated in Section 5.2 to obtain an optimal dual vector from an optimal tableau for the original problem. However, if $M^* = \mathbf{Q}M$ where $\mathbf{Q}$ is a pivot matrix of the form

$$\mathbf{Q} = \begin{bmatrix} 1 & \mathbf{r}^T \\ \mathbf{0} & \mathbf{R} \end{bmatrix}$$

then show that the vector $\mathbf{y}^* = -\mathbf{r}$ is optimal for the dual.

5.6 (a) Find the dual of the refinery problem in Section 2.2.
(b) Solve the dual problem.
(c) What are the shadow prices for each of the three grades of crude oil?

5.7 A certain resource allocation problem of the form

$$\begin{aligned} \text{maximize} \quad & \mathbf{c}^T\mathbf{x} \\ \text{subject to} \quad & \mathbf{A}\mathbf{x} \leq \mathbf{b} \\ & \mathbf{x} \geq \mathbf{0} \end{aligned}$$

has the following initial canonical form tableau M when slack variables s_i are added to each of the three resource constraints. Here, the variables x_j, $j = 1, \ldots, 4$ are the numbers of units of each of four products produced. Pivoting by the simplex algorithm yields the optimal tableau M^*.

		x_1	x_2	x_3	x_4	s_1	s_2	s_3
	0	−2	−5	−3	−7	0	0	0
$M =$	50	1	1	0	3	1	0	0
	150	2	1	2	1	0	1	0
	80	1	1	1	4	0	0	1

$M^* =$

	x_1	x_2	x_3	x_4	s_1	s_2	s_3
340	3	0	0	11	2	0	3
50	1	1	0	3	1	0	0
40	1	0	0	−4	1	1	−2
30	0	0	1	1	−1	0	1

(a) Prior to production, what is the minimum price we should charge for one unit of resource 1 so that our total revenue (from selling the one unit and making products with the available resources) remains at 340?
(b) Answer the same question for resource 2 and then for resource 3.
(c) The per-unit price in part (a) is valid up to what amount of resource 1 sold?

5.8 Find the dual of the following linear programs:

(a)
$$\begin{aligned} \text{minimize} \quad & \mathbf{c}^T\mathbf{x} + \mathbf{a}^T\mathbf{y} \\ \text{subject to} \quad & \mathbf{A}\mathbf{x} = \mathbf{b} \\ & \mathbf{D}\mathbf{x} + \mathbf{B}\mathbf{y} \geq \mathbf{d} \\ & \mathbf{x} \geq \mathbf{0}, \mathbf{y} \text{ free} \end{aligned}$$

(b)
$$\begin{aligned} \text{minimize} \quad & \mathbf{c}^T\mathbf{x} \\ \text{subject to} \quad & \mathbf{A}\mathbf{x} = \mathbf{b} \\ & \mathbf{B}\mathbf{x} \leq \mathbf{a} \\ & \mathbf{x} \geq \mathbf{0} \end{aligned}$$

(c)
$$\begin{aligned} \text{maximize} \quad & \mathbf{c}^T\mathbf{w} + \mathbf{a}^T\mathbf{v} \\ \text{subject to} \quad & \mathbf{A}\mathbf{w} + \mathbf{B}\mathbf{v} = \mathbf{b} \\ & \mathbf{w} \geq \mathbf{0}, \mathbf{v} \text{ free} \end{aligned}$$

(d)
$$\begin{aligned} \text{minimize} \quad & \mathbf{c}^T\mathbf{x} + \mathbf{a}^T\mathbf{y} \\ \text{subject to} \quad & \mathbf{A}\mathbf{x} + \mathbf{B}\mathbf{y} \leq \mathbf{b} \\ & \mathbf{x} \geq \mathbf{0}, \mathbf{y} \text{ free} \end{aligned}$$

5.9 If $\mathbf{x}$ and $\mathbf{y}$ are feasible for the minimization and maximization problems, respectively, in the standard dual pair and also satisfy the complementary slackness conditions, show that they must be optimal vectors.

5.10 Consider the following linear program:

$$\begin{aligned} \text{minimize} \quad & 2x_1 + x_2 + 3x_3 \\ \text{subject to} \quad & \\ & -x_1 - x_2 + 2x_3 \leq -1 \\ & x_1 - x_2 - x_3 \leq -5 \\ & x_j \geq 0 \quad j = 1, 2, 3 \end{aligned}$$

(a) Add slack variables to obtain standard form and solve this problem using the dual simplex method.
(b) Use the method of artificial variables to obtain an initial canonical form and solve the problem using the simplex algorithm.
(c) Use the subproblem technique to obtain an initial canonical form and solve the problem using the simplex algorithm.

5.11 Solve the following problem using the dual simplex method.

$$\begin{array}{ll} \text{minimize} & 3x_1 + x_2 \\ \text{subject to} & \\ & x_1 - x_2 \le -1 \\ & -x_1 - x_2 \le -4 \\ & x_1 + x_2 \le \ \ 1 \\ & x_1 \ge 0, x_2 \ge 0 \end{array}$$

5.12 Suppose that the following optimal tableau is obtained by solving a linear program in standard form.

	x_1	x_2	x_3	x_4	x_5	x_6
−10	0	0	5	4	0	3
4	1	0	−1	2	0	−1
2	0	0	4	−1	1	2
5	0	1	1	1	0	−1

Unfortunately, the constraint that

$$x_1 + x_2 + x_3 \le 8$$

was not included in the original model, but an optimal solution satisfying this constraint is required. Add this constraint to the above tableau and show how to pivot to obtain a new optimal form while maintaining nonnegativity of the cost coefficients.

5.13 For any matrix $\mathbf{A}$, consider the following statement:

either the system $\mathbf{Ax} > \mathbf{0}$ has a solution,
or the system $\mathbf{A}^T\mathbf{y} = \mathbf{0}$, $\mathbf{y} \ge \mathbf{0}$, and $\mathbf{y} \ne \mathbf{0}$ has a solution,
but not both systems have solutions.

(a) Write down a linear program that will determine whether or not the second system has a solution.
(b) Find the dual of the linear program in part (a).
(c) Use the duality relations to show that the above statement is true.

5.14 Consider the linear program

$$\begin{array}{ll} \text{maximize} & -\mathbf{b}^T\mathbf{y} \\ \text{subject to} & -\mathbf{A}^t\mathbf{y} \le \mathbf{c} \\ & \quad \mathbf{y} \text{ free} \end{array}$$

(a) Letting $\mathbf{y} = \mathbf{u} - \mathbf{v}$, where $\mathbf{u} \ge \mathbf{0}$ and $\mathbf{v} \ge \mathbf{0}$, show that the dual of this linear program can be written as

$$\begin{array}{ll} \text{minimize} & \mathbf{c}^T\mathbf{x} \\ \text{subject to} & \mathbf{Ax} = \mathbf{b} \\ & \mathbf{x} \ge \mathbf{0} \end{array}$$

(b) We can also replace the free vector $\mathbf{y}$ by letting $\mathbf{y} = \mathbf{u} - w\mathbf{e}$, where $\mathbf{u}$ and w are nonnegative and $\mathbf{e}$ is a vector with every entry equal to 1 (see Section 3.7). Using this replacement, show that the dual of the linear program can also be written as

$$\begin{array}{ll} \text{minimize} & \mathbf{c}^T\mathbf{x} \\ \text{subject to} & \mathbf{Ax} \le \mathbf{b} \\ & (\mathbf{e}^T\mathbf{A})\mathbf{x} \ge \mathbf{e}^T\mathbf{b} \\ & \mathbf{x} \ge \mathbf{0} \end{array}$$

(c) Therefore, the minimization problem in part (b) is equivalent to our standard form linear program. Using this equivalence, show that if a linear program in standard form has feasible vectors, then an initial canonical form can be obtained by solving only one subproblem in the subproblem technique (or by introducing only one artificial variable if the method of artificial variables is used).

5.15 An important result involving the solution of a system of equations states that for any matrix $\mathbf{A}$,

either the system $\mathbf{Ax} = \mathbf{b}$ has a solution,
or the system $\mathbf{A}^T\mathbf{y} = \mathbf{0}$, $\mathbf{b}^T\mathbf{y} \neq 0$ has a solution,
but not both systems have solutions.

(a) Write down a linear program that will determine whether or not the second system has a solution.
(b) Find the dual of the linear program in part (a) and use the duality relations to show why the above result is true.
(c) Give a geometric interpretation of this result.

5.16 True or False: Give an explanation for your answer.
(a) If a linear program is infeasible, then its dual must be unbounded.
(b) If a slack variable for a resource allocation problem is nonbasic in an optimal tableau, then the shadow price for that resource cannot be zero.
(c) If a minimization problem has a feasible vector, then its dual can never have an unbounded maximum value.
(d) If a minimization problem in the standard dual pair has an optimal vector with the first component positive, then the first constraint in its dual must be active.

5.17 Show that, if the minimization problem in the standard dual pair has multiple optimal solutions, then the maximization problem is degenerate. Give an example.

5.18 Use the dual simplex method to solve the linear programming problems with the following standard from tableaus:

(a)

0	0	0	5	3	0
−1	1	0	−1	1	0
2	0	0	1	−1	1
−1	0	1	−1	−1	0

(b)

0	0	0	5	3	0
−2	1	0	−1	1	0
2	0	0	1	−1	1
−3	0	1	−1	0	1

(c)

0	0	5	3	2	0
−2	1	−1	−1	0	0
−3	0	1	−1	−1	1

CHAPTER 6

SENSITIVITY ANALYSIS IN LINEAR PROGRAMMING

Many linear programming models have data that changes from time to time and may even have some coefficients that are not precisely known. For example, in a resource allocation problem, the selling price of a product may change and/or the availability of a resource may only be known to lie within some range of values. Analyzing how an optimal solution changes with changes in the data or model is thus extremely important and is usually referred to as **sensitivity analysis.** In this chapter we shall see that sensitivity analysis usually does not require completely re-solving the linear program with the new data. Often an optimal tableau for one problem can be used to obtain an optimal solution for a problem where the data has changed slightly.

6.1 THE BREWERY PROBLEM

Throughout this chapter we use the brewery problem first introduced in Section 2.2 to illustrate the various techniques of sensitivity analysis. For that problem, 50 units of malt, 150 units of hops, and 80 units of yeast are available for the production of four products called Light, Dark, Ale, and Special. Using the data in Section 2.2, we obtained the following linear programming model where the objective is to maximize revenue.

The Brewery Model

maximize $6x_1 + 5x_2 + 3x_3 + 7x_4$
subject to

$$\begin{aligned} x_1 + x_2 + \quad\; + 3x_4 &\le 50 \\ 2x_1 + x_2 + 2x_3 + x_4 &\le 150 \\ x_1 + x_2 + x_3 + 4x_4 &\le 80 \end{aligned}$$

$$x_j \ge 0 \qquad j = 1, \ldots, 4$$

Here,

$$\begin{bmatrix} x_1 \\ x_2 \\ x_3 \\ x_4 \end{bmatrix} = \text{amount of} \begin{bmatrix} \text{Light} \\ \text{Dark} \\ \text{Ale} \\ \text{Special} \end{bmatrix} \text{to be produced}$$

and, for example, one unit of Light uses one unit of malt, two units of hops, and one unit of yeast and sells for \$6.

After adding slack variables s_1, s_2, and s_3, for malt, hops, and yeast, respectively, we obtain the following standard form tableau M:

$M =$

	x_1	x_2	x_3	x_4	s_1	s_2	s_3
0	−6	−5	−3	−7	0	0	0
50	1	1	0	3	1	0	0
150	2	1	2	1	0	1	0
80	1	1	1	4	0	0	1

Pivoting by the simplex algorithm yields the following optimal tableau M^*:

$M^* =$

	x_1	x_2	x_3	x_4	s_1	s_2	s_3
380	0	0	0	7	3	1	1
10	0	1	0	7	0	−1	2
40	1	0	0	−4	1	1	−2
30	0	0	1	1	−1	0	1

Recall from Section 3.9 that given M and M^*, it is easy to write down a matrix $\mathbf{Q}$ such that $M^* = \mathbf{Q}M$. We simply "do unto the identity as we would do unto M." Since M contains all of the columns of an $m + 1$ by $m + 1$ identity matrix excepting the first, and since the first column of $\mathbf{Q}$ is always the first identity column, we see from M^* that

$$\mathbf{Q} = \begin{bmatrix} 1 & 3 & 1 & 1 \\ 0 & 0 & -1 & 2 \\ 0 & 1 & 1 & -2 \\ 0 & -1 & 0 & 1 \end{bmatrix}$$

This pivot matrix plays an important role in sensitivity analysis.

In the following sections we develop techniques for analyzing how an optimal vector changes with

- changes in production requirements,
- changes in available resources,
- changes in the selling price of a product,
- adding a new product, and
- adding a new constraint.

We will be mainly concerned with what happens when only one of the changes occurs at a time. Analyzing what happens to the optimal vector when simultaneous changes occur is usually much more complicated and is sometimes not possible without completely re-solving the problem. We will, however, provide some results for simultaneous changes in all resources and for simultaneous changes in all selling prices.

6.2 CHANGES IN PRODUCTION REQUIREMENTS

Changes in Nonbasic Variables

In our discussion of shadow prices in Section 5.3, we saw how an optimal vector changes with changes in slack variables that are nonbasic in the optimal tableau. Actually, the same technique we used there can be applied to any nonbasic variable in the optimal tableau. For example, in the brewery problem, suppose we wanted one unit of the product Special to be produced; that is, instead of having the nonbasic variable $x_4 = 0$ in the optimal solution, suppose we required that $x_4 = 1$. We would like to move, with the smallest possible increase in the objective function value, from the optimal point $\mathbf{x}^*$ to another feasible point having $x_4 = 1$.

Letting $x_4 = 1$ and combining the x_4 column with the constant column gives the following tableau:

	x_1	x_2	x_3	s_1	s_2	s_3
$380 - 7x_4$	0	0	0	3	1	1
$10 - 7x_4$	0	1	0	0	-1	2
$40 + 4x_4$	1	0	0	1	1	-2
$30 - x_4$	0	0	1	-1	0	1

When $x_4 = 1$ the constant column in the preceding tableau remains nonnegative, and so it is an optimal form tableau. Of course, this constant column remains nonnegative as long as x_4 is less than the minimum ratio for the x_4 column in the original tableau. Thus the new optimal vector becomes

$$\begin{bmatrix} 40 + 4x_4 \\ 10 - 7x_4 \\ 30 - x_4 \\ x_4 \end{bmatrix} = \begin{bmatrix} 44 \\ 3 \\ 29 \\ 1 \end{bmatrix} \quad \text{when } x_4 = 1$$

From this example it is clear that the general situation can be summarized as follows:

> In general, given an optimal tableau M^*,
>
> if the required increase in a nonbasic variable x_k is less than the minimum ratio for the x_k column,
>
> to obtain a new optimal vector
>
> set x_k equal to the required increase and set all other nonbasic variables equal to zero. This determines the values of the basic variables.

As another example, suppose we required that $s_2 = 5$ in an optimal solution so that 5 units of hops would be left over. In the s_2 column of M we see that the minimum ratio equals 40, and so we can simply write down the new optimal vector as

$$\begin{bmatrix} 40 - 1s_2 \\ 10 + 1s_2 \\ 30 \\ 0 \end{bmatrix} = \begin{bmatrix} 35 \\ 15 \\ 30 \\ 0 \end{bmatrix} \quad \text{when } s_2 = 5$$

Increasing Basic Variables

The preceding analysis shows how to deal with changes in a nonbasic variable provided the change is not greater than the minimum ratio. To handle the case where a change in a nonbasic variable is greater than the minimum ratio for its column, we first need to consider how changes in variables that are basic in an optimal tableau affect the optimal solution.

In the brewery problem suppose, for example, that we want the amount of Light produced to be increased to 41 (one more unit than the optimal amount given by the basic variable x_1 in M^*). For convenience, we repeat the optimal tableau M^* here.

$M^* =$

	x_1	x_2	x_3	x_4	s_1	s_2	s_3
380	0	0	0	7	3	1	1
10	0	1	0	7	0	−1	2
40	1	0	0	−4	1	1	−2
30	0	0	1	1	−1	0	1

An increase in a variable that is basic in an optimal tableau M^* is only possible if the row of M^* containing that variable has at least one negative entry. The row containing x_1 in M^* corresponds to the equation

$$40 = x_1 - 4x_4 + s_1 + s_2 - 2s_3$$

(If there had been no negative coefficients in this row then, clearly, all feasible vectors would satisfy $x_1 \leq 40$.) In order to increase x_1 from its current value of 40 in $\mathbf{x}^*$, we need to increase either x_4 or s_3. From the preceding equation we see that x_1 can be increased up to 41 by

$$\text{either letting } x_4 = \tfrac{1}{4}$$

$$\text{or letting } s_3 = \tfrac{1}{2}$$

The change that is optimal is the one that increases the objective function the least. From M^*, the objective function is given by

$$z = -380 + 7x_4 + 3s_1 + s_2 + s_3$$

so the proposed change in x_4 increases z by $\frac{7}{4}$ while the proposed change in s_3 only increases z by $\frac{1}{2}$. Thus the new optimal vector is obtained by letting $s_3 = \frac{1}{2}$ in M^* and keeping all other nonbasic variables equal to zero. Note that s_3 is selected over x_4 because the ratio $\frac{1}{2}$ is less than the ratio $\frac{7}{4}$. Increasing s_3 gives the new optimal vector

$$\begin{bmatrix} 40 + 2s_3 \\ 10 - 2s_3 \\ 30 - s_3 \\ 0 \end{bmatrix} = \begin{bmatrix} 41 \\ 9 \\ 29.5 \\ 0 \end{bmatrix} \quad \text{when } s_3 = \tfrac{1}{2}$$

The general situation is summarized as follows:

An increase in a basic variable in row h of M^* is possible only if some $a_{hj} < 0$.

The new optimal vector is obtained by increasing the nonbasic variable x_k, where k is such that

$$\frac{c_k}{a_{hk}} = \max \left\{ \frac{c_j}{a_{hj}} \,\middle|\, a_{hj} < 0 \right\}$$

provided the increase in x_k necessary to achieve the increase in the basic variable does not exceed the minimum ratio for the x_k column.

In our example s_3 was selected by this rule and $s_3 = \frac{1}{2}$ is less than the minimum ratio for the s_3 column in M^*, and thus it is feasible to allow the proposed increase in the nonbasic variable s_3 to obtain the required increase in the basic variable x_1.

It is interesting to interpret the preceding analysis from a geometric point of view. The feasible region for the brewery problem is a four-dimensional convex polyhedron. Requiring $x_1 = 41$ forces us to consider a new feasible region, namely, the three-dimensional region Y determined by intersecting the hyperplane $x_1 = 41$ with the original feasible set. We want to move from the optimal point $\mathbf{x}^*$ for the original problem with as small an increase in the objective function as possible to a point in Y. From $\mathbf{x}^*$ we have a choice of edges to move along to reach Y, namely, the edges defined by increasing x_4 or s_3. We select the edge defined by increasing s_3 because the initial increase in the objective function is the smallest for this edge. Since the required increase in s_3 is less than the minimum ratio, we know that this edge intersects Y at the new optimal point.

As another example of changes in a variable that is basic in the optimal tableau, suppose we required $x_3 = 35$ instead of the optimal value of 30 given in M^*. In this case there is only one nonbasic variable that can be increased to yield the required change, namely, s_1. Letting $s_1 = 5$, which is less than the minimum ratio for the s_1 column, gives the new optimal vector

$$\begin{bmatrix} 40 - s_1 \\ 10 \\ 30 + s_1 \\ 0 \end{bmatrix} = \begin{bmatrix} 35 \\ 10 \\ 35 \\ 0 \end{bmatrix} \quad \text{when } s_1 = 5$$

Decreasing Basic Variables

We can similarly analyze decreases in variables that are basic in an optimal tableau. For example, suppose we require that $x_2 = 8$ instead of 10. Clearly, moving to a feasible point with x_2 decreasing from its current value of 10 is only possible if the row in M^* containing x_2 has at least one positive entry. In M^* the objective function z is

$$z = -380 + 7x_4 + 3s_1 + s_2 + s_3$$

and the constraint row containing x_2 is

$$10 = x_2 + 7x_4 - s_2 + 2s_3$$

Thus we can decrease x_2 from 10 to 8 either by increasing x_4 to $\frac{2}{7}$ or by increasing s_3 to 1. Keeping the other nonbasic variables equal to zero and letting

$$x_4 = \tfrac{2}{7} \quad \text{gives } z = -380 + 7(\tfrac{2}{7}) = -378$$

but letting

$$s_3 = 1 \quad \text{gives } z = -380 + 1(1) = -379$$

Thus the optimal change is to increase s_3 to 1, and the new optimal vector can be written down from M^* as

$$\begin{bmatrix} 40 + 2s_3 \\ 10 - 2s_3 \\ 30 - s_3 \\ 0 \end{bmatrix} = \begin{bmatrix} 42 \\ 8 \\ 29 \\ 0 \end{bmatrix} \quad \text{when } s_3 = 1$$

The general situation is summarized as follows:

A decrease in a basic variable in row h of M^* is possible only if some $a_{hj} > 0$.

The new optimal vector is obtained by increasing the nonbasic variable x_k, where k is such that

$$\frac{c_k}{a_{hk}} = \min \left\{ \frac{c_j}{a_{hj}} \,\middle|\, a_{hj} > 0 \right\}$$

provided the increase in x_k necessary to achieve the increase in the basic variable does not exceed the minimum ratio for the x_k column.

When a Nonbasic Variable Exceeds Its Minimum Ratio

We conclude this section by combining the technique for changing nonbasic variables with the technique for changing basic variables to show how the optimal vector changes when the change required in a nonbasic variable x_k is greater than the minimum ratio for the x_k column.

For convenience we repeat the optimal tableau M^*.

$M^* =$

	x_1	x_2	x_3	x_4	s_1	s_2	s_3
380	0	0	0	7	3	1	1
10	0	1	0	7	0	−1	2
40	1	0	0	−4	1	1	−2
30	0	0	1	1	−1	0	1

Suppose that we want to find the new optimal vector when $x_4 = 5$, which is a value greater than the minimum ratio for the x_4 column. In this case we cannot simply let $x_4 = 5$ in M^* and combine the x_4 column with the constant column because this would result in some negative constant column entries.

Here, to obtain the new optimal vector, we first pivot on M^* to get x_4 up to the minimum ratio of $\frac{10}{7}$ and to make x_4 a basic variable. We then analyze further increases in x_4 as a basic variable. Performing the minimum ratio pivot on the x_4 column in M^* gives

	x_1	x_2	x_3	x_4	s_1	s_2	s_3
370	0	−1	0	0	3	2	−1
10/7	0	1/7	0	1	0	−1/7	2/7
320/7	1	4/7	0	0	1	3/7	−6/7
200/7	0	−1/7	1	0	−1	1/7	5/7

The variable x_4 is now basic, and a further increase in x_4 from its current value of $\frac{10}{7}$ is possible because the row in the preceding tableau containing x_4 has at least one negative entry, namely, in the s_2 column. Increasing s_2 up to 25 (which is less than the minimum ratio for the s_2 column) gives $x_4 = 5$, and so the new optimal vector is

$$\begin{bmatrix} 320/7 - (3/7)s_2 \\ 0 \\ 200 - (1/7)s_2 \\ 10/7 + (1/7)s_2 \end{bmatrix} = \begin{bmatrix} 245/7 \\ 0 \\ 175/7 \\ 5 \end{bmatrix} \quad \text{when } s_2 = 25$$

6.3 CHANGES IN AVAILABLE RESOURCES

In this section we analyze how the optimal vector changes with a change in the availability of a resource. Of course, if the optimal production program does not use up all of a resource, then an increase in that resource will not change the optimal vector. The optimal vector will also remain unchanged if the resource is decreased by less than the amount that was left over.

In the brewery problem all of the slack variables are zero in the optimal solution $\mathbf{x}^*$, so each resource is used up by the optimal production program. Thus the optimal solution will be affected by a decrease in the availability of any resource. We first consider changes in only one of the resources.

Changes in One Resource

Suppose the amount of yeast available is $80 + a$ units instead of 80 units, where the change, a, can be positive or negative. To analyze how such a change in the availability of yeast affects the optimal solution, consider the following tableau, M_1, which differs from the original tableau M only in that $80 + a$ replaces 80 in the constant column.

$M_1 =$

	x_1	x_2	x_3	x_4	s_1	s_2	s_3
0	−6	−5	−3	−7	0	0	0
50	1	1	0	3	1	0	0
150	2	1	2	1	0	1	0
$80 + a$	1	1	1	4	0	0	1

In Section 6.1 we saw that $M^* = \mathbf{Q}M$, where the pivot matrix $\mathbf{Q}$ is

$$\mathbf{Q} = \begin{bmatrix} 1 & 3 & 1 & 1 \\ 0 & 0 & -1 & 2 \\ 0 & 1 & 1 & -2 \\ 0 & -1 & 0 & 1 \end{bmatrix}$$

If we perform on M_1 the same sequence of pivots that transforms M into M^*, the tableau that results is $M_2 = \mathbf{Q}M_1$. Because M_2 differs from M^* only in the constant column, we really only need to compute that column. Doing so yields

$$\begin{bmatrix} 1 & 3 & 1 & 1 \\ 0 & 0 & -1 & 2 \\ 0 & 1 & 1 & -2 \\ 0 & -1 & 0 & 1 \end{bmatrix} \begin{bmatrix} 0 \\ 50 \\ 150 \\ 80 + a \end{bmatrix} = \begin{bmatrix} 380 + a \\ 10 + 2a \\ 40 - 2a \\ 30 + a \end{bmatrix}$$

Thus M_2 is given by

$M_2 =$

	x_1	x_2	x_3	x_4	s_1	s_2	s_3
$380 + 1a$	0	0	0	7	3	1	1
$10 + 2a$	0	1	0	7	0	−1	2
$40 - 2a$	1	0	0	−4	1	1	−2
$30 + 1a$	0	0	1	1	−1	0	1

This tableau is in optimal form provided that

$$10 + 2a \geq 0 \qquad 40 - 2a \geq 0 \qquad 30 + a \geq 0$$

that is, provided $-5 \leq a \leq 20$. Thus, if $a = 10$ additional units of yeast are available, then from M_2 we can simply write down the new optimal vector, namely,

$$\begin{bmatrix} 40 - 2a \\ 10 + 2a \\ 30 + a \\ 0 \end{bmatrix} = \begin{bmatrix} 20 \\ 30 \\ 40 \\ 0 \end{bmatrix} \quad \text{when } a = 10$$

It is not necessary to use the matrix **Q** to determine the adjusted constant column entries in M_2. Simply note that the coefficients of a in the constant column of M_2 are the coefficients in the s_3 column of M^*. Thus we can use

$M^* =$

	x_1	x_2	x_3	x_4	s_1	s_2	s_3
380	0	0	0	7	3	1	1
10	0	1	0	7	0	−1	2
40	1	0	0	−4	1	1	−2
30	0	0	1	1	−1	0	1

and let $s_3 = -10$ to determine the new optimal vector as

$$\begin{bmatrix} 40 + 2s_3 \\ 10 - 2s_3 \\ 30 - s_3 \\ 0 \end{bmatrix} = \begin{bmatrix} 20 \\ 30 \\ 40 \\ 0 \end{bmatrix} \quad \text{when } s_3 = -10$$

It is true in general that an increase, a, in any single resource can be analyzed by letting the slack variable corresponding to that resource take on the value $-a$. Of course, we have already seen that a decrease in the availability of a resource can be analyzed by letting the corresponding slack variable take on a positive value. Thus, in both cases, the new optimal vector can be written down directly from M^* provided that the change in the slack variable does not produce any negative entries in the adjusted constant column. In order for that to be true, the new value of s_k must satisfy the following condition:

$$\max\left\{\frac{b_i}{a_{ik}} \;\middle|\; a_{ik} < 0\right\} \le s_k \le \min\left\{\frac{b_i}{a_{ik}} \;\middle|\; a_{ik} > 0\right\}$$

As another example, suppose we wish to analyze changes in the amount of hops, the resource corresponding to the slack variable s_2. If we let s_2 take on a fixed value and combine the s_2 column with the constant column, the new constant column is nonnegative as long as s_2 satisfies

$$-10 \le s_2 \le 40$$

Thus, if the amount of hops is increased by 5, we compute the new optimal vector as

$$\begin{bmatrix} 40 - 1s_2 \\ 10 + 1s_2 \\ 30 \\ 0 \end{bmatrix} = \begin{bmatrix} 45 \\ 5 \\ 30 \\ 0 \end{bmatrix} \quad \text{when } s_2 = -5$$

On the other hand, if there were a decrease of 25 in the amount of hops available, then we would let $s_2 = 25$ and the new optimal vector would be given by

$$\begin{bmatrix} 40 - 1s_2 \\ 10 + 1s_2 \\ 30 \\ 0 \end{bmatrix} = \begin{bmatrix} 15 \\ 35 \\ 30 \\ 0 \end{bmatrix} \quad \text{when } s_2 = 25$$

Simultaneous Changes in Several Resources

The preceding analysis concerns a change in only a single resource. Now suppose that simultaneous changes occur in the availability of several resources. For example, suppose that in the brewery problem the amount of each resource changes so that the initial tableau M is replaced by tableau T below.

$T =$

	x_1	x_2	x_3	x_4	s_1	s_2	s_3
0	−6	−5	−3	−7	0	0	0
$50 + a$	1	1	0	3	1	0	0
$150 + b$	2	1	2	1	0	1	0
$80 + c$	1	1	1	4	0	0	1

Here a, b, and c denote changes in the amounts of malt, hops, and yeast, respectively. Using the fact that $M^* = \mathbf{Q}M$, where

$$\mathbf{Q} = \begin{bmatrix} 1 & 3 & 1 & 1 \\ 0 & 0 & -1 & 2 \\ 0 & 1 & 1 & -2 \\ 0 & -1 & 0 & 1 \end{bmatrix}$$

we can perform the same pivots on T that we performed on M, obtaining the following tableau $T^* = \mathbf{Q}T$.

$T^* =$

	x_1	x_2	x_3	x_4	s_1	s_2	s_3
$380 + 3a + b + c$	0	0	0	7	3	1	1
$10 - b + 2c$	0	1	0	7	0	−1	2
$40 + a + b - 2c$	1	0	0	−4	1	1	−2
$30 - a + c$	0	0	1	1	−1	0	1

Of course this tableau is in optimal form provided the vector of changes, $[a, b, c]^T$, yields a nonnegative constant column in T^*, that is, provided the changes satisfy the following system of linear inequalities:

$$\begin{aligned} 10 \qquad - b + 2c &\geq 0 \\ 40 + a + b - 2c &\geq 0 \\ 30 - a + \qquad c &\geq 0 \end{aligned}$$

Usually, the effects of simultaneous changes in the availability of several resources are difficult to analyze in a systematic way using such a system of inequalities. However, for many particular simultaneous changes we can use the above result to write down the new optimal vector. For example, the combination of changes

$$\begin{bmatrix} a \\ b \\ c \end{bmatrix} = \begin{bmatrix} -10 \\ 20 \\ 5 \end{bmatrix}$$

satisfies the preceding system of inequalities, so for these changes we can simply write down the new optimal vector as

$$\begin{bmatrix} 40 + a + b - 2c \\ 10 \quad - b + 2c \\ 30 - a \quad + c \\ 0 \end{bmatrix} = \begin{bmatrix} 40 \\ 0 \\ 45 \\ 0 \end{bmatrix} \quad \text{when} \quad \begin{bmatrix} a \\ b \\ c \end{bmatrix} = \begin{bmatrix} -10 \\ 20 \\ 5 \end{bmatrix}$$

Sometimes the change in the resources can be rather large with the final tableau T still being in optimal form. For example, the simultaneous changes

$$\begin{bmatrix} a \\ b \\ c \end{bmatrix} = \begin{bmatrix} 65 \\ 80 \\ 35 \end{bmatrix} \quad \text{or} \quad \begin{bmatrix} a \\ b \\ c \end{bmatrix} = \begin{bmatrix} 5 \\ -15 \\ -10 \end{bmatrix}$$

yield a nonnegative constant column in T^*, and so the new optimal vector can easily be written down. Note that each individual change actually exceeds the maximum increase allowed when only a single resource is allowed to change.

If a vector of changes does not satisfy the system of inequalities that guarantees that the constant column remains nonnegative, then we can use the dual simplex method to pivot to a new optimal vector if one exists. For example, the vector of simultaneous changes

$$\begin{bmatrix} a \\ b \\ c \end{bmatrix} = \begin{bmatrix} 5 \\ 25 \\ 5 \end{bmatrix}$$

does not satisfy the inequality system, and in this case tableau T^* becomes

	x_1	x_2	x_3	x_4	s_1	s_2	s_3
425	0	0	0	7	3	1	1
−5	0	1	0	7	0	(−1)	2
60	1	0	0	−4	1	1	−2
30	0	0	1	1	−1	0	1

Performing the circled pivot determined by the dual simplex method yields the tableau

	x_1	x_2	x_3	x_4	s_1	s_2	s_3
420	0	1	0	14	3	0	3
5	0	−1	0	−7	0	1	−2
55	1	1	0	3	1	0	0
30	0	0	1	1	−1	0	1

which is in optimal form, and from this tableau the new optimal vector $\mathbf{x}^*$ is

$$\mathbf{x}^* = \begin{bmatrix} 55 \\ 0 \\ 30 \\ 0 \end{bmatrix}$$

Computing Shadow Prices

In Section 5.3 we saw that the optimal values of the dual variables in a resource allocation problem are shadow prices for the resources. In particular, from the optimal tableau

$M^* =$

	x_1	x_2	x_3	x_4	s_1	s_2	s_3
380	0	0	0	7	3	1	1
10	0	1	0	7	0	−1	2
40	1	0	0	−4	1	1	−2
30	0	0	1	1	−1	0	1

for the brewery problem, we see that the shadow prices for malt, hops, and yeast are 3, 1, and 1, respectively.

Recall that the shadow price of a resource is the amount by which the maximum revenue decreases if one less unit of resource is available. Thus the minimum price we should charge for selling one unit of malt is 3, and this revenue plus the revenue from the new optimal production equals 380. Of course, this shadow price is only valid up to 40 units of malt because the minimum ratio for the s_1 column in M^* is 40.

Suppose we wish to sell 45 units of malt. We know that we should charge 3 per unit for the first 40 units, but what is the per-unit price above 40? To answer this question, we first pivot on M^* to get s_1 up to 40, and then we analyze further increases in s_1 as a basic variable. Pivoting to make s_1 basic gives

$R_1 =$

	x_1	x_2	x_3	x_4	s_1	s_2	s_3
260	−3	0	0	19	0	−2	7
10	0	1	0	7	0	−1	②
40	1	0	0	−4	1	1	−2
70	1	0	1	−3	0	1	−1

Thus, if we insist on having 40 units of s_1 left over, then our revenue decreases by 120. If we want $s_1 > 40$, then we need to consider changes in s_1 as a basic variable in R_1.

We can increase s_1 above 40 by increasing either x_4 or s_3, and the change that leads to the new optimal vector is determined using the analysis developed in Section 6.2. A one-unit increase in s_1 obtained by increasing s_3 leads to a $\frac{7}{2}$ decrease in revenue, whereas increasing x_4 to achieve the same increase in s_1 leads to a $\frac{19}{4}$ decrease in revenue. Therefore, we select s_3 as the variable to increase. This also shows that we should charge $\frac{7}{2}$ for each unit in excess of 40 that we sell. This price is valid so long as s_3 is less than its minimum ratio of 5. When $s_3 = 5$, we see from tableau R_1 that $s_1 = 50$. Thus the price of $\frac{7}{2}$ is valid between 40 and 50 units. In particular, the amount we should charge for 45 units is

$$3(40) + (\tfrac{7}{2})(5) = 137.5$$

Performing the circled pivot on R_1 to get s_3 up to 5 gives the tableau:

$R_2 =$

	x_1	x_2	x_3	x_4	s_1	s_2	s_3
225	−3	−7/2	0	−11/2	0	3/2	0
5	0	1/2	0	7/2	0	−1/2	1
50	1	1	0	3	1	0	0
75	1	1/2	1	1/2	0	1/2	0

Only 50 units of malt are available, and tableau R_2 shows that if all 50 units are left over, then the new optimal vector is

$$\mathbf{x}^* = \begin{bmatrix} 0 \\ 0 \\ 75 \\ 0 \end{bmatrix}$$

Note that the third product, Ale, does not use any malt. Selling 75 units of Ale yields 225 in revenue and selling all the malt yields

$$3(40) + (\tfrac{7}{2})10 = 155$$

in revenue. The total revenue, 225 + 155, equals the original maximum revenue of 380.

6.4 CHANGES IN SELLING PRICES

From the graphical representation of a linear program, we have seen that the same extreme point of a feasible region may be optimal for many different objective functions. It is actually quite easy to analytically determine the range in which a single objective function coefficient can vary without changing the optimal vector.

For example, in the brewery problem, if q denotes a change in the selling price of Light so that the new selling price is $6 + q$ instead of 6, then the initial tableau becomes N_1:

$N_1 =$

	x_1	x_2	x_3	x_4	s_1	s_2	s_3
0	$-6 - q$	−5	−3	−7	0	0	0
50	1	1	0	3	1	0	0
150	2	1	2	1	0	1	0
80	1	1	1	4	0	0	1

Performing on N_1 the same pivots that were performed on M to get M^* yields the following tableau N_2, which differs from M^* in only the c_1 entry.

$N_2 =$

	x_1	x_2	x_3	x_4	s_1	s_2	s_3
380	$-q$	0	0	7	3	1	1
10	0	1	0	7	0	−1	2
40	(1)	0	0	−4	1	1	−2
30	0	0	1	1	−1	0	1

To see why this is true, just use the fact that $N_2 = \mathbf{Q}N_1$, where $\mathbf{Q}$ is the pivot matrix such that $M^* = \mathbf{Q}M$. Thus the x_1 column of N_2 is calculated as

$$\begin{bmatrix} 1 & 3 & 1 & 1 \\ 0 & 0 & -1 & 2 \\ 0 & 1 & 1 & -2 \\ 0 & -1 & 0 & 1 \end{bmatrix}\begin{bmatrix} -6 - q \\ 1 \\ 2 \\ 1 \end{bmatrix} = \begin{bmatrix} -q \\ 0 \\ 1 \\ 0 \end{bmatrix}$$

Tableau N_2 is not in canonical form, but performing the circled pivot gives the tableau N_3:

$N_3 =$

	x_1	x_2	x_3	x_4	s_1	s_2	s_3
380 + 40q	0	0	0	7 − 4q	3 + q	1 + q	1 − 2q
10	0	1	0	7	0	−1	2
40	1	0	0	−4	1	1	−2
30	0	0	1	1	−1	0	1

Tableau N_3 is in optimal form if

$$7 - 4q \geq 0 \qquad 3 + q \geq 0 \qquad 1 + q \geq 0 \qquad 1 - 2q \geq 0$$

that is, if

$$\max\{-\tfrac{3}{1}, -\tfrac{1}{1}\} \leq q \leq \min\{\tfrac{7}{4}, \tfrac{1}{2}\}$$

Actually, this range on q can be determined directly from M^*. For convenience, we repeat M^*:

$M^* =$

	x_1	x_2	x_3	x_4	s_1	s_2	s_3
380	0	0	0	7	3	1	1
10	0	1	0	7	0	−1	2
40	1	0	0	−4	1	1	−2
30	0	0	1	1	−1	0	1

Note that for our example the change in selling price corresponds to the basic variable that appears in the second row of M^*. For each nonbasic column in M^*, if we divide the objective function coefficient by the negative of the entry in the second row, then we obtain the ratios used in the above ratio test on q. The following result is a generalization from the preceding example:

> If the change q in the selling price of a product corresponding to the basic variable in row h of M^* satisfies
>
> $$\max\left\{\frac{-c_j}{a_{hj}} \;\middle|\; a_{hj} > 0\right\} \leq q \leq \min\left\{\frac{c_j}{-a_{hj}} \;\middle|\; a_{hj} < 0\right\}$$
>
> then the optimal vector does not change.

For a change in the selling price of a product corresponding to a variable that is nonbasic in M^*, the analysis is straightforward. For example, if $(7 + r)$ is the new selling price of Special, then the cost coefficient in the x_4 column of

M becomes $(-7 - r)$. Performing the pivots to obtain M^* causes this entry to become $(7 - r)$, that is,

$$\begin{bmatrix} 1 & 3 & 1 & 1 \\ 0 & 0 & -1 & 2 \\ 0 & 1 & 1 & -2 \\ 0 & -1 & 0 & 1 \end{bmatrix} \begin{bmatrix} -7 - r \\ 3 \\ 1 \\ 4 \end{bmatrix} = \begin{bmatrix} 7 - r \\ 7 \\ -4 \\ 1 \end{bmatrix}$$

Thus the new tableau is in optimal form provided $7 - r \geq 0$, and so the optimal vector does not change unless the selling price of Special exceeds 14.

Simultaneous Changes in Prices

Just as we dealt with simultaneous changes in the amounts of resources available, we can also deal with simultaneous changes in the selling prices of the products. If we let p, q, r, and s denote the change in the selling price of Light, Dark, Ale, and Special, respectively, then the initial tableau becomes

$P =$

	x_1	x_2	x_3	x_4	s_1	s_2	s_3
0	$-6 - p$	$-5 - q$	$-3 - r$	$-7 - s$	0	0	0
50	1	1	0	3	1	0	0
150	2	1	2	1	0	1	0
80	1	1	1	4	0	0	1

Performing on P the same pivots that were performed on M to get M^* yields the tableau P_1:

$P_1 =$

	x_1	x_2	x_3	x_4	s_1	s_2	s_3
380	$-p$	$-q$	$-r$	$7 - s$	3	1	1
10	0	1	0	7	0	-1	2
40	1	0	0	-4	1	1	-2
30	0	0	1	1	-1	0	1

Performing the necessary pivots to get zeros above the identity columns in P_1 yields the tableau P_2:

$P_2 =$

	x_1	x_2	x_3	x_4	s_1	s_2	s_3
v	0	0	0	a	b	c	d
10	0	1	0	7	0	-1	2
40	1	0	0	-4	1	1	-2
30	0	0	1	1	-1	0	1

where

$$\begin{aligned} a &= 7 - 4p + 7q + r - s \\ b &= 3 + p - r \\ c &= 1 + p - q \\ d &= 1 - 2p + 2q + r \end{aligned}$$

$$v = 380 + 40p + 10q + 30r$$

Tableau P_2 is in optimal form provided *a, b, c,* and *d* are nonnegative, that is, provided the vector $[p, q, r, s]^T$ of simultaneous changes satisfies the system

$$\begin{aligned}
7 - 4p + 7q + r - s &\geq 0 \\
3 + p \qquad - r \quad &\geq 0 \\
1 + p - q \qquad\quad &\geq 0 \\
1 - 2p + 2q + r \quad &\geq 0
\end{aligned}$$

For example, the vector

$$\begin{bmatrix} p \\ q \\ r \\ s \end{bmatrix} = \begin{bmatrix} -1 \\ 0 \\ 2 \\ 1 \end{bmatrix}$$

satisfies the above system of inequalities and so if

the selling price for Light decreases by 1,
the selling price for Dark stays the same,
the selling price for Ale increases by 2, and
the selling price for Special increases by 1,

then the optimal vector does not change.

Of course, if the changes in selling prices are such that at least one of *a, b, c,* or *d* in tableau P_2 is negative, then additional simplex rule pivots need to be performed on P_2 to obtain the new optimal vector.

6.5 ADDING NEW PRODUCTS OR CONSTRAINTS

New Products

Suppose in the brewery problem a new product called Extra-Special is proposed that requires 2 units of malt, 3 units of hops, and 1 unit of yeast, and its proposed per-unit selling price is 9. To determine whether it is profitable to produce any of this new product, we simply add the column

$$\mathbf{u} = \begin{bmatrix} -9 \\ 2 \\ 3 \\ 1 \end{bmatrix}$$

to the tableau *M* and then calculate

$$\mathbf{Qu} = \begin{bmatrix} 1 & 3 & 1 & 1 \\ 0 & 0 & -1 & 2 \\ 0 & 1 & 1 & -2 \\ 0 & -1 & 0 & 1 \end{bmatrix} \begin{bmatrix} -9 \\ 2 \\ 3 \\ 1 \end{bmatrix} = \begin{bmatrix} 1 \\ -1 \\ 3 \\ -1 \end{bmatrix} = \mathbf{u}^*$$

Thus, performing the pivots on *M* to get M^* results in the new column $\mathbf{u}^*$ in M^*. Because the cost coefficient in $\mathbf{u}^*$ is positive, the product Extra-Special would not be produced in an optimal production program.

If p denotes the selling price of Extra-Special, then

$$\mathbf{Q}\begin{bmatrix} -p \\ 2 \\ 3 \\ 1 \end{bmatrix} = \begin{bmatrix} -p + 10 \\ -1 \\ 3 \\ -1 \end{bmatrix}$$

Thus the selling price of Extra-Special must be at least 10 per unit in order for it to be produced in an optimal program. In fact, if p is slightly greater than 10, then the above analysis shows that Extra-Special would replace Light in the optimal production program. The reason for this is that as $(-p + 10)$ becomes negative, then a pivot in the new column of M^* is required. This pivot would cause x_1 to become a nonbasic variable and so Light would no longer be produced.

Technology Changes

The preceding technique can also be used to analyze what happens when the technology for making a product changes. For example, suppose that a new way of making Special is invented that only takes 2 units of yeast instead of 4 and 1 unit of hops instead of 3. To see if this results in Special becoming a competitive product, we simply perform the matrix multiplication

$$\mathbf{Q}\begin{bmatrix} -7 \\ 1 \\ 1 \\ 2 \end{bmatrix} = \begin{bmatrix} 1 & 3 & 1 & 1 \\ 0 & 0 & -1 & 2 \\ 0 & 1 & 1 & -2 \\ 0 & -1 & 0 & 1 \end{bmatrix}\begin{bmatrix} -7 \\ 1 \\ 1 \\ 2 \end{bmatrix} = \begin{bmatrix} -1 \\ 3 \\ 0 \\ 1 \end{bmatrix}$$

Thus with these technology changes the tableau M^* becomes

$M^* =$

	x_1	x_2	x_3	x_4	s_1	s_2	s_3
380	0	0	0	−1	3	1	1
10	0	1	0	(3)	0	−1	2
40	1	0	0	0	1	1	−2
30	0	0	1	1	−1	0	1

Because the cost coefficient of the x_4 column is negative, M^* is no longer in optimal form and Special is now a competitive product. Performing the circled pivot in M^* gives the new optimal tableau:

$M^{**} =$

	x_1	x_2	x_3	x_4	s_1	s_2	s_3
1150/3	0	1/3	0	0	3	2/3	5/3
10/3	0	1/3	0	1	0	−1/3	2/3
40	1	0	0	0	1	1	−2
80/3	0	−1/3	1	0	−1	1/3	1/3

Therefore, the new optimal production program is

$$\mathbf{x}^* = \begin{bmatrix} 40 \\ 0 \\ 80/3 \\ 10/3 \end{bmatrix}$$

with $x_4 = \frac{10}{3}$ units of Special being produced.

New Constraints

We saw in Section 5.5 how the dual simplex method can be used to advantage when new constraints are added to a linear programming problem. We conclude this section with another example of this using the brewery problem.

Suppose that a new constraint is imposed that requires the total amount of Light and Dark produced to be at least twice the total amount of Ale and Special produced. The current optimal vector does not satisfy this constraint, namely, that

$$x_1 + x_2 \geq 2(x_3 + x_4)$$

Adding this constraint to the tableau M^* yields the tableau:

	x_1	x_2	x_3	x_4	s_1	s_2	s_3	s_4
380	0	0	0	7	3	1	1	0
10	0	1	0	7	0	−1	2	0
40	1	0	0	−4	1	1	−2	0
30	0	0	1	1	−1	0	1	0
0	−1	−1	2	2	0	0	0	1

where s_4 is a new slack variable. Performing the three pivots to obtain identity columns yields the tableau:

	x_1	x_2	x_3	x_4	s_1	s_2	s_3	s_4
380	0	0	0	7	3	1	1	0
10	0	1	0	7	0	−1	2	0
40	1	0	0	−4	1	1	−2	0
30	0	0	1	1	−1	0	1	0
−10	0	0	0	3	3	0	−2	1

Now, pivoting by the dual simplex method on the circled entry gives the following optimal tableau for the problem:

	x_1	x_2	x_3	x_4	s_1	s_2	s_3	s_4
375	0	0	0	17/2	9/2	1	0	1/2
0	0	1	0	10	3	−1	0	1
50	1	0	0	−7	−2	1	0	−1
25	0	0	1	5/2	1/2	0	0	1/2
5	0	0	0	−3/2	−3/2	0	1	−1/2

Thus the new optimal production program becomes

$$\mathbf{x}^* = \begin{bmatrix} 50 \\ 0 \\ 25 \\ 0 \end{bmatrix}$$

SELECTED REFERENCES

Bradley, S. P., Hax, A. C., and Magnanti, T. L., *Applied Mathematical Programming,* Addison Wesley, Reading, MA, 1977.
Dantzig, G. B., *Linear Programming and Extensions,* Princeton University Press, Princeton, 1963.
Spivey, A. W., and Thrall, R. M., *Linear Optimization,* Holt, Rinehart, and Winston, New York, 1970.

EXERCISES

6.1 Consider the following optimal tableau for a standard form linear program:

	x_1	x_2	x_3	x_4	x_5	x_6	x_7
−75	0	4	0	3	2	0	0
10	1	−1	0	−2	−1	0	0
5	0	1	0	2	3	0	1
20	0	−2	1	−1	1	0	0
30	0	1	0	−1	0	1	0

(a) Find an optimal vector when the additional requirement $x_2 = 3$ is added to the original model.
(b) If we insist on the requirement that $x_2 = 6$, does the linear program have any feasible solutions?
(c) Using the second constraint equation, find upper bounds on the variables x_2, x_4, x_5, and x_7 that must be satisfied by any feasible solution.
(d) Find an optimal vector when the additional requirement $x_1 = 12$ is added to the original model.
(e) Find an optimal vector when the additional requirement $x_3 = 21$ is added to the original model.

6.2 Suppose on solving a linear program the following optimal form tableau is obtained:

	x_1	x_2	x_3	x_4
−45	0	1	1	0
10	0	1	−1	1
20	1	1	2	0

(a) Find the new optimal vector when the additional requirement $x_2 = 5$ is added.
(b) Construct a graphical representation of the feasible set in x_2x_3 − space, identify the optimal point, and draw a contour line for the objective function passing through the optimal point.
(c) On the graph of part (b) indicate the intersection, Y, of the feasible set with the hyperplane $x_2 = 5$.
(d) Find the point on Y corresponding to the new optimal vector found in part (a) and show graphically that this point is optimal when $x_2 = 5$ is required.

6.3 The following tableau M is the initial canonical form tableau for a resource allocation problem of the form:

$$\text{maximize} \quad \mathbf{c}^T\mathbf{x} \quad \text{subject to} \quad \mathbf{Ax} \leq \mathbf{b}, \mathbf{x} \geq \mathbf{0}$$

$M =$

	x_1	x_2	x_3	x_4	s_1	s_2	s_3
0	−7	−8	−3	−7	0	0	0
50	1	2	1	1	1	0	0
40	2	2	1	1	0	1	0
30	1	2	1	5	0	0	1

Here, x_j denotes the amount of product j produced, and s_i is the slack variable for resource i. For example, product 2 sells for 8 and uses 2 units of each of the resources. The optimal tableau M^* is

$M^* =$

	x_1	x_2	x_3	x_4	s_1	s_2	s_3
150	0	0	1	1	0	3	1
20	0	0	0	−4	1	0	−1
10	1	0	0	−4	0	1	−1
10	0	1	1/2	9/2	0	−1/2	1

(a) Write down an optimal vector for the dual problem.
(b) Find a matrix $\mathbf{Q}$ such that $M^* = \mathbf{Q}M$.
(c) What is the minimum amount we should charge for 8 units of resource 2 so the total revenue from selling these 8 units and selling products will not be less than 150?
(d) What is the minimum amount we should charge for 22 units of resource 2 so the total revenue from selling these 22 units and selling products will not be less than 150?
(e) What is the new optimal vector $\mathbf{x}^*$ if we insist on making 10 units of product 3?
(f) What is the new optimal vector $\mathbf{x}^*$ if we insist on making 25 units of product 3?
(g) What is the new optimal vector $\mathbf{x}^*$ if only 25 units of resource 3 are available instead of the original 30?
(h) What is the new optimal vector $\mathbf{x}^*$ if 35 units of resource 3 are available instead of the original 30?
(i) Give a range in which the selling price of product 1 can vary without changing the optimal basic sequence.
(j) Suppose a new product is introduced that uses 2, 5, and 7 units of resource 1, 2, and 3, respectively. What is the smallest value we could select for its selling price so that it would be produced in an optimal production program?
(k) What is the smallest amount that the selling price of product 3 should be increased by so that product 3 would be produced in an optimal production program?
(l) What is the new optimal vector $\mathbf{x}^*$ if the amount of resource 1 increases by 10 units but the amount of resource 2 and resource 3 each decrease by 10 units?
(m) Write down the inverse of the matrix $\mathbf{Q}$ in part (b).

6.4 The following tableau M is the initial canonical form tableau for a resource allocation problem of the form:

$$\text{maximize} \quad \mathbf{c}^T\mathbf{x} \quad \text{subject to} \quad \mathbf{Ax} \leq \mathbf{b}, \mathbf{x} \geq \mathbf{0}$$

$M =$

	x_1	x_2	x_3	x_4	s_1	s_2	s_3
0	−6	−1	−4	−5	0	0	0
20	2	1	1	1	1	0	0
10	1	0	2	1	0	1	0
5	1	1	1	0	0	0	1

Here, x_j denotes the amount of product j produced and s_i is the slack variable for resource i. For example, product 2 sells for 1 and uses 1 unit each of resource 1 and resource 3. The optimal tableau M^* is

$M^* =$

	x_1	x_2	x_3	x_4	s_1	s_2	s_3
55	0	0	7	0	0	5	1
5	0	0	−2	0	1	−1	−1
5	0	−1	1	1	0	1	−1
5	1	1	1	0	0	0	1

(a) Find a matrix $\mathbf{Q}$ such that $M^* = \mathbf{Q}M$.
(b) What is the minimum amount we should charge for 8 units of resource 2 so the total revenue from selling these 8 units and selling products will not be less than 55?
(c) What is the new optimal vector $\mathbf{x}^*$ if only 4 units of resource 3 are available instead of the original 5?
(d) What is the new optimal vector $\mathbf{x}^*$ if 8 units of resource 3 are available instead of the original 5?
(e) Give a range in which the selling price of product 1 can vary without changing the optimal basic sequence.
(f) What is the smallest amount by which the selling price of product 3 should be increased so that product 3 would be produced in an optimal production program?
(g) What is the new optimal production program if the selling price of product 4 increases to 6?

6.5 Answer the following questions regarding the brewery problem of Section 6.1.
(a) What is the new optimal production program if 40 units of malt are available instead of the original 50 units?
(b) What is the new optimal production program if an additional 15 units of malt are available?
(c) What is the new optimal production program if only 5 units of malt are available?
(d) Suppose that the amount of malt increases by 10 units, the amount of hops decreases by 30 units, and the amount of yeast increases by 10 units. Does this change the basic sequence that is optimal for the original problem? What is the new optimal solution?
(e) Suppose that the amount of malt increases by 20 units and the amount of hops increases by 20 units. Does the original optimal basic sequence change? Use the dual simplex method to obtain the new optimal solution.

6.6 For the brewery problem of Section 6.1, find a function $f(p)$ giving the amount that the company should charge for p units of yeast so that the amount received plus the revenue from production equals the original optimal revenue of \$380.

6.7 Consider the resource allocation problem of Exercise 6.4.
(a) Suppose that the selling price of product 4 increases by 1 unit and the selling

price of product 2 increases by 1 unit. Is the original optimal solution still optimal? Explain.

(b) If the selling prices of products 1, 2, 3, and 4 change by p, q, r, and s, respectively, find a system of inequalities on p, q, r, and s having the property that if p, q, r, and s satisfy the system, then the optimal solution does not change.

6.8 For the brewery problem of Section 6.1, suppose that the amounts of malt, hops, and yeast increase in the proportions $1p:1p:2p$, respectively.

(a) Find the maximum value of p so that the optimal basic sequence given in M^* does not change.

(b) What is the new optimal vector when p takes on its maximum value?

CHAPTER 7

LINEAR PROGRAMMING MODELS FOR NETWORK FLOW PROBLEMS

Many linear programming models have special structure that allows the simplex algorithm to be implemented in a very efficient manner and permits us to solve problems that otherwise would be too large to solve using current computer technology. Even for smaller problems that could be solved using a straightforward application of the simplex algorithm, the savings in computational effort obtained by exploiting the special structure can be enormous. In this chapter we show how the simplex algorithm can be specialized to solve several linear programming models that arise from network flow problems.

7.1 THE TRANSPORTATION PROBLEM

Historically, one of the first examples of exploiting special structure in a linear programming model occurred with the **transportation problem.** In the transportation problem units of a single product are to be shipped from m sources to n destinations where

a_i = number of units available at source i
b_j = number of units required at destination j
c_{ij} = cost of shipping one unit from source i to destination j

The problem is to determine a shipping program that meets the required demands and minimizes the total shipping cost. Initially, we assume that the total supply

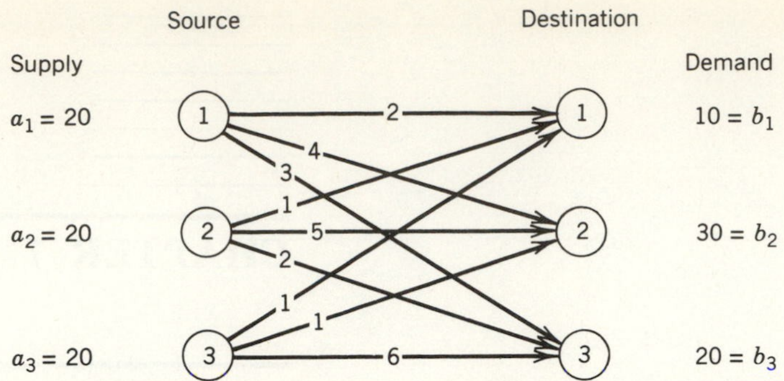

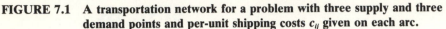

FIGURE 7.1 A transportation network for a problem with three supply and three demand points and per-unit shipping costs c_{ij} given on each arc.

equals the total demand, that is,

$$\sum_{i=1}^{m} a_i = \sum_{j=1}^{n} b_j$$

We shall show later that it is easy to handle the case where this equality does not hold.

The transportation problem is a special case of a network flow problem where the nodes of the network are the supply-and-demand points and the arcs in the network are the shipping routes from each supply point to each demand point. For example, consider a problem with 3 supply and 3 demand points where each supply point has 20 units of the product and the demands are 10, 30, and 20 units, respectively, at the 3 demand points. The network associated with this problem is given in Figure 7.1.

If we define the decision variables by

x_{ij} = number of units shipped from source i to destination j

the transportation problem can be formulated as the following linear program:

THE TRANSPORTATION PROBLEM AS A LINEAR PROGRAM

$$\begin{aligned}
\text{minimize} \quad & \sum_{i=1}^{m}\sum_{j=1}^{n} c_{ij}\, x_{ij} \\
\text{subject to} \quad & \sum_{j=1}^{n} x_{ij} = a_i \qquad i = 1, \ldots, m \quad \text{(supply constraints)} \\
& \sum_{i=1}^{m} x_{ij} = b_j \qquad j = 1, \ldots, n \quad \text{(demand constraints)} \\
& x_{ij} \geq 0 \quad \text{for each } (i, j)
\end{aligned}$$

Note that we can use equality constraints in the model because we have assumed that total supply equals total demand.

The transportation model is already in standard form, and we could, of course, use the simplex algorithm to pivot the initial standard form tableau to optimal form. However, even for modest values of m and n, the resulting linear program can be very large. For example, when $m = 100$ and $n = 100$, there are 200 constraints and 10,000 variables in the model.

To see the special linear programming structure of a transportation problem, consider the example problem summarized by the network in Figure 7.1. The simplex tableau for this problem is as follows:

	x_{11}	x_{12}	x_{13}	x_{21}	x_{22}	x_{23}	x_{31}	x_{32}	x_{33}
0	2	4	3	1	5	2	1	1	6
20	1	1	1	0	0	0	0	0	0
20	0	0	0	1	1	1	0	0	0
20	0	0	0	0	0	0	1	1	1
10	1	0	0	1	0	0	1	0	0
30	0	1	0	0	1	0	0	1	0
20	0	0	1	0	0	1	0	0	1

This tableau is not in canonical form, but obtaining an initial basic feasible solution is very easy for a transportation problem and does not require the use of artificial variables or the subproblem technique. To see how this can be done, consider the network representation of the problem in Figure 7.1. The following illustrates a general procedure, called the **northwest corner rule,** for finding a feasible solution, and as we shall see the solution obtained is actually a basic feasible solution.

Given the network in Figure 7.1, suppose we start with the first source node, S_1 (in the northwest corner), and ship as much as possible to the first destination, D_1. Because S_1 has 20 units of supply and D_1 needs 10 units, we assign $x_{11} = 10$. Then S_1 has 10 units of supply left, and we ship that remaining supply to D_2 by assigning $x_{12} = 10$. We then consider S_2 and ship as much as possible to D_2 because its demand is not yet met. Thus we assign $x_{22} = 20$, which meets the demand at D_2 and uses up the supply at S_2. We then go on to the last supply node, S_3, and ship 20 units from S_3 to D_3 by assigning $x_{33} = 20$, which gives us a feasible solution. Actually, if an assignment, other than the final one, simultaneously meets a demand and uses up a supply, let us agree to assign a shipment of 0 units from the current source to the next destination. Thus, in our example, we also assign $x_{23} = 0$. The reason for doing this will be made clear when we see that this means that the transportation problem is degenerate.

It is not hard to see how this procedure for finding a feasible solution can be generalized to a transportation problem with m supply and n demand points, and we shall give a formal statement of the procedure later. It is important to realize that the feasible solution obtained by the northwest corner rule will always consist of $m + n - 1$ assignments, some of which may be zero. We shall represent such a solution by a subnetwork of the transportation network that consists of the $m + n - 1$ arcs where assignments are made. For our example the $m + n - 1 = 5$ arcs in this subnetwork are given in Figure 7.2.

It is clear that the solution given in Figure 7.2 is a feasible solution, but to see that it is a basic feasible solution we need to examine the simplex tableau for this problem, which we repeat here for convenience.

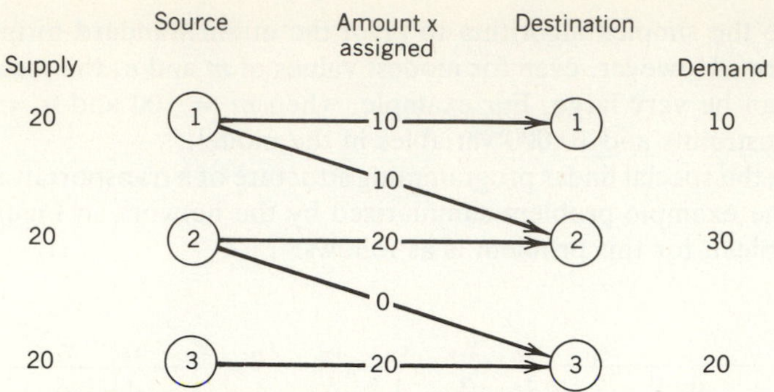

FIGURE 7.2 A subnetwork representing the feasible solution obtained by the northwest corner rule.

	x_{11}	x_{12}	x_{13}	x_{21}	x_{22}	x_{23}	x_{31}	x_{32}	x_{33}
0	2	4	3	1	5	2	1	1	6
20	1	1	1	0	0	0	0	0	0
20	0	0	0	1	1	1	0	0	0
20	0	0	0	0	0	0	1	1	1
10	(1)	0	0	1	0	0	1	0	0
30	0	1	0	0	1	0	0	1	0
20	0	0	1	0	0	1	0	0	1

We need to show that it is possible to pivot on this tableau to obtain a basic feasible solution that is identical to the solution obtained by the northwest corner rule. Note that the sum of the supply constraints (the first three rows) equals the sum of the demand constraints, and so one of the constraint equations is redundant. In particular, if we multiply the last three rows in the above tableau by -1 and then add the first five rows to the last row, we obtain a row with every entry equal to zero. Thus any basic feasible solution will have $m + n - 1$ basic variables.

We saw in constructing the northwest corner rule solution represented in Figure 7.2 that it is possible to obtain a feasible solution by sequentially assigning values to $m + n - 1$ variables so that an assignment does not affect the value of a variable already assigned. We can make these same sequential assignments by pivoting on the preceding simplex tableau. Using the same order in which the assignments were made by the northwest corner rule, we pivot to increase each x_{ij} up to the level it has in the northwest corner rule solution.

For example, from Figure 7.2 we see that $x_{11} = 10$, so we pivot in the x_{11} column of the preceding simplex tableau to increase x_{11} up to 10, which gives the following tableau:

	x_{11}	x_{12}	x_{13}	x_{21}	x_{22}	x_{23}	x_{31}	x_{32}	x_{33}
-20	0	4	3	-1	5	2	-1	1	6
10	0	(1)	1	-1	0	0	-1	0	0
20	0	0	0	1	1	1	0	0	0
20	0	0	0	0	0	0	1	1	1
10	1	0	0	1	0	0	1	0	0
30	0	1	0	0	1	0	0	1	0
20	0	0	1	0	0	1	0	0	1

Continuing down the arcs in Figure 7.2, we now need to increase x_{12} up to 10, which is accomplished by performing the circled pivot in the preceding tableau to give the following tableau:

	x_{11}	x_{12}	x_{13}	x_{21}	x_{22}	x_{23}	x_{31}	x_{32}	x_{33}
−60	0	0	−1	3	5	2	3	1	6
10	0	1	1	−1	0	0	−1	0	0
20	0	0	0	1	1	1	0	0	0
20	0	0	0	0	0	0	1	1	1
10	1	0	0	1	0	0	1	0	0
20	0	0	−1	1	①	0	1	1	0
20	0	0	1	0	0	1	0	0	1

In this tableau we can increase x_{22} up to 20 by pivoting in row 2 or in row 5. Because we want to have $x_{23} = 0$ on the next pivot, we chose row 5 as the pivot row. This (circled) pivot yields the following tableau:

	x_{11}	x_{12}	x_{13}	x_{21}	x_{22}	x_{23}	x_{31}	x_{32}	x_{33}
−160	0	0	4	−2	0	2	−2	−4	6
10	0	1	1	−1	0	0	−1	0	0
0	0	0	1	0	0	①	−1	−1	0
20	0	0	0	0	0	0	1	1	1
10	1	0	0	1	0	0	1	0	0
20	0	0	−1	1	1	0	1	1	0
20	0	0	1	0	0	1	0	0	1

Two more pivots to obtain $x_{23} = 0$ and $x_{33} = 20$ as basic variables give the following two tableaus:

	x_{11}	x_{12}	x_{13}	x_{21}	x_{22}	x_{23}	x_{31}	x_{32}	x_{33}
−160	0	0	2	−2	0	0	0	−2	6
10	0	1	1	−1	0	0	−1	0	0
0	0	0	1	0	0	1	−1	−1	0
20	0	0	0	0	0	0	1	1	①
10	1	0	0	1	0	0	1	0	0
20	0	0	−1	1	1	0	1	1	0
20	0	0	0	0	0	0	1	1	1

	x_{11}	x_{12}	x_{13}	x_{21}	x_{22}	x_{23}	x_{31}	x_{32}	x_{33}
−280	0	0	2	−2	0	0	−6	−8	0
10	0	1	1	−1	0	0	−1	0	0
0	0	0	1	0	0	1	−1	−1	0
20	0	0	0	0	0	0	1	1	1
10	1	0	0	1	0	0	1	0	0
20	0	0	−1	1	1	0	1	1	0
0	0	0	0	0	0	0	0	0	0

Eliminating the row of zeros (corresponding to a redundant constraint) gives a canonical form tableau whose associated basic feasible solution is precisely the northwest corner rule solution obtained earlier. Also, it is now clear why we wanted to include an arc with a zero assignment when supply is used up simultaneously with demand being met. This corresponds to the degenerate case where we need to have a basic variable at level zero.

Thus, at least for this example, the above sequence of pivots shows that the feasible solution obtained by the northwest corner rule is a basic feasible solution. However, given any solution obtained by the northwest corner rule, we can always perform a sequence of pivots to obtain such a solution as a basic feasible solution. The reason for this is that a northwest corner rule solution is obtained by sequentially assigning values to variables so that an assignment does not affect the value of a variable already assigned. Thus we can pivot to sequentially assign the same values to the variables, and we know we must end up with a feasible solution, namely, the northwest corner rule solution.

The Transportation Tableau

We could use a transportation network as a bookkeeping scheme to record the current basic feasible solution, but doing so would be rather cumbersome. It will be much more convenient to use the following **transportation tableau.**

		b_1	b_2	b_3
		10	30	20
a_1	20	2^{10}	4^{10}	3
a_2	20	1	5^{20}	2^{0}
a_3	20	1	1	6^{20}

the (i, j)th entry is $c_{ij}^{x_{ij}}$

In a transportation tableau we use superscripts on the cost coefficients c_{ij} to indicate the values of the current basic variables. The current nonbasic variables are all zero, so there is no need to indicate them as superscripts. For easy reference we have also listed the supplies and demands for the problem. Note that the superscripts in each row add up to the supplies and the superscripts in each column add up to the demands.

Using the transportation tableau, it is easy to give a precise statement of the steps in the northwest corner rule.

THE NORTHWEST CORNER RULE

Step 1. Pick the cost entry in the upper left (northwest) corner and ship as much as possible by that route so that the supply is used up or the demand is met.

Step 2. If the assignment just made uses up the supply for the row, then eliminate that row from further consideration and return to step 1.

If the assignment just made meets the demand for the column, then eliminate that column from further consideration and return to step 1.

If the assignment just made uses up the supply for row i and meets the demand for the column j, then assign $x_{i,j+1} = 0$ to the cost entry in row i and column $j + 1$, unless the assignment just made was the final assignment. Eliminate row i and column j from further consideration and return to step 1.

We illustrate the steps of the northwest corner rule by using our example problem with the following transportation tableau:

	10	30	20
20	2^{10}	4	3
20	1	5	2
20	1	1	6

Here, we assign as much as possible, namely, $x_{11} = 10$ to the northwest corner cost entry. This meets the demand at destination 1, so we eliminate column 1 from further consideration.

	10	30	20
20	2^{10}	4^{10}	3
20	1	5	2
20	1	1	6

The northwest corner entry is now $c_{12} = 4$, so we assign as much as possible, namely, $x_{12} = 10$ to this entry. This uses up the supply, so we eliminate row 1 from further consideration.

	10	30	20
20	2^{10}	4^{10}	3
20	1	5^{20}	2^{0}
20	1	1	6

We now assign $x_{22} = 20$ to the remaining northwest corner entry. This simultaneously uses up the supply and meets the demand, so we assign $x_{23} = 0$ and eliminate both row 2 and column 2.

	10	30	20
20	2^{10}	4^{10}	3
20	1	5^{20}	2^{0}
20	1	1	6^{20}

Only one cost entry remains, and we assign $x_{33} = 20$ as the final assignment.

We call a position (i, j) in the transportation tableau a **basic position** if it is a position corresponding to a basic variable. Thus, the basic positions are those having an assigned superscript. Positions without a superscript are called **nonbasic positions.**

The northwest corner rule usually does not give a particularly good initial feasible solution because it completely ignores the cost coefficients in constructing the solution. In Section 7.4 we present some methods that usually give better initial basic feasible solutions.

7.2 USING THE DUAL TO IMPROVE THE CURRENT SOLUTION

In the last section we showed how to pivot on the simplex tableau to obtain an initial canonical form tableau whose basic feasible solution is the northwest corner rule solution. For our example, after eliminating the redundant constraint row, the initial tableau is

	x_{11}	x_{12}	x_{13}	x_{21}	x_{22}	x_{23}	x_{31}	x_{32}	x_{33}
-280	0	0	2	-2	0	0	-6	-8	0
10	0	1	1	-1	0	0	-1	0	0
0	0	0	1	0	0	1	-1	-1	0
20	0	0	0	0	0	0	1	1	1
10	1	0	0	1	0	0	1	0	0
20	0	0	-1	1	1	0	1	1	0

From this tableau, it is clear that the current solution can be improved by pivoting in the x_{32} column, for example. In this section we show how this simplex tableau pivot can be interpreted and performed by a simple manipulation in the transportation tableau. We shall see that there is no need to use the much larger simplex tableau to do the bookkeeping and that the simple manipulation will require considerably less computational effort than is necessary for a pivot.

Unlike the initial canonical form simplex tableau, the initial transportation tableau does not tell us whether or not the current solution can be improved. If we are going to solve the problem by only using the transportation tableau, we need to know how that tableau can be used to improve the current solution. Fortunately, by considering the dual of the transportation problem, it is easy to see how the current solution can be improved. In Section 5.4 we derived the following dual for the transportation problem.

The Dual of a Transportation Problem

$$\text{maximize} \sum_{i=1}^{m} a_i u_i + \sum_{j=1}^{n} b_j v_j$$

subject to

$$u_i + v_j \leq c_{ij} \quad \text{for each } i \text{ and } j$$

$$\mathbf{u} \text{ and } \mathbf{v} \text{ free,}$$

where $\mathbf{u} = [u_1, \ldots, u_m]^T$ and $\mathbf{v} = [v_1, \ldots, v_n]^T$

The dual objective function can be written as

$$\beta(\mathbf{u}, \mathbf{v}) = \mathbf{a}^T\mathbf{u} + \mathbf{b}^T\mathbf{v}$$

where $\mathbf{a}$ is the vector of supplies and $\mathbf{b}$ is the vector of demands. Also, we shall denote the primal objective function by

$$\alpha(\mathbf{x}) = \sum_{i=1}^{m} \sum_{j=1}^{n} c_{ij} x_{ij}$$

where $\mathbf{x}$ denotes a vector having mn components x_{ij}.

One way to show that a feasible vector $\mathbf{x}$ for the transportation problem is optimal is to show that there is a feasible dual vector $(\mathbf{u}, \mathbf{v})$ for which

$$\alpha(\mathbf{x}) - \beta(\mathbf{u}, \mathbf{v}) = 0$$

Calculating this difference, we obtain

$$\begin{aligned} \alpha(\mathbf{x}) - \beta(\mathbf{u}, \mathbf{v}) &= \sum_{i=1}^{m} \sum_{j=1}^{n} c_{ij} x_{ij} - \sum_{i=1}^{m} a_i u_i - \sum_{j=1}^{n} b_j v_j \\ &= \sum_{i=1}^{m} \sum_{j=1}^{n} c_{ij} x_{ij} - \sum_{i=1}^{m} \sum_{j=1}^{n} x_{ij} u_i - \sum_{j=1}^{n} \sum_{i=1}^{m} x_{ij} v_j \end{aligned}$$

using the fact that

$$a_i = \sum_{j=1}^{n} x_{ij} \quad \text{and} \quad b_j = \sum_{i=1}^{m} x_{ij}$$

because **x** is feasible for the primal problem. Thus, factoring out x_{ij} yields

$$\alpha(\mathbf{x}) - \beta(\mathbf{u}, \mathbf{v}) = \sum_{i=1}^{m} \sum_{j=1}^{n} (c_{ij} - u_i - v_j)x_{ij}$$

and therefore the following result holds.

> If $(\mathbf{u}, \mathbf{v})$ is any dual vector (feasible or infeasible), then
>
> $$\alpha(\mathbf{x}) = \beta(\mathbf{u}, \mathbf{v}) = \sum_{i=1}^{m} \sum_{j=1}^{n} (c_{ij} - u_i - v_j)x_{ij}$$
>
> provided **x** is a feasible solution.

Given a basic feasible solution **x** for the transportation problem, suppose that we find vectors **u** and **v** such that

$$c_{ij} - u_i - v_j = 0 \quad \text{for each basic position } (i, j)$$

that is, for each (i, j) where x_{ij} is a basic variable. Using the preceding result, we then see that $\alpha(\mathbf{x}) = \beta(\mathbf{u}, \mathbf{v})$. If, in addition, the vector $(\mathbf{u}, \mathbf{v})$ is feasible for the dual problem, then the duality relations of Section 5.1 imply that **x** is optimal for the transportation problem. Recall by part (i) of the duality relations that if **x** is feasible for the transportation problem and $(\mathbf{u}, \mathbf{v})$ is feasible for its dual, then

$$\alpha(\mathbf{x}) \geq \beta(\mathbf{u}, \mathbf{v})$$

Thus no feasible point, **x**, can give an objective function value lower than $\beta(\mathbf{u}, \mathbf{v})$. Therefore, if $\alpha(\mathbf{x}) = \beta(\mathbf{u}, \mathbf{v})$, then **x** must be optimal.

Thus, given a basic feasible solution **x** and the associated transportation tableau, we need to find vectors **u** and **v** such that

$$c_{ij} - u_i - v_j = 0 \quad \text{for each basic position } (i, j)$$

To illustrate how this can always be done, consider the initial transportation tableau in our example.

2^{10}	4^{10}	3
1	5^{20}	2^{0}
1	1	6^{20}

We want to find vectors **u** and **v** satisfying the system

$$u_i + v_j = c_{ij} \quad \text{for each basic position } (i, j)$$

Thus, in our example, **u** and **v** must satisfy the system

$$u_1 + v_1 = 2$$

$$u_1 + v_2 = 4$$

$$u_2 + v_2 = 5$$

$$u_2 + v_3 = 2$$

$$u_3 + v_3 = 6$$

Finding a solution to this type of system is very easy because of its structure. To find a solution, we simply

pick an arbitrary value for u_1, say $u_1 = 0$,
and then $v_1 = 2$ and $v_2 = 4$ are uniquely determined,
and then $u_2 = 1$ is uniquely determined,
and then $v_3 = 1$ is uniquely determined,
and then $u_3 = 5$ is uniquely determined.

In general, we can pick an arbitrary value for one of the variables, and this sets off a chain reaction that uniquely determines the other variables. It is not even necessary to write down the preceding linear system because we can determine a solution by inspection from the transportation tableau. For example, in the preceding transportation tableau, if we set $u_1 = 0$ and follow the chain reaction that this sets off in the transportation tableau, then the vector **u** that results is written on the right of the tableau and the vector **v** that results is written below the tableau.

2^{10}	4^{10}	3	0	u_1
1	5^{20}	2^{0}	1	u_2
1	1	6^{20}	5	u_3
2	4	1		
v_1	v_2	v_3		

By construction of **u** and **v**, we know for the current basic feasible solution **x** that

$$\alpha(\mathbf{x}) = \beta(\mathbf{u}, \mathbf{v})$$

Thus, to determine whether or not the current solution **x** is optimal, we only need to check if $(\mathbf{u}, \mathbf{v})$ is feasible for the dual problem. Because the dual constraints are

$$c_{ij} - u_i - v_j \geq 0, \quad \text{for each } (i, j)$$

we can use the above transportation tableau to check dual feasibility by subtracting u_i from every cost coefficient in row i and subtracting v_j from every cost coefficient in column j. Doing this yields the following tableau where we have also indicated the current solution **x**.

0^{10}	0^{10}	2
-2	0^{20}	0^{0}
-6	-8	0^{20}

i, j entry is $(c_{ij} - u_i - v_j)^{x_{ij}}$

Note in particular that $c_{ij} - u_i - v_j = 0$ whenever x_{ij} is basic. From this tableau we see that the dual vector $(\mathbf{u}, \mathbf{v})$ is not feasible. In particular, three of the dual constraints are not satisfied because there are three negative entries in the tableau. Thus we cannot conclude that the current solution **x** is optimal. However, we shall show how to improve the current solution by using this tableau.

It is important to realize that replacing the cost coefficients c_{ij} in the original problem by the **adjusted cost entries,** $c_{ij} - u_i - v_j$, does not change the optimal vector. To see why this is true, note that making the replacement gives the following objective function:

$$\begin{aligned}
\hat{\alpha}(\mathbf{x}) &= \sum_{i=1}^{m}\sum_{j=1}^{n}(c_{ij} - u_i - v_j)x_{ij} \\
&= \sum_{i=1}^{m}\sum_{j=1}^{n} c_{ij}x_{ij} - \sum_{i=1}^{m}\sum_{j=1}^{n} u_i x_{ij} - \sum_{i=1}^{m}\sum_{j=1}^{n} v_j x_{ij} \\
&= \sum_{i=1}^{m}\sum_{j=1}^{n} c_{ij}x_{ij} - \sum_{i=1}^{m} u_i\left(\sum_{j=1}^{n} x_{ij}\right) - \sum_{j=1}^{n} v_j\left(\sum_{i=1}^{m} x_{ij}\right) \\
&= \alpha(\mathbf{x}) - \sum_{i=1}^{m} u_i a_i - \sum_{j=1}^{n} v_j b_j \\
&= \alpha(\mathbf{x}) - \mathbf{a}^T\mathbf{u} - \mathbf{b}^T\mathbf{v}
\end{aligned}$$

Because $\mathbf{a}^T\mathbf{u}$ and $\mathbf{b}^T\mathbf{v}$ are constants, we know that minimizing $\hat{\alpha}(\mathbf{x})$ over the feasible set is equivalent to minimizing $\alpha(\mathbf{x})$ over the feasible set. Of course, the optimal values are different but the optimal vectors are the same.

The preceding transportation tableau with adjusted cost coefficients, $c_{ij} - u_i - v_j$, corresponds to the initial canonical form tableau obtained earlier for this problem. These two tableaus are:

0^{10}	0^{10}	2
-2	0^{20}	0^{0}
-6	-8	0^{20}

	x_{11}	x_{12}	x_{13}	x_{21}	x_{22}	x_{23}	x_{31}	x_{32}	x_{33}
−280	0	0	2	−2	0	0	−6	−8	0
10	0	1	1	−1	0	0	−1	0	0
0	0	0	1	0	0	1	−1	−1	0
20	0	0	0	0	0	0	1	1	1
10	1	0	0	1	0	0	1	0	0
20	0	0	−1	1	1	0	1	(1)	0

Note that the cost coefficients in the simplex tableau are precisely the adjusted cost entries in the transportation tableau and that the basic feasible solution in the simplex tableau is the solution indicated by the superscripts in the transportation tableau.

The simplex tableau is not in optimal form because there is at least one negative cost entry. Performing the circled pivot increases the nonbasic variable x_{32} up to 20 and decreases the total cost by $8(20) = 160$. To see how this pivot can be accomplished by a simple manipulation in the transportation tableau, we first interpret this pivot in terms of the subnetwork representing the current solution. Increasing x_{32} from zero corresponds to introducing a new arc from source 3 to destination 2. In Figure 7.3 this new arc is given by a dotted line.

We would like to let the flow (amount shipped) on the dotted arc be positive, but we need to adjust the other flows so that the solution we obtain will be

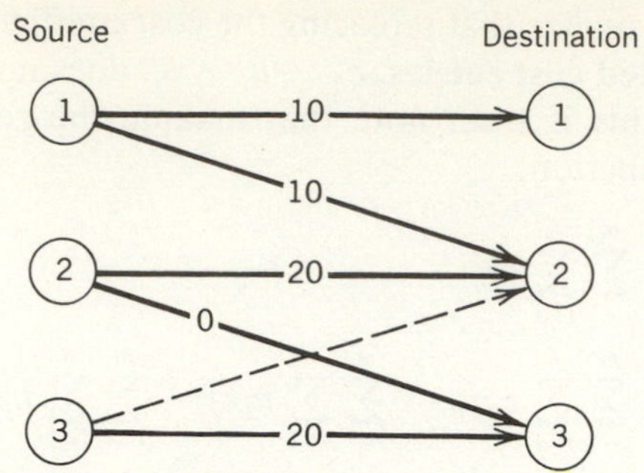

FIGURE 7.3 Introducing a new shipping route.

feasible. It is easy to do this using the transportation network because introducing the dotted arc gives a **loop** in the subnetwork; that is, starting at one node we can follow a path of consecutive arcs (sometimes going against the flow) and return to the starting node. We denote the loop in Figure 7.3 by $S_3D_2S_2D_3S_3$ where S_i refers to source i and D_j refers to destination j.

In general, given a subnetwork with $m + n - 1$ arcs and no loops, it is easy to see that adding an arc always creates a unique loop. Also, it is important to realize that the initial northwest corner rule solution will by its very construction never have a loop in the set of basic positions.

The key observation that allows us to see how a pivot on the canonical form simplex tableau can be interpreted in the network is the following:

> If we start with the dotted arc and alternately add and subtract a positive flow t to the arcs in the loop, then the new solution $\mathbf{x}(t)$ is feasible provided all of its components are nonnegative.

For example, suppose we increase the flow on the dotted arc by letting $x_{32} = t$. Then from the arcs in the loop we see that

$x_{22} = 20 - t$ (subtract t from the arc S_2D_2)
$x_{23} = 0 + t$ (add t to the arc S_2D_3)
$x_{33} = 20 - t$ (subtract t from the arc S_3D_3)

and all the other flows in the current solution remain unchanged because they are not involved in the loop.

By construction, the new solution $\mathbf{x}(t)$ satisfies the supply-and-demand constraints of the problem and so it is feasible provided t satisfies

$$20 - t \geq 0$$

Thus, $\mathbf{x}(t)$ is feasible provided $t \leq 20$.

Going around the loop in Figure 7.3, if we add $t = 20$ when we go with the flow of an arc and subtract $t = 20$ when we go against the flow of an arc, we obtain the new feasible solution and, using the adjusted costs, we see that this new solution decreases the total cost by $20(8) = 160$. In this example shifting 20 units around the loop in the manner specified causes two basic variables to become zero. This is the second way that degeneracy can occur in a transportation problem.

Looking back at the canonical form simplex tableau associated with the current solution, we see that the x_{32} column does not have a unique minimum ratio row. Thus, when we pivot to make $x_{32} = 20$ as a basic variable, we have a choice as to which basic variable it is going to replace, namely, x_{22} or x_{33}. Arbitrarily choosing x_{22} as the basic variable to become nonbasic, we obtain the new basic feasible solution represented by the subnetwork in Figure 7.4, where x_{33} remains as a basic variable at level zero.

The network analysis that we used to determine the new solution represented in Figure 7.4 corresponds exactly with the t analysis introduced in Section 3.2 for finding a better basic feasible solution for a canonical form tableau. In particular, the maximum amount t that we can shift around the loop is simply the minimum ratio for the x_{32} column in the initial canonical form tableau. Thus, the new solution represented by the network in Figure 7.4 is the next basic feasible solution obtained by performing a simplex rule pivot in the x_{32} column of the initial canonical form tableau.

The process of shifting an amount t around a loop in the transportation network of Figure 7.3 can be interpreted in terms of the associated transportation tableau,

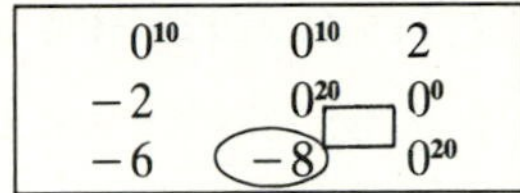

0^{10}	0^{10}	2
-2	0^{20}	0^{0}
-6	-8	0^{20}

We simply start with a position having a negative adjusted cost and see how we can loop back to this starting position by using basic positions. We know that such a loop must exist because adding an arc in the network always creates a unique loop. For example, picking the cost entry of -8 in the tableau gives the loop indicated above.

If we start with the selected negative cost position and alternately add and subtract the amount t to x_{ij} entries associated with the positions making up the loop, then it is clear that t can be no larger than any of the x_{ij}'s from which it is subtracted. From the preceding tableau, we see that at most $t = 20$ units can be shifted around the loop, so we shift 20 units around the loop to obtain the largest decrease in cost. Note that we have the choice of keeping either x_{22} or x_{33} as a basic variable. Here we keep x_{33} as a basic variable at level zero to agree with our choice in Figure 7.4. The new basic feasible solution is given in the

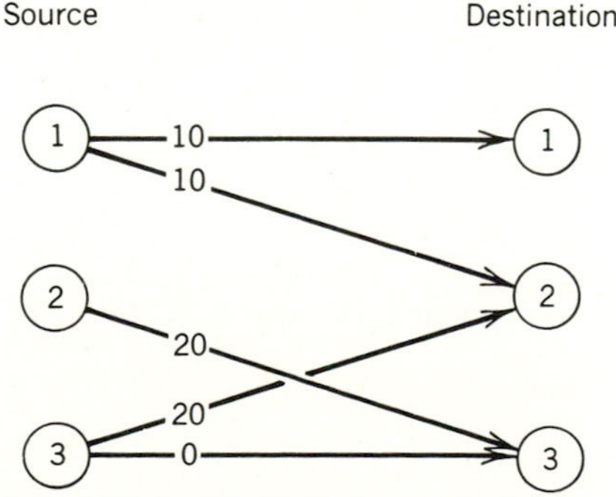

FIGURE 7.4 The subnetwork representing the new solution.

transportation tableau

0^{10}	0^{10}	2
-2	0	0^{20}
-6	-8^{20}	0^{0}

Of course, we could have picked the cost entry equal to -6, and then the loop is the one indicated in the following tableau:

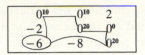

0^{10}	0^{10}	2
-2	0^{20}	0^{0}
(-6)	-8	0^{20}

The maximum amount t that we can shift around this loop is $t = 10$, and the resulting decrease in the total cost is 60. Note that the positions with cost entries of -2 and -8 are not in the loop so we do not change the associated x_{ij}'s.

Obtaining an Optimal Transportation Tableau

To check if the improved basic feasible solution is optimal, we use the transportation tableau that represents the new solution, namely, the following tableau:

0^{10}	0^{10}	2
-2	0	0^{20}
-6	-8^{20}	0^{0}

This tableau uses adjusted cost coefficients, but we have already seen that if a vector is optimal for a transportation problem with these adjusted cost coefficients, then it is optimal for the original problem. Therefore, with the adjusted costs replacing the original costs, we repeat the above process to see if the current solution can be improved. Thus, denoting the adjusted cost coefficients by c_{ij}, we first find a dual vector (u, v) so that $u_i + v_j = c_{ij}$ whenever x_{ij} is basic. Setting $u_1 = 0$ sets off a chain reaction that uniquely determines $\mathbf{u}$ and $\mathbf{v}$ as

0^{10}	0^{10}	2	0	u_1
-2	0	0^{20}	-8	u_2
-6	-8^{20}	0^{0}	-8	u_3
0	0	8		
v_1	v_2	v_3		

Using this dual vector to calculate new adjusted cost coefficients gives the tableau

0^{10}	0^{10}	(-6)
6	8	0^{20}
2	0^{10}	0^{0}

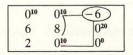

This tableau shows that one dual constraint is violated, so we look for the unique loop starting at the cost entry -6. That loop is indicated in the preceding tableau, and we see that $t = 0$ is the maximum amount that can be alternately added and subtracted to the basic positions in the loop. Performing this degenerate shift of 0 units around the loop only causes x_{13} to replace x_{33} as a basic variable. The new basic feasible solution is given in the following tableau:

0^{10}	0^{10}	-6^{0}	0	u_1
6	8	0^{20}	6	u_2
2	0^{20}	0	0	u_3
0	0	-6		
v_1	v_2	v_3		

To see if this new solution is optimal, we perform another iteration and calculate a dual vector (**u**, **v**) by setting $u_1 = 0$ and letting the usual chain reaction determine the other components. The dual vector (**u**, **v**) obtained is given with the preceding tableau. Calculating adjusted cost coefficients gives the following tableau:

0^{10}	0^{10}	0^{0}
0	2	0^{20}
2	0^{20}	6

This tableau shows that (**u**, **v**) is dual feasible and, therefore, the current **x** is optimal.

A Final Comparison with the Simplex Algorithm

It is instructive to compare the transportation tableaus obtained in the preceding section with the simplex tableaus obtained by solving the problem directly with the simplex algorithm. The following sequence of simplex tableaus results if we start with the initial canonical form and always select the most negative cost column as the pivot column. For comparison purposes, we also give the transportation tableaus obtained previously.

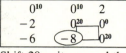

0^{10}	0^{10}	2
-2	0^{20}	0^{0}
-6	(-8)	0^{20}

Shift 20 units around the loop.

	x_{11}	x_{12}	x_{13}	x_{21}	x_{22}	x_{23}	x_{31}	x_{32}	x_{33}
-280	0	0	2	-2	0	0	-6	-8	0
10	0	1	1	-1	0	0	-1	0	0
0	0	0	1	0	0	1	-1	-1	0
20	0	0	0	0	0	0	1	1	1
10	1	0	0	1	0	0	1	0	0
20	0	0	-1	1	1	0	1	(1)	0

Pivot to get x_{32} up to 20.

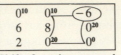

0^{10}	0^{10}	-6
6	8	0^{20}
2	0^{20}	0^{0}

Shift 0 units around the loop.

	x_{11}	x_{12}	x_{13}	x_{21}	x_{22}	x_{23}	x_{31}	x_{32}	x_{33}
−160	0	0	−6	6	8	0	2	0	0
10	0	1	1	−1	0	0	−1	0	0
20	0	0	0	1	1	1	0	0	0
0	0	0	(1)	−1	−1	0	0	0	1
10	1	0	0	1	0	0	1	0	0
20	0	0	−1	1	1	0	1	1	0

Perform the degenerate pivot to make x_{13} basic.

0^{10}	0^{10}	0^{0}
0	2	0^{20}
2	0^{20}	6

Optimal form

	x_{11}	x_{12}	x_{13}	x_{21}	x_{22}	x_{23}	x_{31}	x_{32}	x_{33}
−160	0	0	0	0	2	0	2	0	6
10	0	1	0	0	1	0	−1	0	−1
20	0	0	0	1	1	1	0	0	0
0	0	0	1	−1	−1	0	0	0	1
10	1	0	0	1	0	0	1	0	0
20	0	0	0	0	0	0	1	1	1

Optimal form

It is clear from this sequence of tableaus that the transportation tableaus contain essentially the same information as the simplex tableaus. Instead of performing a pivot on the simplex tableau in a column with a negative cost entry, we simply find a loop in the transportation tableau starting at a negative adjusted cost and shift as much as possible around the loop. As we noted earlier, the amount shifted around the loop is equal to the minimum ratio associated with the pivot column in the simplex tableau. If no negative cost entries exist, the simplex tableau is in optimal form, and the transportation tableau is also in optimal form, as we argued earlier using the duality relations.

7.3 THE TRANSPORTATION ALGORITHM

The computational effort required to solve a transportation problem using the algorithm developed above is much less than that required by the simplex algorithm. We summarize the steps of the algorithm as follows:

Step 1. Find an initial basic feasible solution **x**.

Step 2. For the current solution **x** and the current cost coefficients c_{ij}, find a dual vector (**u**, **v**) such that

$$u_i + v_j = c_{ij} \quad \text{for each } (i, j) \text{ with } x_{ij} \text{ basic}$$

and calculate the adjusted cost coefficients

$$c_{ij} - u_i - v_j \quad \text{for each } (i, j)$$

Step 3. If each adjusted cost coefficient is nonnegative, stop; the current **x** is optimal.

Otherwise, pick a position with a negative adjusted cost and find the unique loop starting at that position with all other positions in the loop being those associated with basic variables.

Step 4. Shift as much as possible around the loop to obtain a new basic feasible solution **x** and return to step 2 with the adjusted costs as the current costs.

We have seen how the transportation algorithm is just the simplex algorithm applied to the transportation problem but with the bookkeeping done using a transportation tableau instead of the usual simplex tableau. From this point of view we know that the algorithm will converge in a finite number of iterations provided the problem is nondegenerate. We could alter the algorithm to guarantee that it converges even in the degenerate case, but we shall not present those details here.

Actually, if the original data of the problem is integer valued, then there is an easy way to see why this algorithm always converges in the nondegenerate case. If each a_i and each b_j is integer valued, then it is not hard to find an initial feasible **x** that has each x_{ij} integer valued. The northwest corner rule, for example, gives such a solution. If the initial **x** has all components integer, then the preceding algorithm preserves this property for all subsequent feasible solutions because the maximum amount shifted around a loop in each iteration must be an integer. Also, if each of the original cost coefficients is integer, then the dual vector (**u**, **v**) will have integer components. This means that the adjusted cost coefficients will be integer. Therefore, the decrease in total cost at each iteration will always be integer valued. Thus only a finite number of iterations of the transportation algorithm can be performed because the minimum cost can be no smaller than

$$K\left(\sum_{i=1}^{m} a_i\right) \quad \text{where } K = \underset{(i,j)}{\text{minimum}} \{c_{ij}\}$$

For example, from the initial tableau for our example,

	10	30	20
20	2	4	3
20	1	5	2
20	1	1	6

we see that the minimum cost can be no smaller than 60 because the minimum shipping cost over any route is 1, and 60 units need to be shipped.

7.4 FINDING AN INITIAL BASIC FEASIBLE SOLUTION

By taking the cost coefficients into account, it is usually possible to find an initial feasible solution with a lower total cost than a solution obtained by using the northwest corner rule. In this section we present two other methods for finding an initial basic feasible solution that can be used in step 1 of our algorithm. Again we assume that the total supply equals the total demand.

The Smallest Cost Entry Method

Step 1. Pick the smallest remaining cost entry in the transportation tableau and ship as much as possible by that route so that the supply is used up or the demand is met.

Step 2. If the assignment just made uses up the supply for the row, then eliminate that row from further consideration and return to step 1.

If the assignment just made meets the demand for the column, then eliminate that column from further consideration and return to step 1.

If the assignment just made uses up the supply for the row and meets the demand for the column and it is not the final assignment, then assign $x_{ij} = 0$ to the next smallest cost entry in the column. Eliminate both the row and the column from further consideration and return to step 1.

As we have seen before, if a shipment uses up the supply and simultaneously meets the demand, then the transportation problem is degenerate, unless, of course, it was the last shipment. Therefore, in step 2 we make an assignment of 0 so that we end up with $m + n - 1$ basic variables. We shall see that the $\mathbf{x}$ obtained by the minimum cost entry method is a basic feasible solution.

As an example of obtaining a feasible solution by the smallest cost entry method, consider the following tableau with three cost entries that could be chosen as the smallest entry. We arbitrarily select the one in column 1 to start the method.

	15	10	10	5	30
10	3	6	8	11	5
15	1 (15)	9	3	2	7
30	4	2	8	25	15
5	9	1	4	9	8
10	2 (0)	4	2	11	1

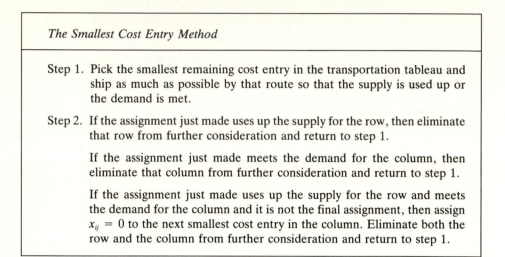

Assigning $x_{21} = 15$ uses up the supply and meets the demand, so we eliminate both row 2 and column 1 after assigning $x_{51} = 0$ to the next smallest entry in column 1.

	15	10	10	5	30
10	3	6	8	11	5
15	1 (15)	9	3	2	7
30	4	2	8	25	15
5	9	1 (5)	4	9	8
10	2 (0)	4	2	11	1

Here we assign $x_{42} = 5$ and eliminate row 4 from further consideration.

	15	10	10	5	30
10	3	6	8	11	5
15	1^{15}	9	3	2	7
30	4	2	8	25	15
5	9	1^{5}	4	9	8
10	2^{0}	4	2	11	1^{10}

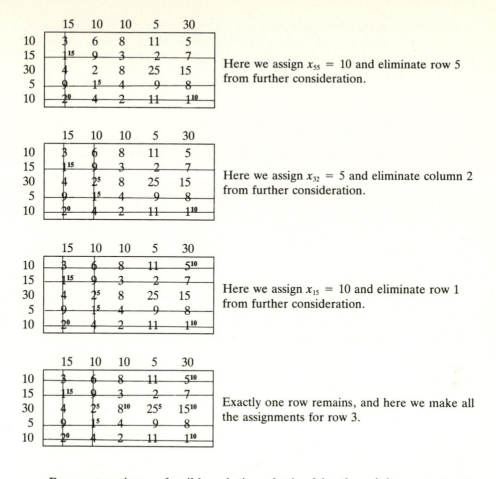

Here we assign $x_{55} = 10$ and eliminate row 5 from further consideration.

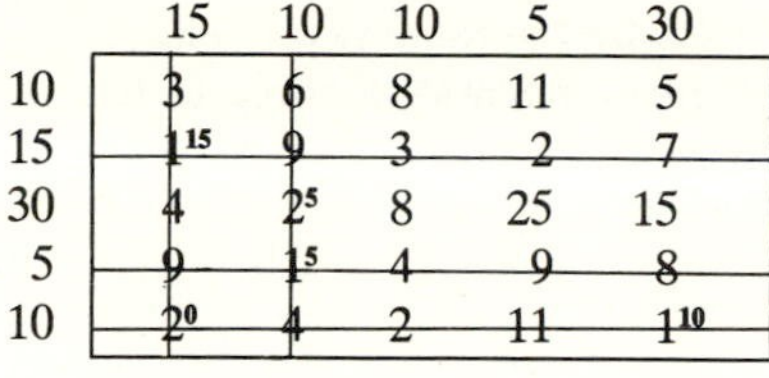

Here we assign $x_{32} = 5$ and eliminate column 2 from further consideration.

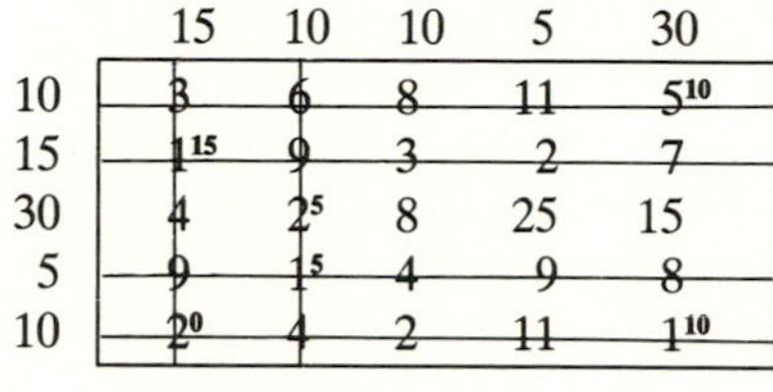

Here we assign $x_{15} = 10$ and eliminate row 1 from further consideration.

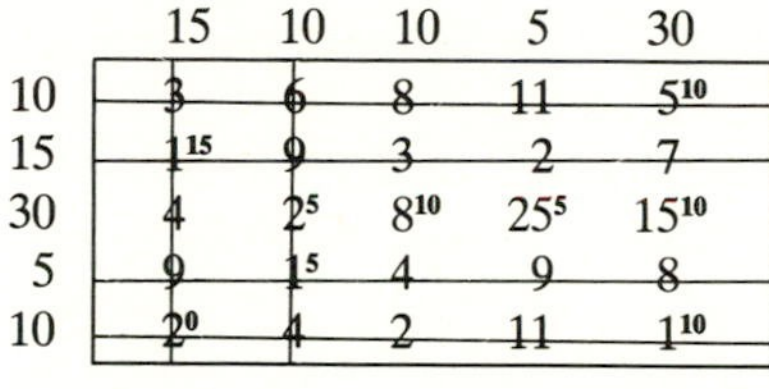

Exactly one row remains, and here we make all the assignments for row 3.

By construction, a feasible solution obtained by the minimum cost entry method will always have exactly $m + n - 1$ superscripts because every time we make an assignment we use up a supply or meet a demand. The last assignment made, however, uses up a supply and meets a demand, so we make exactly $m + n - 1$ assignments. Also, there can be no loops in a feasible solution obtained by the minimum cost entry rule because every time we make an assignment we prevent the row or column from receiving another assignment that could be used in a loop, so there cannot be a loop starting from that position.

In general, given a feasible solution with no loops and exactly $m + n - 1$ assignments (some of which may be zero in the degenerate case), it is always possible to pivot in the associated simplex tableau to obtain this solution as a basic feasible solution. For example, a feasible solution obtained by the northwest corner rule always has $m + n - 1$ assignments and no loops. We saw that it is possible to pivot to obtain the northwest corner rule solution as a basic feasible solution. Similarly, we can also show that the feasible solution obtained by the minimum cost entry method is a basic feasible solution. Using the same order that the assignments were made, we can successively pivot in the associated simplex tableau to increase each X_{ij} up to the level it has in the given solution. We only have to take care not to change any basic variable already assigned.

Although the smallest cost entry method does take into account cost data in finding a feasible solution, it does not "look ahead." For example, the last two assignments made above were relatively costly assignments. We now present a method for finding an initial basic feasible solution that tries to avoid getting trapped into making costly assignments.

Vogel's Advanced Start Method

Step 1. Compute the difference between the two smallest remaining cost entries in each row and in each column and assign as much as possible to the smallest cost entry in the row or column with the largest difference so that the supply is used up or the demand is met.

Step 2. If the assignment just made uses up the supply for the row, then eliminate that row from further consideration and return to step 1.

If the assignment just made meets the demand for the column, then eliminate that column from further consideration and return to step 1.

If the assignment just made uses up the supply for the row and meets the demand for the column and it is not the final assignment, then assign $x_{ij} = 0$ to the next smallest cost entry in the column. Eliminate both the row and the column from further consideration and return to step 1.

Again, in step 2, we make an assignment of zero when the problem is degenerate and supply and demand are simultaneously met.

To illustrate the method, we consider the same problem used previously to find a starting solution by the smallest cost entry method. We give the differences between the two smallest entries in each row on the right of the tableau and the differences between the two smallest entries in each column on the bottom of the tableau.

	15	10	10	5	30	
10	3	6	8	11	5	2
15	1	9	3	2^5	7	1
30	4	2	8	25	15	2
5	9	1	4	9	8	3
10	2	4	2	11	1	1
	1	1	1	7	4	

Here we assign $x_{14} = 5$ to the smallest entry in column 4 and eliminate column 4 from further consideration.

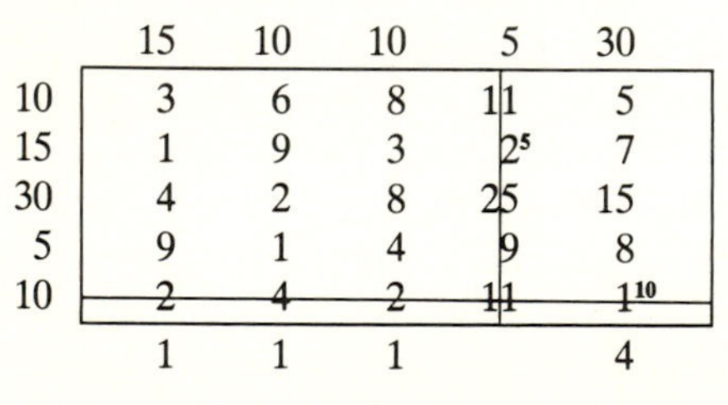

	15	10	10	5	30	
10	3	6	8	11	5	2
15	1	9	3	2^5	7	2
30	4	2	8	25	15	2
5	9	1	4	9	8	3
10	2	4	2	11	1^{10}	1
	1	1	1		4	

After updating the row differences, we assign $x_{55} = 10$ to the smallest entry in column 5 and eliminate row 5 from further consideration.

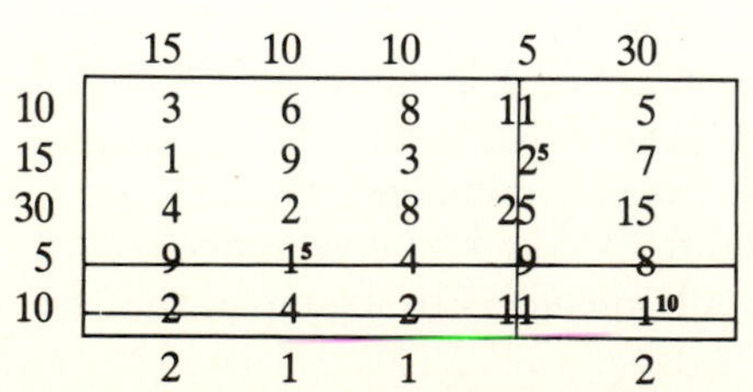

	15	10	10	5	30	
10	3	6	8	11	5	2
15	1	9	3	2^5	7	2
30	4	2	8	25	15	2
5	9	1^5	4	9	8	3
10	2	4	2	11	1^{10}	
	2	1	1		2	

After updating the column differences, we assign $x_{42} = 5$ to the smallest entry in row 4 and eliminate row 4 from further consideration.

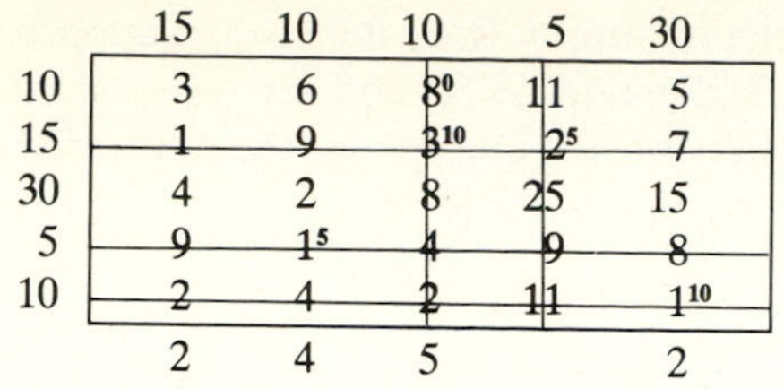

After updating the column differences, we assign $x_{23} = 10$ to the smallest entry in column 3, which meets the column 3 demand and uses up all the row 2 supply. Thus, we assign $x_{13} = 0$ to the next smallest entry in column 3 and eliminate row 2 and column 3 from further consideration.

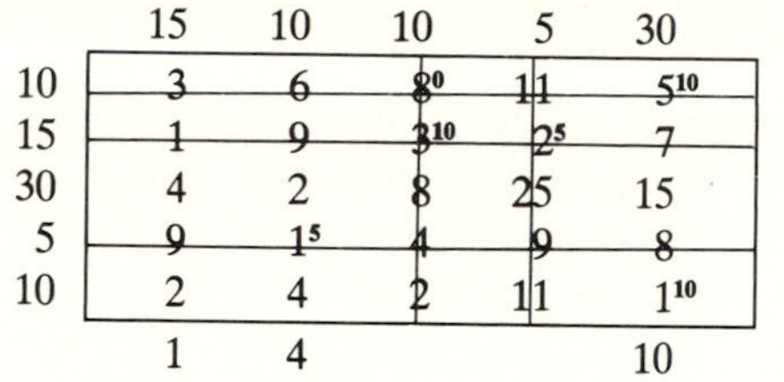

After updating the row and column differences, we assign $x_{15} = 10$ to the smallest entry in column 5 and eliminate row 1 from further consideration.

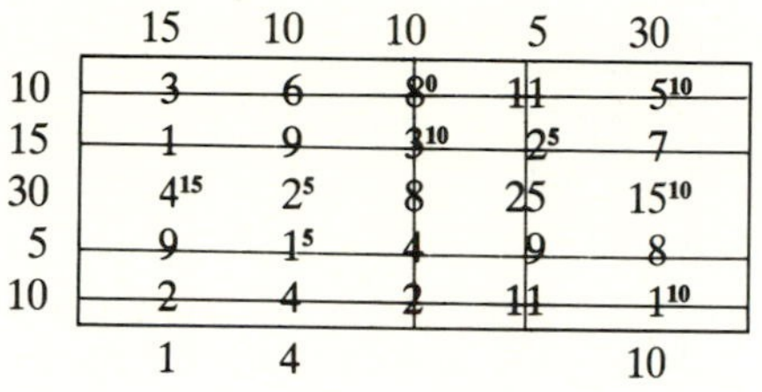

Here, only row 3 remains, and we are forced to assign $x_{31} = 15$, $x_{32} = 5$, and $x_{35} = 10$.

For this example it is interesting to compare the cost associated with the feasible solutions obtained by the three different methods we have discussed. The northwest corner rule solution has a total cost of 595, the smallest cost entry solution has a total cost of 445, and the Vogel advanced start solution has a total cost of 325. Typically, the extra calculations required in the Vogel advanced start method are worthwhile.

7.5 VARIATIONS OF THE TRANSPORTATION PROBLEM

Unequal Supply and Demand

In developing the transportation algorithm, we assumed that the total supply was equal to the total demand, that is,

$$\sum_{i=1}^{m} a_i = \sum_{j=1}^{n} b_j$$

If this condition does not hold, it is easy to formulate an equivalent transportation problem for which total supply equals total demand and whose solution solves the original problem. We illustrate how this can always be done by considering an example problem with the following transportation tableau:

	20	30	20
50	5	3	8
70	2	9	1

In this example there are 50 more units of supply than demand. Suppose we add a fictitious destination and give it a demand of 50 and let the cost of shipping from any source to that destination be 0. This gives the following transportation tableau:

	20	30	20	50
50	5	3	8	0
70	2	9	1	0

This new problem has total supply equal to total demand, so we can apply our transportation algorithm. If **x** is optimal for this new problem, we can obtain an optimal solution for the original problem by simply deleting the fictitious column. To see why this is true, consider the following optimal tableau for the reformulated problem.

	20	30	20	50
50	5	3^{30}	8	0^{20}
70	2^{20}	9	1^{20}	0^{30}

Suppose it were possible to adjust the x_{ij}'s in the original part of the tableau to obtain a solution satisfying demand and having a lower cost than the above solution. We could then simply adjust the x_{ij}'s in the fictitious part of the tableau (so that the row sums add up to the supplies) to give a solution that has a lower cost than the optimal cost for the reformulated problem, which is impossible. Thus, the x_{ij}'s in the original part of the tableau must be optimal for the original problem.

We can similarly reformulate a problem if total supply is less than total demand. As an example, consider the problem with the following transportation tableau:

	20	20	20
10	3	1	5
20	2	9	6
15	1	4	2

Here, we add a fictitious source and give it a supply equal to the excess demand, namely, 15, and we let the cost of shipping from this source to any destination be 0. This results in the following transportation tableau:

	20	20	20
10	3	1	5
20	2	9	6
15	1	4	2
15	0	0	0

Solving this problem by the transportation algorithm gives the following optimal tableau:

	20	20	20
10	3	1^{10}	5
20	2^{20}	9	6^{0}
15	1	4	2^{15}
15	0	0^{10}	0^{0}

Again, it is clear that if the x_{ij}'s could be adjusted in the original part of the tableau to obtain a lower cost solution using the available supplies, then we could simply adjust the x_{ij}'s in the fictitious part of the tableau to give a feasible solution for the reformulated problem with a lower cost than the optimal cost, which is impossible. Thus, the x_{ij}'s in the original part of the tableau must be optimal for the original problem.

The Transshipment Problem

In some applications the sources and destinations in a transportation problem may also serve as **transshipment points,** that is, units may be shipped through an intermediate source or destination on the way to their final destination. We will show how to solve such a problem by formulating a transportation problem whose optimal solution can be used to obtain an optimal solution to the original transshipment problem.

We illustrate how this can always be done by considering an example. Suppose we are given a transportation problem with the following transportation tableau:

	20	20	20
30	9	8	1
30	1	7	8

In addition, suppose that all the sources and destinations can serve as transshipment points. For example, depending on the costs involved, instead of shipping units directly from source 1 to destination 1, it may be cheaper to ship the units from source 1 to source 2 and then ship the units on to destination 1. In this case source 2 serves as a transshipment point.

The key in solving a transshipment problem is to formulate a larger transportation problem that allows for all possible transshipments. To this end consider the problem where each destination in the original problem is also considered a source and each source in the original problem is also considered a destination. For our examle we would then construct the following transportation tableau:

		S_1	S_2	D_1	D_2	D_3
		0	0	20	20	20
S_1	30	0	1	9	8	1
S_2	30	2	0	1	7	8
D_1	0	9	1	0	2	1
D_2	0	8	7	2	0	9
D_3	0	1	8	1	9	0

Here we have used shipping costs from D_j to S_i that are the same as the shipping costs from S_i to D_j, but this need not be the case. We do, however, need to know all the possible shipping costs. For this problem we assume that the costs are those given in the preceding tableau.

Note that with the above costs it is cheaper to ship units from S_1 to S_2 to D_1 than it is to ship units directly from S_1 to D_1. In this tableau we have initially set the demands for the new "destinations" S_1 and S_2 to 0, and, similarly, we have set the supplies for the new "sources" D_1, D_2, and D_3 to 0. Therefore, if we were to apply our algorithm to this larger problem, the transshipment from S_1 to S_2 to D_1, for example, would not be possible. We need to alter the preceding tableau so that all such transshipments are possible.

Even though we do not know how much should be transshipped through each source and destination in the optimal solution, we do know an upper bound on these values, namely, the total supply. In our example suppose we add 60 (the total supply) to each of the supplies and demands to obtain the following tableau:

		S_1	S_2	D_1	D_2	D_3
		60	60	80	80	80
S_1	90	0	1	9	8	1
S_2	90	2	0	1	7	8
D_1	60	9	1	0	2	1
D_2	60	8	7	2	0	9
D_3	60	1	8	1	9	0

With these new supply-and-demand values, if it is cheaper to transship an amount through a source or destination, it will be possible to do so. However, the question of how we can use an optimal solution to this larger problem to obtain an optimal solution to the original transshipment problem still remains. We illustrate how this can be done by using the transportation algorithm to solve the enlarged problem starting with the feasible solution in the following tableau:

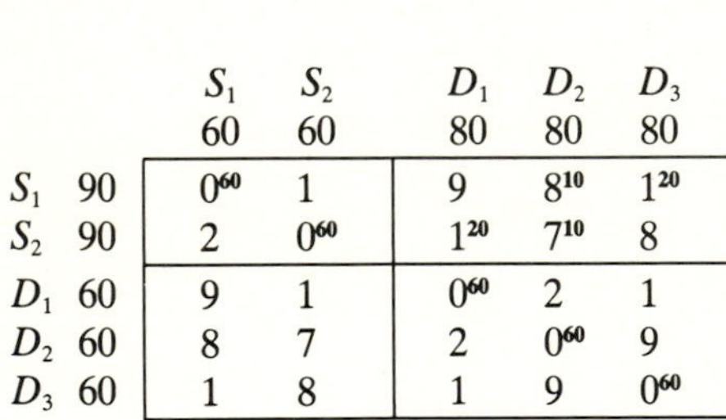

		S_1	S_2	D_1	D_2	D_3
		60	60	80	80	80
S_1	90	0^{60}	1	9	8^{10}	1^{20}
S_2	90	2	0^{60}	1^{20}	7^{10}	8
D_1	60	9	1	0^{60}	2	1
D_2	60	8	7	2	0^{60}	9
D_3	60	1	8	1	9	0^{60}

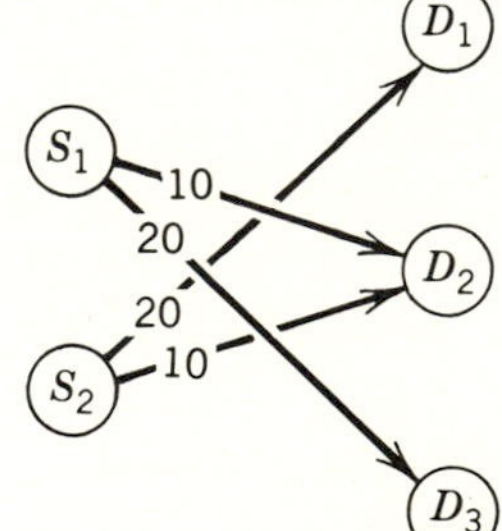

Associated network

For this particular feasible solution, none of the fictitious 60 units added to each supply and demand point are shipped anywhere. It is convenient to think of the 60 units on each diagonal entry as a "buffer capacity" for handling transshipments. The transportation network shows all the real shipments in this initial feasible solution.

Applying the transportation algorithm to this problem gives the dual vector (**u**, **v**) indicated below.

		S_1	S_2	D_1	D_2	D_3		
		60	60	80	80	80		
S_1	90	0^{60}	1	9	8^{10}	1^{20}	0	u_1
S_2	90	2	0^{60}	1^{20}	7^{10}	8	−1	u_2
D_1	60	9	1	0^{60}	2	1	−2	u_3
D_2	60	8	7	2	0^{60}	9	−8	u_4
D_3	60	1	8	1	9	0^{60}	−1	u_5
		0	1	2	8	1		
		v_1	v_2	v_3	v_4	v_5		

Calculating the adjusted costs gives the following tableau:

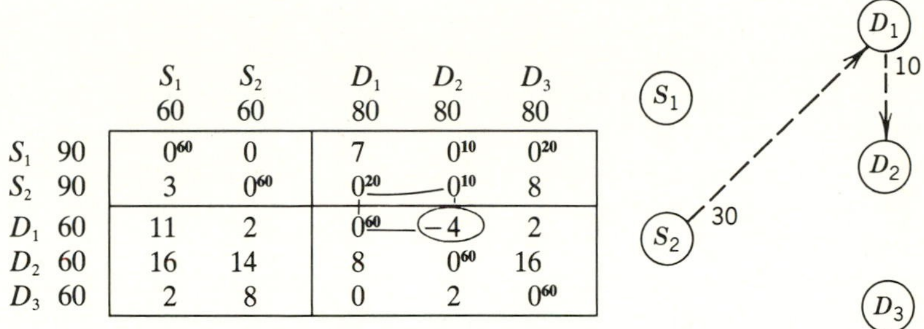

		S_1	S_2	D_1	D_2	D_3
		60	60	80	80	80
S_1	90	0^{60}	0	7	0^{10}	0^{20}
S_2	90	3	0^{60}	0^{20}	0^{10}	8
D_1	60	11	2	0^{60}	(−4)	2
D_2	60	16	14	8	0^{60}	16
D_3	60	2	8	0	2	0^{60}

Transship through D_1

This tableau shows that the current solution is not optimal. In particular, the loop shows that it is advantageous to transship 10 units from S_2 through D_1 to D_2 rather than ship them directly from S_2 to D_2. Using 10 units of buffer capacity to shift 10 units around the loop gives the feasible solution in the following tableau:

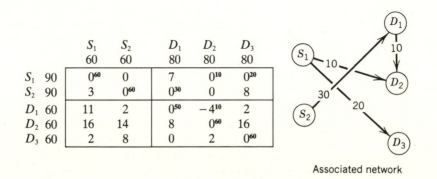

		S_1	S_2	D_1	D_2	D_3
		60	60	80	80	80
S_1	90	0^{60}	0	7	0^{10}	0^{20}
S_2	90	3	0^{60}	0^{30}	0	8
D_1	60	11	2	0^{50}	-4^{10}	2
D_2	60	16	14	8	0^{60}	16
D_3	60	2	8	0	2	0^{60}

D_1 10 S_1 10 D_2 30 20 S_2 D_3

Associated network

One more iteration of the transportation algorithm yields the following optimal tableau for the enlarged problem:

		S_1	S_2	D_1	D_2	D_3
		60	60	80	80	80
S_1	90	0^{60}	0^{10}	7	4	0^{20}
S_2	90	3	0^{50}	0^{40}	0	8
D_1	60	11	2	0^{40}	0^{20}	2
D_2	60	12	10	4	0^{60}	12
D_3	60	2	8	0	2	0^{60}

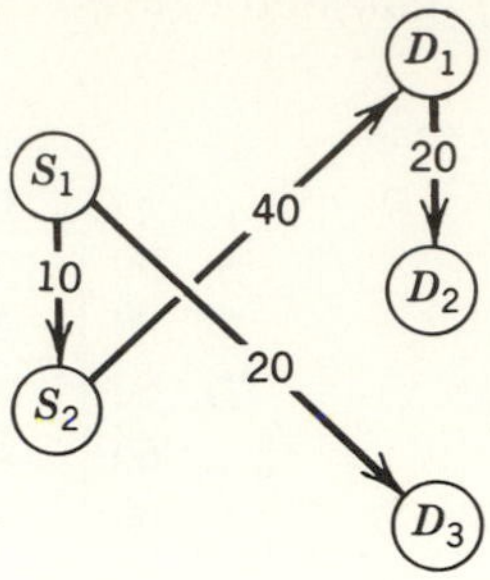

Associated network

The optimal solution transships a total of 30 units, 10 through S_2 to D_1 and 20 through D_1 to D_2. The only way these transshipments can be made is to use some of the buffer capacity on the diagonal cost entries. For example, if we had insisted on having each superscript on the diagonal remain at 60, then none of the transshipments in the solution would have been possible. Note that a total of exactly 30 units are taken away from the diagonal superscripts in the S_2 and the D_1 rows because these are the only nodes through which transshipping is advantageous.

In general, if the buffer capacity that we initially add to each supply and demand is greater than the amount that could be transshipped through each node, then in the optimal tableau there will always be a positive x_{ij} on each diagonal entry, so the diagonal cost entries will remain at zero throughout the solution process.

The x_{ij}'s in the off-diagonal positions of the preceding tableau give a feasible solution to the transshipment problem. To see why this solution is, in fact, optimal for the transshipment problem, suppose that there was a better way to meet the demands while allowing transshipments. We could then simply write down that better solution in the off-diagonal positions of the tableau and adjust the diagonal superscripts to obtain a feasible solution with a cost smaller than the minimum cost obtained, because all of the diagonal cost entries are zero. This, of course, would contradict the optimality of the preceding solution for the enlarged transportation problem.

Multiple Optimal Solutions

In Section 4.3 we saw how to use the optimal simplex tableau to see if there is more than one optimal solution. We can also use an optimal transportation tableau to determine whether or not an optimal solution is unique. We simply check to see if there are any zero cost entries in the positions of an optimal transportation tableau corresponding to the nonbasic variables in the optimal solution. For each nonbasic position with a zero cost entry, there is another optimal basic feasible solution.

For example, suppose the following optimal tableau is obtained by applying our algorithm to a certain transportation problem.

	15	30	15	30
20	(0)	2	3	0^{20}
15	6	0^{15}	1	0
10	0^{10}	0^{0}	9	7
15	3	0^{15}	4	2
30	0^{5}	5	0^{15}	0^{10}

This tableau has an optimal solution with $m + n - 1 = 8$ basic positions, and two of the nonbasic positions have 0 adjusted cost entries. We know that if we add an arc to the transportation network representing the current basic feasible solution, a unique loop will be created. Therefore, for each nonbasic position with a zero cost entry, there must be a unique loop starting at that position, with all other positions in the loop being basic positions. If we simply shift as much as possible around each such loop, we obtain other optimal basic feasible solutions. Of course, in the resulting tableaus, we need to check again for nonbasic positions with zero costs and continue to perform the appropriate shifts if we wish to find all optimal basic feasible solutions.

In the preceding tableau we indicate the loop corresponding to the nonbasic position with $c_{11} = 0$. Shifting the maximim amount possible around the loop gives the optimal basic feasible solution shown in the following tableau:

	15	30	15	30
20	0^{5}	2	3	0^{15}
15	6	0^{15}	1	(0)
10	0^{10}	0^{0}	9	7
15	3	0^{15}	4	2
30	0	5	0^{15}	0^{15}

In this tableau we have $c_{24} = 0$ in a nonbasic position and the loop starting at c_{24} is indicated. Shifting 10 units around this loop gives a third optimal basic feasible solution. Actually, for this problem there are 4 distinct optimal basic feasible solutions. You should find the other one.

Of course, there are optimal solutions that are not basic feasible solutions. For example, in the preceding tableau, we could shift one unit around the loop, and this would certainly give a feasible solution but not a basic feasible solution.

7.6 THE ASSIGNMENT PROBLEM

In this section we show that our transportation algorithm can be used to solve an important class of problems called **assignment problems.** The prototype assignment problem is usually stated as follows:

> Given n people to be assigned to n jobs, with c_{ij} being the cost of assigning person i to job j, find an assignment of minimum total cost.

A mathematical model for an assignment problem can be formulated by introducing decision variables x_{ij} defined by

$$x_{ij} = \begin{cases} 1, & \text{if person } i \text{ is assigned to job } j \\ 0, & \text{otherwise} \end{cases}$$

Using these variables, the assignment problem is formulated as the following integer programming problem.

THE ASSIGNMENT PROBLEM

$$\text{minimize} \quad \sum_{i=1}^{n} \sum_{j=1}^{n} c_{ij} x_{ij}$$

subject to

$$\sum_{j=1}^{n} x_{ij} = 1 \qquad i = 1, \ldots, n$$

$$\sum_{i=1}^{n} x_{ij} = 1 \qquad j = 1, \ldots, n$$

$$x_{ij} = 0 \text{ or } 1 \quad \text{for each } i, j$$

It is important to realize that if each x_{ij} is restricted to be 0 or 1, then the first set of constraints force each person i to be assigned to one job, and the second set of constraints force each job j to have one person assigned to it.

If the restriction that each x_{ij} be 0 or 1 were replaced by $x_{ij} \geq 0$, then the assignment problem would be a transportation problem with the number of sources equal to the number of destinations, and with each supply and demand equal to 1.

We have already remarked that for the transportation algorithm if the initial solution has each x_{ij} integer valued, then an optimal solution also has this property because every time the current solution is improved an integer amount is shifted around a loop. In particular, if the initial solution has a value of 0 or 1 for each x_{ij}, then each current solution will also have this property. This is true because any loop found in the algorithm will have x_{ij} values that alternate between 0 and 1, so we would never add 1 to a position with $x_{ij} = 1$ because then we would have to subtract 1 from the next position, which has $x_{ij} = 0$.

Therefore, the transportation algorithm can be used to obtain an optimal solution to the assignment problem. We simply relax the 0–1 integer restriction to $x_{ij} \geq 0$ and solve the relaxed problem starting with a feasible solution that has each x_{ij} equal to 0 or 1. The minimum value to the assignment problem can be no smaller than the minimum value of the relaxed problem because any $\mathbf{x}$ feasible for the assignment problem is also feasible for the relaxed problem. Thus, if an optimal solution to the relaxed problem is feasible for the assignment problem, it must also be optimal for the assignment problem.

We conclude this section by solving an assignment problem by our transportation algorithm. Suppose the costs of assignments are given by the following assignment tableau. Of course, an initial basic feasible solution that has only n positive x_{ij}'s is surely degenerate. As usual, we indicate the basic positions with a superscript (of 0 or 1) and for assignment problems there must be exactly one 1 in each row and in each column. The feasible solution in the following tableau was obtained by using the smallest cost entry method.

9	1^1	7	3	0
3	5	2	1^1	1
8^1	4^0	4^0	5	3
7	5	3^1	2^0	2
5	1	1	0	

In step 2 of the transportation algorithm we calculate the dual vector ($\mathbf{u}$, $\mathbf{v}$) indicated in the preceding tableau and this gives the adjusted cost entries in the following tableau:

4	0^1	6	3
(−3)	3	0	0^1
0^1	0^0	0^0	2
0	2	0^1	0^0

Because there is exactly one 1 in each row and in each column, the positions in the loop have x_{ij} values that alternate between 0 and 1. Therefore, the next solution will have each x_{ij} equal to 0 or 1 because either 0 or 1 will be alternately added and subtracted to the loop positions, and we cannot add 1 to a position with $x_{ij} = 1$ because then we would have to subtract 1 from the next position, which has $x_{ij} = 0$.

In our example, shifting as much as possible, namely 1, around the indicated loop gives the improved solution in the following tableau:

4	0^1	6	3	0
-3^1	3	0	0^0	0
0	0^0	0^1	2	0
0	2	0^0	0^1	0
−3	0	0	0	

As usual, we have to arbitrarily choose one of the basic positions to become a nonbasic position, and above we selected x_{31} to become nonbasic. Continuing with the transportation algorithm, we calculate $\mathbf{u}$ and $\mathbf{v}$ and then calculate the adjusted costs in the following tableau:

7	0^1	6	3
0^1	3	0	0^0
3	0^0	0^1	2
3	2	0^0	0^1

This final tableau shows that the current solution is optimal for the relaxed problem and is therefore optimal for the assignment problem.

We should mention that there are many other methods for solving the assignment problem that are more efficient than simply using the transportation algorithm as we have done here. One particularly elegant and efficient method that uses the dual simplex method was recently discovered by M. L. Balinski; see the selected references at the end of this chapter.

7.7 GENERAL NETWORK MODELS

All the models considered in this chapter are examples of network flow models. In this section we discuss a general setting for such problems and show how our algorithm for solving the transportation problem can be generalized to provide an algorithm for solving minimum cost network flow problems. As is the case for our transportation algorithm, the more general algorithm we present in this section is a streamlined version of the simplex algorithm.

As an example of a general network model, consider the network in Figure 7.5 with five nodes and nine possible shipping routes with the per-unit cost, c_{ij}, of shipping from node i to node j indicated on each arc.

The network in Figure 7.5 has a supply of 50 at node 1, a supply of 10 at node 4, a demand of 40 units at node 2, and a demand of 20 units at node 5. Node 3 is an intermediate node that only serves as a transshipment node. From the structure of the network, we see that supply nodes and demand nodes may also serve as transshipment nodes. Also, this network is **connected** in that every pair of nodes is connected by arcs in the network.

In a general network flow problem, we want to know how much to ship on each arc in the network so that demand is satisfied with the available supplies and total shipping costs are minimized. Again we assume that total supply equals total demand. If we let

$$x_{ij} = \text{the flow from node } i \text{ to node } j$$

then there will be one variable for each arc in the network.

The constraints for a general network flow problem come from the following observations.

For supply nodes

$$\text{flow into the node} + \text{supply} = \text{flow out of the node}$$

For demand nodes

$$\text{flow into the node} - \text{demand} = \text{flow out of the node}$$

For intermediate nodes

$$\text{flow into the node} = \text{flow out of the node}$$

For the network represented in Figure 7.5, there are five constraints, one for each node. The linear programming model for this network is as follows:

$$\text{minimize} \quad 5x_{12} + x_{13} + 2x_{24} + 6x_{25} + 3x_{32} + 4x_{34} + 5x_{35} + 2x_{43} + 7x_{45}$$

subject to

$$\begin{aligned}
x_{12} + x_{13} &= 50 \\
x_{24} + x_{25} - x_{12} - x_{32} &= -40 \\
x_{32} + x_{34} + x_{35} - x_{13} - x_{43} &= 0 \\
x_{43} + x_{45} - x_{24} - x_{34} &= 10 \\
- x_{25} - x_{35} - x_{45} &= -20 \\
x_{ij} \geq 0 \qquad \text{for each } (i, j)
\end{aligned}$$

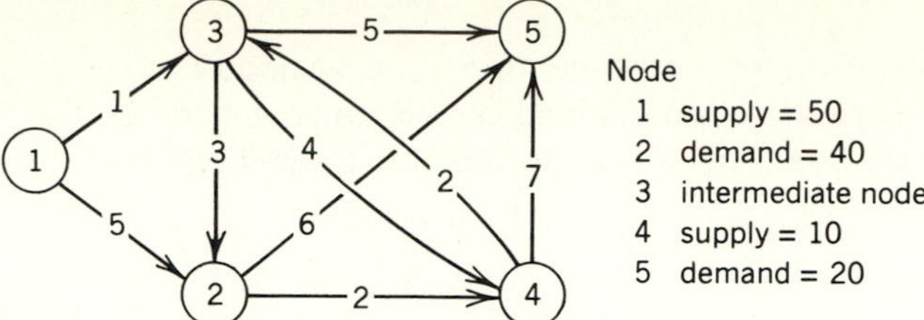

FIGURE 7.5 A general network flow problem.

The above constraints are written in the form

$$(\text{flow out}) - (\text{flow in}) = \begin{cases} \text{supply}, & \text{for supply nodes} \\ 0, & \text{for intermediate nodes} \\ -\text{demand}, & \text{for demand nodes} \end{cases}$$

It is important to realize that these constraints allow for all possible transshipments in the network.

The preceding linear program is in standard form and has the following simplex tableau:

	x_{12}	x_{13}	x_{24}	x_{25}	x_{32}	x_{34}	x_{35}	x_{43}	x_{45}
0	5	1	2	6	3	4	5	2	7
50	1	1	0	0	0	0	0	0	0
−40	−1	0	1	1	−1	0	0	0	0
0	0	−1	0	0	1	1	1	−1	0
10	0	0	−1	0	0	−1	0	1	1
−20	0	0	0	−1	0	0	−1	0	−1

Note that each column of this tableau has exactly one 1 and exactly one −1, with all other column entries equal to 0. This is always the case for this type of network flow model because each arc goes out of exactly one node and into exactly one other node. For example, the x_{13} column has a 1 in row 1 and a −1 in row 3 because the variable x_{13} represents the flow out of node 1 into node 3. A consequence of this special structure is that if we add each of the first $n - 1$ equations to the last equation, then the last row in the simplex tableau becomes a row of zeros. Thus, just as in transportation problems, these problems also have one redundant constraint equation, so a basic feasible solution has $n - 1$ basic variables.

The General Network Flow Model

A network flow problem with a total of n supply, demand, and intermediate nodes has the following form:

$$\text{minimize} \quad \sum_{i=1}^{n} \sum_{j=1}^{n} c_{ij} x_{ij}$$

subject to

$$\sum_{j=1}^{n} x_{ij} - \sum_{k=1}^{n} x_{ki} = b_i \quad \text{for each node } i$$

$$x_{ij} \geq 0 \quad \text{for each } (i, j)$$

where

$b_i > 0$ means there is a supply of b_i units at node i
$b_i < 0$ means there is a demand of $-b_i$ units at node i
$b_i = 0$ means that node i serves only for transshipment

In this model it is to be understood that the only variables x_{ij} that appear are those for which there is an arc from node i to node j in the network.

From the structure of the model, especially as revealed by the simplex tableau, we see that the dual of this standard form linear program has the following form:

$$\text{maximize } \sum_{i=1}^{n} b_i y_i$$

subject to

$$y_i - y_j \le c_{ij} \quad \text{for each } (i, j) \text{ corresponding to an arc in the network}$$
$$\mathbf{y} \text{ free}$$

Given a feasible solution to the network flow problem, we can use the dual variables, y_i for $i = 1, \ldots, n$, to see whether or not a current basic feasible solution can be improved. This dual problem will play the same role in the network flow algorithm as the transportation dual plays in our transportation algorithm. Letting

$$\alpha(\mathbf{x}) = \sum_{i=1}^{n} \sum_{j=1}^{n} c_{ij} x_{ij}$$

$$\beta(\mathbf{y}) = \sum_{i=1}^{n} b_i y_i$$

the difference $\alpha(\mathbf{x}) - \beta(\mathbf{y})$ can be calculated (see Exercise 7.18) as

$$\alpha(\mathbf{x}) - \beta(\mathbf{y}) = \sum_{i=1}^{n} \sum_{j=1}^{n} (c_{ij} - y_i + y_j) x_{ij} + \sum_{i=1}^{n} y_i \left(\sum_{j=1}^{n} x_{ij} - \sum_{k=1}^{n} x_{ki} - b_i \right)$$

If $\mathbf{x}$ is feasible, then the last term in parentheses is zero, and thus we obtain the following key result:

> If $\mathbf{y}$ is any dual vector (feasible or infeasible), then
>
> $$\alpha(\mathbf{x}) - \beta(\mathbf{y}) = \sum_{i=1}^{n} \sum_{j=1}^{n} (c_{ij} - y_i + y_j) x_{ij}$$
>
> provided $\mathbf{x}$ is a feasible solution.

We shall use this result in the same way we used its counterpart in the transportation algorithm. Given a basic feasible $\mathbf{x}$ for the network problem, we first calculate a dual vector $\mathbf{y}$ satisfying

$$c_{ij} - y_i + y_j = 0 \quad \text{whenever } x_{ij} \text{ is basic}$$

Such a dual vector guarantees that $\alpha(\mathbf{x}) = \beta(\mathbf{y})$ and so if, in addition, $\mathbf{y}$ is feasible, then we know from the duality relations that $\mathbf{x}$ is optimal for the network flow problem. The vector $\mathbf{y}$ is dual feasible if the **adjusted costs** are nonnegative, that is, if

$$c_{ij} - y_i + y_j \geq 0 \quad \text{for each } (i, j) \text{ in the network}$$

Just as we saw for the transportation problem, replacing the original costs with the adjusted costs does not change the optimal vector. To see why this holds here, we simply calculate a new objective function $\hat{\alpha}(\mathbf{x})$ using these adjusted costs:

$$\begin{aligned}
\hat{\alpha}(\mathbf{x}) &= \sum_{i=1}^{n} \sum_{j=1}^{n} (c_{ij} - y_i + y_j)x_{ij} \\
&= \alpha(\mathbf{x}) + \sum_{i=1}^{n} y_i \left(\sum_{j=1}^{n} x_{ij} - \sum_{k=1}^{n} x_{ki} \right) \\
&= \alpha(\mathbf{x}) + \sum_{i=1}^{n} y_i b_i
\end{aligned}$$

Because $\hat{\alpha}(\mathbf{x})$ and $\alpha(\mathbf{x})$ only differ by the constant $\mathbf{b}^T\mathbf{y}$, minimizing $\hat{\alpha}(\mathbf{x})$ over the feasible set is equivalent to minimizing $\alpha(\mathbf{x})$ over the feasible set. Therefore, if one of the adjusted costs is negative, the current solution can be improved by shifting some flow to that route. Before showing how such a shift is always possible, we need to generalize some of the procedures used in solving the transportation problem.

Spanning Trees and Basic Feasible Solutions

As we saw earlier in the network representation of a transportation problem with a total of $m + n$ supply and demand nodes, a basic feasible solution always corresponds to a subset of $n + m - 1$ arcs having no loops.

In general, given a network with n nodes, a connected subset of arcs having no loops and joining together all n nodes is called a **spanning tree.** For example, a subnetwork representing a basic feasible solution for a transportation problem is a spanning tree (see Figures 7.1 and 7.2). For connected networks the following is an important characterization of spanning trees:

> In a connected network with n nodes, a subset of arcs is a spanning tree if and only if it consists of $n - 1$ arcs and has no loops.

For a proof of this result, see almost any elementary graph theory text.

As we saw earlier for transportation networks, there is a one-to-one correspondence between basic feasible solutions and subnetworks that consist of

$m + n - 1$ arcs and have no loops. This, along with the preceding result, establishes the following:

> For transportation networks, basic feasible solutions correspond to spanning trees.

For general network flow problems, the situation is slightly more complicated. Unlike transportation problems, not every spanning tree corresponds to a basic feasible solution. To illustrate this point, consider the spanning tree indicated by the solid arcs in Figure 7.6 for our example network flow problem.

From Figure 7.6 it is clear that by only using shipping routes corresponding to arcs in the spanning tree, it is not possible to meet the demand. In particular, there is no way to ship the 10 units at supply node 4 out of the node unless we go against the flow. We need to construct a **feasible spanning tree,** namely, a spanning tree for which it is possible to meet the demand using the arcs of the tree.

Later we will give a general procedure for finding an initial feasible spanning tree, but for our example it is fairly easy to find such a tree by examining the network. In Figure 7.7 we give a feasible spanning tree with the flow indicated on each arc.

In general,

> For network flow problems, basic feasible solutions correspond to feasible spanning trees.

For example, the feasible spanning tree in Figure 7.7 corresponds to the basic feasible solution with basic variables

$$x_{12} = 50 \qquad x_{13} = 0, \qquad x_{25} = 10, \qquad x_{45} = 10$$

We shall not establish this general result here, but we do use it to show how the current basic feasible solution can be improved. In particular, given a feasible spanning tree, adding an arc to the tree gives a unique loop that can be used to improve the current solution.

Because a feasible spanning tree corresponds to a basic feasible solution, an arc in a feasible spanning tree is called a **basic arc** and all other arcs in the network are called **nonbasic arcs.**

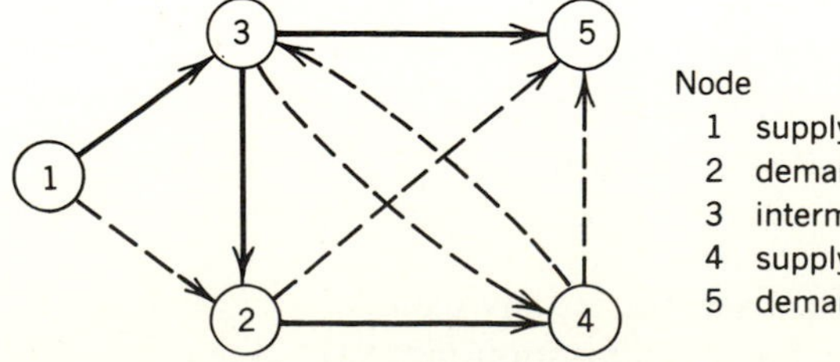

FIGURE 7.6 A spanning tree that is not feasible.

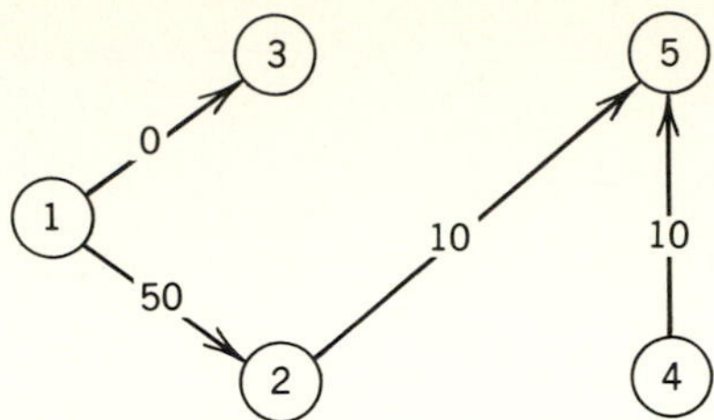

FIGURE 7.7 A feasible spanning tree.

We shall illustrate how the general network flow algorithm works by using it to solve our example problem. Unfortunately, the only bookkeeping scheme that can be conveniently used here is the network itself.

Solving a General Network Flow Problem

Given the initial feasible spanning tree (basic feasible solution) in Figure 7.7, we first calculate a dual vector $\mathbf{y}$ satisfying

$$c_{ij} - y_i + y_j = 0 \quad \text{for each basic arc}$$

For convenience, we repeat our example network in Figure 7.8 and indicate the basic arcs as solid arcs and nonbasic arcs as dotted arcs. The shipping cost c_{ij} is given on each arc.

Using the four basic (solid) arcs in the spanning tree to calculate a dual vector $\mathbf{y}$ gives the linear system

$$\begin{aligned} y_1 - y_2 &= 5 \\ y_1 - y_3 &= 1 \\ y_2 - y_5 &= 6 \\ y_4 - y_5 &= 7 \end{aligned} \quad \text{which yields} \quad \begin{bmatrix} y_1 \\ y_2 \\ y_3 \\ y_4 \\ y_5 \end{bmatrix} = \begin{bmatrix} 0 \\ -5 \\ -1 \\ -4 \\ -11 \end{bmatrix} \quad \text{when } y_1 = 0$$

To check whether or not the current solution $\mathbf{x}$ is optimal, we calculate the adjusted costs to see if the current $\mathbf{y}$ is dual feasible. Of course, the adjusted costs are zero for all basic arcs, so we need only check if

$$c_{ij} - y_i + y_j \geq 0 \quad \text{for all nonbasic arcs.}$$

There are five nonbasic (dotted) arcs in the spanning tree of Figure 7.8, which

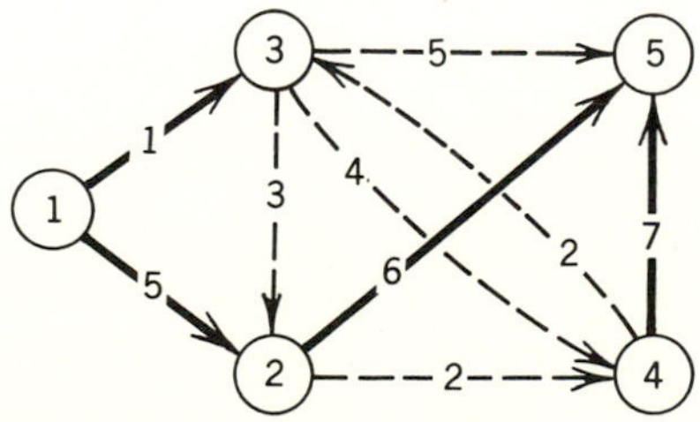

FIGURE 7.8 Basic and nonbasic arcs with associated costs.

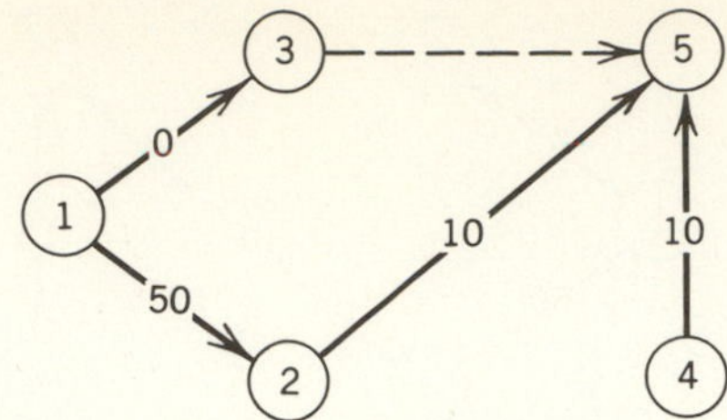

FIGURE 7.9 Adding an arc to the spanning tree creates a unique loop.

give the following adjusted costs:

$$c_{24} - y_2 + y_4 = 2 - (-5) + (-4) = 3$$

$$c_{32} - y_3 + y_2 = 3 - (-1) + (-5) = -1$$

$$c_{34} - y_3 + y_4 = 4 - (-1) + (-4) = 1$$

$$c_{35} - y_3 + y_5 = 5 - (-1) + (-11) = -5 \longleftarrow$$

$$c_{43} - y_4 + y_3 = 2 - (-4) + (-1) = 5$$

Here, two dual constraints are not satisfied, and we cannot conclude that the current $\mathbf{x}$ is optimal. Adding the arc with the most negative adjusted cost to the current spanning tree creates the unique loop indicated in Figure 7.9.

To obtain an improved solution, we simply add as much as possible to the dotted arc in the loop of Figure 7.9 and adjust the flow around the loop to make sure that the solution remains feasible. However, to maintain feasibility, we cannot simply add and subtract an amount t to the alternate arcs in the loop as we did in the transportation algorithm. We need to pay closer attention to the direction of flow on each arc of the loop. In Figure 7.10, we show how to shift an amount t around the loop to maintain feasibility.

Note in Figure 7.10 that while going around the loop in the direction of the dotted arc, if we add t when we go with the flow of an arc and subtract t when we go against the flow of an arc, then we always end up with a feasible solution provided t is not too large. Here we can shift a maximum of $t = 10$ around the loop, which results in a decrease of 50 in the total cost because the adjusted cost for the new arc is -5. Performing this shift drives x_{25} to zero, and thus the arc from node 2 to node 5 becomes a nonbasic arc. The new basic feasible solution (feasible spanning tree) is given by the solid arcs in Figure 7.11. For convenience, we have repeated the original shipping costs c_{ij} on each arc. Given this new improved basic feasible solution, we are now ready to repeat the process.

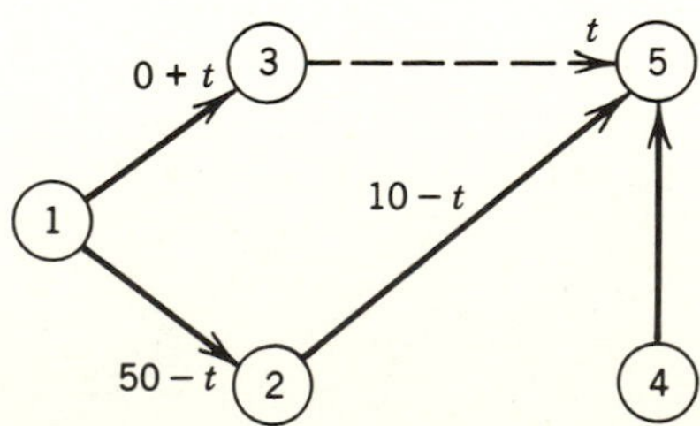

FIGURE 7.10 Shifting an amount *t* around the loop to maintain feasibility.

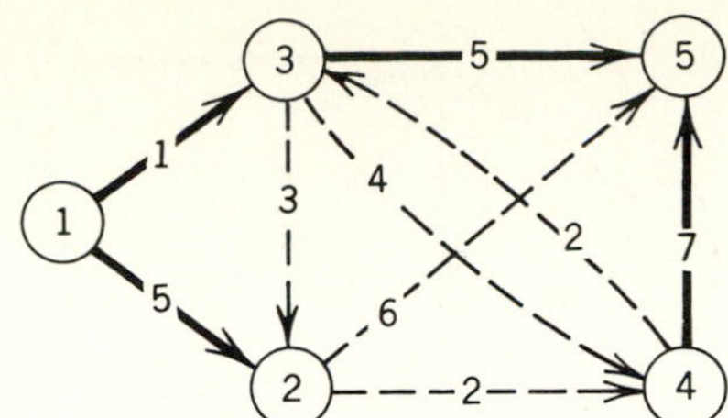

FIGURE 7.11 The improved feasible spanning tree.

To see if the current solution is optimal, we first find a dual vector **y** so that

$$c_{ij} - y_i + y_j = 0 \quad \text{for each basic arc}$$

Using the original costs c_{ij}, we obtain the system

$$\begin{aligned} y_1 - y_2 &= 5 \\ y_1 - y_3 &= 1 \\ y_3 - y_5 &= 5 \\ y_4 - y_5 &= 7 \end{aligned} \quad \text{which yields} \quad \begin{bmatrix} y_1 \\ y_2 \\ y_3 \\ y_4 \\ y_5 \end{bmatrix} = \begin{bmatrix} 0 \\ -5 \\ -1 \\ 1 \\ -6 \end{bmatrix} \quad \text{when } y_1 = 0$$

This **y** is dual feasible if

$$c_{ij} - y_i + y_j \geq 0 \quad \text{for each nonbasic arc}$$

in the current network. From Figure 7.11, we obtain

$$\begin{aligned} c_{24} - y_2 + y_4 &= 2 - (-5) + (1) = 8 \\ c_{25} - y_2 + y_5 &= 6 - (-5) + (-6) = 5 \\ c_{32} - y_3 + y_2 &= 3 - (-1) + (-5) = -1 \longleftarrow \\ c_{34} - y_3 + y_4 &= 4 - (-1) + (1) = 6 \\ c_{43} - y_4 + y_3 &= 2 - (1) + (-1) = 0 \end{aligned}$$

Thus the current **y** is not dual feasible, and we increase the flow on the arc from node 3 to node 2 with the negative adjusted cost. Adding that arc to the current spanning tree gives the unique loop indicated in Figure 7.12.

Shifting as much as possible, namely, $t = 40$, around the loop gives the new basic feasible solution represented by the new feasible spanning tree in Figure 7.13, where we have again recorded the original costs.

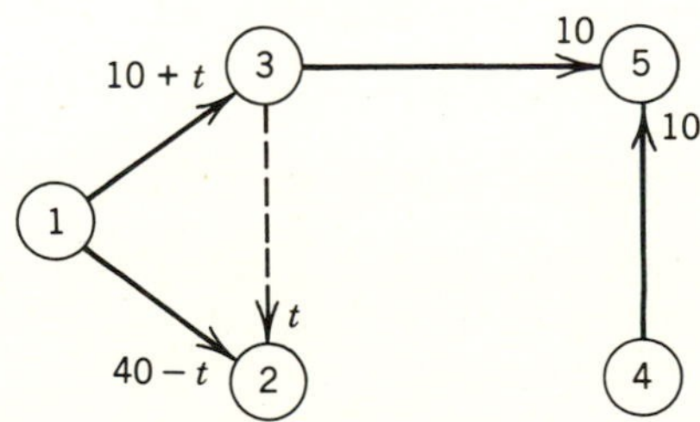

FIGURE 7.12 Adding an arc creates a unique loop.

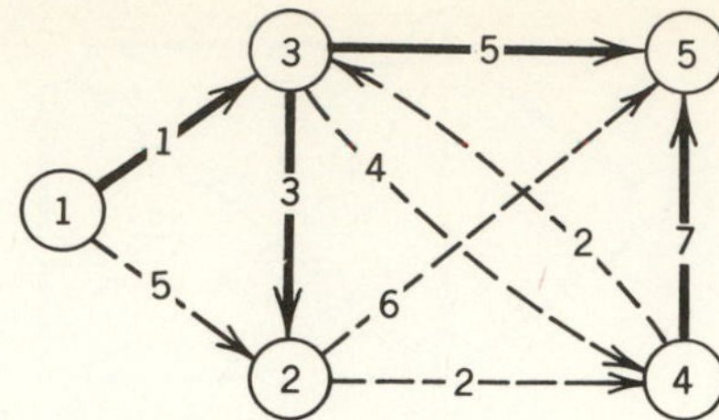

FIGURE 7.13 A new feasible spanning tree.

We now repeat this process using the spanning tree and costs given in Figure 7.13 to calculate a dual vector **y** so that $c_{ij} - y_i + y_j = 0$ for each basic arc. This gives the system

$$\begin{aligned} y_1 - y_3 &= 1 \\ y_3 - y_2 &= 3 \\ y_3 - y_5 &= 5 \\ y_4 - y_5 &= 7 \end{aligned} \quad \text{which yields} \quad \begin{bmatrix} y_1 \\ y_2 \\ y_3 \\ y_4 \\ y_5 \end{bmatrix} = \begin{bmatrix} 0 \\ -4 \\ -1 \\ 1 \\ -6 \end{bmatrix} \quad \text{when } y_1 = 0$$

Using the costs c_{ij} given in Figure 7.13, we calculate

$$c_{ij} - y_i + y_j \quad \text{for each nonbasic arc}$$

in the current network. This yields the adjusted costs

$$\begin{aligned} c_{12} - y_1 + y_2 &= 5 - (0) \quad + (-4) = 1 \\ c_{24} - y_2 + y_4 &= 2 - (-4) + (1) \quad = 7 \\ c_{25} - y_2 + y_5 &= 6 - (-4) + (-6) = 4 \\ c_{34} - y_3 + y_4 &= 4 - (-1) + (1) \quad = 6 \\ c_{43} - y_4 + y_3 &= 2 - (1) \quad + (-1) = 0 \end{aligned}$$

Because all of the adjusted costs are nonnegative, the current **y** is dual feasible and therefore the current **x** is optimal.

At each iteration of this process we used the adjusted costs to determine how the flow on the network could be improved. We did not use the current adjusted costs as the starting costs for the next iteration as we did with the transportation algorithm. Because of the bookkeeping scheme being used, there is no particular advantage to updating the costs at each iteration. In fact, if the original costs are used at each iteration, then the adjusted costs need only be calculated until the first negative is obtained.

Summary of the General Network Flow Algorithm

It should now be clear that the algorithm illustrated for solving a general network flow problem is a generalization of our transportation algorithm. You should compare the following statement of the algorithm to that of the transportation algorithm given in Section 7.3.

Step 1. Find an initial feasible spanning tree.

Step 2. For the current spanning tree and the current cost coefficients c_{ij}, find a dual vector $\mathbf{y}$ such that

$$c_{ij} - y_i + y_j = 0 \quad \text{for each basic arc}$$

and calculate the adjusted cost coefficients

$$c_{ij} - y_i + y_j \quad \text{for each nonbasic arc}$$

Step 3. If each adjusted cost coefficient is nonnegative, stop; the current $\mathbf{x}$ is optimal.

Otherwise, pick an arc with a negative adjusted cost and add that arc to the current spanning tree to create a unique loop.

Step 4. Shift as much as possible around the loop to obtain an improved feasible $\mathbf{x}$ and a new feasible spanning tree. Return to step 2 with this new spanning tree and with the original costs.

Finding an Initial Feasible Spanning Tree

As mentioned earlier, finding an initial basic feasible solution for a general network flow problem is not as easy as it is for transportation problems because not every spanning tree corresponds to a basic feasible solution. We need a procedure for finding a feasible spanning tree.

It is important to realize that some care has to be taken in constructing a general network flow model that actually has feasible solutions. Unlike the transportation problem, it is not enough to have total supply equal total demand. We need to be sure that enough arcs are included so it is possible to ship the supply to the various demand nodes.

The key to constructing a feasible spanning tree is to introduce artificial arcs so there is an arc between each supply point and every demand point. Ignoring all the intermediate nodes, we would then have a network flow problem that is just like a transportation problem. Using the artificial arcs along with the original arcs, we could then use a method for finding an initial basic feasible solution for the transportation problem to find an initial spanning tree for the supply and demand nodes. This would not provide a spanning tree for the entire network because the intermediate nodes would not be included. However, if there are p intermediate nodes, they must be connected to the supply-and-

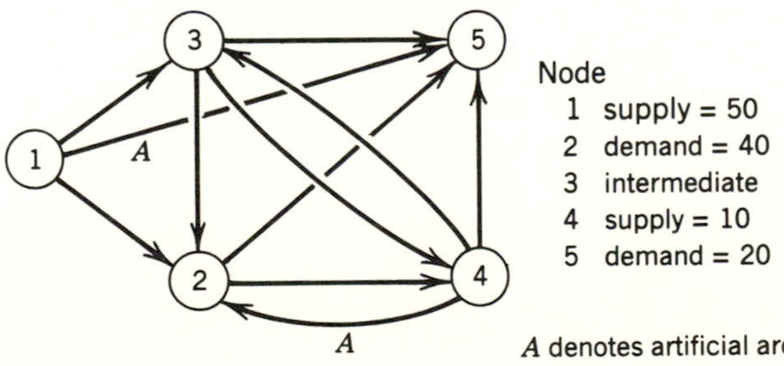

FIGURE 7.14 Adding artificial arcs to allow shipping from each source to each destination.

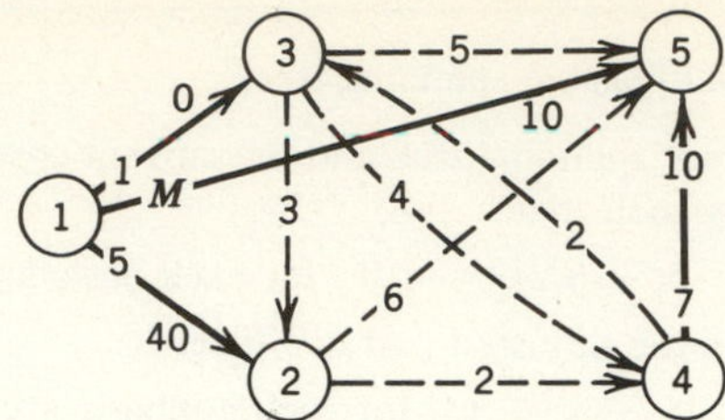

FIGURE 7.15 A feasible spanning tree for the artificial problem.

demand nodes, and we can use p existing arcs with 0 flow to obtain a feasible spanning tree. The costs on the artificial arcs would be set to some large value, and, in applying the above algorithm to find the minimum cost solution, the artificial arcs would become nonbasic and could be eliminated unless no feasible spanning tree existed using only the original arcs.

We illustrate this procedure by using the example of this section. Consider the network in Figure 7.14 where two artificial arcs are added so that every supply node can ship to every demand node.

The smallest cost entry method for a transportation problem with nodes 1 and 4 as supply nodes and nodes 2 and 5 as demand nodes gives a basic feasible solution with $x_{12} = 40$, $x_{15} = 10$, and $x_{45} = 10$. In Figure 7.15 we construct a feasible spanning tree for the artificial problem by using the arcs corresponding to x_{12}, x_{15}, and x_{45} and by using, for intermediate node 3, the arc from node 1 to node 3 with a flow of $x_{13} = 0$. Note that the artificial arc from node 4 to node 2 has been dropped because it is not needed in the initial feasible spanning tree for the artificial problem.

We can now apply our algorithm to find the minimum cost solution and, if a feasible spanning tree exists for the original problem, the remaining artificial arc will become nonbasic and can then be eliminated from the problem. Setting $M = 500$, for example, and using the feasible spanning tree in Figure 7.15, we calculate a dual vector y such that

$$c_{ij} - y_i + y_j = 0 \quad \text{for each basic arc}$$

This gives the system

$$\begin{aligned} y_1 - y_2 &= 3 \\ y_1 - y_3 &= 1 \\ y_1 - y_5 &= 500 \\ y_4 - y_5 &= 7 \end{aligned} \quad \text{which yields} \quad \begin{bmatrix} y_1 \\ y_2 \\ y_3 \\ y_4 \\ y_5 \end{bmatrix} = \begin{bmatrix} 0 \\ -3 \\ -1 \\ -493 \\ -500 \end{bmatrix} \quad \text{when } y_1 = 0$$

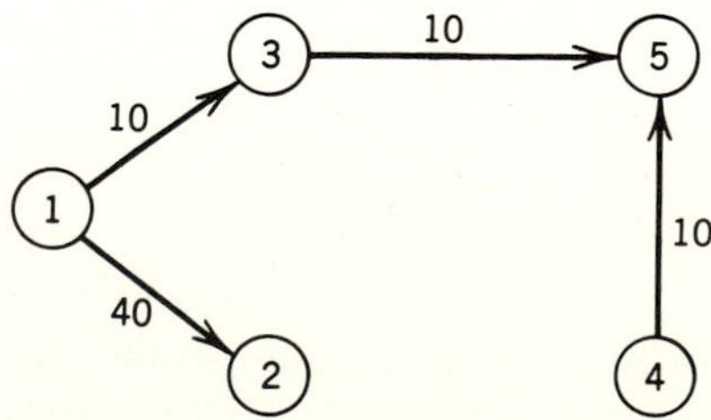

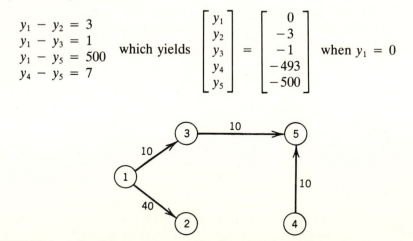

FIGURE 7.16 A feasible spanning tree for the original problem.

Using the costs c_{ij} given in Figure 7.15, we then use $\mathbf{y}$ to calculate

$$c_{ij} - y_i + y_j \quad \text{for each nonbasic arc}$$

in the current spanning tree. This yields the adjusted costs

$$\begin{aligned}
c_{24} - y_2 + y_4 &= 2 - (-3) \quad + (-493) = -488 \\
c_{25} - y_2 + y_5 &= 6 - (-3) \quad + (-500) = -491 \\
c_{32} - y_3 + y_2 &= 3 - (-1) \quad + (-3) \quad = \quad 1 \\
c_{34} - y_3 + y_4 &= 4 - (-1) \quad + (-493) = -488 \\
c_{35} - y_3 + y_5 &= 5 - (-1) \quad + (-500) = -495 \\
c_{43} - y_4 + y_3 &= 2 - (-493) + (-1) \quad = \quad 494
\end{aligned}$$

Adding the arc from node 3 to node 5 (corresponding to the most negative adjusted cost) gives the loop from node 3 to node 5 to node 1 to node 3. Shifting 10 units around this loop causes the artificial arc to become nonbasic and gives the feasible spanning tree in Figure 7.16 for the original problem.

We can now use the spanning tree in Figure 7.16 as the initial feasible spanning tree and proceed with the usual steps of the algorithm to solve the original problem. Actually, the feasible spanning tree given in Figure 7.16 is the one used in Figure 7.11, and thus only one more iteration is required to find the optimal solution.

SELECTED REFERENCES

Balinski, M. L., A Competitive (Dual) Simplex Method for the Assignment Problem, *Mathematical Programming* **34**:125–141 (1986).

Bazaraa, M. S., and Jarvis, J. J., *Linear Programming and Network Flows*, Wiley, New York, 1977.

Bradley, S. P., Hax, A. C., and Magnanti, T. L., *Applied Mathematical Programming*, Addison-Wesley, Reading MA, 1977, Chapter 8.

Jensen, P. A., and Barnes, J. W., *Network Flow Programming,* Wiley, New York, 1980.

Minieka, E., *Optimization Algorithms for Networks and Graphs,* Marcel Dekker, New York, 1978.

Wagner, H. M., *Principles of Operations Research,* Prentice-Hall, Englewood Cliffs, NJ, 1969, Chapters 6 and 7.

EXERCISES

7.1 A company has 3 warehouses that are used to supply 3 stores with a product. Each warehouse has 20 units of the product and store 1, store 2, and store 3 need 20, 15, and 15 units, respectively. The per-unit costs of shipping from warehouse 1 to stores 1 and 2 are 5 and 7, respectively. Warehouse 1 cannot ship units to store 3. The per unit costs of shipping from warehouse 2 to stores 1, 2, and 3 are 6, 2, and 9, respectively. The per-unit costs of shipping from warehouse 3 to stores 1, 2, and 3 are 1, 1, and 5, respectively.

(a) Write down a transportation tableau with supplies, demands, and costs specified so that an optimal solution will provide the shipping schedule of minimum total cost.
(b) Find an optimal shipping schedule.

7.2 Use the transportation algorithm to solve the transportation problems with the following tableaus and with supplies and demands as indicated. Be sure to check whether total supply equals total demand.

(a)

	20	5	5	10
10	9	3	4	5
15	2	5	1	2
15	3	9	1	5

(b)

	10	15	10	15
20	7	3	8	5
20	4	1	6	2

(c)

	20	15	20
20	7	1	2
20	1	5	5
25	4	1	4

7.3 Consider the following transportation tableau with an initial basic feasible solution indicated by the superscripts.

1^5	1^5	1
2^{10}	4	3^5
4	4^{10}	1

(a) Starting with the indicated basic feasible solution, apply the algorithm developed in the text to obtain an optimal transportation tableau.
(b) Write down the optimal vector $\mathbf{x}$ and the minimum cost.
(c) Suppose that it is not possible to use the shipping route from source 1 to destination 2. Find an optimal solution subject to this added restriction.

7.4 For the following optimal transportation tableaus, find all optimal basic feasible solutions.

(a)

0^5	2	0^5	4	7	8
4	0^{10}	6	0^5	2	1
5	9	0^{10}	6	0^{10}	5
4	0^5	9	7	6	0^{10}
0	1	5	0^{15}	4	9
2	7	1	5	0^{15}	0^{10}

(b)

2	0^5	0^5	2	0^{10}
0^{15}	0	1	1	0^5
0	4	0^5	4	1
3	0^5	1	0^{10}	1

7.5 Consider the following transportation tableau, with supplies and demands indicated as usual.

	15	10	15	20
20	9	8	1	3
30	2	6	5	4
10	0	7	9	8

(a) Find an initial basic feasible solution by the northwest corner rule.
(b) Find an initial basic feasible solution by the smallest cost entry method.
(c) Find an initial basic feasible solution by the Vogel advanced start method.

7.6 Use our transportation algorithm to solve the example problem given in Section 7.4 using the feasible solution found by the Vogel advanced start method as the initial basic feasible solution.

7.7 True or False. Give an explanation for your answer.
(a) Any feasible **x** for a transportation problem with exactly $m + n - 1$ positive components is a basic feasible solution.
(b) If **x** is feasible for a transportation problem and (**u**, **v**) is any dual vector, then the difference between the primal and dual objective function values is

$$\sum_i \sum_j (c_{ij} - u_i - v_j)x_{ij}$$

(c) If **x** is the current basic feasible solution for a transportation problem and if a dual vector (**u**, **v**) is calculated in the algorithm with at least one negative adjusted cost, $c_{ij} - u_i - v_j$, then **x** cannot be optimal for the problem.
(d) If **x** is feasible for a transportation problem and (**u**, **v**) is a dual vector for which

$$(c_{ij} - u_i - v_j)x_{ij} = 0$$

then **x** must be optimal for the problem.
(e) In our transportation algorithm, if we start with an initial basic feasible solution having some noninteger components, then an optimal solution obtained by the algorithm must also have some noninteger components.

7.8 Suppose that during exam week sharpened pencils are required at the start of each exam day. The daily requirements call for 60, 50, 80, 40, and 50 sharpened pencils for Monday through Friday, respectively. Sharp pencils can be bought for 15 cents each at the bookstore. Pencils used during one exam day can be resharpened at Bud's 1-day service for 2 cents each or at Mac's 2-day service for 1 cent each. For example, a pencil used on Monday can be sharpened by Bud on Tuesday and be ready for use on Wednesday, but if it is sent to Mac, it will not be ready for use until Thursday. At the end of the week, all pencils can be sold for 5 cents each.
(a) Formulate a transportation problem that will determine how to supply pencils at minimum total cost. Write down a transportation tableau with supplies and demands indicated. Be sure to define the supply and demand points.
(b) Find an optimal solution to this problem.

7.9 The production manager of a large company is in charge of 3 separate plants that make turbo-encabulators. The plants produce the same quality of product, but they have different production capacities. Plants 1, 2, and 3 can produce 1500, 2500, and 2000 turbo-encabulators, respectively, each month. There are 4 outlet stores that need this product. Unfortunately, not each plant can ship to each outlet because of interstate trade restrictions. Plant 1 can ship to all 4 stores with per-unit shipping costs of 4, 3, 2, and 5 to stores 1, 2, 3, and 4, respectively. Plant 2 can only ship to stores 1, 2, and 3 with per-unit shipping costs of 2, 1, and 4, respectively. Plant 3 can only ship to stores 2, 3, and 4 with per-unit shipping costs of 4, 2, and 7, respectively. Store 1 needs exactly 1000 turbo-encabulators; store

2 needs at least 1000 but would like all it can get; store 3 needs exactly 1000; and store 4 needs at least 500 but will take more if any are available.
(a) Formulate (but do not solve) a transportation problem whose solution will yield a minimum cost shipping schedule. You may simply write down a transportation tableau with the supplies and demands indicated. Be sure that total supply equals total demand.
(b) From the formulation of the problem, show without actually solving the problem that store 2 will get 3500 turbo-encabulators in an optimal solution.

7.10 Consider a transportation problem with the following transportation tableau.

	10	10	10	10
10	2	4	3	2
10	1	1	4	3
20	3	6	9	5

Suppose that each source can be used as a transshipment point, but that destination points cannot transship. Let the cost of shipping from source i to source j be $i + j$ if $i \neq j$ and let it be zero otherwise. Formulate a transportation problem that will yield a least-cost shipping schedule. You may simply give a transportation tableau with appropriate supplies, demands, and costs.

7.11 Consider the following initial and optimal transportation tableaus for a problem with two supply points and three demand points.

	15	15	10
20	1	3	2
20	2	2	6

Initial tableau

0^{10}	2	0^{10}
0^{5}	0^{15}	3

Optimal tableau

(a) Write down an initial simplex tableau for this problem when it is formulated as a linear program.
(b) Perform four pivots on the initial simplex tableau to obtain the optimal solution indicated in the optimal transportation tableau.

7.12 An oil company has three refineries and five marketing areas. The production costs per gallon, wholesale price per gallon, and the cost of transporting one gallon from the ith refinery to the jth marketing area are given below, along with production capacities and demands.

Refinery	*Production Cost per Gallon*	*Production Capacity per Day*
1	12	200,000 gal
2	10	300,000 gal
3	14	400,000 gal

Market Area	*Wholesale Price per Gallon*	*Demand per Day*
1	20	200,000 gal
2	24	150,000 gal
3	18	400,000 gal
4	22	100,000 gal
5	16	200,000 gal

Per-unit shipping costs from refinery i to market area j are given as the entries A_{ij} in the following matrix $\mathbf{A}$:

$$\mathbf{A} = \begin{bmatrix} 5 & 4 & 3 & 4 & 3 \\ 3 & 8 & 9 & 5 & 6 \\ 7 & 7 & 3 & 9 & 4 \end{bmatrix}$$

Assume that the company will not ship from refinery i to area j if a net profit cannot be realized.
(a) Suppose we wish to determine the amount to be shipped from refinery i to area j so as to maximize profit while meeting demands. Formulate this problem as a transportation problem by giving an initial transportation tableau with appropriate supplies, demands, and costs.
(b) Find an initial feasible $\mathbf{x}$ by the Vogel advanced start method and perform one iteration of the transportation algorithm to see if this solution is optimal.

7.13 In a national epidemic, 4 cities need a certain vaccine that can be shipped from 3 laboratories. Laboratories 1, 2, and 3 have supplies in the amounts of 15, 25, and 5 units, respectively. The 4 cities need 12, 8, 15, and 10 units, respectively. Travel time between laboratory i and city j is given as the entry T_{ij} in the following matrix $\mathbf{T}$:

$$\mathbf{T} = \begin{bmatrix} 10 & 2 & 20 & 11 \\ 1 & 7 & 9 & 20 \\ 12 & 14 & 16 & 18 \end{bmatrix}$$

Here, time is more important than the cost of shipping, and we seek a shipping schedule so that the maximum shipping time is as small as possible.
(a) Describe a general method for solving similar least-time transportation problems.
(b) Use your method to find a feasible solution $\mathbf{x}$ so that $\max\{T_{ij} | x_{ij} > 0\}$ is as small as possible.

7.14 For a transportation problem, suppose we have found a feasible $\mathbf{x}$. However, due to a misunderstanding by the people who collected the data and provided us with the feasible solution, we only know the costs c_{ij} associated with the positive components in $\mathbf{x}$. This information is given in the following partial transportation tableau:

1^{10}	*	*	*	*	*
*	3^{10}	*	*	*	2^{10}
*	*	*	4^{10}	*	*
*	*	*	*	3^{5}	4^{10}
*	*	2^{10}	*	*	*
*	2^{5}	*	*	4^{10}	*

Here, * indicates that the corresponding x_{ij} equals zero and so the cost is unknown.

The feasible $\mathbf{x}$ indicated above has a cost of 215. In this situation of limited information, find a feasible solution with as small a cost as possible.

7.15 Find an optimal assignment for the assignment problems with the following cost matrices.

(a)

8	2	8	3	2
3	9	2	1	1
7	9	7	7	3
5	3	4	5	1
1	1	4	4	9

(b)

8	11	4	3
9	3	9	8
8	7	3	5
8	3	9	2

7.16 Consider a network flow problem with the per-unit cost of shipping given on each arc and with the node specifications given on the right.

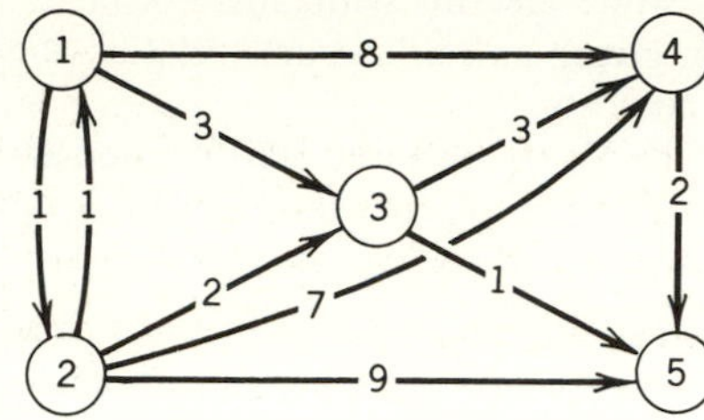

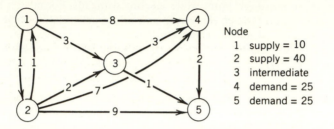

Node
1 supply = 10
2 supply = 40
3 intermediate
4 demand = 25
5 demand = 25

(a) Write down the simplex tableau for this problem when it is formulated as a linear program.
(b) Using the initial feasible spanning tree corresponding to the flow $x_{14} = 10$, $x_{13} = 0$, $x_{24} = 15$, and $x_{25} = 25$, apply the general network flow algorithm to obtain an optimal solution.

7.17 Consider the following network flow problem with the per-unit cost of shipping given on each arc and with the node specifications given on the right.

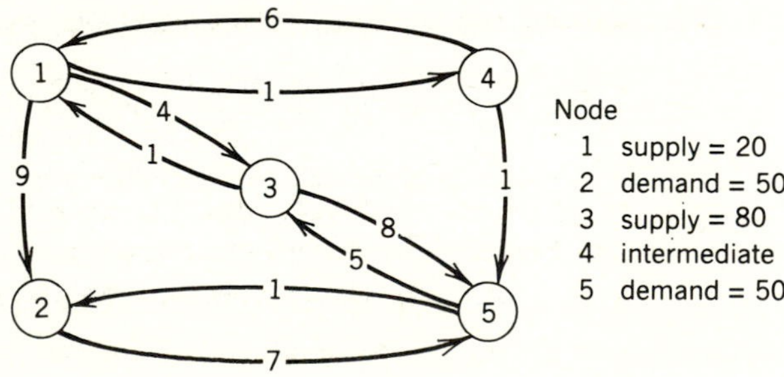

Node
1 supply = 20
2 demand = 50
3 supply = 80
4 intermediate
5 demand = 50

(a) Find an infeasible spanning tree (that is, a spanning tree for which the demands cannot be met using the arcs of the tree).
(b) Starting with the feasible spanning tree corresponding to the flow $x_{12} = 50$, $x_{14} = 0$, $x_{31} = 30$, and $x_{35} = 50$, apply the network flow algorithm to obtain an optimal solution.

7.18 (a) Show that

$$\sum_{i=1}^{n}\sum_{j=1}^{n} y_i x_{ij} - \sum_{i=1}^{n}\sum_{j=1}^{n} y_j x_{ij} - \sum_{i=1}^{n} b_i y_i$$
$$= \sum_{i=1}^{n}\sum_{j=1}^{n} y_i x_{ij} - \sum_{i=1}^{n}\sum_{k=1}^{n} y_i x_{ki} - \sum_{i=1}^{n} b_i y_i$$
$$= \sum_{i=1}^{n} y_i \left(\sum_{j=1}^{n} x_{ij} - \sum_{k=1}^{n} x_{ki} - \sum_{i=1}^{n} b_i \right)$$

(b) Using the result in part (a), show for the general network model of Section 7.7 that the difference between the primal objective function, $\alpha(\mathbf{x})$, and the dual objective function, $\beta(\mathbf{y})$, can be calculated as

$$\alpha(\mathbf{x}) - \beta(\mathbf{y}) = \sum_{i=1}^{n} \sum_{j=1}^{n} (c_{ij} - y_i + y_j)x_{ij} + \sum_{i=1}^{n} y_i \left(\sum_{j=1}^{n} x_{ij} - \sum_{k=1}^{n} x_{ki} - b_i \right)$$

PART TWO

INTEGER, NONLINEAR, AND DYNAMIC PROGRAMMING

CHAPTER 8

INTEGER PROGRAMMING

The linear programming models of Part One embody three important assumptions that are clearly inappropriate in some situations:

- The decision variables must be permitted to take on arbitrary real values.
- The objective and constraint functions are linear.
- The functional relationships are deterministic.

In Part Two we consider models that relax the first two of these assumptions to permit integer constraints and nonlinear functions. In Part Three we shall introduce some models that are appropriate in situations when chance occurrences are important.

In this chapter we introduce **integer programs,** optimization problems in which some or all of the decision variables are restricted to take on only certain discrete values. The discussion will be framed mostly in terms of problems having linear functions, but the general technique outlined here can also be used to solve nonlinear integer programs.

8.1 THE INTEGER PROGRAMMING PROBLEM

Consider the following simple optimization problem, which is quite similar to the linear programming (LP) problems discussed in Part One.

IP: minimize $z(\mathbf{x}) = -x_1 - x_2$
subject to

$$-2x_1 + 2x_2 \leq 1$$
$$16x_1 - 14x_2 \leq 7$$

$$x_1 \geq 0 \qquad x_2 \geq 0$$
$$x_1 \text{ and } x_2 \text{ integers}$$

This problem would be a linear program except for the restriction that the variables take on only integer values. One might therefore suppose that such an integer program could be solved by simply ignoring the integrality requirement, solving the linear program, and rounding off any noninteger solution components to the nearest integers. The linear program that results from ignoring the integer constraint is called the **linear programming relaxation** of the integer program. For the above integer program (IP), the linear programming relaxation has the optimal vector

$$\mathbf{x}^*_{\text{LP}} = [7, 7.5]^T \quad \text{with optimal value} \quad z^*_{\text{LP}} = -14.5$$

Rounding off the noninteger solution component would yield either

$$\hat{\mathbf{x}} = [7, 7]^T \quad \text{with} \quad \hat{z} = -14$$

or

$$\bar{\mathbf{x}} = [7, 8]^T \quad \text{with} \quad \bar{z} = -15$$

Unfortunately, neither of these points is feasible for the inequality constraints. In fact, as shown in Figure 8.1, only four points (⊕) having integer components $[x_1, x_2]^T$ satisfy the inequality constraints, and the true optimal solution to IP,

$$\mathbf{x}^*_{\text{IP}} = [3, 3]^T \quad \text{with} \quad z^*_{\text{IP}} = -6$$

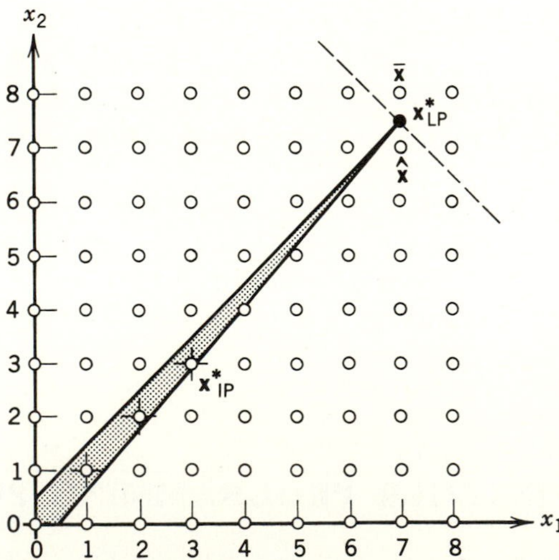

FIGURE 8.1 An integer programming problem.

is far from either rounded linear programming solution in both solution components and objective function value. Thus, the example shows that an integer program cannot in general be solved by simply rounding off the solution of the linear programming relaxation. The linear programming solution does provide some information about the optimal integer solution, however. Adding extra restrictions to an optimization problem makes the feasible set smaller (or perhaps does not change the feasible set). Reducing the feasible set can never result in a better optimal objective function value, and usually makes it worse. Therefore, adding integer constraints to a linear program can never make the solution better, and

> if z^*_{IP} is the minimum value for an integer program,
> and z^*_{LP} is the minimum value for its LP relaxation,
> then $z^*_{\text{IP}} \geq z^*_{\text{LP}}$.

In other words the linear programming solution provides a **lower bound** on the optimal objective value for the integer problem.

Figure 8.1 suggests another naive approach that might seem plausible for solving integer programs. The feasible set for a linear programming problem contains infinitely many points because the variables are allowed to take on arbitrary real values. If the variables must have integer values, however, it may be that there are only a finite number of possible solutions, and then they could be examined one by one in search of the optimal point. In the preceding problem, for example, we could conclude from the inequality constraints that there are no feasible points with either coordinate bigger than 8. Because x_1 and x_2 must be nonnegative, the only integer points that could possibly be feasible for IP are those for which

$$x_1 \in \{0, 1, 2, 3, 4, 5, 6, 7, 8\}$$

and

$$x_2 \in \{0, 1, 2, 3, 4, 5, 6, 7, 8\}$$

Each of these $9 \cdot 9 = 81$ points could be checked to see if it satisfies the inequality constraints. The objective value could then be calculated for each feasible point, and the optimal solution could finally be identified as the feasible point having the lowest objective function value.

This method of solution is called **exhaustive enumeration.** The method requires that we specify the set of integer values that can be taken on by each variable. In the preceding example we could see from Figure 8.1 that any integer point feasible for the inequality constraints must have $x_1 \leq 8$ and $x_2 \leq 8$, and from this observation it was easy to write down the sets of possible values. In most problems it is not so easy, however, because bounds on the variables may not be available by inspection. The inequality constraints of some problems define unbounded regions, in which case it is not possible to state bounds on the variables at all.

Even when it is possible to write down finite sets of possible values for the variables, exhaustive enumeration is computationally intractable for all but the simplest of integer programming problems. Practical integer programs commonly have many variables, and each variable may be capable of taking on many integer values. If a problem had only 10 variables, and if each variable could take on only 9 values, as in our example, the number of points to be examined in an exhaustive enumeration would be $9^{10} = 3486784401$. Thus, even if the calculations are performed on a high-speed computer, exhaustive enumeration is usually out of the question.

A practical algorithm for solving integer programs must somehow consider every possible solution point without explicitly enumerating them all. The next section shows how that can be done by using a sequence of linear programming relaxations.

8.2 IMPLICIT ENUMERATION

The example of Section 8.1 shows that the optimal solution to an integer program cannot, in general, be obtained by simply rounding off the solution of the linear programming relaxation. However, if the linear programming solution happened to have all integer components, it would of course also be optimal for the integer problem. This suggests that we might somehow construct a linear program whose solution is optimal for the integer program we wish to solve. In this section we shall show that this is indeed possible and introduce a systematic method for constructing the required linear program. To illustrate the method, we use the following example.

$$\begin{aligned}
\text{minimize} \quad z(\mathbf{x}) &= 4x_1 - 6x_2 \\
\text{subject to} \quad -x_1 + x_2 &\leq 1 \\
x_1 + 3x_2 &\leq 9 \\
3x_1 + x_2 &\leq 15 \\
x_1 \geq 0 \quad & x_2 \geq 0 \\
x_1 \text{ and } x_2 & \text{ integers}
\end{aligned}$$

It will be convenient to let F denote the feasible set for the linear programming relaxation of the problem; that is, we let F be the set of all nonnegative vectors $\mathbf{x} \in R^2$ satisfying the inequality constraints. Then the linear programming relaxation of the example problem can be concisely written in the following form:

$$\min_{\mathbf{x} \in F} z(\mathbf{x})$$

The optimal solution of this linear program is

$$x_1 = 1.5 \qquad x_2 = 2.5 \quad \text{with} \quad z = -9$$

Figure 8.2 shows a graph of the feasible region F, the solution point $\mathbf{x}^*_{\text{LP}}$ of the LP relaxation, and the objective function contour passing through that point. The figure also shows some of the points (○) in the x_1x_2-plane that have integer values for both of their coordinates; such points are called **lattice points.** Because the solution to the LP relaxation has components that are noninteger, it is obviously not feasible for the integer program IP.

Note that because x_1 must be an integer in the optimal solution to IP, the

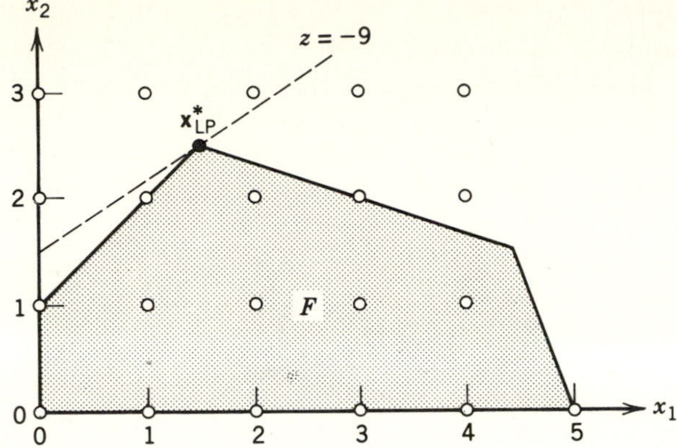

FIGURE 8.2 The linear programming problem $\min_{\mathbf{x} \in F} z(\mathbf{x})$.

solution to IP must satisfy

$$\text{either} \quad \begin{array}{l} \mathbf{x} \in F \\ x_1 \leq 1 \end{array} \qquad \text{or} \qquad \begin{array}{l} \mathbf{x} \in F \\ x_1 \geq 2 \end{array}$$

Moreover, if we construct two new linear programming problems having these constraint sets, the unwanted noninteger solution component $x_1 = 1.5$ will be excluded from the solutions of both linear programs. The two new problems and their solutions are

$$\begin{array}{c} \min z(\mathbf{x}) \\ \mathbf{x} \in F \\ x_1 \leq 1 \end{array} \qquad \begin{array}{c} \min z(\mathbf{x}) \\ \mathbf{x} \in F \\ x_1 \geq 2 \end{array}$$

$$\boxed{\begin{array}{rl} x_1 &= 1 \\ x_2 &= 2 \\ z &= -8 \end{array}} \qquad \boxed{\begin{array}{rl} x_1 &= 2 \\ x_2 &= \frac{7}{3} \\ z &= -6 \end{array}}$$

The constraint sets for these two problems are shown in Figure 8.3, with their solution points (•) and the objective function contours passing through those points. The process of constructing new problems by adding constraints is called **branching,** and because the constraints we added were on x_1 we would say that we **branched** on x_1. As illustrated in Figure 8.3, branching on x_1 has the effect of partitioning the constraint set F into the two subsets

$$F \cap \{\mathbf{x} \mid x_1 \leq 1\} \quad \text{and} \quad F \cap \{\mathbf{x} \mid x_1 \geq 2\}$$

The linear programs that result from the branching are called **subproblems,** as distinct from IP itself, which is referred to as the **master problem.**

The solution to the left-hand subproblem happens to satisfy the integer constraints, and it is therefore feasible for IP. Of course, it would still not be optimal for IP if there were some other integer solution with a lower objective value, so now we must investigate whether some other integer-feasible point might be better.

The optimal values of the subproblems are lower bounds on the objective function over their respective constraint sets. For example, in the right-hand

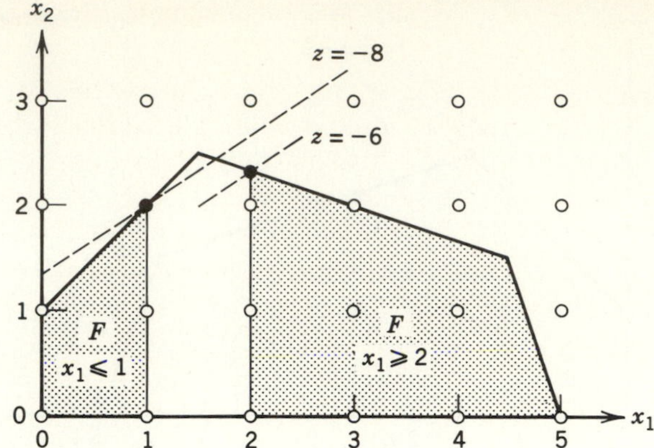

FIGURE 8.3 The subproblems and their solutions.

subproblem no point in the set defined by the constraints $\mathbf{x} \in F$, $x_1 \geq 2$, could possibly yield a better (that is, a more negative) objective value than $z = -6$. Therefore, no integer point in the set could yield a better value than that. Because we already know of an integer solution having a better objective value, namely, $\mathbf{x} = [1, 2]^T$ with $z = -8$, it must be that the optimal solution to IP does not have $x_1 \geq 2$.

The optimal value of the left-hand subproblem is a lower bound on the objective function over the set defined by the constraints $\mathbf{x} \in F$, $x_1 \leq 1$, so no point in that set could possibly yield a better objective value than $z = -8$. Thus no integer point having $x_1 \leq 1$ could be better than the point $\mathbf{x}^* = [1, 2]^T$. Therefore, $\mathbf{x}^*$ must be the optimal solution to IP.

We could just as well have solved the problem by branching on x_2. The following two subproblems result, and their solutions are given in the boxes.

$\min z(\mathbf{x})$ $\mathbf{x} \in F$ $x_2 \leq 2$	$\min z(\mathbf{x})$ $\mathbf{x} \in F$ $x_2 \geq 3$
$x_1 = 1$ $x_2 = 2$ $z = -8$	Infeasible

The infeasibility of the right-hand subproblem shows that an optimal solution to IP cannot have $x_2 \geq 3$. This is consistent with Figure 8.3 in which it is clear that no points in F, integer or otherwise, have $x_2 \geq 3$. The solution to the left-hand subproblem is the best that could possibly be done over the subset of F for which $x_2 \leq 2$, so there are surely no integer solutions in that subset better than $[1, 2]^T$. Thus, we again conclude that $[1, 2]^T$ is optimal for IP.

The method employed in solving this example problem is called **branch and bound** because it proceeds by branching and then reasoning, on the basis of objective function bounds, that some subset of the possible integer solutions cannot contain the optimal point. The process of deciding that a subset of the possible solutions can be excluded from further consideration is called **fathoming.** Fathoming produces the same result as would be obtained by explicitly enumerating all of the possible solutions in the subset and discovering that none of them is optimal, but it does so without actually examining all of the points. For this reason the branch-and-bound method is called an **implicit enumeration** scheme.

In this example it was necessary to solve a linear program to find a lower bound on the objective function over each of the various subsets of possible solutions. However, the logic of branch and bound does not depend on how the lower bounds are obtained, so the method can be used to solve nonlinear integer programs as well as linear ones. The subproblems are then nonlinear programming problems, of course, and must be solved by methods such as those discussed in Chapter 9. In some cases, one of which is described in Section 8.4, the subproblems may be simple enough so that they can be solved by inspection.

8.3 SOLUTION BY BRANCH AND BOUND

As illustrated by the example of Section 8.2, the branch-and-bound method generates a sequence of subproblems that differ from one another only in the bounds on the variables. The solutions of the subproblems are used to deduce that certain subsets of integer points could not be optimal for the original integer program, and these subsets are systematically eliminated from further consideration. The method is sure eventually to generate a subproblem whose solution can be deduced to be optimal for the original integer program.

The Branch-and-Bound Algorithm

To simplify our formal statement of the branch-and-bound algorithm we shall use the notation introduced in Section 8.2, in which the integer program has the following form:

THE GENERAL INTEGER PROGRAMMING PROBLEM

IP: minimize $z(\mathbf{x})$
subject to
$\mathbf{x} \in F$
x_j integer, $j = 1, \ldots, n$

The following statement of the algorithm, using this notation, is just a careful description of the method illustrated in Section 8.2.

A BRANCH-AND-BOUND ALGORITHM FOR GENERAL INTEGER PROGRAMS

0. **Initialize**
   Solve the linear programming relaxation of the original problem; if the solution satisfies the integer constraints, STOP (the solution is optimal for the integer program).

   Find an upper bound z_U on the optimal objective value, equal to the objective value at some point that is feasible for the integer program or $+\infty$ if no such feasible point is known.

1. **Branch**
   Select a remaining subset of feasible solutions (on the first iteration, select F). Select a noninteger component of the solution to the corresponding subproblem,

and partition the subset into two smaller subsets by adding constraints to exclude the noninteger value of the chosen component.

2. **Bound**
   For each new subset obtain a lower bound z_L on the objective value over that subset.
3. **Fathom**
   Examine every subset that might still contain the optimal point, and exclude a subset from further consideration if

   (a) $z_L \geq z_U$
   or (b) the subset has no feasible points
   or (c) z_L is attained at an integer-feasible point in the subset and $z_L < z_U$

   In case (c),

   call the integer-feasible point the **incumbent solution**
   let $z_U = z_L$
   GO TO 3 to see if other subsets can now be fathomed
4. **Test**
   If no subsets remain unfathomed, STOP (the incumbent solution is optimal for IP). Otherwise, GO TO 1.

There are many ways to actually implement the branch-and-bound algorithm, differing in the precise rules for deciding which variable to branch on next, what order to use in following the branches, and how to determine the lower bounds on the objective value. The example of Section 8.2 suggested the possibility of such variations because we were able to get the solution in two ways by branching on different variables. For the expository purposes of this section, it is useful to permit some flexibility by omitting precise implementation details from the statement of the algorithm; thus, both solution processes given in Section 8.2 follow the algorithm given above. However, in designing a computer program to implement the branch-and-bound algorithm for a certain class of problems, it would be necessary to specify in advance exactly what rules to use at each point in the algorithm where a choice is necessary. We shall see in Section 8.4 that special variations in the implementation of the branch-and-bound method can be used to increase its efficiency when the integer program has a special structure, and then also it will be important to use a more precise algorithm statement.

A Branch-and-Bound Example in Detail

To illustrate the branch-and-bound algorithm stated above, we will use it to solve the following example problem:

$$\begin{aligned}
\text{IP: min} \quad & z(\mathbf{x}) = 3x_1 - 7x_2 - 12x_3 \\
\text{subject to} \quad & \\
& -3x_1 + 6x_2 + 8x_3 \leq 12 \\
& 6x_1 - 3x_2 + 7x_3 \leq 8 \\
& -6x_1 + 3x_2 + 3x_3 \leq 5 \\
& x_1, x_2, x_3 \text{ nonnegative integers}
\end{aligned}$$

Here, F is the set of all nonnegative vectors $\mathbf{x} \in R^3$ such that all three linear inequality constraints are satisfied.

Step 0 (initialize).

If we ignore the requirement that each x_j be integer and use the simplex method to solve the resulting linear programming relaxation of IP, we get

$$\mathbf{x} = [0, \tfrac{10}{33}, \tfrac{14}{11}]^T$$

This solution has noninteger components, so it is clearly not optimal for IP, and we cannot stop yet. To establish an upper bound on the objective function value we note that $\mathbf{x} = \mathbf{0}$ is feasible for IP and yields an objective value of $z(\mathbf{x}) = 0$. Thus the minimum value of IP is surely no bigger than $z_U = 0$ because we can do at least that well by simply picking $\mathbf{x} = \mathbf{0}$. We could use $z_U = +\infty$ instead, and the algorithm would still work, but because of the way z_U is employed in the bounding step, it sometimes speeds things up to start with a tighter upper bound. In this case it is very easy to get the tighter upper bound of $z_U = 0$, so we use that value and declare $\mathbf{x} = [0, 0, 0]^T$ to be the corresponding incumbent solution.

Step 1 (branch).

According to the algorithm statement, we can choose either x_1 or x_2 as the variable on which to branch. Arbitrarily selecting x_2, we partition the set of prospective solution points into two parts having, respectively, $x_2 \le 0$ and $x_2 \ge 1$. Because x_2 must be integer, the optimal solution to IP must be in

$$\text{either} \quad F \cap \{\mathbf{x} \mid x_2 \le 0\} \quad \text{or} \quad F \cap \{\mathbf{x} \mid x_2 \ge 1\}$$

Because the variables are constrained to be nonnegative, every point in the left-hand subset has $x_2 = 0$.

Step 2 (bound).

We must now establish lower bounds on the IP objective value over each of the subsets produced by the branch. This can be done by solving the two linear programs that are obtained by appending the extra constraints to the original linear programming relaxation. The resulting linear programs and their solutions are as follows:

$\underset{\mathbf{x}\in F}{\text{minimize}}\ z(\mathbf{x})$ subject to $x_2 \le 0$	$\underset{\mathbf{x}\in F}{\text{minimize}}\ z(\mathbf{x})$ subject to $x_2 \ge 1$
$\mathbf{x} = [0, 0, \tfrac{8}{7}]^T$, $z = -\tfrac{96}{7}$	$\mathbf{x} = [\tfrac{2}{3}, 1, 1]^T$, $z = -17$

Step 3 (fathom).

Checking the fathoming conditions in the algorithm statement shows that neither of the preceding subsets can be excluded from further consideration:

$$\left.\begin{array}{l} -\tfrac{96}{7} \text{ is not more than } z_U = 0 \\ -17 \text{ is not more than } z_U = 0 \end{array}\right\} \Rightarrow \text{cannot fathom by (a)}$$

neither subproblem is infeasible $\Rightarrow$ cannot fathom by (b)

neither subproblem solution is attained at an integer point $\Rightarrow$ cannot fathom by (c)

Step 4 (test).

Both subsets remain unfathomed, so we cannot stop yet and instead return to step 1 of the algorithm.

Each complete pass through steps 1 to 4 of the algorithm is called an **iteration,** so we have now performed one iteration of the algorithm. At this point in the solution process it becomes helpful to track the progress of the algorithm by means of a diagram. Figure 8.4*a* shows the initialization step, the branching, the subproblems and their answers, and the results of the fathoming check performed so far. Solution vector coordinates and objective function values are shown rounded off to the nearest tenth. In working a problem by hand, it is often helpful to organize the process by drawing such a **branching diagram** as the work progresses, and we will complete the diagram started in Figure 8.4*a* as we explain the remainder of the solution process for the example problem.

In a branching diagram the subproblems appear as the nodes of a binary tree, connected by links that show how the branchings are chosen. We shall therefore often refer to a subproblem as a **node** of the branching diagram. When step 3 of the algorithm leads to the fathoming of a subset of possible solutions, in referring to the branching diagram we shall say that the corresponding node has been fathomed. The process of selecting a subset of feasible solutions, referred to in algorithm step 1, corresponds to selecting an unfathomed node to branch from next.

As shown in Figure 8.4*a*, for the example problem the first iteration yields two unfathomed nodes. Suppose we select the left node to begin the second iteration. The subproblem at the left node has the solution $\mathbf{x} = [0, 0, 1.1]^T$, so we must choose x_3 as the variable on which to branch. To exclude the noninteger value, we introduce the additional constraints $x_3 \geq 2$ and $x_3 \leq 1$ to form, respectively, the left and right subproblems shown at the bottom of Figure 8.4*b*.

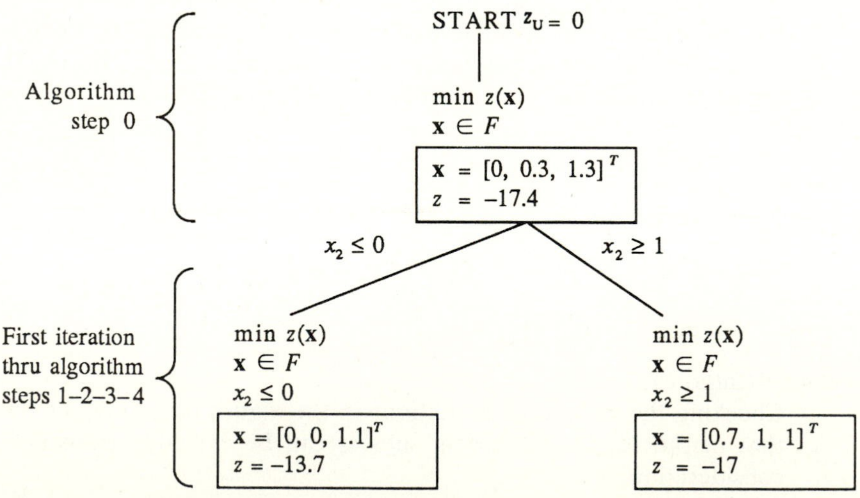

FIGURE 8.4*a* Branching diagram through iteration 1 for the example problem.

The leftmost new node in Figure 8.4*b* is fathomed because the subproblem is infeasible. Recall that we can fathom a node corresponding to an infeasible subproblem because there are no points at all that satisfy both the original constraints and those added in branching. In this subproblem

$$F \cap \{\mathbf{x} \mid x_2 \leq 0, x_3 \geq 2\} = \emptyset$$

Because the constraint set is empty, it certainly cannot contain the optimal point for IP. We therefore exclude this subset of F from further consideration and note on the branching diagram that it has been fathomed by condition (b) of step 3 in the algorithm statement.

The other new subproblem at iteration 2 yields the integer point $\mathbf{x} = [0, 0, 1]^T$ and has $z = -12 < z_U = 0$, so it is fathomed by condition (c). Recall that a node is fathomed when we can deduce that there is nothing to be gained by examining it further. The minimum for that subproblem occurs at the integer point $[0, 0, 1]^T$, so it must be that there are no integer points in that subset having an objective value lower than $z = -12$, and it is thus unnecessary to consider the subset further. If it turns out that the subset contains the optimal point for the integer program, then it must be the point $\mathbf{x} = [0, 0, 1]^T$ that we have already found.

Because an integer point has been found with objective value less than the current upper bound of $z_U = 0$ on the optimal value of the integer program,

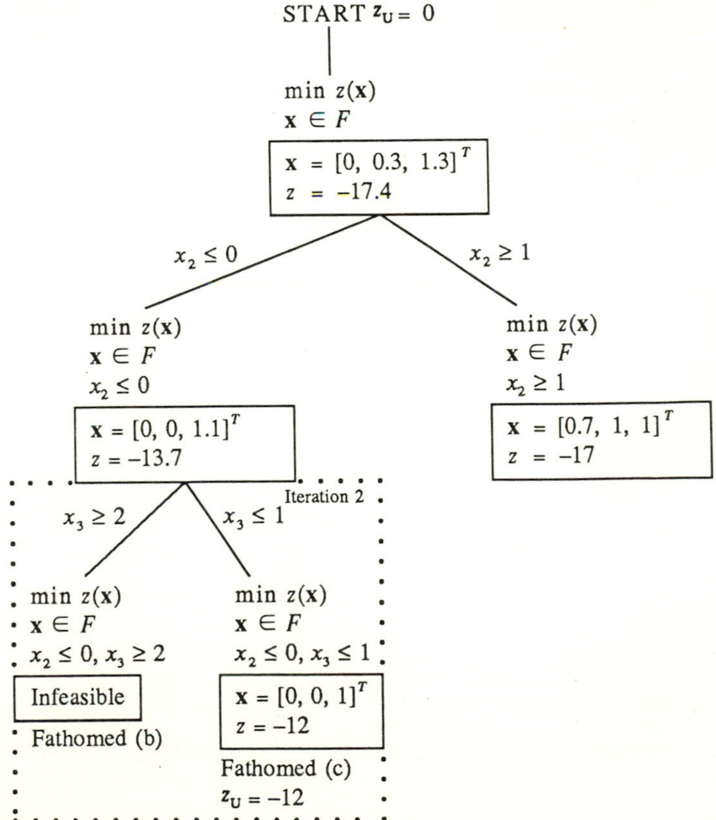

FIGURE 8.4*b* Branching diagram through iteration 2 for the example problem.

we update the bound to $z_U = -12$ and declare $\mathbf{x} = [0, 0, 1]^T$ to be the new incumbent solution. Recall that the minimum value for the integer program cannot be any larger than $z = -12$ because $\mathbf{x} = [0, 0, 1]^T$ is feasible and yields an objective function value of $z = -12$.

At this point it is necessary to check the only other remaining unfathomed node again to see if that node can be fathomed now that z_U has been revised. If this node can now be fathomed, we can stop and declare the current incumbent solution optimal for the integer program. Checking the fathoming conditions, we see that $z = -17 < z_U = -12$, so the node still cannot be fathomed and another branching is required. If there had been other unfathomed nodes, it would have been necessary to check them again also.

As shown in Figure 8.4*c*, the next iteration begins with a branching on x_1 from the single unfathomed node. This yields an infeasible subproblem, which is fathomed by condition (b), and a feasible subproblem having a noninteger optimal point. The node with the feasible subproblem cannot be fathomed because it has $z = -16.8 < z_U = -12$, so another branching is required. Both x_2 and x_3 have noninteger values in the subproblem solution, so we can branch on either of those variables. Arbitrarily selecting x_3, we continue the solution process as shown in Figure 8.4*d*.

One of the iteration 4 subproblems is infeasible, so that node is fathomed. The other new subproblem has a noninteger solution, and it cannot be fathomed

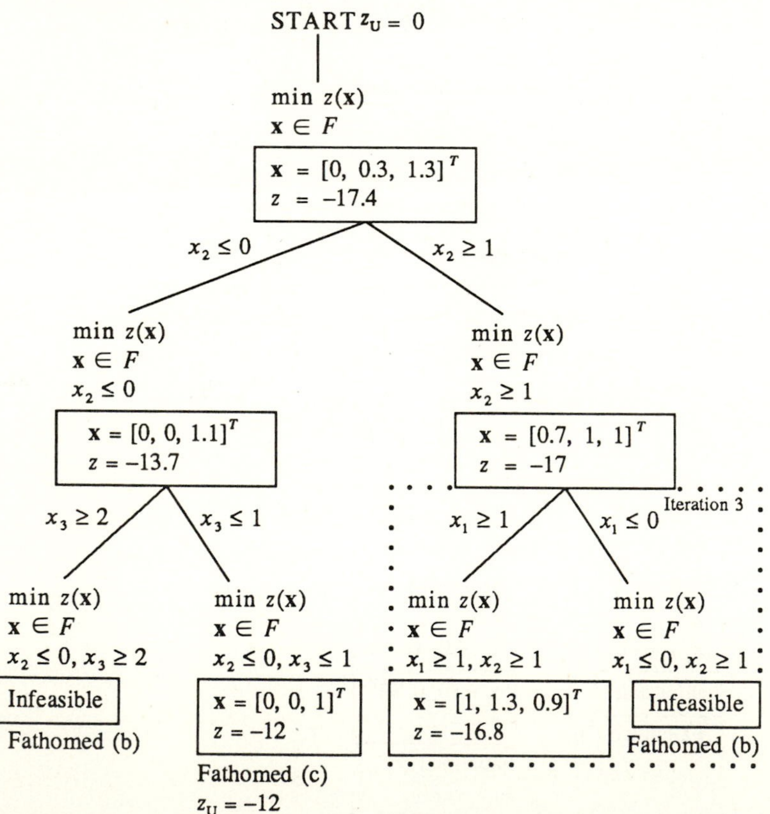

FIGURE 8.4*c* Branching diagram through iteration 3 for the example problem.

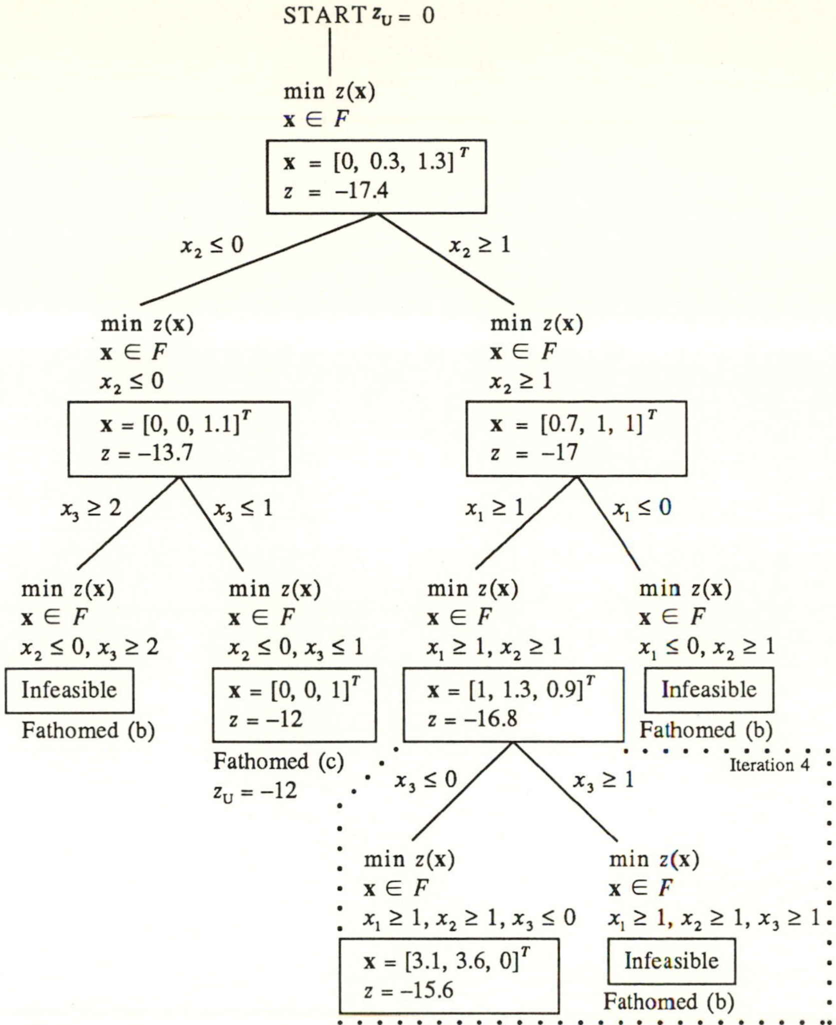

FIGURE 8.4*d* Branching diagram through iteration 4 for the example problem.

because it has $z = -15.6 < z_U = -12$, so another iteration is required, starting with a branch on either x_1 or x_2. Arbitrarily selecting x_2, we continue the solution process as shown in Figure 8.4*e*.

Solving the iteration 5 subproblems shows that one of them is infeasible and the other has $z = -15$ at the integer point $\mathbf{x} = [2, 3, 0]^T$. Thus both new nodes are fathomed and no further branchings are required. Next we find that $z = -15 < z_U = -12$, so we update the upper bound to $z_U = -15$ and make $\mathbf{x} = [2, 3, 0]^T$ the new incumbent solution. Finally the convergence test, step 4 of the algorithm, stops the algorithm because no unfathomed subsets remain. Thus the optimal solution to the integer program is the final incumbent solution $\mathbf{x}_{IP}^* = [2, 3, 0]^T$ with $z_{IP}^* = -15$.

In this example problem the optimal point happened to be the solution to a subproblem generated in the final iteration, but that does not always happen. As we began the last iteration we had no way of knowing whether the current

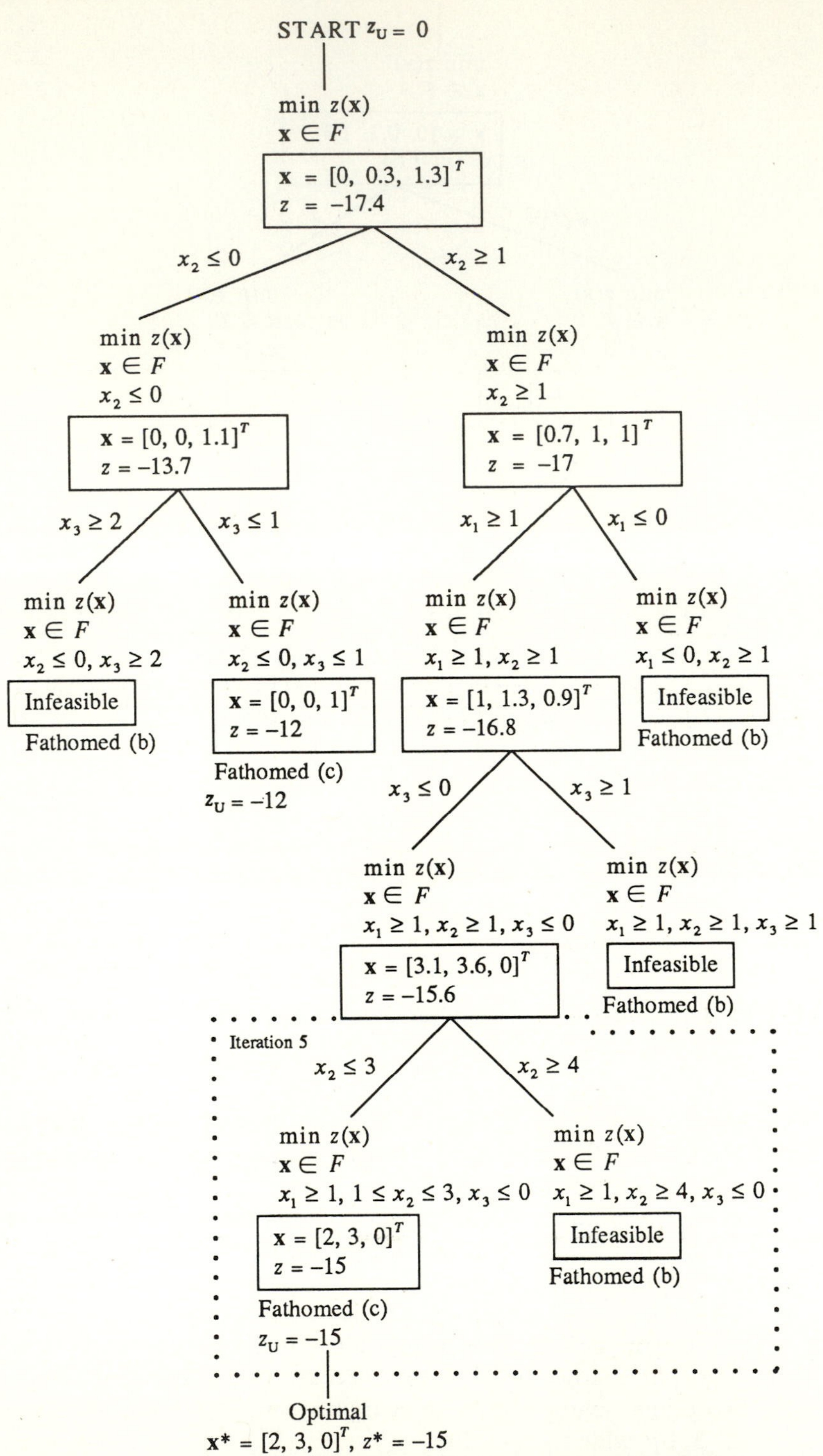

FIGURE 8.4e Complete branching diagram for the example problem.

incumbent solution (it was then $\mathbf{x} = [0, 0, 1]^T$) would turn out to be the optimal solution to the integer program, or whether one of the new subproblems would yield an integer point with a better objective value. In some problems the point that turns out to be optimal is discovered and made the incumbent at an iteration prior to the final one, and the subsequent nodes are all fathomed without discovering a better point. It is still necessary to complete the process and fathom all the nodes, of course, because until that is done we have not ruled out the possibility that the incumbent will be replaced by a better point.

The Order of Selecting Unfathomed Nodes

In the preceding example, there was only one occasion on which we had a choice about which unfathomed node to select for consideration next. That was at the end of the first iteration, shown in Figure 8.4*a*, when we chose the left subproblem to work on first. This choice was arbitrary because algorithm step 1 requires only that we select some unfathomed subset and does not specify how this subset is to be chosen. It would also have been in keeping with the algorithm statement if we had chosen the right subproblem instead. We could then have followed the right branchings part or all of the way down the diagram before returning to the left subproblem that was generated in the first iteration.

Figure 8.4*f* shows the branching diagram that results from following the right-hand branchings as far as possible. When we reach the bottom of the diagram, iteration 4 yields $z_U = -15$ at the incumbent solution $\mathbf{x} = [2, 3, 0]^T$. In completing the fathoming step of iteration 4, we check the one remaining unfathomed node, namely, the left node generated in iteration 1, and discover that it can now be fathomed because $z_U = -15 < z = -13.7$. Thus, using this strategy it turns out to be unnecessary to branch at this node.

Branching to generate all of the nodes at the current depth of the diagram before generating any that are further down, as we did in Figures 8.4*a* through 8.4*e*, is called a **breadth-first** subset selection strategy. Extending the diagram as far down as possible before considering nodes to the left or right, as shown in Figure 8.4*f*, is called a **depth-first** strategy. For a very small problem there might not be any choices in the order of branching. For larger problems it is usually possible to use either a breadth-first or a depth-first strategy, or to switch between them and extend the branching diagram downwards or sideways in whatever pattern seems convenient. For the solution process to remain faithful to the algorithm, it is of course necessary always to solve both new subproblems resulting from a branch before branching further, no matter what node of the diagram is chosen to branch from next.

As illustrated in Figures 8.4*e* and 8.4*f*, the order in which subsets are selected can affect the total number of subproblems to be solved, and thus the work required to solve a problem. It is usually impossible to tell ahead of time what subset selection strategy will work best for any particular problem, but at some points in the solution process it might be possible to make a reasonable guess whether to go sideways or downwards. In the example the right subproblem generated in iteration 1 has a much lower optimal value than the left subproblem, so it would be reasonable to expect that following the right-hand branch might yield an integer point with an objective value low enough to later fathom the left node.

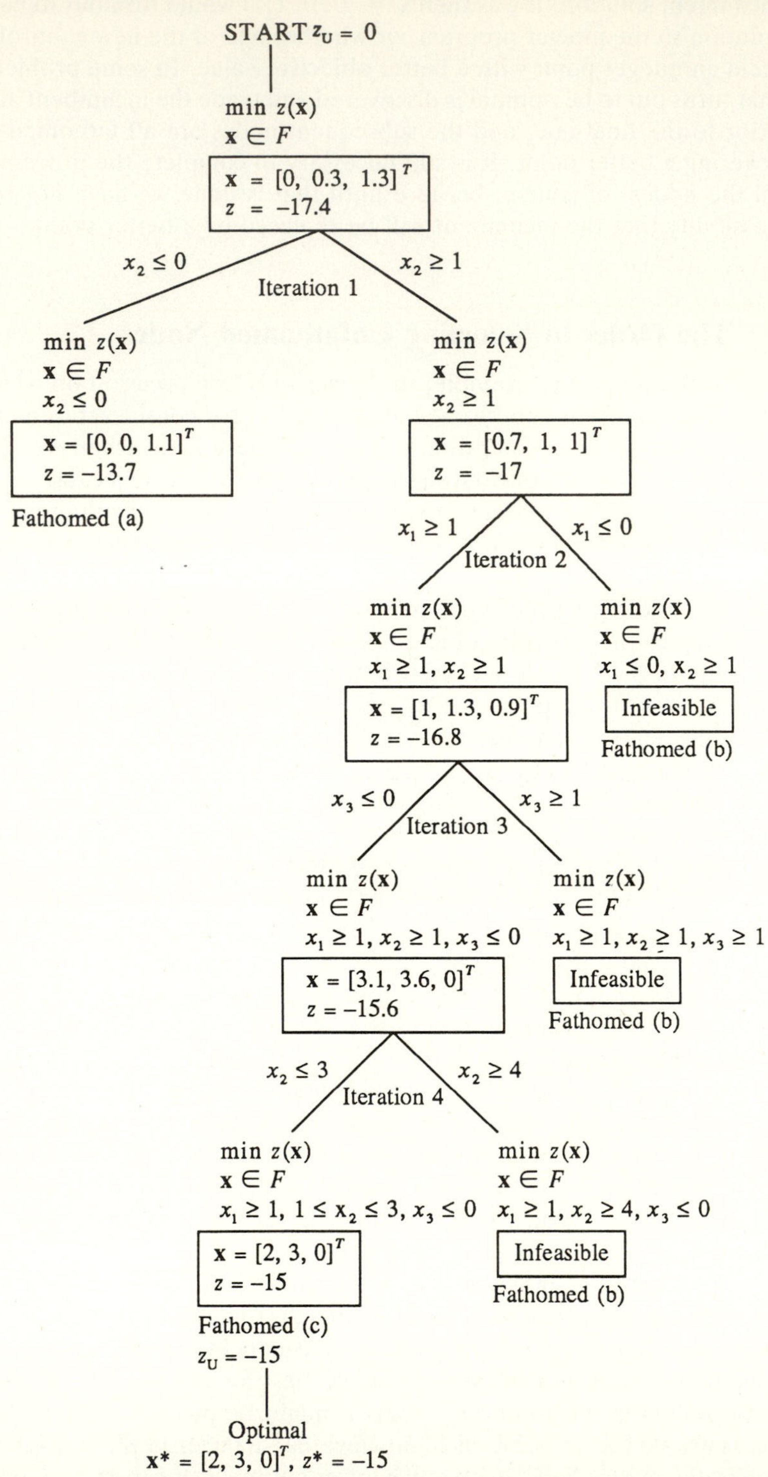

FIGURE 8.4f Complete branching diagram for the example problem, using the depth-first subset selection strategy.

Practical Considerations in Using Branch and Bound

At first it might seem that the branch-and-bound algorithm is rather cumbersome, but the underlying logic is actually quite simple. The preceding exposition examines the workings of the algorithm in minute detail, to clarify the reasoning and convince the reader of its correctness. Once the logic of the algorithm is understood, this degree of introspection is no longer necessary, and the solution process can be performed much more quickly. After a little practice the method can be used to solve small problems quite easily by hand, provided the subproblem solutions are easy to obtain. It is also straightforward to program a computer to implement the algorithm, although some care is required in choosing a sensible data structure to represent the branching diagram.

In many applications it might be satisfactory to stop the branch-and-bound algorithm when a solution has been found that is not optimal but only sufficiently close. Thus, if at some stage in the solution process z_U is only slightly higher than the lowest objective value among the nodes that remain unfathomed, little further improvement can be obtained, and the current incumbent solution might be declared close enough to optimal.

In the example problem we were able to fathom many of the nodes as the solution progressed, so the total number of nodes we needed to consider was quite small. Unfortunately, this does not always happen. Depending on the data of a problem, it can turn out that many branch-and-bound iterations are performed without fathoming many nodes, leading to a dramatic growth in the number of subproblems to be solved. In fact, the total amount of work needed to solve an integer program can grow exponentially with the size of the problem (which depends on the number of variables and the number of values they can take on). This phenomenon is thought to be possible no matter what algorithm is used, and makes integer programming fundamentally more difficult than linear programming. Although the behavior of problems from real applications is never as bad as it could be in the worst case, the proliferation of subproblems imposes a practical limit on the size of problems that can be solved on a computer of given memory capacity and speed.

Usually, most of the computational work in the branch-and-bound algorithm is devoted to solving the subproblems. In the preceding example we ignored this point by simply providing the solutions of the subproblems at each stage, but the work in the bounding step of the algorithm is typically much greater than the work involved in deciding how to branch, in fathoming nodes, and in checking for unfathomed nodes. As we shall see in Chapter 9, this is particularly true when the subproblems are nonlinear programming programs. However, there is also an important special case in which the bounding step is very easy, and that is the subject of the next section.

8.4 0–1 INTEGER PROGRAMS

Many integer programming problems that arise in practical settings have the special property that some or all of their variables are restricted to take on only the values 0 or 1. Such variables are called **0–1 variables,** and, as we shall see in Section 8.6, they often arise naturally in the formulation of problems that involve yes-or-no decisions. If an integer programming problem has only 0–1 variables, the bounding step in the branch-and-bound algorithm can be simplified considerably. Because most of the work of the branch-and-bound algorithm is in the bounding step, this simplification can make the algorithm very much faster.

A 0–1 Example

To see how the branch-and-bound algorithm can be speeded up when an integer program contains only 0–1 variables, consider the following example:

$$\begin{aligned} \text{min} \quad & z(\mathbf{x}) = 2x_1 + 3x_2 + 7x_3 + 7x_4 \\ \text{subject to} \quad & \\ & x_1 + x_2 - 2x_3 - 5x_4 \geq 2 \\ & -x_1 + 2x_2 + x_3 + 4x_4 \geq -3 \\ & x_j = 0 \text{ or } 1, \; j = 1, \ldots, 4 \end{aligned}$$

The first step in the branch-and-bound algorithm of Section 8.3 is to solve the linear programming relaxation of the original problem. We used the linear programming relaxation to get a point that satisfies the linear inequalities but probably has some noninteger components. From among the noninteger variables, we selected one on which to branch at the beginning of iteration 1 of the algorithm. The same approach could be used when the variables take on only the values 0 and 1, but it is much faster to get started by solving a different relaxation of the original problem instead. The relaxation that we use for 0–1 problems ignores the inequalities instead of ignoring the integrality requirement, and yields a point that has integer components but probably does not satisfy the linear inequalities of the original problem. It will be convenient to let I denote the feasible set for this relaxation of the problem; that is, if there are n variables in the problem,

$$I = \{ [x_1, x_2, \ldots, x_n]^T \mid x_j = 0 \text{ or } 1, \; j = 1, \ldots, n\}$$

Then the required relaxation can be concisely written in the following form:

$$\min_{\mathbf{x} \in I} z(\mathbf{x})$$

For the preceding example problem, ignoring the inequalities gives the relaxation below.

$$\begin{aligned} \min z(\mathbf{x}) &= 2x_1 + 3x_2 + 7x_3 + 7x_4 \\ x_j &= 0 \text{ or } 1, \; j = 1 \ldots, 4 \end{aligned}$$

This relaxation is clearly an easy problem to solve. The objective function coefficients happen all to be positive, so the smallest we could possibly make $z(\mathbf{x})$ would be $z(\mathbf{x}) = 0$ when $\mathbf{x} = [0, 0, 0, 0]^T$. If that point were feasible for the inequality constraints, it would be optimal and no further work would be needed. Of course we can see by inspection of the original problem that $\mathbf{x} = [0, 0, 0, 0]^T$ is not feasible. We will see below that a 0–1 problem can always be reformulated in such a way that the objective function has all nonnegative cost coefficients. Thus the initial relaxed problem can always be solved by inspection, and deciding if its solution is optimal for the original problem only requires checking whether $\mathbf{x} = \mathbf{0}$ is feasible for the inequality constraints.

Next we need to find an upper bound on the optimal objective function value. This is also easy to get by inspection. The largest $z(\mathbf{x})$ could possibly be in the example, if each x_j is either 0 or 1, is clearly $2 + 3 + 7 + 7 = 19$ when $\mathbf{x} = [1, 1, 1, 1]^T$, so for our example problem the branch-and-bound process starts with $z_U = 19$. We can always take the starting z_U to be just the sum of the (nonnegative) objective function cost coefficients.

The results of this process are shown at the top of the branching diagram in Figure 8.5*a*. Our goal in applying the branch-and-bound algorithm will be to systematically eliminate from further consideration subsets of I that can be determined not to contain the optimal point for the original problem. This is precisely analogous to the process we used in previous sections in which we eliminated subsets of F known not to contain the optimal point. The only difference is that earlier we generated subproblem solutions in F and checked them for integer feasibility, whereas here we generate subproblem solutions in I and check whether they are also in F. A branch on any of the variables will partition I in the required way, so we arbitrarily begin iteration 1 with a branch on x_1, as shown in Figure 8.5*a*.

Next we must obtain a lower bound on the objective value over each of the subproblem constraint sets. To find a lower bound on the optimal value of the objective function over the subset of I in the right subproblem, we examine the objective function with $x_1 = 1$, which is

$$z(\mathbf{x}) = 2 + 3x_2 + 7x_3 + 7x_4$$

The lowest the objective value could be over the right subset is clearly $z(\mathbf{x}) = 2$, at the point $\mathbf{x} = [1, 0, 0, 0]^T$. This point is infeasible for the inequality constraints, but that is unimportant for fixing a lower bound on the objective value. Requiring $\mathbf{x}$ to satisfy the inequality constraints (that is, to be in F as well as being in I) could never result in a decrease of the objective function value, and no 0–1 point having $x_1 = 1$ could yield a lower objective value than $z_L = 2$. By a similar line of reasoning, it is easy to establish that a lower bound over the left subset is $z_L = 0$ (actually, we already knew that from the initialization step).

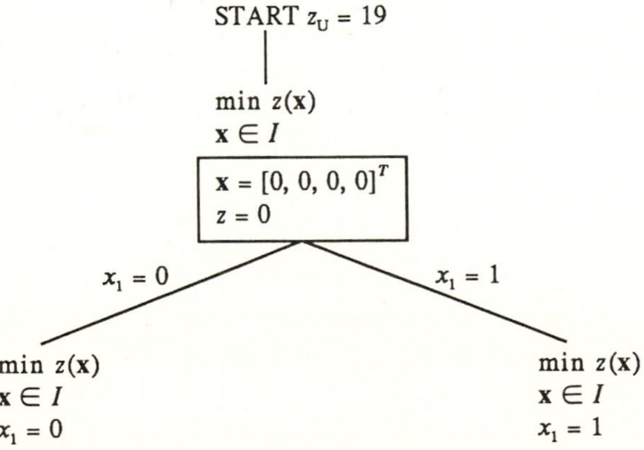

FIGURE 8.5*a* Branching diagram after first branch step for the example problem.

The next step in the algorithm is to see if either of the new nodes can be fathomed. In both cases z_L is less than the current upper bound of $z_U = 19$, so neither node can be fathomed on the basis of condition (a). To check fathoming condition (b), we must determine the feasibility of the subproblems. Before showing how this is done, it is convenient to introduce some new terminology.

All solutions obtained by branching from the left node will have $x_1 = 0$. There are $2^3 = 8$ such solutions, listed here in tabular form:

x_1	x_2	x_3	x_4
0	0	0	0
0	0	0	1
0	0	1	0
0	0	1	1
0	1	0	0
0	1	0	1
0	1	1	0
0	1	1	1

The choice of $x_1 = 0$ is called a **partial solution** because the values of x_2, x_3, and x_4 are not specified. As a reminder of the variables that are yet to be determined, we write the partial solution $x_1 = 0$ as $\mathbf{x} = 0\_\_\_$. When a particular combination of x_2, x_3, and x_4 is assumed to fill the blanks in a partial solution, the resulting vector is called a **completion** of the partial solution. Thus, the entries in the preceding table are all completions of the partial solution $\mathbf{x} = 0\_\_\_$. A completion that has all 0's after the partial solution is called the **zero completion.** Thus, for example, $1\underline{000}$ is the zero completion of the partial solution $1\_\_\_$. A completion that satisfies the inequality constraints is called a **feasible completion.**

Returning to the example, we can fathom the left node in Figure 8.5*a* if the subproblem is infeasible, so we would like to know if there are feasible points in the subset of I having $x_1 = 0$. This is equivalent to the question "are there any feasible completions for the partial solution $\mathbf{x} = 0\_\_\_$?" A naive way of answering this question would be to exhaustively enumerate all of the completions to $\mathbf{x} = 0\_\_\_$, as in the table, and check them in the original inequality constraints. Unfortunately, this usually requires too much work. In Section 8.1 we saw that exhaustive enumeration is not practical for problems of realistic size when the variables can take on arbitrary nonnegative integer values. This is also true in the case of 0–1 integer programs, even though the calculations are somewhat easier because the variable values are all either 0 or 1. Therefore, rather than trying to decide whether a 0–1 subproblem contains any feasible points, we instead merely eliminate subproblems that are easily discovered to be infeasible.

To see how this is possible, consider the inequality constraints of the example problem with $x_1 = 0$ for the left subproblem. With this substitution the constraints are as follows:

$$f_1(\mathbf{x}) = (0) + x_2 - 2x_3 - 5x_4 \geq 2$$

$$f_2(\mathbf{x}) = (0) + 2x_2 + x_3 + 4x_4 \geq -3$$

The largest value that the first constraint function could take on is obtained by setting the variables with positive coefficients to 1 and all the others to zero, giving the completion $\mathbf{x} = 0\underline{100}$ and $f_1(\mathbf{x}) = 1$. Thus, no completion of $\mathbf{x} = 0\_\_\_$ will make $f_1(\mathbf{x}) \geq 2$, so there are no feasible completions to $\mathbf{x} = 0\_\_\_$ and the left node is fathomed by condition (b). In practice we always examine the constraints one at a time, so that it is not necessary to check them all except in the case when none of them show the subproblem to be infeasible.

Next we need to perform the same sort of analysis for the right subproblem. With $x_1 = 1$ the inequality constraints reduce to the following:

$$f_1(\mathbf{x}) = 1 + x_2 - 2x_3 - 5x_4 \geq 2$$

$$f_2(\mathbf{x}) = -1 + 2x_2 + x_3 + 4x_4 \geq -3$$

The largest possible value of the first constraint function is $f_1(\mathbf{x}) = 2$, attained at the completion $\mathbf{x} = 1\underline{100}$. This is not inconsistent with the constraint $f_1(\mathbf{x}) \geq 2$, so we cannot conclude that the right subproblem is infeasible. Checking the second constraint function, we find that its largest possible value is $f_2(\mathbf{x}) = 6$, attained at the completion $\mathbf{x} = 1\underline{111}$, and this is not inconsistent with $f_2(\mathbf{x}) \geq -3$ so once again we cannot conclude that the right subproblem is infeasible. Of course, this does not guarantee that the right subproblem is feasible, because it might be that no completion satisfies both constraints. As noted, that condition would be much harder to check. Fortunately, however, checking for simultaneous feasibility is not required because if it turns out that no completion satisfies all of the constraints, we shall discover that fact at a later iteration of the algorithm. At this iteration all we need to conclude is that the right node cannot be fathomed on the basis that some single constraint is violated by all possible completions.

To finish the fathoming check for the right node, we need to determine whether the zero completion used to obtain the lower bound is feasible for the right subproblem. Checking the constraints, we find that it is not. Thus condition (c) fails as well; the right node of Figure 8.5*a* cannot be fathomed; and another branching is required. Arbitrarily selecting x_2 as the variable on which to branch, we get the new subproblems shown at the bottom of Figure 8.5*b*.

The original problem is reproduced here for reference in discussing the next iteration of the algorithm.

$$\begin{aligned} \text{min} \quad & z(\mathbf{x}) = 2x_1 + 3x_2 + 7x_3 + 7x_4 \\ \text{subject to} \quad & \\ & x_1 + x_2 - 2x_3 - 5x_4 \geq 2 \\ & -x_1 + 2x_2 + x_3 + 4x_4 \geq -3 \\ & x_j = 0 \text{ or } 1, \quad j = 1, \ldots, 4 \end{aligned}$$

After branching to form the new subproblems, the next step is to find a lower bound over each new subset. Remember that at this step we ignore feasibility and simply look for the lowest possible objective value. For the right subproblem, with the partial solution $11\_\_$, the best we could do would be with the completion $11\underline{00}$, for a lower bound of $z_L = 5$. For the left subproblem $z_L = 2$ with the completion $10\underline{00}$.

Neither node can be fathomed on the basis of its lower bound because both lower bounds are less than the current upper bound, which is still $z_U = 19$. We

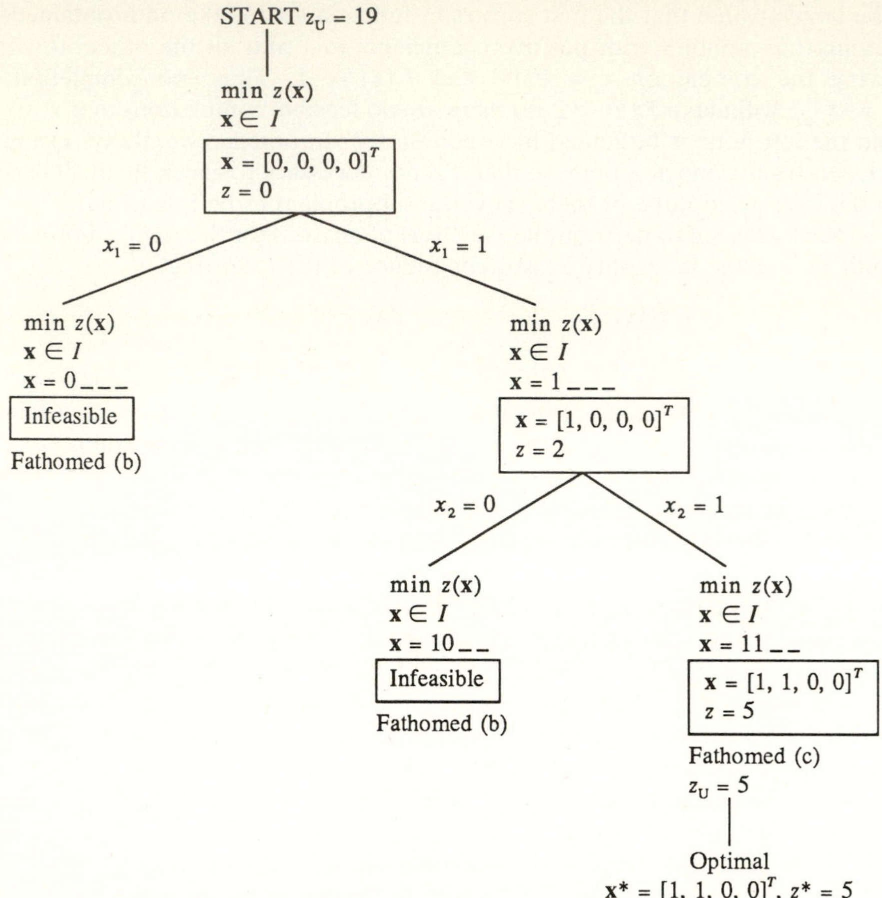

FIGURE 8.5*b* Complete branching diagram for the example problem.

therefore proceed to check whether the subproblems are infeasible. With the partial solution 10__ it is impossible to satisfy the first inequality, so the left subproblem is infeasible, and we fathom that node. Each constraint has feasible completions to the partial solution 11__, however, so we cannot fathom the right node on the basis of subproblem infeasibility. To complete the fathoming check for the right node, we investigate the feasibility of the point yielding the lower bound. It turns out that $\mathbf{x} = [1, 1, 0, 0]^T$ satisfies both inequality constraints, so we declare that point the new incumbent solution and $z_U = 5$ the new upper bound, and fathom the node.

Finally, because no unfathomed nodes remain, we declare $\mathbf{x}^* = [1, 1, 0, 0]^T$ optimal, with $z^* = 5$.

Note that we never attempt in any single step of the algorithm to find a best feasible completion; in checking for infeasibility we pay no attention to the objective function value, and in finding a completion that minimizes the objective value, we pay no attention to whether the resulting solution satisfies the inequality constraints. This makes each step in the branch-and-bound algorithm for 0–1 programs easy enough to be performed by inspection (for problems that are small enough to work by hand) or by trivial arithmetic calculations (when using a computer). In particular, of course, the bounding step is very much easier than in the case of general integer programs.

Looking Ahead

Recall that a node is fathomed when $z_L \geq z_U$; that is, when the lower bound on the objective function for the corresponding subset of possible solutions is equal to or higher than the current upper bound on the optimal objective value for the problem. There is no reason to explore a branch any further if the best solution we could possibly find is no better than a solution we already know (namely, the current incumbent solution). The higher a subset's z_L, the sooner its node is likely to be fathomed, perhaps preventing subsequent branching and thus saving work. It is therefore possible to limit the growth of the branching diagram and speed convergence of the algorithm by establishing tighter (that is, higher) lower bounds z_L than those obtained by using the zero completion. Of course, finding a tighter lower bound requires more work in the bounding step, but sometimes a small amount of extra work results in a significant improvement in speed of convergence. As an illustration of the idea consider the following problem, which differs from the first example introduced in this section only in the sign of one term.

$$\begin{aligned}
\text{min} \quad & z(\mathbf{x}) = 2x_1 + 3x_2 + 7x_3 + 7x_4 \\
\text{subject to} \quad & \\
& x_1 + x_2 - 2x_3 + 5x_4 \geq 2 \\
& -x_1 + 2x_2 + x_3 + 4x_4 \geq -3 \\
& x_j = 0 \text{ or } 1, \ j = 1, \ldots, 4
\end{aligned}$$

As shown in Figure 8.6*a*, changing the "-5" in the original problem to a "$+5$" yields a new problem that branches rather extravagantly when it is solved using the approach described.

The initialization step of the Figure 8.6*a* solution consists of discovering that $\mathbf{x} = [0, 0, 0, 0]^T$ is infeasible and setting $z_U = 19$ to start the branch-and-bound process. Then the branch on x_1 produces the two subproblems shown. For each subproblem it turns out that (a) the lower bound obtained with the zero completion does not exceed the current upper bound $z_U = 19$ on the optimal value for the problem, (b) it is not possible to rule out feasible completions for the partial solution constructed so far, and (c) the zero completion is not itself feasible. Thus all three fathoming conditions fail for both of the subproblems, and neither of the nodes can be fathomed.

Using a breadth-first strategy, the branchings of iterations 2 and 3 (labeled at the branch points in Figure 8.6*a*) produce four new subproblems. For the rightmost subproblem the partial solution is 11__ and the zero completion yields the point $\mathbf{x} = [1, 1, 0, 0]^T$, which is feasible for the inequality constraints with $z_L = 5$. Thus we fathom that node and revise the upper bound to $z_U = 5$. Even with this decrease of z_U, the other nodes cannot be fathomed.

Iterations 4, 5, and 6 yield six new subproblems. Half of these fail the fathoming tests and the others are fathomed because they yield lower bounds that exceed the current upper bound of $z_U = 5$. Even though there is only one variable remaining to branch on, there are still three unfathomed nodes from

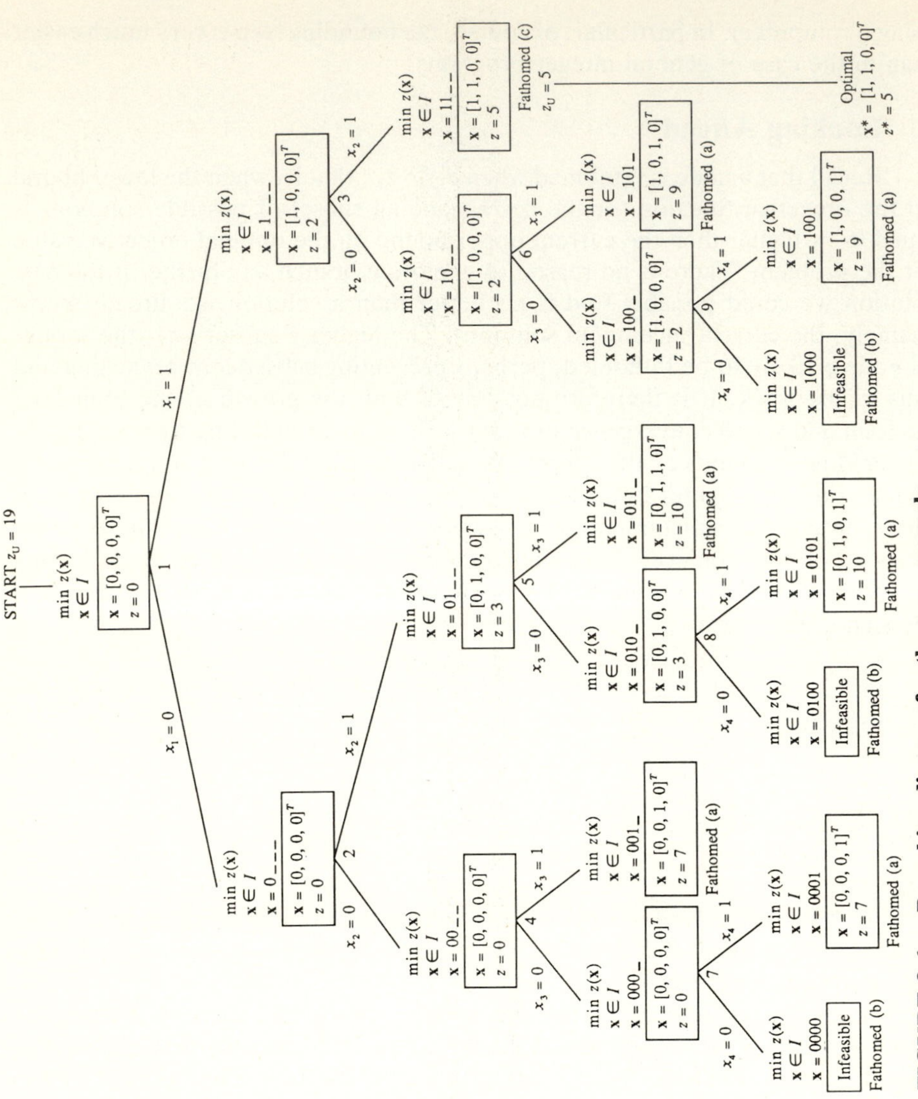

FIGURE 8.6a Branching diagram for the new example.

which new subproblems are produced in iterations 7, 8, and 9. The final six subproblems must of course all be fathomed by one condition or another because no more branching is possible. Half of them turn out to be infeasible, and the others are fathomed because their lower bounds exceed the current upper bound, which is still $z_U = 5$.

Finally, the solution $\mathbf{x}^* = [1, 1, 0, 0]^T$ is declared optimal, but only after 10 out of the $2^4 = 16$ possible solutions, or almost two thirds of them, have been examined. The process also required the construction and solution of 19 subproblems, and thus can hardly be considered preferable to exhaustive enumeration. This example suggests that some improvement to the algorithm is badly needed. Moreover, inspection of Figure 8.6*a* reveals that considerable work was really wasted in examining the solution $\mathbf{x} = [0, 0, 0, 0]^T$ over and over again as the branching diagram was followed down its left side.

Now consider the starting node in Figure 8.6*a* where the point $\mathbf{x} = [0, 0, 0, 0]^T$ was found to be infeasible and a lower bound of $z = 0$ on the optimal value was given. The statement of the problem is reproduced here for easy reference:

$$\begin{array}{lrl} \min & z(\mathbf{x}) = & 2x_1 + 3x_2 + 7x_3 + 7x_4 \\ \text{subject to} & & x_1 + x_2 - 2x_3 + 5x_4 \geq 2 \\ & & -x_1 + 2x_2 + x_3 + 4x_4 \geq -3 \\ & & \\ & & x_j = 0 \text{ or } 1, \quad j = 1, 2, 3, 4 \end{array}$$

At the starting node a higher lower bound than $z = 0$ can be obtained by making the following simple observation.

> Because the point $\mathbf{x} = [0, 0, 0, 0]^T$ is infeasible,
> any feasible point must have at least one $x_j = 1$.

Because all of the cost coefficients c_j are positive, this means that the optimal value must be at least as large as the smallest c_j, namely, $c_1 = 2$. Thus, at the starting node we can use a lower bound of $z_L = 2$ corresponding to the point $\mathbf{x}_L = [1, 0, 0, 0]^T$. This information is recorded at the starting node in Figure 8.6*b*. If $\mathbf{x}_L$ were feasible, we could conclude that it is optimal. However, the point $\mathbf{x}_L = [1, 0, 0, 0]^T$ is not feasible, so we partition the feasible set into two subsets by branching on x_1.

Now consider Figure 8.6*b* again, and the determination of the lower bound z_L for the left subproblem generated at iteration 1 by branching on x_1. The same type of reasoning used in obtaining a higher lower bound at the starting node can be used to obtain a higher lower bound for the subproblems.

To obtain a lower bound over completions of the partial solution $\mathbf{x} = 0\_\_\_$, in Figure 8.6*a* we used the zero completion and calculated

$$z = 2(0) + 3(0) + 7(0) + 7(0) = 0$$

This is surely a lower bound. However, because the zero completion was found to be infeasible at the previous node, any feasible completion to $\mathbf{x} = 0\_\_\_$ must have at least one of the other variables equal to 1. That is, one or more of x_2, x_3, or x_4 must be 1. From this observation and the coefficients c_j of the objective function, we can deduce that any feasible completion to $\mathbf{x} = 0\_\_\_$ will have an

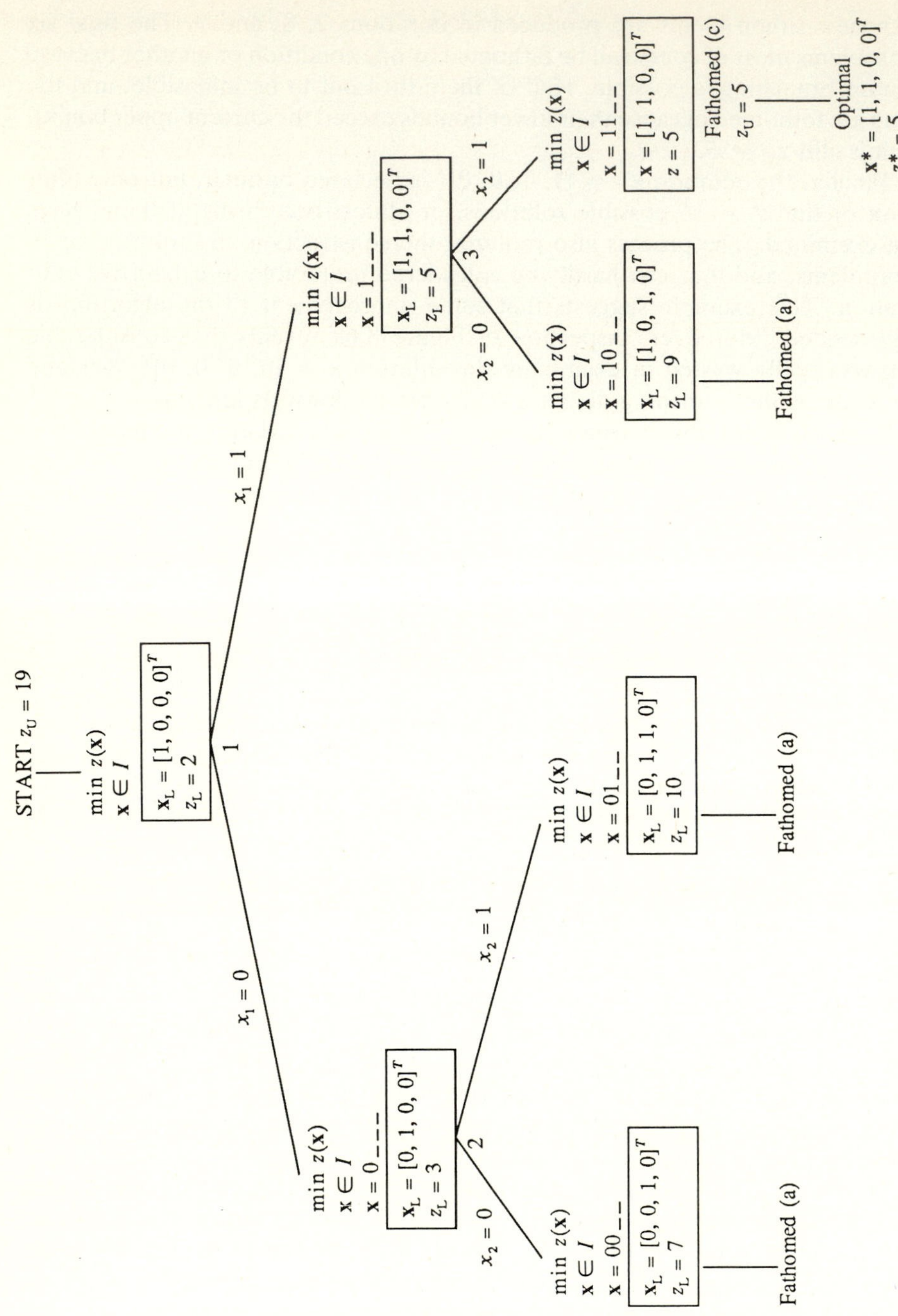

FIGURE 8.6*b* **Branching diagram with look-ahead.**

objective value of at least 3 because $c_2 = 3$ is the smallest objective function coefficient among those corresponding to variables that are not yet in the partial solution.

The same sort of argument can be used to get a tighter lower bound for the right subproblem of Figure 8.6*b*. At the previous node we already discovered that the zero completion $\mathbf{x} = 1\underline{000}$ is infeasible, so once again we know that at least one of x_2, x_3, or x_4 must be 1. The smallest objective coefficient remaining is again $c_2 = 3$, so we could use a lower bound of $z_L = 2 + 3 = 5$.

This reasoning generalizes to the following **look-ahead rule:**

> Because the zero completion is not feasible,
> find the completion having exactly one previously unassigned variable $x_k = 1$, where k is chosen so that c_k is as small as possible,
> and compute the lower bound z_L using that completion.

Figure 8.6*b* shows the complete solution process when this look-ahead rule is used. In previous branching diagrams the lower bound z_L at each node has been the subproblem objective value z, so we have denoted both by that symbol. From now on the lower bounds given in branching diagrams will be denoted z_L when the look-ahead rule gives a value different from the subproblem objective value. The points listed will still be those at which z_L occurs, but will now be denoted $\mathbf{x}_L$ when they differ from the subproblem optimal points $\mathbf{x}$. Now the lower bounds obtained from the subproblems are higher than those used in Figure 8.6*a*, with the effect that nodes are fathomed earlier in the solution process on several occasions. Note in Figure 8.6*b* that when the new upper bound of $z_U = 5$ is obtained at iteration 3, the two nodes generated at iteration 2 can be fathomed because $z_L > z_U$. The result of this improvement to the algorithm is that the branching diagram is considerably smaller and fewer subproblems are solved in finding the optimal point.

The process of constructing the completion required by the look-ahead rule is simple for this example because the c_j's happen to be in increasing order. Whenever this is true, the look-ahead rule can be modified to say "find the completion having exactly one variable $x_k = 1$, where k is the index of the next variable to be included in the partial solution." Because of this simplification, having the c_j's in ascending order makes it a little easier to solve small problems by hand. Of course, if the c_j's are not in increasing order, we can simply rearrange the order of the variables in the problem to make that true. In the formal statement of our branch-and-bound algorithm for 0–1 programming, we shall assume that the c_j's are in increasing order.

How Far To Look Ahead

The success of the simple look-ahead rule suggests that more elaborate rules might improve the basic algorithm further by providing still sharper lower bounds z_L. For instance, if the completion with a single 1 in the leftmost position is infeasible, one could move that 1 to the next position and check that completion, and so forth. If no completions with a single 1 were found to be feasible, one could check completions containing two 1's, and so forth. Of course, it would always be necessary to avoid doing anything approaching an exhaustive enumeration because then the extra work required to solve the subproblems would offset any improvement in the overall performance of the algorithm resulting from having the sharper lower bounds.

The trade-off between doing more work to sharpen the lower bounds and doing more work in solving more subproblems is itself a sort of optimization problem in which the objective is to design an algorithm that solves integer programs as fast as possible. This problem has been the subject of much research, and several methods have been devised (based on far better ideas than the one suggested in the previous paragraph) for quickly finding very good lower bounds. The reader who is interested in learning more about this topic should consult the research literature and the selected references for this chapter that deal specifically with integer programming.

Getting Nonnegative, Increasing Cost Coefficients

We have claimed that any linear 0–1 program can be written in such a way that its objective function coefficients are all nonnegative, and we suggested rearranging the variables, if necessary, to get the objective function coefficients in increasing order. To see how both transformations can be easily accomplished by one set of substitutions, consider the following example.

$$\begin{array}{ll} \text{min} & z(\mathbf{y}) = y_1 + 8y_2 - 3y_3 + 7y_4 - 2y_5 + 8y_6 \\ \text{subject to} & \\ & 5y_1 - y_2 - 2y_3 - 3y_4 + 3y_5 + 2y_6 \geq -4 \\ & -y_1 + 3y_2 - 2y_3 + y_4 - 3y_6 \geq -3 \\ & y_1 + 5y_2 + y_3 - 2y_5 - y_6 \geq 2 \\ & y_j = 0 \text{ or } 1 \end{array}$$

First rearrange the variables so that the objective function coefficients are in increasing order of magnitude, ignoring their signs. This yields the following:

$$\begin{array}{ll} \text{min} & z(\mathbf{y}) = y_1 - 2y_5 - 3y_3 + 7y_4 + 8y_2 + 8y_6 \\ \text{subject to} & \\ & 5y_1 + 3y_5 - 2y_3 - 3y_4 - y_2 + 2y_6 \geq -4 \\ & -y_1 - 2y_3 + y_4 + 3y_2 - 3y_6 \geq -3 \\ & y_1 - 2y_5 + y_3 + 5y_2 - y_6 \geq 2 \\ & y_j = 0 \text{ or } 1 \end{array}$$

If we now let

$$\begin{aligned} y_1 &= x_1 \\ y_5 &= 1 - x_2 \\ y_3 &= 1 - x_3 \\ y_4 &= x_4 \\ y_2 &= x_5 \\ y_6 &= x_6 \end{aligned}$$

the objective becomes

$$z(\mathbf{x}) = x_1 - 2(1 - x_2) - 3(1 - x_3) + 7x_4 + 8x_5 + 8x_6$$

Making the same substitutions in the constraints and simplifying, we get the problem

$$\begin{array}{ll} \min & z(\mathbf{x}) = x_1 + 2x_2 + 3x_3 + 7x_4 + 8x_5 + 8x_6 - 5 \\ \text{subject to} & \\ & 5x_1 - 3x_2 + 2x_3 - 3x_4 - x_5 + 2x_6 \geq -5 \\ & -x_1 + 2x_3 + x_4 + 3x_5 - 3x_6 \geq -1 \\ & x_1 + 2x_2 - x_3 + 5x_5 - x_6 \geq 3 \end{array}$$

$$x_j = 0 \text{ or } 1$$

All of the objective function coefficients are now nonnegative, but this is still a 0–1 program because

$$x_2 = 0 \text{ or } 1 \quad \text{as} \quad y_5 = 1 \text{ or } 0$$

$$x_3 = 0 \text{ or } 1 \quad \text{as} \quad y_3 = 1 \text{ or } 0$$

and the other x_j's are 0 or 1 as the corresponding y_j's are 0 or 1. The constant -5 in the new objective function does not affect the minimization, and we can omit it from our calculations so long as we remember to make the appropriate adjustment in reporting the optimal objective value. However, the constants that are introduced into the constraint inequalities cannot be ignored.

If the variables are in the desired order, the general procedure for removing negative coefficients c_j from the objective function of a linear 0–1 program is simply

> If $c_j < 0$, make the substitution $x_j = 1 - y_j$ in both the objective and the constraints.

The branch-and-bound method can be used to solve 0–1 programs in which the objective function coefficients are not increasing or are not all nonnegative, but the details of the algorithm then become more complicated and more difficult to remember. To simplify the formal algorithm statement given below, we shall always use the preceding technique to transform the problem, if necessary, so that the objective function coefficients are increasing and nonnegative.

The Branch-and-Bound Algorithm for 0–1 Programs

We are now in a position to give a precise statement of the branch-and-bound algorithm for 0–1 integer programming, corresponding to the algorithm for general integer programs given in Section 8.3. After stating the algorithm, we shall illustrate its use by completing the solution of the example introduced in the previous subsection. In this statement of the algorithm, we are assuming that $0 \leq c_1 \leq c_2 \leq \ldots \leq c_n$.

A BRANCH-AND-BOUND ALGORITHM FOR 0–1 INTEGER PROGRAMS

0. **Initialization**
   Check whether $\mathbf{x} = \mathbf{0}$ is feasible; if it is STOP ($\mathbf{x} = \mathbf{0}$ is optimal). Otherwise set the upper bound z_U on the optimal value equal to the sum of the c_j's, set the lower bound on the optimal value to $z_L = c_1$, and set $\mathbf{x}_L = [1, 0, \ldots, 0]^T$. Check whether $\mathbf{x}_L$ is feasible; if it is STOP ($\mathbf{x}_L$ is optimal). Set the iteration counter to $k = 1$.

1. **Branch**
   Select a remaining subset of feasible solutions (on the first iteration, select the feasible set), and partition the subset into two smaller subsets by adding constraints $x_k = 0$ and $x_k = 1$.

2. **Bound**
   For each new subset, set $\mathbf{x}_L$ to the completion having its $(k + 1)$st component equal to 1 and all subsequent components equal to 0, and use $\mathbf{x}_L$ to determine a lower bound z_L on the objective value over that subset.

3. **Fathom**
   Examine every unfathomed subset and fathom it if

   (a) $z_L \geq z_U$
   or (b) there is at least one constraint that cannot be satisfied by any completion in the subset
   or (c) $\mathbf{x}_L$ is feasible

   In case (c),
   declare $\mathbf{x}_L$ the incumbent solution
   let $z_U = z_L$
   GO TO 3 to see if other subsets can now be fathomed

4. **Test**
   If no subsets remain unfathomed, STOP (the incumbent solution is optimal for IP). Otherwise, set $k = k + 1$ and GO TO 1.

We now apply the algorithm to the preceding example where we made the necessary substitutions to get the c_j's increasing and nonnegative. We repeat the reformulated problem here for easy reference.

$$
\begin{array}{ll}
\min & z(\mathbf{x}) = x_1 + 2x_2 + 3x_3 + 7x_4 + 8x_5 + 8x_6 - 5 \\
\text{subject to} & \\
& 5x_1 - 3x_2 + 2x_3 - 3x_4 - x_5 + 2x_6 \geq -5 \\
& -x_1 + 2x_3 + x_4 + 3x_5 - 3x_6 \geq -1 \\
& x_1 + 2x_2 - x_3 + 5x_5 - x_6 \geq 3 \\
& \\
& x_j = 0 \text{ or } 1
\end{array}
$$

The "initialization" consists of noting that $\mathbf{x} = \mathbf{0}$ is infeasible for the third inequality, setting $z_U = 1 + 2 + 3 + 7 + 8 + 8 = 29$, and setting $z_L = 1$ with $\mathbf{x}_L = [1, 0, 0, 0, 0, 0]^T$. A simple check shows that $\mathbf{x}_L$ is not feasible so we set $k = 1$, which completes step 0. Next we branch on x_1, obtaining the two subproblems shown in Figure 8.7*a*.

For the left subproblem we find the lower bound $z_L = 2$ using the com-

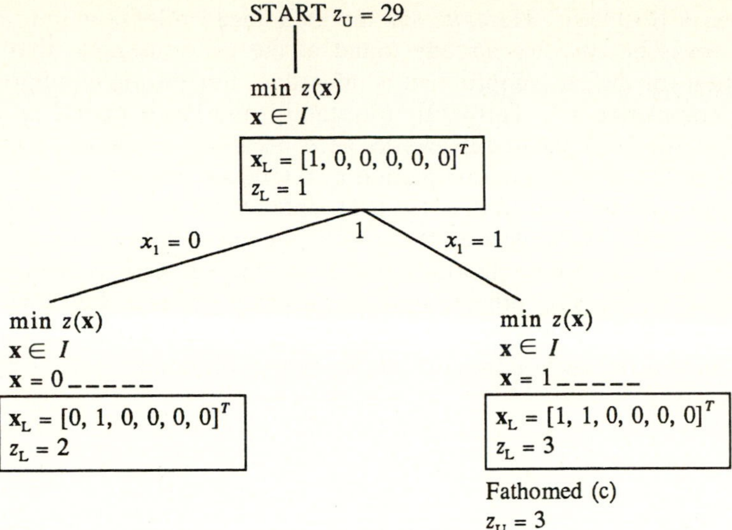

FIGURE 8.7*a* Branching diagram through iteration 1 for the third example.

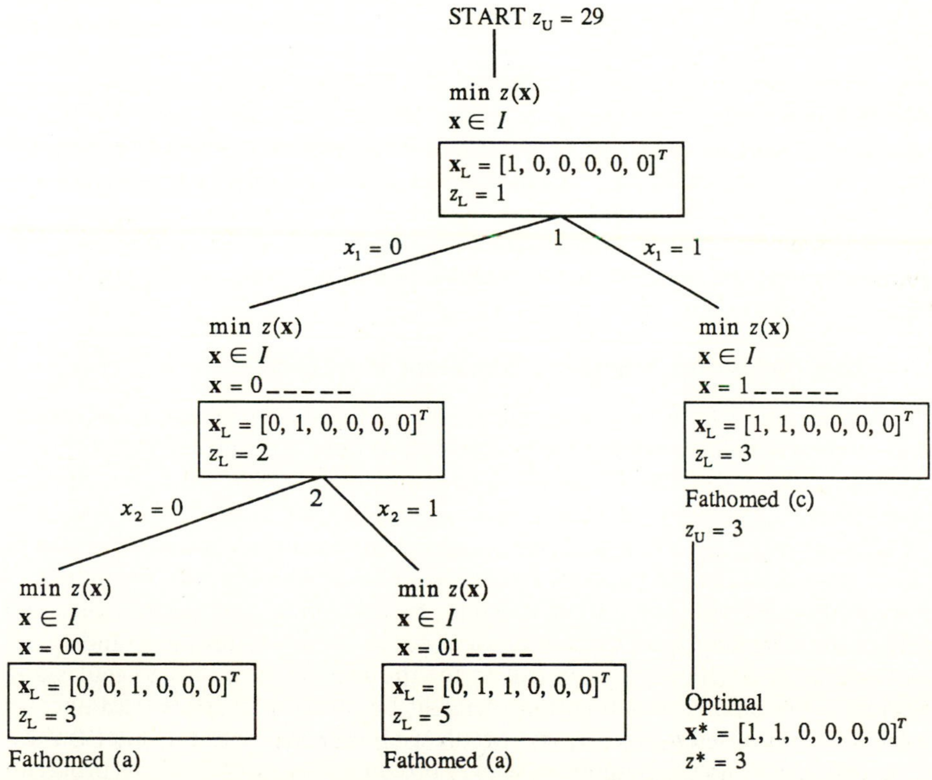

FIGURE 8.7*b* Complete branching diagram for the third example.

pletion $\mathbf{x} = [0\underline{10000}]^T$. Here we see the look-ahead rule in action, and recall that it works because we already found at the previous node that the zero completion for the left subproblem is infeasible. For the right subproblem we use the completion $x = [11\underline{0000}]^T$ to establish the lower bound $z_L = 3$, and recall that the look-ahead rule works here because we already found at the previous node that the zero completion $\mathbf{x} = [1\underline{00000}]^T$ is infeasible.

The point $\mathbf{x} = [1, 1, 0, 0, 0, 0]^T$ is feasible for the right subproblem, so we fathom that node and update z_U to $z_U = 3$. The left node remains unfathomed, so we set $k = 2$, return to algorithm step 1, and branch on x_2 as shown in Figure 8.7*b*. The resulting subproblems are both fathomed because their lower bounds are not less than $z_U = 3$. Testing in algorithm step 4, we find that no unfathomed subsets remain and declare the current incumbent solution $\mathbf{x}^* = [1, 1, 0, 0, 0, 0]^T$ optimal with $z^* = 3$.

8.5 INTEGER PROGRAMMING FORMULATION EXAMPLES

The examples in this section illustrate some applications of integer programming and demonstrate the formulation process. The best way to learn how to formulate integer programming models is to attempt your own formulations and to study formulations made by others. It is therefore important for the reader to be an active participant in the study of these examples, by stopping after the statement of each problem to ponder what the decision variables should be and to attempt a formulation. Only after you have done this should you go on to study the formulation we provide. Every problem seems obvious once its solution is revealed, and the value of the examples will be greatly reduced if you simply read the solutions first. (This advice should sound familiar because we gave it at the beginning of Section 2.2 on linear programming formulations, but it is sufficiently important to bear repetition here.) Like linear programming problems, integer programming problems typically have several alternative formulations that are mathematically equivalent and therefore equally correct.

The Oakwood Furniture Problem Revisited

In Section 2.1 we used a simple resource allocation problem to introduce linear programming models. The problem concerned a furniture company with 12.5 units of wood on hand from which to manufacture tables and chairs. Making a table uses 2 units of wood and making a chair uses 1 unit. The company's distributor will pay $20 for each table and $15 for each chair but will not accept more than 8 chairs and requires at least two chairs per table. We formulated a linear program with continuous variables to determine how many tables and chairs the company should produce so as to maximize its revenue, and found an optimal production program calling for the manufacture of 2.25 tables and 8 chairs. We pointed out then that it might be difficult to sell a fraction of a table, and promised to take up the problem of integer constraints later (in this chapter). Let us now formulate an integer programming model for this problem.

We must decide how many tables and chairs to make, so it is natural to choose as decision variables

x_1 = number of tables to make
x_2 = number of chairs to make

just as we did in formulating the continuous-variable linear programming model. Then the integer programming model is simply

$$\begin{aligned}
&\text{maximize } 20x_1 + 15x_2 \\
&\text{subject to} \\
&\qquad 2x_1 + x_2 \le 12.5 \\
&\qquad 2x_1 \le x_2 \\
&\qquad x_2 \le 8 \\
&\qquad x_j \ge 0 \text{ and integer}
\end{aligned}$$

This is a general integer program that could be solved using the branch-and-bound algorithm of Section 8.3. Unfortunately, as shown by the remainder of the examples in this section, most integer programming formulations require more than just writing down a linear program and appending the constraint that some or all of the variables take on only integer values.

The Knapsack Problem

Suppose some items are to be selected for carrying in a knapsack. There are n items to choose from, and item j has value v_j and weight w_j. How should the items to pack be chosen, from among the n items that are available, so that their total value is maximized while their total weight does not exceed W?

For each item we must decide whether or not it is to be included in the knapsack. Thus the decision variables are

$$x_j = \begin{cases} 1 & \text{if item } j \text{ is included} \\ 0 & \text{if item } j \text{ is left behind} \end{cases}$$

With these integer variables the model becomes

$$\begin{aligned}
&\text{maximize } \sum_{j=1}^{n} v_j x_j \\
&\text{subject to} \\
&\qquad \sum_{j=1}^{n} w_j x_j \le W \\
&\qquad x_j = 0 \text{ or } 1 \qquad j = 1, \ldots, n
\end{aligned}$$

Because the variables are restricted to be 0 or 1, this is a 0–1 integer program of the kind discussed in Section 8.4. Many variations on the knapsack problem have been studied, and there is a voluminous research literature on the subject.

Capital Budgeting

A generalization of the knapsack problem is to decide which subset of a set of possible projects should be undertaken. This type of problem is called a **capital budgeting problem** because it involves budgeting available capital (resources) among competing projects. Suppose that m resources are available and that n different revenue-earning projects are being considered. Not all the projects can be undertaken because not enough resources are available. Let

r_j = revenue returned if project j is undertaken
b_i = total amount of resource i available
a_{ij} = amount of resource i required for project j

Which of the projects should be undertaken to maximize the total revenue returned?

Fractional projects cannot be undertaken; that is, if the exact amount of each resource required for a project is not available, then no revenue can be earned. Thus the decision variables are

$$x_j = \begin{cases} 1 & \text{if project } j \text{ is undertaken} \\ 0 & \text{otherwise} \end{cases}$$

The capital budgeting model then becomes

$$\text{maximize} \sum_{j=1}^{n} r_j x_j$$

subject to

$$\sum_{j=1}^{n} a_{ij} x_j \le b_i \qquad i = 1, \ldots, m$$

$$x_j = 0 \text{ or } 1 \qquad j = 1, \ldots, n$$

Many capital budgeting models have additional constraints on the projects. For example, suppose that project 3 can be undertaken only if project 5 is also undertaken. Because the variable values must be 0 or 1, this additional restriction can be modeled by simply adding the constraint $x_5 \ge x_3$, so if $x_5 = 0$, then $x_3 = 0$ must also hold. Also, it may be that some projects are mutually exclusive. For example, suppose that at most one of the projects 2, 5, or 7 can be undertaken. This requirement is modeled by adding the constraint

$$x_2 + x_5 + x_7 \le 1$$

Facility Location

Suppose that facilities for distributing a product to n customers can be placed at m possible locations. If location i is used as a distribution point, there is a fixed cost F_i for setting up the facility and c_{ij} will be the cost of having location i send one unit of the product to customer j. This cost c_{ij} includes the per-unit operating costs at location i and the per-unit shipping cost from location i to customer j. Customer j demands d_j units of the product. At which of the m possible locations should distribution facilities be placed, and how much should be shipped from each facility to each customer so as to meet the demands and minimize total cost?

To model the selection of facility locations, let

$$y_i = \begin{cases} 1 & \text{if a facility is placed at location } i \\ 0 & \text{otherwise} \end{cases}$$

To model the decision about how much to ship from each facility to each customer, let

$$x_{ij} = \text{amount shipped from location } i \text{ to customer } j$$

We assume that the variables x_{ij} are not constrained to be integer valued.

The main difficulty in formulating the constraints is to model the condition that it is only possible to ship from an existing facility. If a variable $y_i = 0$, then nothing can be shipped from location i to any of the customers. Thus we want

$$\sum_{j=1}^{n} x_{ij} = 0 \quad \text{for those } i \text{ with } y_i = 0$$

However, if $y_i = 1$, then the most that could possibly be shipped out of location i is the total demand of all the customers. Thus the following condition is always satisfied by any shipping program that meets customer demand.

$$\sum_{j=1}^{n} x_{ij} \le \sum_{j=1}^{n} d_j \quad \text{for those } i \text{ with } y_i = 1$$

These two conditions can be modeled with the single set of constraints

$$\sum_{j=1}^{n} x_{ij} \le y_i \sum_{j=1}^{n} d_j \qquad i = 1, \ldots, m$$

along with the conditions that $x_{ij} \ge 0$ and $y_i = 0$ or 1. Note that when $y_i = 0$ this constraint implies that

$$\sum_{j=1}^{n} x_{ij} \le 0$$

and so each $x_{ij} = 0$ because each x_{ij} is nonnegative. Also when $y_i = 1$, the constraint becomes redundant because it is automatically satisfied by any shipping program that meets customer demand. Thus the final model becomes

$$\text{minimize} \quad \underbrace{\sum_{i=1}^{m} \sum_{j=1}^{n} c_{ij} x_{ij}}_{\text{cost of shipments}} + \underbrace{\sum_{i=1}^{m} F_i y_i}_{\text{cost of facilities}}$$

subject to

meet customer demand

$$\sum_{i=1}^{m} x_{ij} = d_j \qquad j = 1, \ldots, n$$

only ship from an existing facility

$$\sum_{j=1}^{n} x_{ij} \le y_i \sum_{j=1}^{n} d_j \qquad i = 1, \ldots, m$$

$$x_{ij} \ge 0 \qquad i = 1, \ldots, m \text{ and } j = 1, \ldots, n$$

$$y_i = 0 \text{ or } 1 \qquad i = 1, \ldots, m$$

This model is an example of a mixed 0–1 program; that is, it contains both continuous and 0–1 variables.

The Traveling Salesman Problem

Consider the problem of a traveling salesman who must visit each of n cities exactly once, minimizing the total cost of travel and returning to the city from which he began. If the cost of travel from city i to city j is c_{ij}, in what order should he visit the cities? This harmless-sounding problem is both subtle of formulation and notorious for its computational difficulty.

It is natural to consider the cities as nodes in a network, connected by links that are the individual trips that make up the salesman's route or **tour.** Because the cost of a tour depends on the link costs c_{ij}, it is helpful to think about the links in the tour when selecting decision variables, rather than about the cities themselves. Let

$$x_{ij} = \begin{cases} 1 & \text{if the tour uses a link from city } i \text{ to city } j \\ 0 & \text{otherwise} \end{cases}$$

The total cost of a tour is then

$$\sum_{i=1}^{n} \sum_{j=1}^{n} c_{ij} x_{ij}$$

In order for the salesman's tour to visit each city exactly once, it must be that he arrives at each city exactly once and leaves each city exactly once. In terms of our decision variables, this means that

$$\sum_{j=1}^{n} x_{ij} = 1 \qquad i = 1, \ldots, n \quad \text{leave each city exactly once}$$

$$\sum_{i=1}^{n} x_{ij} = 1 \qquad j = 1, \ldots, n \quad \text{enter each city exactly once}$$

So far, this is just an assignment problem, and could be solved (quickly) by the transportation algorithm of Chapter 7. Unfortunately, the formulation is not yet complete because it does not guarantee that the tour will be connected. A tour that is not connected is said to consist of **subtours,** as shown in Figure 8.8 for the case of $n = 9$. Of course, many other combinations of subtours could be used to connect the cities in Figure 8.8 while still ensuring that each city is entered exactly once and exited exactly once.

Extra constraints are clearly needed to guarantee that the traveling salesman's tour will be connected rather than a collection of subtours. For example,

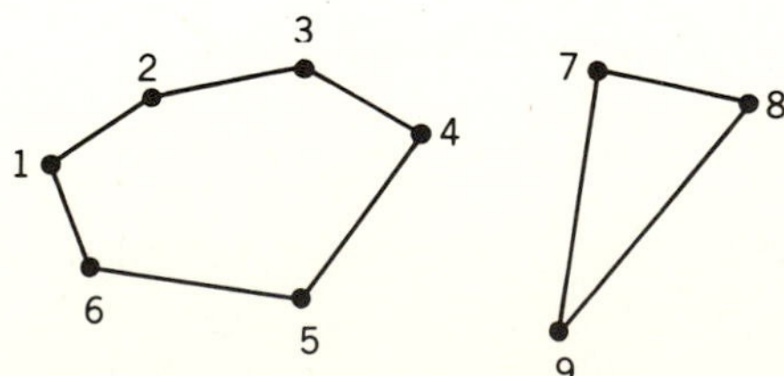

FIGURE 8.8 Subtours in the Traveling Salesman Problem for n = 9.

to rule out the two subtours shown in Figure 8.8, we could include the following constraints:

$$x_{97} + x_{78} + x_{89} \leq 2$$

$$x_{12} + x_{23} + x_{34} + x_{45} + x_{56} + x_{61} \leq 5$$

Such added constraints make the problem no longer an assignment problem, and a large number of them (see Exercise 8.15) would be necessary to rule out all possible subtours in the preceding $n = 9$ example.

It is possible to rule out all subtours in the traveling salesman problem with n cities by introducing $n - 1$ auxiliary variables u_i for $i = 2, \ldots, n$ and including the constraints

$$u_i - u_j + nx_{ij} \leq n - 1 \qquad i = 2, \ldots, n, j = 2, \ldots, n, i \neq j$$

but it is not obvious how this artifice works (see the book by Wagner listed in the selected references at the end of this chapter) and we do not consider it further here.

A Scheduling Problem

A small air freight company provides service between the five cities shown in the route map of Figure 8.9. The route map shows the flights that the company provides, their flight numbers, and (in parentheses) the time in hours that each flight takes, including time in transit and average landing delays. The company guarantees to its customers that each flight on the route map will be flown at least once every day, but does not specify the precise times of the flights. The company employs four pilots. Pilots Able and Baker live in Plattsburgh, while Smith and Jones live in Jamestown.

The company enforces three special policies regarding its flight operations.

- Each day, each pilot flies a sequence of flights starting and ending at his or her home base.

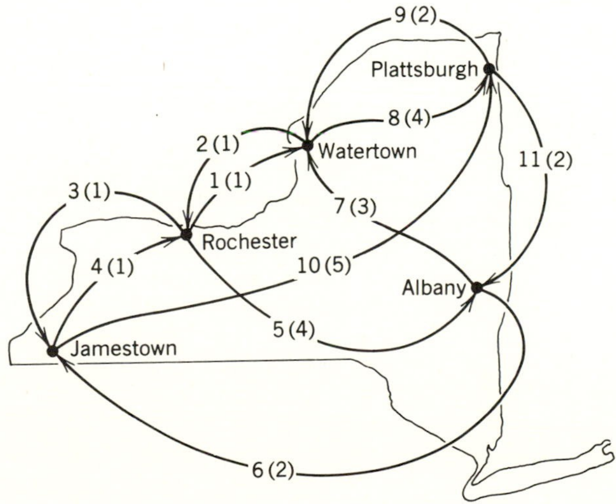

FIGURE 8.9 Air Freight Route Map.

- No pilot is permitted to fly more than 10 hr per day.
- The sequence of flights flown by a pilot must not include any flight more than once. This rules out flight sequences such as 9-2-1-2-1-8.

The freight company pays a pilot $10 per flying hour plus $10 per takeoff. The problem is to assign the pilots to the flights in a way that minimizes the total cost to the company for pilots' pay.

Because of the requirement that each pilot start from and return to his or her home base city, it seems natural to consider the various routes that it would be possible for each pilot to take. For example, either pilot Able or pilot Baker could make the sequence of flights 11-6-10 because that route starts and ends at Plattsburgh and takes less than 10 hr to fly. The cost of that route to the company would be

$$\$10 \cdot (2 + 2 + 5 \text{ hr}) + \$10 \cdot (3 \text{ takeoffs}) = \$120$$

Similarly, either pilot Smith or pilot Jones could make the sequence of flights 10-11-6, which would also cover flights 6, 10, and 11 for the day and would cost the company the same amount as having Able or Baker fly the sequence 11-6-10. Covering the flights and minimizing cost are the only things that matter to the company in choosing between feasible assignments of pilots to flights. The flight sequences 11-6-10 and 10-11-6 are thus equivalent from the perspective of the company and can be thought of as merely different ways of describing a single route to which any of the four pilots could be assigned. The problem of assigning pilots to flights then reduces to the problem of assigning pilots to routes that are feasible for them.

Of course, not all routes could be flown by all pilots. For example, the route consisting of flights 9-2-1-8 does not include Jamestown, and therefore cannot be flown by either Smith or Jones. There are also routes, such as 10-9-8-11-6, that cannot be flown by anyone because they are more than 10 hr long. There are few enough feasible routes that we can explicitly enumerate them all and summarize them in Table 8.1. Having summarized the information in this

TABLE 8.1 Feasible Routes in the Air Freight Network

Route Number, j	*Feasible for Pilots*	*Flights in the Route*	*Cost* c_j
1	A, B	9, 8	80
2	A, B	11, 7, 8	120
3	A, B	9, 2, 1, 8	120
4	A, B, S, J	9, 2, 3, 4, 1, 8	160
5	A, B, S, J	11, 6, 4, 1, 8	150
6	A, B, S, J	11, 6, 10	120
7	A, B, S, J	9, 2, 3, 10	130
8	S, J	4, 3	40
9	S, J	4, 5, 6	100
10	S, J	4, 1, 2, 3	80
11	S, J	4, 1, 2, 5, 6	140
12	S, J	4, 5, 7, 2, 3	150

way, it is natural to choose as decision variables x_j, $j = 1, \ldots, 11$, where

$$x_j = \begin{cases} 1 & \text{if route } j \text{ is used} \\ 0 & \text{if not} \end{cases}$$

With this choice of decision variables, the objective function to be minimized is just the total cost of the routes used, or

$$\sum_{j=1}^{12} c_j x_j$$

Inspection of Table 8.1 shows that if route 3 is used, or route 4, or route 5, or route 10, or route 11, then flight 1 will be flown. Thus, to enforce the requirement that flight 1 occur, we can use the constraint

$$x_3 + x_4 + x_5 + x_{10} + x_{11} \geq 1$$

Repeating this process for the other flights and summarizing the constraints in matrix form, the requirement that each flight occur at least once a day can be expressed as follows.

$$\begin{array}{c} \begin{array}{cccccccccccc} 1 & 2 & 3 & 4 & 5 & 6 & 7 & 8 & 9 & 10 & 11 & 12 \end{array} \\ \text{(route)} \\ \begin{bmatrix} 0 & 0 & 1 & 1 & 1 & 0 & 0 & 0 & 0 & 1 & 1 & 0 \\ 0 & 0 & 1 & 1 & 0 & 0 & 1 & 0 & 0 & 1 & 1 & 1 \\ 0 & 0 & 0 & 1 & 0 & 0 & 1 & 1 & 0 & 1 & 0 & 1 \\ 0 & 0 & 0 & 1 & 1 & 0 & 0 & 1 & 1 & 1 & 1 & 1 \\ 0 & 0 & 0 & 0 & 0 & 0 & 0 & 0 & 1 & 0 & 1 & 1 \\ 0 & 0 & 0 & 0 & 1 & 1 & 0 & 0 & 1 & 0 & 1 & 0 \\ 0 & 1 & 0 & 0 & 0 & 0 & 0 & 0 & 0 & 0 & 0 & 1 \\ 1 & 1 & 1 & 1 & 1 & 0 & 0 & 0 & 0 & 0 & 0 & 0 \\ 1 & 1 & 1 & 1 & 0 & 0 & 1 & 0 & 0 & 0 & 0 & 0 \\ 0 & 0 & 0 & 0 & 0 & 1 & 1 & 0 & 0 & 0 & 0 & 0 \\ 0 & 0 & 0 & 0 & 1 & 1 & 0 & 0 & 0 & 0 & 0 & 0 \end{bmatrix} \end{array} \begin{bmatrix} x_1 \\ x_2 \\ x_3 \\ x_4 \\ x_5 \\ x_6 \\ x_7 \\ x_8 \\ x_9 \\ x_{10} \\ x_{11} \\ x_{12} \end{bmatrix} \geq \begin{bmatrix} 1 \\ 1 \\ 1 \\ 1 \\ 1 \\ 1 \\ 1 \\ 1 \\ 1 \\ 1 \\ 1 \end{bmatrix} \begin{array}{c} \text{(flight)} \\ 1 \\ 2 \\ 3 \\ 4 \\ 5 \\ 6 \\ 7 \\ 8 \\ 9 \\ 10 \\ 11 \end{array}$$

Or, more briefly,

$$Ax \geq e \quad \text{where } a_{ij} = \begin{cases} 1 & \text{if flight } i \text{ is in route } j \\ 0 & \text{if not} \end{cases}$$

and e is the vector of all 1's. In addition to the requirement that all of the flights be flown, we must not use more routes than there are pilots to cover them. Because of the way in which the route numbers are assigned in Table 8.1, we can state these requirements as follows:

at most 2 of the first 3 routes can be used

$$x_1 + x_2 + x_3 \leq 2$$

at most 2 of the last 5 routes can be used

$$x_8 + x_9 + x_{10} + x_{11} + x_{12} \leq 2$$

at most 4 of all 12 routes can be used

$$x_1 + x_2 + \cdots + x_{12} \leq 4$$

8.6 INTEGER PROGRAMMING FORMULATION TECHNIQUES

In Section 8.5 we used examples to illustrate some general approaches to the process of formulating integer programming problems. In this section we elaborate on a few specific techniques that sometimes turn out to be useful in the formulation process.

Writing General Integer Programs as 0–1 Programs

In Section 8.4 we saw that a 0–1 program can be solved with much less computational work than an integer programming problem of comparable size in which the variables are not restricted to take on only the values of 0 and 1. It is therefore sometimes desirable to be able to reformulate a general integer program as a 0–1 problem. To see how this can often be done, consider the following example:

$$\begin{aligned} \text{min} \quad & z(\mathbf{x}) = -x_1 - 3x_2 \\ \text{subject to} \quad & \\ & -x_1 + x_2 \le 10 \\ & x_1 + x_2 \le 14.5 \\ & x_j \ge 0 \text{ and integer} \end{aligned}$$

From the second constraint, we can see that $x_1 \le 14$ and $x_2 \le 14$. Now suppose we represent x_1 and x_2 each as a sum of powers of 2, as follows:

$$\begin{aligned} & \quad\;\; 2^3 \quad\;\; 2^2 \quad\;\; 2^1 \quad\;\; 2^0 \\ x_1 &= 8w_1 + 4w_2 + 2w_3 + 1w_4 \\ x_2 &= 8w_5 + 4w_6 + 2w_7 + 1w_8 \\ & \text{where } w_j = 0 \text{ or } 1 \end{aligned}$$

In this scheme, $w_1, \ldots, w_4$ are the bits of the **binary representation** of x_1, and $w_5, \ldots, w_8$ are the bits of the binary representation of x_2. Thus, for example, a value of $x_1 = 9$ would correspond to $[w_1, w_2, w_3, w_4]^T = [1, 0, 0, 1]^T$ because $9_{10} = 1001_2$. With 4 bits it is actually possible to count up to $2^4 - 1 = 15$, so we can surely count up to the maximum values of 14 required. Substituting the expressions for x_1 and x_2 in the original problem statement yields the following equivalent 0–1 program:

$$\begin{aligned} \text{min} \quad & z(\mathbf{x}) = -(8w_1 + 4w_2 + 2w_3 + w_4) - 3(8w_5 + 4w_6 + 2w_7 + w_8) \\ \text{subject to} \quad & \\ & -(8w_1 + 4w_2 + 2w_3 + w_4) + (8w_5 + 4w_6 + 2w_7 + w_8) \le 10 \\ & 8w_1 + 4w_2 + 2w_3 + w_4 + 8w_5 + 4w_6 + 2w_7 + w_8 \le 14.5 \\ & w_j = 0 \text{ or } 1 \qquad j = 1, \ldots, 8 \end{aligned}$$

This method always works if the original integer program has **bounded variables,** that is, if it can be deduced from the constraints that $x_j \le U_j$ for $j = 1, \ldots, n$.

To rewrite a general integer program as a 0–1 program:

> Find the smallest k so that $2^k - 1 \geq U_1$.
> Write $x_1 = 2^{k-1}w_1 + 2^{k-2}w_2 + \cdots + 2^1 w_{k-1} + 2^0 w_k$.
> Repeat the process for $x_2, \ldots, x_n$ using $U_2, \ldots, U_n$, introducing new variables w_j as required.

If the bounds U_j are too large, it may not be worthwhile to do the conversion because the equivalent 0–1 program may have so many variables that it is actually more difficult to solve than the original, general integer program. In the preceding example the 0–1 program has so many variables compared to the original that it is not clear which would be easier to solve. If the bounds are small, however, the speed of the 0–1 algorithm can often be obtained at the expense of increasing the number of variables by only a modest amount.

Mixed-Integer Programs

It may be convenient (or unavoidable) to use some decision variables that are real and others that are restricted to be integers. An optimization problem in which some of the variables are real and others are integer valued, and having linear objective and constraint functions, is called a **mixed-integer linear program,** or simply a mixed-integer program.

A naive way of dealing with mixed-integer problems is to use the branch-and-bound method for general integer programs but never branch on a real variable. The real variables participate in the linear programming subproblems, of course, so the solution found by the branch-and-bound algorithm is sure to be optimal for the mixed problem.

Enforcing Logical Conditions

Mixed-integer problems often involve logical conditions that can be modeled by switching constraints on and off. To see how a 0–1 variable can be used as a switch to effectively remove a constraint from a problem, suppose that we want to enforce the following conditions:

y is a 0–1 variable

x_1 and x_2 are nonnegative real variables

$x_1 \leq 4.5$

$x_2 \leq 5.5$

$f(\mathbf{x}) = x_1 + x_2 \leq 3.5 \quad$ only if $y = 0$

(if $y = 1$, ignore the constraint)

Because of the fixed constraints on x_1 and x_2, the largest $f(x)$ can ever be is

$$\text{maximum } [f(\mathbf{x})] = 4.5 + 5.5 = 10.$$

Now consider the following two new constraints:

$$f(\mathbf{x}) \leq 3.5 + 0 \cdot (10 - 3.5) = 3.5$$

$$f(\mathbf{x}) \leq 3.5 + 1 \cdot (10 - 3.5) = 3.5 + 6.5 = 10$$

The first constraint is the one to be enforced only when $y = 0$. The second constraint, however, is always satisfied because as stated the fixed constraints $x_1 \leq 4.5$ and $x_2 \leq 5.5$ ensure that $f(\mathbf{x})$ will never be greater than 10. Thus, to achieve the desired effect we can use the single constraint

$$f(\mathbf{x}) \leq 3.5 + y(10 - 3.5) = 3.5 + 6.5y$$

When $y = 0$, the original constraint is in the problem and must be satisfied at the solution, but when $y = 1$ the constraint is modified in such a way that it is always satisfied, so that it is effectively taken out of the problem. In general, the method is:

> Given the constraint $f(\mathbf{x}) \leq b$ to be switched on and off, find a large value F so that $f(\mathbf{x}) \leq F$ holds for all points $\mathbf{x}$ feasible for the problem. Introduce a 0–1 variable y to serve as the switch and replace the constraint $f(\mathbf{x}) \leq b$ with
>
> $$f(\mathbf{x}) \leq b + y(F - b)$$
>
> When $y = 0$, the constraint $f(\mathbf{x}) \leq b$ is enforced (turned on). When $y = 1$, the constraint $f(\mathbf{x}) \leq b$ is replaced by a redundant constraint [so $f(\mathbf{x}) \leq b$ is turned off].

By introducing more than one 0–1 switch variable $y_j \in \{0, 1\}$, more complicated logical conditions can be enforced on constraints. For example,

$$\left.\begin{array}{l} f_1(\mathbf{x}) \leq b_1 \\ f_2(\mathbf{x}) \leq b_2 \end{array}\right\} \text{at least one must hold}$$

is equivalent to

$$\left.\begin{array}{l} f_1(\mathbf{x}) \leq b_1 + y_1(F_1 - b_1) \\ f_2(\mathbf{x}) \leq b_2 + y_2(F_2 - b_2) \end{array}\right\} y_1 + y_2 \leq 1$$

In order for $y_1 + y_2 \leq 1$, it must be that either $y_1 = 0$ (in which case the first constraint is enforced), or $y_2 = 0$ (in which case the second constraint is enforced), or both.

Another use of switch variables is to let a variable take on one of a finite number of real values. Thus, for example,

$$x \in \{1.4, 2.7, 4.2, 5.6\}$$

is equivalent to

$$x = 1.4y_1 + 2.7y_2 + 4.2y_3 + 5.6y_4$$
$$\text{with } y_1 + y_2 + y_3 + y_4 = 1$$

The constraint on the y_j's ensures that only one of them will be 1, and x will be equal to the coefficient of the y_j that is 1.

In addition to the techniques outlined, many other ways might be found to use integer variables in modeling logical conditions that arise in practical problems. In devising any such scheme, it is very desirable to preserve the linearity of the mathematical programming problem because the solution techniques that we have discussed so far rely on that property. Thus, for example, a construction such as

$$y\cdot[f(\mathbf{x}) - b] \leq 0$$

would switch the constraint $f(\mathbf{x}) \leq b$ on for $y = 1$ and off for $y = 0$, but would introduce a nonlinearity because of the multiplication of variables in $y\cdot f(\mathbf{x})$. Nonlinear optimization problems are much more difficult to solve than linear ones, and nonlinear integer or mixed-integer problems are especially difficult, so it is important to avoid nonlinearities if at all possible. Sometimes, considerable ingenuity may be required to achieve that goal.

SELECTED REFERENCES

Balas, E., An Adaptive Algorithm for Solving Linear Programs with Zero-One Variables, *Operations Research* **13:**517–546 (1965).

Bradley, S. P., Hax, A. C., and Magnanti, T. L., *Applied Mathematical Programming,* Addison-Wesley, Reading, MA, 1977, Chapter 9.

Conley, W., *Computer Optimization Techniques,* Petrocelli, New York, 1980.

Garfinkel, R., and Nemhauser, G., *Integer Programming,* Wiley, New York, 1972.

Geoffrion, A., Integer Programming by Implicit Enumeration and Balas' Method, *SIAM Review* **9:**178–190 (1967).

Geoffrion, A., and Marsten, R., Integer Programming Algorithms: A Framework and State-of-the-Art Survey, *Management Science* **18:**465–491 (1972).

Hillier, F. S., and Lieberman, G. J., *Introduction to Operations Research,* Holden-Day, San Francisco, 1980, Chapter 18.

Nemhauser, G., and Ullman, Z., A Note on the Generalized Lagrange Multiplier Solution to an Integer Programming Problem, *Operations Research* **16:**450–453 (1968).

Wagner, H. M., *Principles of Operations Research,* Prentice-Hall, Englewood Cliffs, NJ, 1969, Chapter 13.

EXERCISES

8.1 For a 0–1 integer program, checking all of the integer points immediately adjacent to an LP solution is the same as exhaustive enumeration. Why? In a 0–1 problem with 37 variables, how many lattice points are adjacent to an LP solution?

8.2 Review the assignment problem discussed in Chapter 7 and explain how it can be formulated as an integer linear programming problem. Why can't all integer linear programs be solved as assignment problems by using the transportation algorithm?

8.3 Use the branch-and-bound algorithm to solve the example of Section 8.1.

8.4 Re-solve the example of Section 8.3 (see Figure 8.4*d*) by
(a) branching on x_1 instead of x_2 in iteration 5
(b) branching on x_2 instead of x_3 in iteration 4
(c) branching on x_3 instead of x_2 in iteration 1

8.5 Use the branch-and-bound algorithm to solve the following integer programming problems:

(a) minimize $z(\mathbf{x}) = -x_1 - 2x_2$

subject to

$$-x_1 + x_2 \le 10$$
$$15x_1 + 16x_2 \le 240$$
$$x_1 \text{ and } x_2 \text{ nonnegative integers}$$

(b) minimize $z(\mathbf{x}) = -3x_1 - 3x_2 + 13x_3$

subject to

$$-3x_1 + 6x_2 + 7x_3 \le 8$$
$$6x_1 - 3x_2 + 7x_3 \le 8$$
$$x_j \ge 0 \text{ and integer} \qquad j = 1, \ldots, 3$$

(c) minimize $z(\mathbf{x}) = -3x_1 - 5x_2 - 7x_3$

subject to

$$-3x_1 + 6x_2 + 7x_3 \le 8$$
$$6x_1 - 3x_2 + 7x_3 \le 8$$
$$x_j \ge 0 \text{ and integer} \qquad j = 1, \ldots, 3$$

8.6 In applying the branch-and-bound algorithm, can it ever happen that after a branch both subproblems are infeasible? If yes, construct an example; if no, explain why not.

8.7 If an integer programming problem has two optimal vectors, $\mathbf{x}$ and $\mathbf{y}$, must the optimal vectors satisfy the property that for at least one index j, $x_j - y_j = +1$ or -1? If yes, explain why; if no, construct a counterexample.

8.8 Find all the optimal points for the following integer program.

minimize $z(\mathbf{x}) = -15x_1 + 10x_2$

subject to

$$x_1 + x_2 \le 13$$
$$17x_1 - 12x_2 \le -12$$
$$x_j \text{ integer}, \quad j = 1, \ldots, 3$$

8.9 Use the algorithm of Section 8.4 to solve the following 0–1 programs. Be sure to show the fathoming conditions.

(a) minimize $-2x_1 - 10x_2 - x_3$
subject to

$$
\begin{aligned}
5x_1 + 2x_2 + x_3 &\le 7 \\
2x_1 + x_2 + 7x_3 &\le 9 \\
x_1 + 3x_2 + 2x_3 &\le 5 \\
x_j = 0 \text{ or } 1 &\qquad j = 1, \ldots, 3
\end{aligned}
$$

(b) minimize $10x_1 + 2x_2 - 11x_3$
subject to

$$
\begin{aligned}
2x_1 - 7x_2 + x_3 &\le 2 \\
5x_1 - 8x_2 + 2x_3 &\le 1 \\
x_j = 0 \text{ or } 1 &\qquad j = 1, \ldots, 3
\end{aligned}
$$

(c) minimize $2x_1 + 4x_2 - 5x_3 + 7x_4$
subject to

$$
\begin{aligned}
-x_1 - 2x_2 - 3x_3 - 3x_4 &\ge -8 \\
-2x_1 + 3x_2 + x_3 + 2x_4 &\ge 2 \\
x_j = 0 \text{ or } 1 &\qquad j = 1, \ldots, 4
\end{aligned}
$$

(d) minimize $0x_1 + 4x_2 + 5x_3 + 7x_4$
subject to

$$
\begin{aligned}
x_1 - 2x_2 - 3x_3 + 3x_4 &\le 5 \\
2x_1 - 3x_2 + x_3 - 2x_4 &\le -1 \\
x_j = 0 \text{ or } 1 &\qquad j = 1, \ldots, 4
\end{aligned}
$$

(e) maximize $-2x_1 - 4x_2 - 6x_3 - 8x_4$
subject to

$$
\begin{aligned}
x_1 + 2x_2 - x_3 + x_4 &\le 5 \\
2x_1 - x_2 - x_3 &\le -2 \\
x_j = 0 \text{ or } 1 &\qquad j = 1, \ldots, 4
\end{aligned}
$$

(f) minimize $5x_1 - 2x_2 + 3x_3 + 6x_4$
subject to

$$
\begin{aligned}
2x_1 - x_2 + 3x_3 - x_4 &\ge -2 \\
-x_1 + 2x_3 + x_4 &\ge 2 \\
x_j = 0 \text{ or } 1 &\qquad j = 1, \ldots, 4
\end{aligned}
$$

(g) minimize $2x_1 + 3x_2 - 4x_3 + 7x_4$
subject to

$$
\begin{aligned}
x_1 - 2x_2 + x_3 - 4x_4 &\ge 1 \\
-x_1 + 2x_2 - 2x_3 + x_4 &\ge -1 \\
x_j = 0 \text{ or } 1 &\qquad j = 1, \ldots, 4
\end{aligned}
$$

(h) minimize $-x_1 + 2x_2 - 3x_3 - x_4 + 2x_5$
subject to

$$
\begin{aligned}
-3x_1 + 2x_2 + x_3 + x_4 &\ge 1 \\
-x_1 - 3x_2 + x_3 + x_4 + x_5 &\ge -1 \\
x_j = 0 \text{ or } 1 &\qquad j = 1, \ldots, 5
\end{aligned}
$$

8.10 Consider the integer version of the Oakwood Furniture Company problem formulated in Section 8.5.
(a) Use the branch-and-bound algorithm of Section 8.3 to find an optimal integer production program.
(b) Comment on the relative merits of continuous-variable linear programming models and integer programming models. Should an integer model be used in every instance where the decision variables really take on only integer values?

8.11 Fire stations are to be established providing service to six different areas of a city. Eight possible locations can be used, providing the area coverages given in the following table:

Location	*Areas Served*
A	1, 2, 6
B	1, 2, 3, 4
C	5, 6
D	3, 4, 5
E	1, 3, 5
F	1, 4, 6
G	2, 3, 5
H	1, 4, 5, 6

(a) Formulate a 0–1 program whose solution will tell which locations to use so as to cover all the areas with the fewest fire stations.
(b) Find an optimal choice of locations.

8.12 Six different items are being considered for shipment. The weight and value of each item is given in the following table:

Item	*Weight*	*Value*
1	10	5
2	9	2
3	15	7
4	2	4
5	11	1
6	6	6

Formulate and solve a 0–1 integer program whose solution will give the shipment of maximum value when the total weight of the shipment can be no larger than 33.

8.13 An insurance company is considering the purchase of new high-, medium-, and low-capacity computers. The purchase price per machine would be \$3.5 million, \$2.5 million, and \$1.5 million, respectively. The maximum commitment the company can make for these purchases is \$15 million. It is estimated the net annual profit utilizing these computers would be \$350,000 per high-capacity, \$280,000 per medium-capacity, and \$120,000 per low-capacity machine. There will be 9 trained operators available, and 1 will be required for each new computer. If only low-capacity computers are purchased, the maintenance facilities will only be able to

handle 15 new computers. Each medium-capacity computer has maintenance requirements equivalent to 1.25 low-capacity computers, and the maintenance requirements of each high-capacity computer are equivalent to those of 1.75 low-capacity computers. How many computers of which sizes should be purchased to maximize net profit after maintenance expenses?
(a) Formulate an integer linear programming model for this problem.
(b) Use the branch-and-bound algorithm to solve the problem.

8.14 A gear train is to be designed consisting of 4 gears having x_1, x_2, x_3, and x_4 teeth. Gear 1 engages gear 2, which is fastened to a common axle with gear 3. Gear 3 engages gear 4. When gear 1 is turned through angle a, gear 4 turns in the same direction through angle b. The actual gear ratio b/a is to be made as close as possible to the desired value r, by selecting x_1, x_2, x_3, and x_4 from the set of available gear sizes. Gears are available in 21 sizes, having 16, 18, . . . , $(16 + 2n)$, . . . , 56 teeth. Formulate a nonlinear integer program to determine the best values for x_1, x_2, x_3, and x_4.

8.15 In Section 8.5 it is shown how added constraints can rule out subtours in the traveling salesman problem.
(a) What conditions on the variables x_{ij} can be imposed to eliminate subtours consisting of a simple loop from a city to itself?
(b) For the case of $n = 5$ cities, how many ways can the five cities be partitioned into subtours, not counting the simple subtours ruled out by the constraints in part (a)?
(c) What is the smallest number of constraints that can be added to those of part (a) to rule out all subtours in the case of $n = 5$ cities?

8.16 Reformulate the following integer program as a 0–1 program having each objective function coefficient nonnegative.

minimize $2x_1 - 3x_2$

subject to

$$x_1 + x_2 \leq 6$$

$$x_1 \geq 0 \qquad x_2 \geq 0 \qquad x_j \text{ integer} \quad j = 1, 2$$

8.17 Reformulate the following problems as mixed-integer linear programs.

(a) minimize $2x_1 - 7x_2$

subject to

$$0 \leq x_1 \leq 10 \qquad 0 \leq x_2 \leq 10$$

and at least one of the following must hold

$$-2x_1 + 3x_2 \geq 0$$

$$5x_1 - 4x_2 \geq 0$$

(b) minimize $\mathbf{c}^T\mathbf{x}$

subject to

$$\mathbf{Ax} = \mathbf{b} \qquad \mathbf{x} \geq \mathbf{0} \qquad x_1 \in \{r_1, r_2, \ldots, r_9\}$$

(c) minimize $3x_1 + 2x_2 - 5x_3$

subject to

$$0 \leq x_1 \leq 5 \qquad x_2 \in \{2, 3, 6\} \qquad 0 \leq x_3 \leq 5$$

at least one of the following must hold

$$-3x_1 + 2x_2 - x_3 \geq 0$$

$$2x_1 + 2x_2 + 2x_3 \geq 0$$

and exactly two of the following must hold

$$
\begin{aligned}
x_1 - x_2 \phantom{{}+x_3} &\geq 0 \\
x_1 \phantom{{}-x_2} - x_3 &\geq 0 \\
x_2 + x_3 &\geq 2 \\
x_1 + x_2 + x_3 &\geq 4
\end{aligned}
$$

8.18 Suppose that 0–1 variables $y_1, \ldots, y_m$ have been used in some integer programming problem to serve as switches that turn the constraints $f_i \leq 0$, $i = 1, \ldots, m$, on and off as described in Section 8.6. Write down a (linear) constraint or constraints that could be added to the problem in order to enforce the following conditions.

(a) At least k out of the m constraints must hold.

(b) Exactly k out of the m constraints must hold.

(c) If the additional 0–1 variable w has the value 0, all the even-numbered constraints must hold, and if w has the value 1, all the odd-numbered constraints must hold.

CHAPTER 9

NONLINEAR PROGRAMMING

In this chapter we consider mathematical programming problems in which the decision variables are real but the objective and constraint functions are not necessarily linear. Such problems are called continuous-variable nonlinear programs, or simply **nonlinear programs.** We shall see below that any nonlinear program can be written as

$$\begin{aligned} &\underset{\mathbf{x}\in R^n}{\text{minimize}} \;\; f_0(\mathbf{x}) \\ &\text{subject to } f_i(\mathbf{x}) \le 0, \qquad i = 1, \ldots, m \end{aligned}$$

Problems of this form arise unavoidably in engineering applications such as process design and optimal control, in scientific applications such as parameter estimation and the approximation of functions, and as subproblems in nonlinear integer programs. In addition, many problems commonly formulated as linear programs are really nonlinear, and the simplifying approximation of linearity may be justified for only some values of problem data. Thus many of the linear programming applications described in Part I give rise to nonlinear programming formulations when economies of scale and other effects cause departures from strict proportionality between inputs and outputs.

9.1 A NONLINEAR PROGRAMMING PROBLEM

To see how nonlinearities can enter into problems of the sort we have studied before, recall the Oakwood Furniture problem of Section 2.1. Suppose the company has expanded its operation and now has 700 units of wood on hand. Making one table still uses 2 units of wood, and making one chair still uses 1 unit, with some of the wood going to waste as sawdust, shavings, and scraps. If tables and chairs are manufactured in larger quantities, however, less wood is wasted because scraps can be used in place of new wood for some parts. At the larger quantities it also becomes worthwhile to use more precise manufacturing methods. The result of these **economies of scale** is that the amount of wood used to make a table or chair depends on the numbers of tables and chairs being made. Suppose it has been determined that, if x_1 is the number of tables produced and x_2 is the number of chairs, then

$$\text{units of wood used per table} = 1.5 + \frac{0.5}{0.998 + 0.002x_1}$$

and

$$\text{units of wood used per chair} = 0.5 + \frac{0.5}{0.998 + 0.004x_1 + 0.002x_2}$$

Oakwood's distributor still requires at least twice as many chairs as tables, and he will still pay \$20 for one table and \$15 for one chair, but as an inducement to order larger quantities he expects a chain discount of 0.1% for every table or chair, so that

$$\text{price per table} = \$20 \cdot (1 - 0.001)^{(x_1+x_2)}$$
$$\text{price per chair} = \$15 \cdot (1 - 0.001)^{(x_1+x_2)}$$

The problem of maximizing Oakwood's revenue can now be formulated as the following nonlinear program:

$$\underset{x_1, x_2}{\text{maximize}}\ (20x_1 + 15x_2) \cdot 0.999^{(x_1+x_2)}$$

subject to

$$x_1\left(1.5 + \frac{0.5}{0.998 + 0.002x_1}\right) + x_2\left(0.5 + \frac{0.5}{0.998 + 0.004x_1 + 0.002x_2}\right) \le 700$$

$$x_2 \ge 2x_1 \qquad x_1 \ge 0 \qquad x_2 \ge 0$$

Recasting the problem in the standard form given at the beginning of the chapter,

$$\underset{x_1, x_2}{\text{minimize}}\ f_0(\mathbf{x}) = -(20x_1 + 15x_2) \cdot 0.999^{(x_1+x_2)}$$

subject to

$$f_1(\mathbf{x}) = x_1\left(1.5 + \frac{0.5}{0.998 + 0.002x_1}\right) + x_2\left(0.5 + \frac{0.5}{0.998 + 0.004x_1 + 0.002x_2}\right) - 700 \le 0$$

$$f_2(\mathbf{x}) = 2x_1 - x_2 \le 0 \qquad f_3(\mathbf{x}) = -x_1 \le 0 \qquad f_4(\mathbf{x}) = -x_2 \le 0$$

This problem has $n = 2$ variables and $m = 4$ constraints.

This formulation includes nonnegativity constraints on the decision variables because it is impossible to manufacture a negative number of tables or chairs. From the form of the objective function it appears that these constraints will be slack at the optimal point, but, as in linear programming, it is usually not possible to tell which constraints in a nonlinear programming problem will be binding at optimality except by solving the problem. The formal statement of any mathematical programming problem should thus include explicit nonnegativity constraints if they are appropriate. Most methods for solving nonlinear programs do not implicitly assume nonnegative variables as the simplex method does for linear programs, and the standard form for a nonlinear program given at the beginning of this chapter permits variables of arbitrary sign.

The verbal description of the nonlinear Oakwood problem is more complicated than that of the original linear example in Section 2.1, but the formulation process is essentially the same. In practice, of course, it might be more difficult in the nonlinear case to find equations that closely approximate the actual variation of wood consumption or product price with quantity, but once those equations are known the formulation process is straightforward. Nonlinear programming formulations are often like linear programming formulations except that some or all of the functions are not linear. This is in contrast to the integer programming formulations considered in Chapter 8, where special techniques frequently come into play. The difficulty in nonlinear programming is usually not in formulating the problem but in solving it, and the remainder of this chapter is therefore devoted to solution techniques rather than to additional examples of problem formulation.

Contour Plots

The graphical method used in Section 2.1 to solve the linear version of the Oakwood problem can also be used for nonlinear problems having only two variables. Recall that in the linear case we graphed, on x_1, x_2 − coordinates, the constraint hyperplanes and the (straight-line) objective function contours for the linear program. Then, by inspecting the graph, we found the extreme point of the feasible region for which the objective value was lowest.

In the nonlinear case the locus of points $\mathbf{x}$ for which a constraint function or the objective function has a given value is also called a **contour** of the function, but it is a curved line unless the function happens to be linear. Figure 9.1 shows a **contour plot** for the nonlinear Oakwood problem. A **feasible point** for the nonlinear programming problem is a point that satisfies all the constraints, and, as in linear programming, the set of all feasible points is called the **feasible set** or **feasible region.** Thus the feasible region S can be described algebraically by

$$S = \{\mathbf{x} \mid f_i(\mathbf{x}) \leq 0 \qquad i = 1, \ldots, m\}$$

The feasible region S in Figure 9.1 is bounded by the zero contours of the constraint functions $f_1(\mathbf{x})$, $f_2(\mathbf{x})$, and $f_3(\mathbf{x})$, so in this problem $f_4(\mathbf{x}) \leq 0$ is a redundant constraint. As in linear programming, it is usually not possible to identify redundant constraints in a nonlinear programming problem except by solving the problem. The feasible set in Figure 9.1 happens to be bounded, but it is possible for a nonlinear programming problem to have an unbounded feasible set. It is also possible for a nonlinear program to be infeasible, so that S is the empty set, and this too cannot usually be determined except by solving the

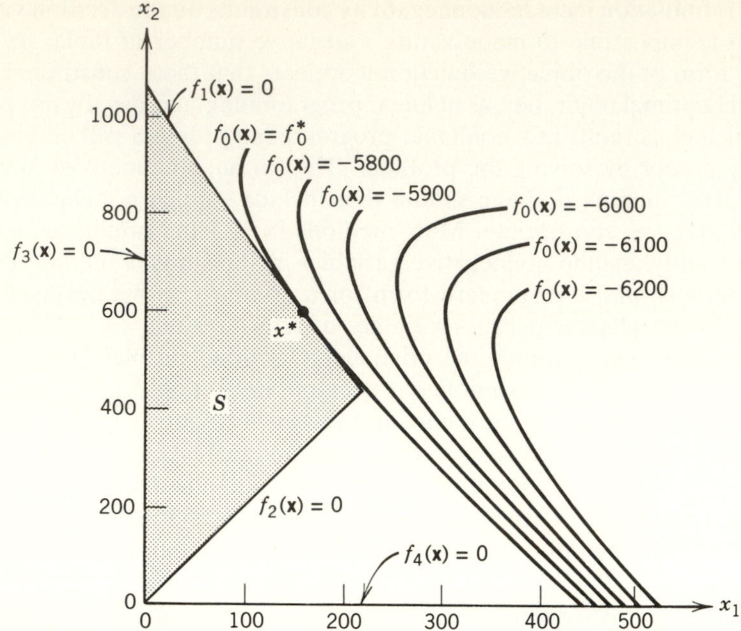

FIGURE 9.1 Contour plot for the nonlinear Oakwood problem.

problem. Although the contour $f_1(\mathbf{x}) = 0$ may at first appear to be a straight line in Figure 9.1, it is actually curved because of the nonlinear form of the function $f_1(\mathbf{x})$. The feasible set for a nonlinear programming problem need not be polyhedral, as in the case of linear programming, nor even connected. The objective function contours in a nonlinear programming problem are also curved lines, like those shown for this example, unless the objective function happens to be linear.

From inspection of Figure 9.1 it is clear that the optimal point $\mathbf{x}^*$ is somewhere on the contour $f_1(\mathbf{x}) = 0$ because every other feasible point lies on an objective function contour having a larger objective value than f_0. We can read off the coordinates of $\mathbf{x}^*$ from the graph as

$$\mathbf{x}^* \approx [150, 600]^T$$

However, it may not be obvious how to find the optimal point analytically. In a linear program that has a unique optimal solution, there are always just as many active constraints as there are variables in the problem, and the optimal solution is at the intersection of the zero contours of those constraints. In contrast, the optimal point for a nonlinear programming problem need not necessarily be where the zero contours of constraint functions intersect one another, and the optimal point can be unique even if there are fewer active constraints than there are variables. In our example constraint 1 is active because relaxing it would permit some improvement in the objective value, while constraints 2, 3, and 4 are inactive because $f_i(\mathbf{x}^*) < 0$ for $i = 2, 3, 4$. Actually, the optimal solution is approximately

$$\mathbf{x}^* = [157.625, 595.825]^T \quad \text{with} \quad f_0^* = -5689.04$$

Assumption of Continuity

In the nonlinear Oakwood example, just as in the linear example of Section 2.1, the company would never actually manufacture a fraction of a table or chair. If the solution to a nonlinear programming problem turns out to be integers, then it is the optimal integer solution. However, if the nonlinear programming solution turns out to be noninteger, then finding the best integer solution usually requires the use of integer programming. One approach is to use the branch-and-bound algorithm of Chapter 8 with continuous nonlinear programming subproblems, solving each subproblem by the techniques described in this chapter. Another way of solving nonlinear integer programs is by means of **dynamic programming,** which will be discussed in Chapter 10. In trivial cases such as the preceding example, it may be possible to use the graphical method to find the best integer solution (Oakwood should actually make 157 tables and 596 chairs).

For the purposes of this chapter we shall ignore the problem of integer variables and assume that the decision variables can take on real values.

Algorithms for Nonlinear Programming

How might we have solved the nonlinear Oakwood problem without graphing contours? How might one solve a nonlinear program having more than two or three variables, for which the graphical method is not practical? Ideally we would like to have a single algorithm that could be applied in some uniform way to every nonlinear program, and that would yield information similar to that provided by the simplex algorithm for linear programs.

Recall that a linear program can be classified as

infeasible

or

feasible with an unbounded optimal value

or

feasible with a finite optimal value
that is attained at an optimal point

The simplex algorithm determines, in a finite number of pivots, which of these three cases holds for a given linear program and yields an optimal point if one exists. These three cases are also possible for a nonlinear program, but a fourth case is needed to classify all nonlinear programs. This is because a nonlinear program could be

feasible with a bounded objective function value
but no optimal point exists

For example, the nonlinear program

$$\text{minimize} \quad \frac{1}{x} \qquad \text{subject to} \qquad x \geq 0$$

has feasible points, and zero is the greatest lower bound on its objective function value. However, this bound is never attained at a feasible point so there is no minimizing point for this nonlinear program. This is one of the many ways in which nonlinear programs are more complicated than linear programs.

Because the general nonlinear programming problem is fundamentally much more difficult than linear programming, no universal algorithm for solving every nonlinear program has yet been devised. However, the search for such an algorithm has given rise to a powerful theory of nonlinear programming that provides useful results in certain circumstances, and effective methods have been developed for the numerical solution of many nonlinear programming problems. It is the aim of this chapter to introduce the theory of nonlinear programming and to describe a few of the computational approaches now in common use.

9.2 UNCONSTRAINED MINIMIZATION

Just as the theory of linear programming is based on linear algebra with several variables, the theory of nonlinear programming is based on calculus with several variables. A convenient and familiar place to begin is therefore with the problem of finding the maximum or minimum of a nonlinear function in the absence of constraints.

Maxima and Minima in One Dimension

Consider the elementary calculus problem of finding the minimum value of the following function of one variable:

$$f(x) = x^2 - 2x + 3$$

A graph of this function is shown in Figure 9.2. From the graph it is clear that the minimum occurs at $x^* = 1$, with $f(x^*) = 2$. To get this result analytically, we would use the fact that the tangent line to the graph of the function is horizontal at the minimizing point. In other words the derivative of the function

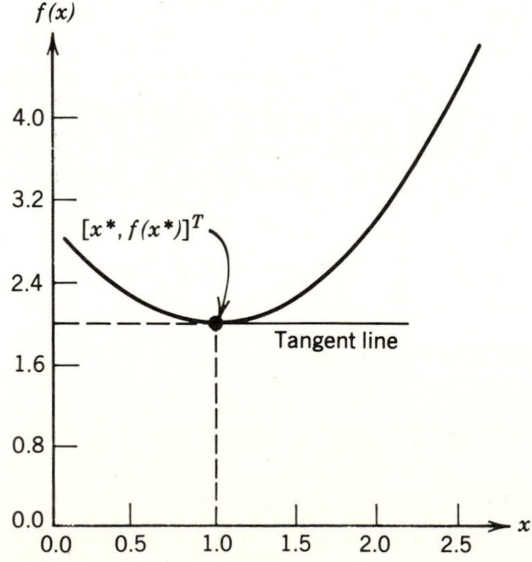

FIGURE 9.2 Graph of $f(x) = x^2 - 2x + 3$.

is zero at the minimizing point, or

$$\frac{df(x)}{dx} = 2x - 2 = 0 \Rightarrow x^* = 1$$

The notation $A \Rightarrow B$ means "A implies B."

With this procedure in mind, consider the more complicated problem of finding the minimum value of

$$f(x) = 65x^6 + 71x^5 - 322x^4 - 401x^3 + 1000$$

When x is large enough in magnitude, the first term dominates in this expression for $f(x)$ and the function behaves like x^6, growing without bound for x greater than about 2 and less than about -2. Between these values of x, the function must have a minimum. Setting the derivative equal to zero yields

$$\frac{df(x)}{dx} = 390x^5 + 355x^4 - 1288x^3 - 1203x^2 = 0$$

which is satisfied by four different values of x. The values are approximately

$$a = -1.79267$$
$$b = -0.94275$$
$$c = 0$$
$$d = 1.82517$$

Which of these is the minimizing point? The behavior of this function is actually rather complicated for values of x between -2 and 2, as shown by the graph of Figure 9.3

The point $x = a$ is called a **local minimizing point** because $f(x) \geq f(a)$ for each x in some neighborhood of the point a. The point $x = d$ is also a local minimizing point, but it is called a **global minimizing point** because $f(x) \geq f(d)$

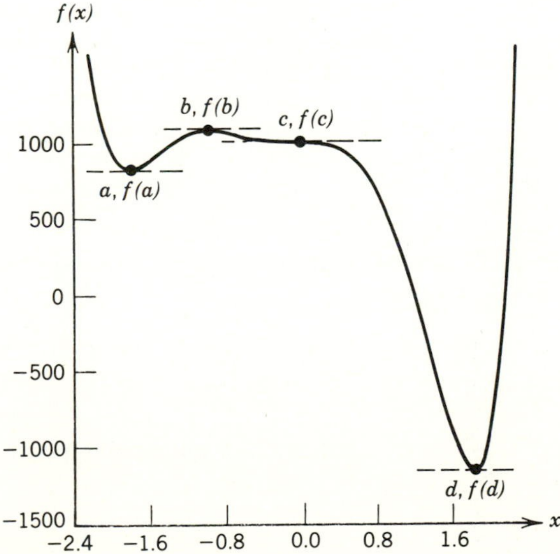

FIGURE 9.3 Graph of $f(x) = 65x^6 + 71x^5 - 322x^4 - 401x^3 + 1000$.

for all points x. Similarly, the point $x = b$ is called a **local maximizing point** because $f(x) \le f(b)$ for each x in some neighbood of the point b. A point $x = e$ is called a **global maximizing point** if $f(x) \le f(e)$ for all points x. In this example there are no global maximizing points, and we say informally that the maximum value is $+\infty$. The point $x = c$ is called an **inflection point** because the derivative of the function is zero there but it is not a local minimizing point or a local maximizing point.

This example illustrates a classical result from elementary calculus, namely, for any differentiable function f of a scalar variable x, in the absence of constraints,

> $\bar{x}$ is a minimizing point or a maximizing point or an inflection point of $f(x)$ if and only if
>
> $$\frac{df(\bar{x})}{dx} = 0$$

The points where the derivative is zero are called **stationary points** of the function. The stationary points can sometimes be sorted analytically into minimizing, maximizing, and inflection points by examining the second derivative of the function. For the preceding example,

$$\frac{d^2f(x)}{dx^2} = 1950x^4 + 1420x^3 - 3864x^2 - 2406x$$

The slope of the tangent line to the graph of this function is negative to the left of a minimizing point and positive to the right, so at the minimum the slope is increasing with increasing x and the second derivative must be positive. Similarly, the second derivative of the function is negative at a maximum and zero at an inflection point. The following table classifies the points at which the first derivative is zero for the example function of Figure 9.3:

Point	$\bar{x}$	$d^2f(\bar{x})/dx^2$	*Classification*
a	−1.79267	3853.8	Minimizing point
b	−0.94275	−815.4	Maximizing point
c	0.0	0.0	Inflection point
d	1.82517	13010.0	Minimizing point

This illustrates another classical result, namely, that for any twice-differentiable function $f(x)$ of a scalar variable x,

> if $\bar{x}$ is a stationary point of $f(x)$, then
>
> $$\frac{d^2f(\bar{x})}{dx^2} > 0 \quad \text{implies } \bar{x} \text{ is a minimizing point}$$
>
> and
>
> $$\frac{d^2f(\bar{x})}{dx^2} < 0 \quad \text{implies } \bar{x} \text{ is a maximizing point}$$

This **second-derivative test** is often useful in classifying stationary points, but it is inconclusive if the second derivative is zero. A stationary point at which the second derivative is zero might be a horizontal inflection point (as in the example of Figure 9.3), or a minimizing point or a maximizing point. The following example shows that the second derivative can be zero at a minimizing point:

$$f(x) = x^4$$

$$\frac{df}{dx} = 4x^3 = 0 \quad \Rightarrow \quad \bar{x} = 0 \text{ is a stationary point}$$

$$\frac{d^2f}{dx^2}(\bar{x}) = 12\bar{x}^2 = 12[0]^2 = 0$$

The second derivative is zero at $\bar{x}$, but $\bar{x}$ is obviously the minimizing point. Throughout the remainder of this chapter, we shall repeatedly encounter similar difficulties in the classification of stationary points for most nonlinear programs.

Minimizing or maximizing a function of one variable is a problem of considerable practical importance in and of itself because it arises in the form of a subproblem called a **line search** in some algorithms for solving nonlinear optimization problems when $\mathbf{x}$ is a vector. We shall therefore return to one-dimensional optimization when we study line searching in Section 9.7. In addition, many aspects of the n-dimensional unconstrained optimization problem can be viewed as generalizations from the one-dimensional case, as we shall see in the following subsection.

Maxima and Minima in *n* Dimensions

We have seen that the stationary points of a function of one variable occur where the tangent line to the graph of the function is horizontal. In the case of a function of several variables, the stationary points occur where the **tangent hyperplane** to the graph of the function is horizontal. This situation is pictured in Figure 9.4 for the function

$$f(\mathbf{x}) = (x_1 - 1)^2 + x_2^2 + 1$$

In order for the tangent hyperplane in Figure 9.4 to be horizontal, it is clear that it must intersect both the $x_1,f(\mathbf{x})$ − plane and the $x_2,f(\mathbf{x})$ − plane in horizontal lines. In other words the partial derivatives of f with respect to both x_1 and x_2 must be zero. This is precisely analogous to the first-derivative condition in the case when x is a scalar. Thus, to find the stationary points of the function

$$f(\mathbf{x}) = (x_1 - 1)^2 + x_2^2 + 1$$

we must solve a system of algebraic equations as follows:

$$\left.\begin{aligned} \frac{\partial f(\mathbf{x})}{\partial x_1} &= 2(x_1 - 1) = 0 \\ \frac{\partial f(\mathbf{x})}{\partial x_2} &= 2x_2 = 0 \end{aligned}\right\} \Rightarrow \text{stationary point } \mathbf{x}^* = \begin{bmatrix} 1 \\ 0 \end{bmatrix}$$

The vector of partial derivatives of a function $f(\mathbf{x})$ with respect to $x_1, \ldots, x_n$

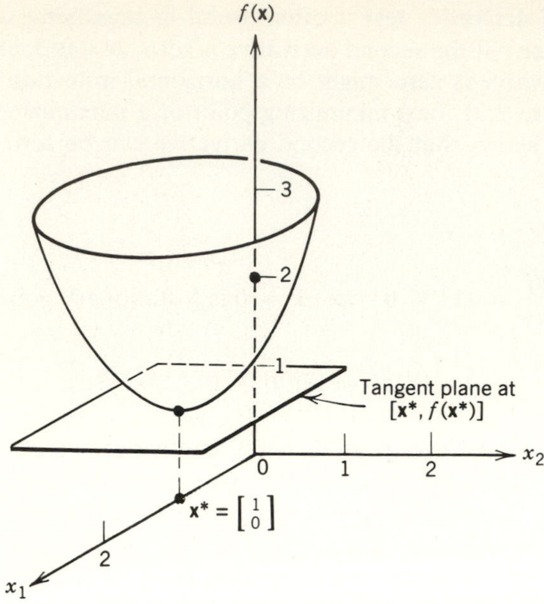

FIGURE 9.4 Graph of $f(\mathbf{x}) = (x_1 - 1)^2 + x_2^2 + 1$.

is called the **gradient vector** of $f(\mathbf{x})$ and is denoted $\nabla f(\mathbf{x})$. That is,

$$\nabla f(\mathbf{x}) = \begin{bmatrix} \dfrac{\partial f(\mathbf{x})}{\partial x_1} \\ \dfrac{\partial f(\mathbf{x})}{\partial x_2} \\ \vdots \\ \dfrac{\partial f(\mathbf{x})}{\partial x_n} \end{bmatrix}$$

For the preceding example

$$\nabla f(\mathbf{x}) = \begin{bmatrix} 2(x_1 - 1) \\ 2x_2 \end{bmatrix}$$

and at the stationary point $\nabla f(\mathbf{x}^*) = [0, 0]^T$. Thus a concise way of stating the requirement that all the partial derivatives be zero is

> $\bar{\mathbf{x}}$ is a stationary point of $f(\mathbf{x})$ if and only if $\nabla f(\bar{\mathbf{x}}) = \mathbf{0}$.

For $\mathbf{x} \in R^n$ the system represented by $\nabla f(\mathbf{x}) = \mathbf{0}$ will contain n equations, and in general the equations will be nonlinear. The simple form of f in our example permitted us to solve analytically for the unique vector $\mathbf{x}^*$, but in general the system $\nabla f(\mathbf{x}) = \mathbf{0}$ might have no solution, or one solution, or more than one solution, and numerical methods might be required to find them.

At $\mathbf{x}^* = [1, 0]^T$ we find that $f(\mathbf{x}) = 1$, and in this particular example we can easily conclude that the point must be a minimizing point because $f(\mathbf{x}) \geq 1$

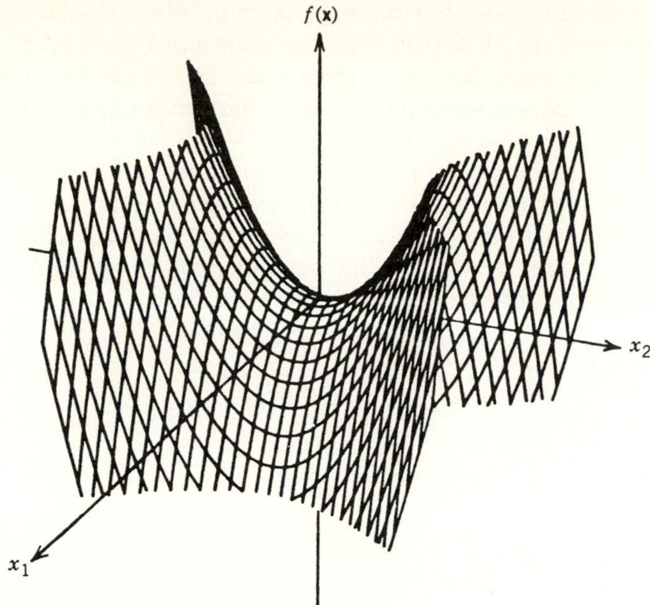

FIGURE 9.5 Saddle Point in $f(\mathbf{x}) = -x_1^2/2 + x_2^2/2$.

for all values of **x**. In general, of course, it might not be possible to use that sort of argument to decide whether a stationary point yields a maximum, a minimum, or neither for the function f. For $n = 1$ a stationary point that is neither a maximizing point nor a minimizing point must be a horizontal inflection point, but for $n \geq 2$ there are other possibilities; for $n = 2$, for example, a stationary point might be a saddle point as shown in Figure 9.5.

It is possible to distinguish two different kinds of local minimum or maximum depending on the behavior of the function near the stationary point. If the function value is higher at all points in the neighborhood of the stationary point no matter what direction one moves away from the point, then the point is a **strict local minimizing point** as shown in Figure 9.6*a*. If the function value

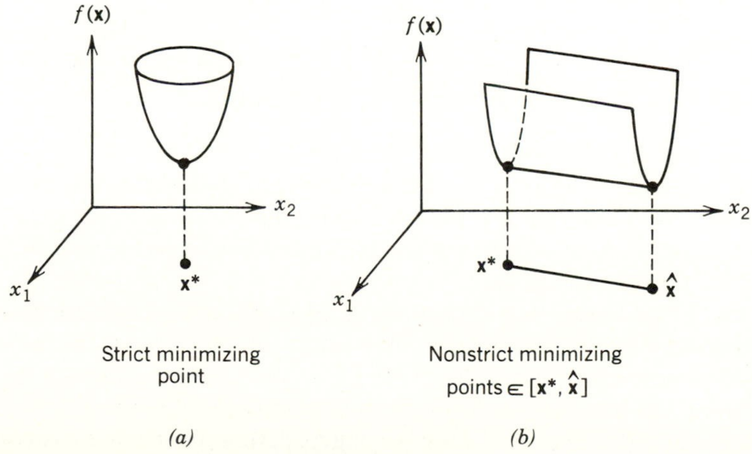

FIGURE 9.6 Strict and nonstrict minimizing points.

increases in some directions but stays the same in others as one moves a small distance away from the stationary point, as shown in Figure 9.6*b*, the point is a local minimizing point but not a strict one. The same distinction is made between a **strict local maximizing point** and a local maximizing point that is not strict.

The second-derivative test for the n-dimensional case is more complicated than the test for the case when x is a scalar because it is not sufficient to check the second partial derivatives $\partial^2 f/\partial x_i \partial x_j$ individually. Because of the way in which these derivatives are related to the curvature of a function's graph, the test must instead be performed using the **Hessian matrix H(x)**, which is defined as follows:

$$\mathbf{H}(\mathbf{x}) = \begin{bmatrix} \dfrac{\partial^2 f}{\partial x_1^2}(\mathbf{x}) & \dfrac{\partial^2 f}{\partial x_1\, \partial x_2}(\mathbf{x}) & \cdots & \dfrac{\partial^2 f}{\partial x_1\, \partial x_n}(\mathbf{x}) \\ \dfrac{\partial^2 f}{\partial x_2\, \partial x_1}(\mathbf{x}) & \dfrac{\partial^2 f}{\partial x_2^2}(\mathbf{x}) & \cdots & \\ \vdots & & & \vdots \\ \dfrac{\partial^2 f}{\partial x_n\, \partial x_1}(\mathbf{x}) & \cdots & & \dfrac{\partial^2 f}{\partial x_n^2}(\mathbf{x}) \end{bmatrix}$$

If f has continuous second partials, as we always assume for the purposes of this chapter, then

$$\frac{\partial^2 f}{\partial x_i\, \partial x_j}(\mathbf{x}) = \frac{\partial^2 f}{\partial x_j\, \partial x_i}(\mathbf{x}) \quad \text{for } i = 1, \ldots, n \text{ and } j = 1, \ldots, n$$

so the Hessian matrix is symmetric.

Checking the sign of the second derivative when $n = 1$ corresponds to checking the **definiteness** of the Hessian matrix when $n > 1$. The different kinds of definiteness are defined as follows, where $\mathbf{M}$ is any $n \times n$ matrix and $\mathbf{z}$ is a vector of length n:

$$\mathbf{M} \text{ is } \left\{ \begin{matrix} \text{positive definite} \\ \text{positive semidefinite} \\ \text{negative semidefinite} \\ \text{negative definite} \end{matrix} \right\} \Longleftrightarrow \mathbf{z}^T\mathbf{M}\mathbf{z} \text{ is } \left\{ \begin{matrix} > 0 \\ \geq 0 \\ \leq 0 \\ < 0 \end{matrix} \right\} \text{ for } \mathbf{z} \neq \mathbf{0}$$

The notation

$$A \Longleftrightarrow B \text{ means ``} A \text{ if and only if } B\text{''};$$

in other words, "A implies B" and "B implies A."

If **M** is the zero matrix (so that it is both positive semidefinite and negative semidefinite) or if the sign of $\mathbf{z}^T\mathbf{M}\mathbf{z}$ varies with the choice of **z**, we shall say that **M** is **indefinite.** We shall sometimes use the abbreviations **pd** for positive definite and **psd** for positive semidefinite.

When a matrix depends on the value of **x**, like the Hessian matrix **H(x)**, its definiteness might vary from one value of **x** to another. To test the definiteness of a Hessian matrix at a particular point $\bar{\mathbf{x}}$, it is necessary to evaluate the matrix at that point and test the resulting constant matrix $\mathbf{M} = \mathbf{H}(\bar{\mathbf{x}})$.

The relationships between Hessian matrix definiteness and the classification of stationary points are summarized for minima:

If $\bar{\mathbf{x}}$ is a stationary point of $f(\mathbf{x})$, then the following implications hold:	
$\mathbf{H}(\bar{\mathbf{x}})$ is positive definite	$\Rightarrow \bar{\mathbf{x}}$ is a strict minimizing point
$\mathbf{H}(\mathbf{x})$ is positive semidefinite for all $\mathbf{x}$ in some neighborhood of $\bar{\mathbf{x}}$	$\Rightarrow \bar{\mathbf{x}}$ is a minimizing point
$\bar{\mathbf{x}}$ is a minimizing point	$\Rightarrow \mathbf{H}(\bar{\mathbf{x}})$ is positive semidefinite

The corresponding relationships for maxima can be obtained by replacing "positive" by "negative" and "minimizing" by "maximizing." Just as the second-derivative test for a function of one variable is inconclusive when the second derivative is zero, this test is inconclusive if $\mathbf{H}(\bar{\mathbf{x}})$ is indefinite or if $\mathbf{H}(\bar{\mathbf{x}})$ is psd at $\bar{\mathbf{x}}$ but not at all points in a neighborhood of $\bar{\mathbf{x}}$. Then the function might have a maximum, or a minimum, or neither at $\bar{\mathbf{x}}$.

For the two-variable example,

$$f(\mathbf{x}) = (x_1 - 1)^2 + x_2^2 + 1$$

$$\nabla f(\mathbf{x}) = \begin{bmatrix} 2(x_1 - 1) \\ 2x_2 \end{bmatrix}$$

$$\mathbf{H}(\mathbf{x}) = \begin{bmatrix} 2 & 0 \\ 0 & 2 \end{bmatrix}$$

To check the definiteness of $\mathbf{H}$, we need to examine the quantity $\mathbf{z}^T\mathbf{H}\mathbf{z}$. In terms of the components of $\mathbf{z}$, this is just

$$\mathbf{z}^T\mathbf{H}\mathbf{z} = [z_1, z_2] \begin{bmatrix} 2 & 0 \\ 0 & 2 \end{bmatrix} \begin{bmatrix} z_1 \\ z_2 \end{bmatrix} = 2z_1^2 + 2z_2^2$$

For $\mathbf{z} \neq \mathbf{0}$, $2(z_1^2 + z_2^2) > 0$, so the Hessian matrix is positive definite and the stationary point $\bar{\mathbf{x}}$ found earlier is a strict local minimizing point.

The function pictured in Figure 9.5 apparently has a stationary point at $\mathbf{x} = [0, 0]^T$, which is confirmed by solving analytically for the point at which the gradient vector is zero, as follows:

$$f(\mathbf{x}) = -\tfrac{1}{2}x_1^2 + \tfrac{1}{2}x_2^2$$

$$\nabla f(\mathbf{x}) = \begin{bmatrix} -x_1 \\ x_2 \end{bmatrix} = 0 \;\Rightarrow\; \bar{\mathbf{x}} = [0, 0]^T$$

The Hessian matrix for this function is

$$\mathbf{H}(\bar{\mathbf{x}}) = \begin{bmatrix} -1 & 0 \\ 0 & 1 \end{bmatrix}$$

so repeating the calculation for the previous example yields $\mathbf{z}^T\mathbf{H}\mathbf{z} = -z_1^2 + z_2^2$. The sign of this quantity clearly depends on the particular values taken on by z_1 and z_2, so the Hessian matrix is indefinite and the point $\bar{\mathbf{x}}$ cannot be classified on the basis of this test.

9.3 EQUALITY CONSTRAINTS

The study of equality-constrained problems is important for two reasons. First, as in the case of linear programming, many nonlinear optimizations are most conveniently formulated in such a way that they contain equality constraints. Second, the theory of equality-constrained nonlinear programming leads naturally to a more complete theory that encompasses both equality and inequality constraints.

To see how equality constraints can be dealt with, reconsider the final example of Section 9.2, with an equality constraint added, as follows:

$$\text{minimize } f_0(\mathbf{x}) = (x_1 - 1)^2 + x_2^2 + 1$$
$$\text{subject to}$$
$$f_1(\mathbf{x}) = x_1 - \tfrac{1}{4}x_2^2 = 0$$

One way to find the optimal point for this problem is from the contour plot of Figure 9.7. The equality constraint requires that the optimal point $\mathbf{x}^*$ lies on the contour $x_1 = \frac{1}{4}x_2^2$, and it is obvious by inspection of the picture that the lowest value $f_0(\mathbf{x})$ can take on along this line is at the point $\mathbf{x}^* = [0, 0]^T$. Of course, for problems having more than two or three variables, using a contour plot is out of the question.

Another approach to handling the equality constraint would be to substitute $x_1 = \frac{1}{4}x_2^2$ into the expression for $f_0(\mathbf{x})$, to obtain

$$f_0(\mathbf{x}) = (\tfrac{1}{4}x_2^2 - 1)^2 + x_2^2 + 1 = \tfrac{1}{16}x_2^4 + \tfrac{1}{2}x_2^2 + 2$$

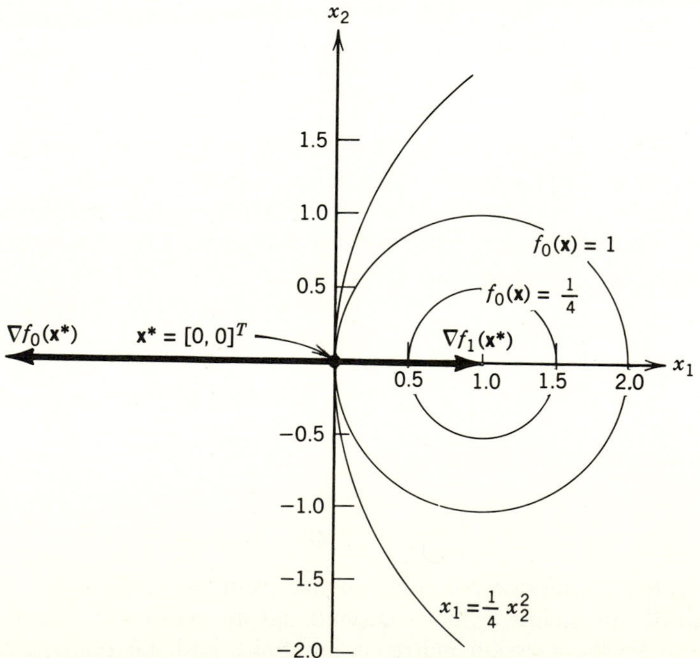

FIGURE 9.7 Contour plot for the equality-constrained problem.

This is called the **reduced objective function** because the dimension of the problem space has been reduced from $n = 2$ to $(n - m) = 2 - 1 = 1$ by the elimination of x_1. The resulting one-dimensional minimization can then be solved as follows:

$$\frac{df_0}{dx_2} = \tfrac{1}{4}x_2^3 + x_2 = 0 \;\Rightarrow\; x_2^* = 0$$

The equality constraint can then be used to find $x_1^* = \frac{1}{4}[0]^2 = 0$, so that $\mathbf{x}^* = [0, 0]^T$. In general, of course, it might not be possible to solve analytically for some variables in terms of the others, particularly if there are several nonlinear constraints.

A third approach is based on the observation that, at the optimal point for this problem, the zero contour of the constraint is tangent to the optimal contour of the objective function, as shown in Figure 9.7. This means that, at the optimal point, the gradient vectors of the two functions are collinear. In this example $\nabla f_0(\mathbf{x}^*) = -u\,\nabla f_1(\mathbf{x}^*)$ for a nonnegative scalar u because the gradients point in opposite directions. Thus,

$$\nabla f_0(\mathbf{x}^*) + u\,\nabla f_1(\mathbf{x}^*) = \mathbf{0}, \quad \text{for some scalar } u$$

Using this observation and the fact that the optimal point satisfies the equality constraint, we obtain the following system that must be satisfied by the optimal point, namely,

$$\begin{bmatrix} 2(x_1 - 1) \\ 2x_2 \end{bmatrix} + u \begin{bmatrix} 1 \\ -\frac{1}{2}x_2 \end{bmatrix} = 0$$

$$x_1 - \tfrac{1}{4}x_2^2 = 0$$

or, equivalently,

$$2(x_1 - 1) + u = 0 \qquad \text{(a)}$$

$$x_2(2 - \tfrac{1}{2}u) = 0 \qquad \text{(b)}$$

$$x_1 - \tfrac{1}{4}x_2^2 = 0 \qquad \text{(c)}$$

Equation (b) implies that either $u = 4$ or $x_2 = 0$. If $u = 4$, then $x_1 = -1$ from (a) and $x_2^2 = -4$ from (c). The square of a real number cannot be negative, so $u \neq 4$ and $x_2 = 0$ must hold. This implies from (c) that $x_1 = 0$ and so $u = 2$ from (a). Thus the point $\mathbf{x}^* = [0, 0]^T$ with $u = 2$ is the unique point satisfying this system. Because the optimal point satisfies the system, we can conclude that $\mathbf{x}^* = [0, 0]^T$ is the optimal point.

Next, we shall show how this approach generalizes when there are $m > 1$ equality constraints.

Parametric Representation of Equality Constraints

The second and third approaches described in the preceding subsection (that is, variable substitution and the use of a relationship between the gradients) are actually related. To see how, it is necessary to consider a different way of using equality constraints to eliminate variables.

In principle, an equality constraint contains the extra information required to permit reformulation of a nonlinear programming problem having n variables into a new problem having $n - 1$ variables. Thus, if we could solve the m equality constraints of a problem analytically, we could eliminate m of the variables to produce an equivalent unconstrained problem having $n - m$ variables. In practice, this is usually not possible. However,

> in most cases, it is possible to represent the feasible set,
>
> $$S = \{\mathbf{x} \in R^n \mid f_i(\mathbf{x}) = 0 \qquad i = 1, \ldots, m\}$$
>
> by n parametric equations,
>
> $$x_j = g_j(\mathbf{t}) \qquad j = 1, \ldots, n$$
>
> where $\mathbf{t}$ is a vector of $n - m$ scalar parameters, and in most cases, it is possible to show that any local minimizing point $\bar{\mathbf{x}}$ must satisfy
>
> $$\nabla f_0(\bar{\mathbf{x}}) + \sum_{i=1}^{m} u_i \nabla f_i(\bar{\mathbf{x}}) = \mathbf{0} \quad \text{for some vector } \mathbf{u} \in R^m$$

A sum of scalar multiples of vectors is called a **linear combination** of those vectors. Thus the last equation implies that the gradient of the objective function is a linear combination of the gradients of the constraints. This important **gradient condition** plays a crucial role in the **Method of Lagrange** for solving equality-constrained nonlinear programs. Before discussing this method, we shall consider some examples where a parameterization of the feasible set can be used to derive the preceding gradient equation.

Following these examples, we shall give precise conditions that assure such a parameterization exists and assure that the gradient condition holds.

Consider the following example, in which it is not possible to solve the constraint equation $f_1(\mathbf{x}) = 0$ algebraically for either of the variables in terms of the other.

$$\underset{\mathbf{x}\in R^2}{\text{minimize}} \quad f_0(\mathbf{x}) = -x_2(x_1 + 2)$$

subject to

$$f_1(\mathbf{x}) = x_1 e^{x_1 x_2} + x_2 - 2 = 0$$

For this example $n - m = 2 - 1 = 1$ so only one parameter is required. If we let

$$t = e^{x_1 x_2}$$

then after a little algebra we find that in place of

$$f_1(x) = x_1 e^{x_1 x_2} + x_2 - 2 = 0$$

we can write

$$x_1 = g_1(t) = \frac{\ln(t)}{1 + \sqrt{1 - t \cdot \ln(t)}}$$

$$x_2 = g_2(t) = 1 + \sqrt{1 - t \cdot \ln(t)}$$

where $\ln(t)$ is the natural logarithm function.

These formulas for x_1 and x_2 are called a **parametric representation** or

parameterization of the constraint $f_1(\mathbf{x}) = 0$. It might not be obvious how to select t as a function of x_1 and x_2, and if we used a different function a different parameterization would result. However, it is easy to verify that the parameterization represents the constraint. When the expressions given are substituted for x_1 and x_2, the equation $f_1(\mathbf{x}) = 0$ is satisfied for all values of t such that $t \cdot \ln(t) \leq 1$.

A contour plot for the problem is shown in Figure 9.8*a*. The feasible set $S = \{\mathbf{x} \mid f_i(\mathbf{x}) = 0\}$ is the curve marked S, that is, the locus of points $[x_1, x_2]^T$ satisfying the constraint equation

$$f_1(\mathbf{x}) = x_1 e^{x_1 x_2} + x_2 - 2 = 0$$

The solid part of the curve is given by the preceding parameterization, and the part that shows in the figure is traced out from left to right as t increases from about 0.08 to about 1.76. For example, the point $\mathbf{x}^*$ corresponds to $t = 1$, a point farther down on the curve at about $[0.25, 1.63]^T$ corresponds to $t = 1.5$, and so forth. Over the solid part of the constraint contour, the parametric representation of the constraint contains the same information as the original constraint equation $f_1(\mathbf{x}) = 0$ because they both describe the same locus of points.

When t has the value that makes $t \cdot \ln(t) = 1$, the parameterization yields

$$x_1 = \frac{\ln(t)}{1 + \sqrt{1 - t \cdot \ln(t)}} \approx 0.566$$

$$x_2 = 1 + \sqrt{1 - t \cdot \ln(t)} = 1$$

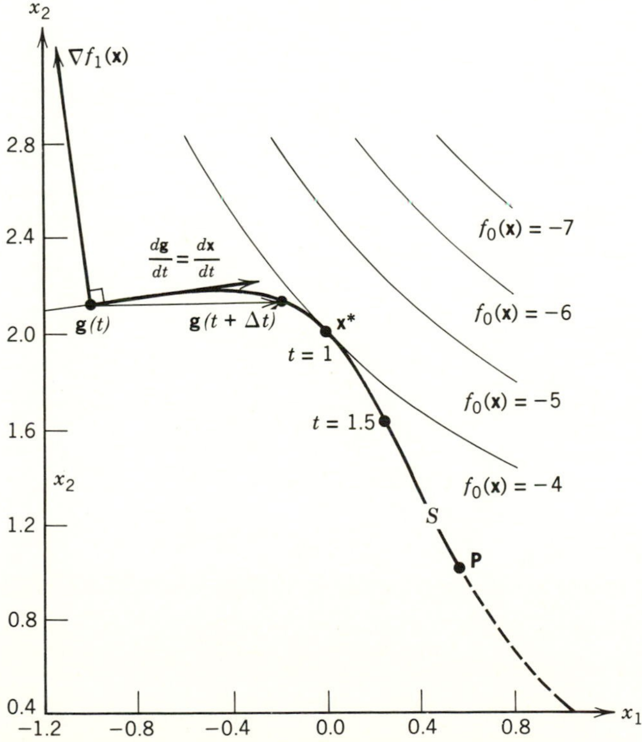

FIGURE 9.8*a* Contour plot for the equality-constrained problem.

which is the point **P** on the solid part of the curve. For larger values of t the square root in the formulas for x_1 and x_2 is undefined, so a different parameterization must be used for the dashed part of the constraint contour. What is important for our purposes is that all points $[x_1, x_2]^T$, near $\mathbf{x}^*$ on the constraint contour $f_1(\mathbf{x}) = 0$ can be represented by the two equations $x_1 = g_1(t)$ and $x_2 = g_2(t)$.

Using this parametric representation of the feasible set, we can rewrite the objective function as a function of the single variable t,

$$\theta(t) = f_0(g_1(t), g_2(t))$$

The function θ has an unconstrained minimum at $t = t^*$, where t^* is the value of t for which $\mathbf{x}^* = [g_1(t^*), g_2(t^*)]^T$, and the derivative of θ must be zero at t^*. Using the chain rule for differentiation, we obtain

$$\begin{aligned}\frac{d\theta}{dt} &= \frac{\partial f_0}{\partial x_1}\frac{dg_1}{dt} + \frac{\partial f_0}{\partial x_2}\frac{dg_2}{dt} \\ &= \left[\frac{\partial f_0}{\partial x_1}, \frac{\partial f_0}{\partial x_2}\right]\begin{bmatrix}\dfrac{dg_1}{dt} \\ \dfrac{dg_2}{dt}\end{bmatrix}\end{aligned}$$

Thus, because $d\theta/dt$ is zero at t^*, we see that

$$\nabla f_0(\mathbf{x}^*)^T \begin{bmatrix}\dfrac{dg_1(t^*)}{dt} \\ \dfrac{dg_2(t^*)}{dt}\end{bmatrix} = 0$$

If the dot product of two nonzero vectors is zero, then the vectors are orthogonal, that is, the angle between them is a right angle.

Thus, the last equation above says that the vector $\nabla f_0(\mathbf{x}^*)$ is orthogonal to the other vector. The other vector has an important geometric interpretation, namely, it is tangent to the contour $f_1(\mathbf{x}) = 0$ at $\mathbf{x}^*$. To see why this is true, let $\mathbf{g}(t)$ be the vector of function values

$$\mathbf{g}(t) = \begin{bmatrix} g_1(t) \\ g_2(t)\end{bmatrix}$$

and let

$$\frac{d\mathbf{g}}{dt} \quad \text{denote the vector} \quad \begin{bmatrix}\dfrac{dg_1}{dt} \\ \dfrac{dg_2}{dt}\end{bmatrix}$$

Consider the points $\mathbf{g}(t)$ and $\mathbf{g}(t + \Delta t)$ in Figure 9.8*a* and the limit

$$\underset{\Delta t \to 0}{\text{limit}} \frac{\mathbf{g}(t + \Delta t) - \mathbf{g}(t)}{\Delta t}$$

From the definition of the derivative, this limit equals the vector $d\mathbf{g}/dt$. Also from Figure 9.8*a*, it is clear that this limiting vector $d\mathbf{g}/dt$ is tangent to the contour $f_1(\mathbf{x}) = 0$. This means that the vector $d\mathbf{g}/dt$ is orthogonal to $\nabla f_1(\mathbf{x})$ at every point $\mathbf{x}$.

We have already seen that the vector $d\mathbf{g}/dt$ is orthogonal to $\nabla f_0(\mathbf{x})$ at the optimal point $\mathbf{x}^*$. Thus, at the optimal point the two gradient vectors, $\nabla f_0(\mathbf{x})$ and $\nabla f_1(\mathbf{x})$, are both orthogonal to the vector $d\mathbf{g}/dt$ and therefore must be collinear.

Because $dg_j/dt = dx_j/dt$, we will also refer to the tangent vector as $d\mathbf{x}/dt$. Figure 9.8*b* shows the tangent vector $d\mathbf{x}/dt$ at the optimal point $\mathbf{x}^*$. For this problem the gradients point in opposite directions, so at the optimal point there is some nonnegative scalar u such that

$$\nabla f_0(\mathbf{x}) = -u \, \nabla f_1(\mathbf{x})$$

This is the same gradient condition observed in Figure 9.7 for the previous example. For that example we were able to solve the system

$$\begin{aligned} \nabla f_0(\mathbf{x}) + u \, \nabla f_1(\mathbf{x}) &= 0 \\ f_1(\mathbf{x}) &= 0 \end{aligned}$$

analytically. The current example illustrates an important point, namely, that it is not always possible to solve such a nonlinear system analytically. Here

$$\nabla f_0(\mathbf{x}) = \begin{bmatrix} -x_2 \\ -(x_1 + 2) \end{bmatrix} \quad \text{and} \quad \nabla f_1(\mathbf{x}) = \begin{bmatrix} x_1 x_2 e^{x_1 x_2} + e^{x_1 x_2} \\ x_1^2 e^{x_1 x_2} + 1 \end{bmatrix}$$

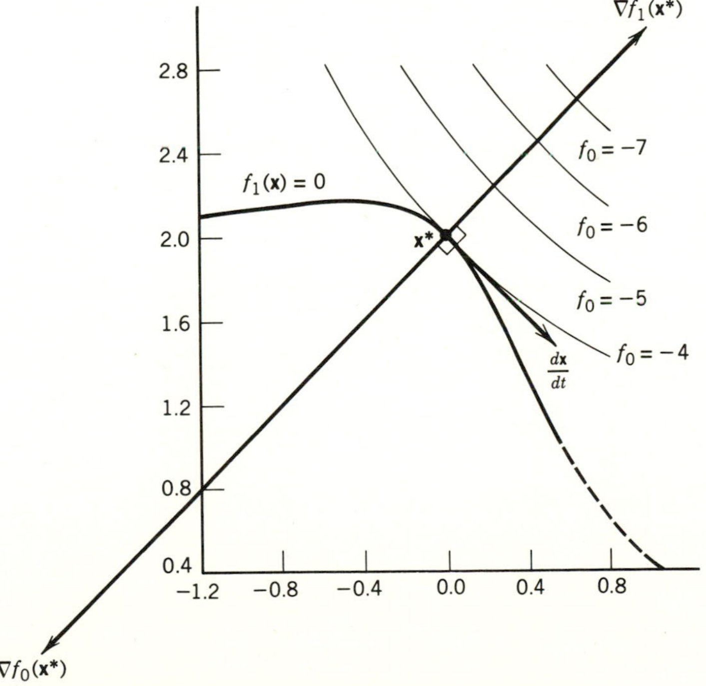

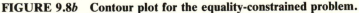

FIGURE 9.8*b* Contour plot for the equality-constrained problem.

so the system $\nabla f_0(\mathbf{x}) + u \nabla f_1(\mathbf{x}) = \mathbf{0}$ and $f_1(\mathbf{x}) = 0$ can be written as

$$-x_2 + ux_1x_2e^{x_1x_2} + ue^{x_1x_2} = 0$$

$$-x_1 - 2 + ux_1^2e^{x_1x_2} + u = 0$$

$$x_1e^{x_1x_2} + x_2 - 2 = 0$$

We challenge the reader to try to solve this system analytically for all solutions. It is not hard to see that $x_1 = 0$ implies that $x_2 = 2$ and $u = 2$. Thus, the point $\mathbf{x}^* = [0, 2]^T$ and $u^* = 2$ is one solution, but showing analytically that it is the only solution is another matter. Inspection of Figure 9.8*b* confirms that $\mathbf{x}^*$ is the optimal point.

Several Equality Constraints

The parametric analysis outlined above can easily be extended to the case in which there are several equality constraints. Consider the problem

$$\underset{x \in R^3}{\text{minimize}} \quad f_0(\mathbf{x}) = (x_1 - \tfrac{13}{3})^2 + (x_2 - \tfrac{1}{2})^2 - x_3$$

subject to

$$f_1(\mathbf{x}) = x_1 + \tfrac{5}{3}x_2 - 10 = 0$$

$$f_2(\mathbf{x}) = (x_2 - 2)^2 + x_3 - 4 = 0$$

A (three-dimensional) contour plot for this problem is shown in Figure 9.9*a*. Suppose that it is possible to represent the set of points

$$S = \{\mathbf{x} \mid f_i(\mathbf{x}) = 0, i = 1, \ldots, m\}$$

parametrically by equations $x_j = g_j(t)$, $j = 1, \ldots, n$. Then, just as we did previously, we can write the objective as a function of the scalar parameter t, namely,

$$\theta(t) = f_0(g_1(t), g_2(t), \ldots, g_n(t))$$

Just as in the previous analysis, the function θ has an unconstrained minimum at the point t corresponding to the optimal point for the constrained problem. Here the chain rule gives

$$\frac{d\theta}{dt} = \frac{\partial f_0}{\partial x_1}\frac{dg_1}{dt} + \frac{\partial f_0}{\partial x_2}\frac{dg_2}{dt} + \cdots + \frac{\partial f_0}{\partial x_n}\frac{dg_n}{dt}$$

But $d\theta/dt$ is zero at the unconstrained minimum, so again

$$\nabla f_0(\mathbf{x})^T \frac{d\mathbf{x}}{dt} = 0$$

at the optimal point. In this example the vector $d\mathbf{x}/dt$ is tangent to the feasible set $S = \{\mathbf{x} \mid f_i(\mathbf{x}) = 0, i = 1, \ldots, m\}$, so it is orthogonal to each of the gradients

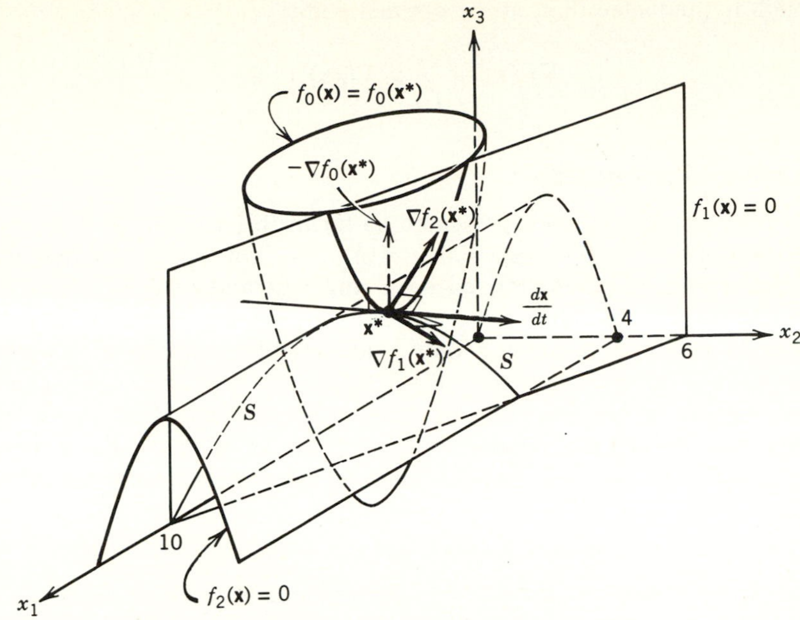

FIGURE 9.9*a* A problem with two equality constraints.

$\nabla f_i(\mathbf{x})$ at every point $\mathbf{x}$. Thus,

> at the optimal point, the gradient vectors $\nabla f_0(\mathbf{x})$ and $\nabla f_i(\mathbf{x})$, $i = 1, \ldots, m$ are orthogonal to the same vector $d\mathbf{x}/dt$ at the same point, and therefore must lie in the same hyperplane.

This situation is pictured for the example in Figure 9.9*a*, where S is a curve in R^3 and the vectors $\nabla f_0(\mathbf{x}^*)$, $\nabla f_1(\mathbf{x}^*)$, and $\nabla f_2(\mathbf{x}^*)$ lie in the same plane.

The fact that these three vectors lie in the same plane is not enough to guarantee that $\nabla f_0(\mathbf{x}^*)$ can be written as a linear combination of the other two. For example, suppose that $\nabla f_1(\mathbf{x}^*)$ and $\nabla f_2(\mathbf{x}^*)$ were collinear and that $\nabla f_0(\mathbf{x}^*)$ were orthogonal to both of them as pictured in Figure 9.9*b*. Note here that $\nabla f_0(\mathbf{x}^*)$ cannot be written as a linear combination of the other two vectors. However, if the gradients of the constraints were **linearly independent** at $\mathbf{x}^*$, then the fact that $\nabla f_0(\mathbf{x}^*)$ and the constraint gradients all lie in the same plane

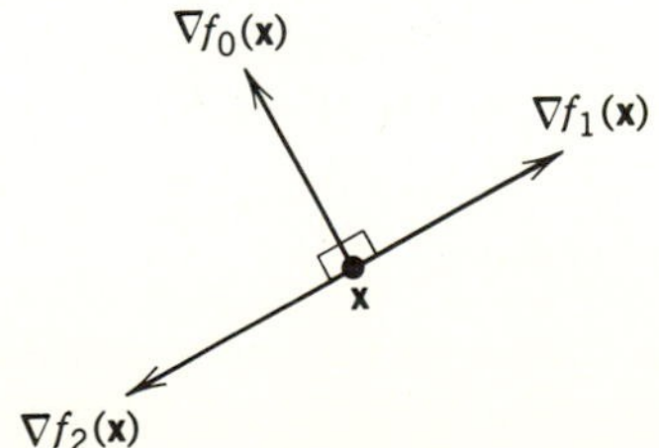

FIGURE 9.9*b* $\nabla f_0(\mathbf{x})$, $\nabla f_1(\mathbf{x})$, $\nabla f_2(\mathbf{x})$ lie in the same plane, but $\nabla f_0(\mathbf{x}) \neq u_1\nabla f_1(\mathbf{x}) + u_2\nabla f_2(\mathbf{x})$ for any u_1 and u_2.

is enough to guarantee that, at the optimal point,

$$\nabla f_0(\mathbf{x}) + \sum_{i=1}^{m} u_i \nabla f_i(\mathbf{x}) = \mathbf{0}$$

Several Parameters

In both of the examples presented so far in this section, $n - m = 1$, so the feasible set $S = \{\mathbf{x} \mid f_i(\mathbf{x}) = 0, i = 1, \ldots, m\}$ was a one-dimensional curve and could be parameterized by a single parameter t. In other words, by parameterizing the constraints, we could use them to construct a new unconstrained problem containing only one variable, namely, t. In general, of course, $n - m > 1$, and then the feasible set is a surface of dimension $n - m$, and must be described by $n - m$ parameters. The unconstrained problem that we get will contain $n - m$ variables, namely, the parameters t_j, $j = 1, \ldots, n - m$. We can still think of writing the objective as a function of $\mathbf{t}$,

$$\theta(\mathbf{t}) = f_0[g_1(\mathbf{t}), g_2(\mathbf{t}), \ldots, g_n(\mathbf{t})]$$

but now $\mathbf{t}$ is a vector. At the point $\mathbf{t}^*$ corresponding to the optimal value for the constrained problem, the function θ has an unconstrained minimum, which requires that all the partial derivatives $\partial\theta/\partial t_j$ be zero, or

$$\frac{\partial\theta}{\partial t_1} = \frac{\partial f_0}{\partial g_1}\frac{\partial g_1}{\partial t_1} + \frac{\partial f_0}{\partial g_2}\frac{\partial g_2}{\partial t_1} + \cdots + \frac{\partial f_0}{\partial g_n}\frac{\partial g_n}{\partial t_1} = 0$$

$$\frac{\partial\theta}{\partial t_2} = \frac{\partial f_0}{\partial g_1}\frac{\partial g_1}{\partial t_2} + \frac{\partial f_0}{\partial g_2}\frac{\partial g_2}{\partial t_2} + \cdots + \frac{\partial f_0}{\partial g_n}\frac{\partial g_n}{\partial t_2} = 0$$

$$\vdots$$

$$\frac{\partial\theta}{\partial t_{(n-m)}} = \frac{\partial f_0}{\partial g_1}\frac{\partial g_1}{\partial t_{(n-m)}} + \cdots + \frac{\partial f_0}{\partial g_n}\frac{\partial g_n}{\partial t_{(n-m)}} = 0$$

These $n - m$ equations give rise to the $n - m$ orthogonality conditions

$$\nabla f_0(\mathbf{x})^T\left[\frac{\partial\mathbf{x}}{\partial t_j}\right] = 0 \qquad j = 1, \ldots, n - m$$

The vectors $[\partial\mathbf{x}/\partial t_j]$ are all tangent to the feasible surface S, just as the vector $[d\mathbf{x}/dt]$ was tangent to the feasible curve in the examples above, so each vector $[\partial\mathbf{x}/\partial t_j]$ is orthogonal to all of the constraint gradients $\nabla f_i(\mathbf{x})$. Thus, at a minimizing point the constraint gradients $\nabla f_i(\mathbf{x})$ and the objective function gradient $\nabla f_0(\mathbf{x})$ all lie in the same m-dimensional hyperplane, and we can once again write $\nabla f_0(\mathbf{x}^*)$ as a linear combination of the $\nabla f_i(\mathbf{x}^*)$.

The Method of Lagrange

The preceding analysis assumes that it is possible to represent the set of points $S = \{\mathbf{x} \mid f_i(\mathbf{x}) = 0, i = 1, \ldots, m\}$ parametrically by the equations $x_j = g_j(t)$, $j = 1, \ldots, n$, and that $\nabla f_0(\mathbf{x})$ can be written as a linear combination of the $\nabla f_i(\mathbf{x})$ at the optimal point. Under precisely what circumstances are these assumptions met? The necessary conditions are given by the **Lagrange multiplier theorem.**

THE LAGRANGE MULTIPLIER THEOREM

Given the nonlinear programming problem

$$\underset{\mathbf{x} \in R^n}{\text{minimize}}\ f_0(\mathbf{x})$$

subject to

$$f_i(\mathbf{x}) = 0 \qquad i = 1, \ldots, m$$

if

$\bar{\mathbf{x}}$ is a local minimizing point for the nonlinear programming problem,
and $n > m$ (there are more variables than constraints),
and the f_i have continuous first partial derivatives with respect to the x_j,
and the $\nabla f_i(\bar{\mathbf{x}})$ are linearly independent vectors,

then

there is a vector $\mathbf{u} = [u_1, u_2, \ldots, u_m]^T$ such that

$$\nabla f_0(\bar{\mathbf{x}}) + \sum_{i=1}^{m} u_i \nabla f_i(\bar{\mathbf{x}}) = \mathbf{0}$$

The numbers u_i are called **Lagrange multipliers.**

In the first example of this section it turned out that the Lagrange multiplier was positive, but the theorem does not require that, and, in general, a Lagrange multiplier for an equality-constrained problem can be of either sign.

The requirement that the $\nabla f_i(\bar{\mathbf{x}})$ be linearly independent is called a **constraint qualification.**

The Lagrange multiplier theorem suggests a method for solving equality-constrained nonlinear programming problems, as follows:

THE METHOD OF LAGRANGE

1. Verify that $n > m$ and each f_i has continuous first partials.
2. Form the **Lagrangian function**

$$L(\mathbf{x}, \mathbf{u}) = f_0(\mathbf{x}) + \sum_{i=1}^{m} u_i f_i(\mathbf{x})$$

3. Find all of the solutions $(\bar{\mathbf{x}}, \bar{\mathbf{u}})$ to the following system of nonlinear algebraic equations:

$$\nabla L(\mathbf{x}, \mathbf{u}) = \nabla f_0(\mathbf{x}) + \sum_{i=1}^{m} u_i \nabla f_i(\mathbf{x}) = \mathbf{0} \qquad (n \text{ equations})$$

$$\frac{\partial L(\mathbf{x}, \mathbf{u})}{\partial u_i} = f_i(\mathbf{x}) = 0 \qquad i = 1, \ldots, m \quad (m \text{ equations})$$

These equations are called the **Lagrange conditions.** The points $(\bar{\mathbf{x}}, \bar{\mathbf{u}})$ are called **Lagrange points.**

4. Examine each solution $(\bar{\mathbf{x}}, \bar{\mathbf{u}})$ to see if it is a minimizing point.

To show how the Lagrange method can be used in practice, we shall solve the two-constraint example pictured in Figure 9.9. The statement of the problem is reproduced here for convenience:

$$\underset{x \in R^3}{\text{minimize}}\ f_0(\mathbf{x}) = (x_1 - \tfrac{13}{3})^2 + (x_2 - \tfrac{1}{2})^2 - x_3$$

subject to

$$f_1(\mathbf{x}) = x_1 + \tfrac{5}{3}x_2 - 10 = 0$$
$$f_2(\mathbf{x}) = (x_2 - 2)^2 + x_3 - 4 = 0$$

1. Check that the method can be used. For this example $n = 3$ and $m = 2$, so $n > m$ as required. The gradients of the constraints are

$$\nabla f_1(\mathbf{x}) = \begin{bmatrix} 1 \\ 5/3 \\ 0 \end{bmatrix} \quad \text{and} \quad \nabla f_2(\mathbf{x}) = \begin{bmatrix} 0 \\ 2x_2 - 4 \\ 1 \end{bmatrix}$$

so the $\partial f_i / \partial x_j$ are continuous functions.

2. Form the Lagrangian function.

$$L(\mathbf{x}, \mathbf{u}) = (x_1 - \tfrac{13}{3})^2 + (x_2 - \tfrac{1}{2})^2 - x_3 + u_1[x_1 + \tfrac{5}{3}x_2 - 10] + u_2[(x_2 - 2)^2 + x_3 - 4]$$

3. Set the derivatives equal to zero and solve.

$$\frac{\partial L}{\partial x_1} = 2\left(x_1 - \frac{13}{3}\right) + u_1 = 0 \qquad \text{(a)}$$

$$\frac{\partial L}{\partial x_2} = 2\left(x_2 - \frac{1}{2}\right) + \frac{5u_1}{3} + 2u_2(x_2 - 2) = 0 \qquad \text{(b)}$$

$$\frac{\partial L}{\partial x_3} = -1 + u_2 = 0 \qquad \text{(c)}$$

$$\frac{\partial L}{\partial u_1} = x_1 + \frac{5x_2}{3} - 10 = 0 \qquad \text{(d)}$$

$$\frac{\partial L}{\partial u_2} = (x_2 - 2)^2 + x_3 - 4 = 0 \qquad \text{(e)}$$

Note that this system is nonlinear because of the product of variables in (b) and the squared term in (e). In many problems the Lagrange conditions cannot be solved analytically, on account of the particular nonlinearities they contain. In other cases an analytical solution is possible, but some ingenuity might be needed to find it.

This system is actually quite easy to solve, and one approach is as follows:

$$\text{(c)} \Rightarrow \quad u_2 = 1 \qquad \text{(f)}$$

$$\text{(f) \& (b)} \Rightarrow \quad 2(x_2 - \tfrac{1}{2}) + \tfrac{5}{3}u_1 + 2[1](x_2 - 2) = 0$$

$$\Rightarrow \quad u_1 = -\tfrac{12}{5}x_2 + 3 \qquad \text{(g)}$$

$$\text{(d)} \Rightarrow \quad x_1 = 10 - \tfrac{5}{3}x_2 \qquad \text{(h)}$$

$$\text{(h) \& (a)} \Rightarrow \quad 2([10 - \tfrac{5}{3}x_2] - \tfrac{13}{3}) + u_1 = 0$$

$$\Rightarrow \quad u_1 = -\tfrac{34}{3} + \tfrac{10}{3}x_2 \qquad \text{(k)}$$

$$(k) \;\&\; (g) \Rightarrow \quad -\tfrac{34}{3} + \tfrac{10}{3}x_2 = \tfrac{12}{5}x_2 + 3$$

$$\Rightarrow \quad x_2 = \tfrac{5}{2} \qquad \text{(p)}$$

$$(p) \;\&\; (g) \Rightarrow \quad u_1 = \frac{-12[5/2]}{5} + 3 = -3$$

$$(p) \;\&\; (h) \Rightarrow \quad x_1 = \frac{10 - 5[5/2]}{3} = \tfrac{35}{6}$$

$$(p) \;\&\; (e) \Rightarrow \quad ([\tfrac{5}{2}] - 2)^2 + x_3 - 4 = 0$$

$$\Rightarrow \quad x_3 = \tfrac{15}{4}$$

Thus, one solution to the Lagrange conditions is

$$\bar{\mathbf{x}} = \begin{bmatrix} 35/6 \\ 5/2 \\ 15/4 \end{bmatrix} \quad \text{and} \quad \bar{\mathbf{u}} = \begin{bmatrix} -3 \\ 1 \end{bmatrix}$$

The instructions for the Lagrange method require that we find all solutions to the system of equations because there might be more than one. In this example the solution can be shown to be unique, but in most problems it is not and multiple solutions must be sought.

4. Examine each solution to see if it is a minimizing point. From inspection of Figure 9.9a, it appears that the unique solution point $\bar{\mathbf{x}}$ is indeed a minimizing point for the example problem.

At the solution point, $\bar{x}_2 = \frac{5}{2}$ so the gradients of the constraints are

$$\nabla f_1(\mathbf{x}) = \begin{bmatrix} 1 \\ 5/3 \\ 0 \end{bmatrix} \quad \text{and} \quad \nabla f_2(\mathbf{x}) = \begin{bmatrix} 0 \\ 1 \\ 1 \end{bmatrix}$$

and these are linearly independent vectors. Thus, the conditions of the theorem are satisfied at $(\bar{\mathbf{x}}, \bar{\mathbf{u}})$. From this example it is clear that the Lagrange method can be used without finding an explicit parameterization of the constraints in the problem.

Classifying Lagrange Points

The Lagrange multiplier theorem does not guarantee that all solutions of the Lagrange conditions will be minimizing points, or even stationary points, so in general it is necessary to check each solution point.

One way of determining whether a point that satisfies the Lagrange conditions is a minimum is by inspecting a contour plot for the problem, as we did for the preceding examples. Of course, that method is of no use when there are more than two or three variables or in other circumstances when a contour plot is not available.

A second approach makes use of the fact that all local minimizing points satisfy the Lagrange conditions. If the solution to the equations is unique, and if it is known that the problem has a minimizing point, then that point is surely the solution to the equations. Each of the preceding example problems is feasible and has a finite optimal value that is taken on, so each has a minimizing point.

Therefore, in each case the fact that $\bar{\mathbf{x}}$ is a unique solution to the Lagrange conditions ensures that it is the minimizing point.

A third approach is to find an explicit parameterization of the constraints, write the objective as a function of the parameters, and use the second-derivative test to check whether the point in question is an unconstrained local minimum of that function. In the final example above, one parameterization is

$$x_1 = g_1(t) = 10 - \tfrac{5}{3}t$$

$$x_2 = g_2(t) = t$$

$$x_3 = g_3(t) = 4 - (t - 2)^2$$

Using this parameterization, we could write the objective function for that problem in reduced form as

$$\begin{aligned} f_0(t) &= ([10 - \tfrac{5}{3}t] - \tfrac{13}{3})^2 + ([t] - \tfrac{1}{2})^2 - [4 - (t - 2)^2] \\ &= (\tfrac{43}{9})t^2 - (\tfrac{215}{9})t + (\tfrac{1165}{36}) \end{aligned}$$

Then

$$\frac{df_0}{dt} = \left(\frac{86}{9}\right)t - \left(\frac{215}{9}\right) = 0 \ \Rightarrow \ t = \frac{5}{2}$$

$$\frac{d^2f_0}{dt^2} = \frac{86}{9} > 0 \ \Rightarrow \ \bar{\mathbf{x}} \text{ is a minimizing point}$$

In cases when $n - m > 1$, it is of course necessary to test the definiteness of the Hessian matrix rather than the sign of a single second derivative. A parameterization might not be known or easily discovered, so this method is not always applicable either.

Finally, there is a **local method** based on approximating the feasible set $S = \{\mathbf{x} \mid f_i(\mathbf{x}) = 0, i = 1, \ldots, m\}$ in the neighborhood of $\bar{\mathbf{x}}$ by a hyperplane tangent to the feasible set at the point $\bar{\mathbf{x}}$. For the final example above, this hyperplane is the line passing through the vector $d\mathbf{x}/dt$ in Figure 9.9a. Any point on this line can be written as

$$\mathbf{x} = \bar{\mathbf{x}} + s\left[\frac{d\mathbf{x}}{dt}\right]$$

where $[d\mathbf{x}/dt]$ is evaluated at $\bar{\mathbf{x}}$ and s is a scalar parameter that measures distance from $\bar{\mathbf{x}}$ along the line. Using the parameterization

$$\mathbf{x} = \begin{bmatrix} 10 - 5t/3 \\ t \\ 4 - (t - 2)^2 \end{bmatrix}$$

we find

$$\frac{d\mathbf{x}}{dt} = \begin{bmatrix} -5/3 \\ 1 \\ -2(t - 2) \end{bmatrix}$$

At the point $\bar{\mathbf{x}} = [\tfrac{35}{6}, \tfrac{5}{2}, \tfrac{15}{4}]^T$, $t = \tfrac{5}{2}$ and the straight-line approximation is given by

$$\mathbf{x} = \begin{bmatrix} 35/6 \\ 5/2 \\ 15/4 \end{bmatrix} + s \begin{bmatrix} -5/3 \\ 1 \\ -1 \end{bmatrix}$$

The following sufficient condition can sometimes be used to classify a solution $(\bar{\mathbf{x}}, \bar{\mathbf{u}})$ to the Lagrange conditions:

> if $(\bar{\mathbf{x}}, \bar{\mathbf{u}})$ satisfies the Lagrange conditions and at $\bar{\mathbf{x}}$ the Hessian, $\mathbf{H}_L(\mathbf{x})$, of the Lagrangian function
>
> $$L(\mathbf{x}, \bar{\mathbf{u}}) = f_0(\mathbf{x}) + \sum_{i=1}^{m} \bar{u}_i f_i(x) \quad \text{with } \mathbf{u} = \bar{\mathbf{u}}$$
>
> satisfies $\mathbf{x}^T \mathbf{H}_L(\bar{\mathbf{x}})\mathbf{x} > 0$ for all vectors $\mathbf{x}$ such that
>
> $$\mathbf{x}^T \nabla f_i(\bar{\mathbf{x}}) = 0 \qquad i = 1, \ldots, m$$
>
> then $\bar{\mathbf{x}}$ is a local minimizing point.

Note that the set of all vectors $\mathbf{x}$ such that $\mathbf{x}^T \nabla f_i(\bar{\mathbf{x}}) = 0$ for each constraint defines a hyperplane orthogonal to each of the constraint gradients and tangent to the feasible set at $\bar{\mathbf{x}}$. For our example, this hyperplane is the line in the direction of the tangent vector $d\mathbf{x}/dt$ in Figure 9.9*a*. Using $\bar{\mathbf{u}} = [-3, 1]^T$, we obtain

$$L(\mathbf{x}, \bar{\mathbf{u}}) = f_0(\mathbf{x}) - 3f_1(\mathbf{x}) + 1f_2(\mathbf{x})$$

For this function the Hessian of the Lagrangian evaluated at $\bar{\mathbf{x}}$ is

$$\mathbf{H}_L(\bar{\mathbf{x}}) = \begin{bmatrix} 2 & 0 & 0 \\ 0 & 4 & 0 \\ 0 & 0 & 0 \end{bmatrix}$$

Note that this matrix is not positive definite over all of R^3 because there are nonzero vectors $\mathbf{x}$ such that $\mathbf{x}^T\mathbf{H}_L(\bar{\mathbf{x}})\mathbf{x} = 0$. For example, $\mathbf{x} = [0, 0, 1]^T$ is such a vector. However, this matrix is positive definite over the hyperplane tangent at $\bar{\mathbf{x}}$. To see why this is true, we use the fact that any point $\mathbf{x}$ on the tangent hyperplane has

$$x_1 = \tfrac{35}{6} - \tfrac{5}{3}s \qquad x_2 = \tfrac{5}{2} + s \qquad x_3 = \tfrac{15}{4} - s$$

For such a point $\mathbf{x}$, $\mathbf{x}^T\mathbf{H}_L(\bar{\mathbf{x}})\mathbf{x}$ can be calculated as a function of s, namely,

$$h(s) = 2(\tfrac{35}{6} - \tfrac{5}{3}s)^2 + 4(\tfrac{5}{2} + s)^2 + 0(\tfrac{15}{4} - s)^2$$

It is not hard to see that $h(s)$ is positive for all s because it has a global minimizing point at $\bar{s} = \frac{10}{44}$ and $h(\bar{s}) > 0$. Thus, $\mathbf{H}_L(\bar{\mathbf{x}})$ is positive definite over the hyperplane orthogonal to each of the constraint gradients at $\bar{\mathbf{x}}$. Therefore, from the above sufficient condition, $\bar{\mathbf{x}}$ is a local minimizing point.

In this analysis we have again used a parametric representation of the feasible set, but this method can be used even if an explicit parameterization is not known. Once the vectors $\nabla f_i(\bar{\mathbf{x}})$ have been calculated, simple techniques from linear algebra can be used to find the hyperplane orthogonal to all of the constraint gradients $\nabla f_i(\bar{\mathbf{x}})$. In our example we have

$$\nabla f_1(\bar{\mathbf{x}}) = \begin{bmatrix} 1 \\ 5/3 \\ 0 \end{bmatrix} \quad \text{and} \quad \nabla f_2(\bar{\mathbf{x}}) = \begin{bmatrix} 0 \\ 1 \\ 1 \end{bmatrix}$$

Vectors $\mathbf{x}$ orthogonal to both these gradients must satisfy

$$\nabla f_1(\bar{\mathbf{x}})^T \mathbf{x} = 0 \quad \text{or} \quad x_1 + (\tfrac{5}{3})x_2 = 0$$

and

$$\nabla f_2(\bar{\mathbf{x}})^T \mathbf{x} = 0 \quad \text{or} \quad x_2 + x_3 = 0$$

Solving this system for x_1 and x_2 in terms of x_3, we find that it is satisfied by all vectors $\mathbf{x}$ such that

$$\begin{aligned} x_1 &= -(\tfrac{5}{3})x_2 = (\tfrac{5}{3})x_3 \\ x_2 &= -x_3 \\ x_3 &= x_3 \end{aligned} \qquad \text{or} \quad \mathbf{x} = s\begin{bmatrix} -5/3 \\ 1 \\ -1 \end{bmatrix}$$

where s is an arbitrary scalar. Then the hyperplane orthogonal to $\nabla f_1(\bar{\mathbf{x}})$ and $\nabla f_2(\bar{\mathbf{x}})$ is the line

$$\mathbf{x} = \bar{\mathbf{x}} + s\begin{bmatrix} -5/3 \\ 1 \\ -1 \end{bmatrix}$$

which is the same expression we obtained earlier by using the explicit parameterization.

As we shall see in Section 9.5, it is possible for inequality-constrained problems to have a special property that makes it easy to classify their solution points, and in that case none of the complications discussed here arise.

An Important Use of the Lagrange Multiplier Theorem

The theorem of Lagrange gives a condition that must necessarily be satisfied by any minimizing point $\bar{\mathbf{x}}$, namely,

$$\nabla f_0(\bar{\mathbf{x}}) + \sum_{i=1}^{m} u_i \nabla f_i(\bar{\mathbf{x}}) = \mathbf{0} \quad \text{for some } \mathbf{u} \in R^m$$

For fixed $\bar{\mathbf{x}}$, this vector equation is simply a system of linear equations in the variables u_i, $i = 1, \ldots, m$. If the assumptions of the Lagrange theorem hold, and

> if there is no $\mathbf{u}$ such that the preceding gradient equation holds at $\bar{\mathbf{x}}$, then the point $\bar{\mathbf{x}}$ cannot be a minimizing point.

To illustrate this observation, consider the following equality-constrained problem:

$$\text{minimize } f_0(\mathbf{x}) = x_1^2 + (x_2 - 2)^2 + x_3^2$$
$$\text{subject to}$$
$$f_1(\mathbf{x}) = x_1^2 + \tfrac{1}{4}x_2^2 + x_3^2 - 1 = 0$$
$$f_2(\mathbf{x}) = -x_2 + 2x_3 = 0$$

Calculating the gradients of f_0, f_1, and f_2, the equation

$$\nabla f_0(\mathbf{x}) + u_1 \nabla f_1(\mathbf{x}) + u_2 \nabla f_2(\mathbf{x}) = 0$$

becomes

$$\begin{bmatrix} 2x_1 \\ 2(x_2 - 1) \\ 2x_3 \end{bmatrix} + u_1 \begin{bmatrix} 2x_1 \\ \frac{1}{2}x_2 \\ 2x_3 \end{bmatrix} + u_2 \begin{bmatrix} 0 \\ -1 \\ 2 \end{bmatrix} = \begin{bmatrix} 0 \\ 0 \\ 0 \end{bmatrix}$$

Suppose that the feasible point

$$\bar{\mathbf{x}} = \begin{bmatrix} 2/\sqrt{6} \\ 2/\sqrt{6} \\ 1/\sqrt{6} \end{bmatrix}$$

is suspected to be a minimizing point. It is not hard to show that the assumptions of the theorem hold for this example at this point $\bar{\mathbf{x}}$. At $\bar{\mathbf{x}}$ the system of gradient equations is

$$\begin{bmatrix} 4/\sqrt{6} \\ 4/\sqrt{6} - 4 \\ 4/\sqrt{6} \end{bmatrix} + u_1 \begin{bmatrix} 4/\sqrt{6} \\ 1/\sqrt{6} \\ 4/\sqrt{6} \end{bmatrix} + u_2 \begin{bmatrix} 0 \\ -1 \\ 2 \end{bmatrix} = \begin{bmatrix} 0 \\ 0 \\ 0 \end{bmatrix}$$

or equivalently

$$4/\sqrt{6} + (4/\sqrt{6})u_1 + 0u_2 = 0$$
$$4/\sqrt{6} - 4 + (1/\sqrt{6})u_1 - u_2 = 0$$
$$4/\sqrt{6} + (4/\sqrt{6})u_1 + 2u_2 = 0$$

From the first equation $u_1 = -1$, and so from the third equation we see that $u_2 = 0$. But these values of u_1 and u_2 do not satisfy the second equation. Thus, there is no $\mathbf{u}$ such that this system has a solution so $\bar{\mathbf{x}}$ cannot be a minimizing point. Of course, it is important to remember that these equations are only necessary conditions for a minimizing point. If a solution $\mathbf{u}$ does exist for a given $\mathbf{x}$, then that point $\mathbf{x}$ may or may not be a minimizing point.

9.4 INEQUALITY CONSTRAINTS

Any equality constraint can be written as a pair of inequality constraints by using the fact that

$$f_i(\mathbf{x}) = 0 \iff f_i(x) \le 0 \quad \text{and} \quad f_i(x) \ge 0$$
$$\text{or}$$
$$f_i(x) \le 0 \quad \text{and} \quad -f_i(x) \le 0$$

Thus, as noted in the introduction to this chapter, any nonlinear program can be written in the following **canonical form:**

$$\underset{\mathbf{x} \in \mathrm{R}^n}{\text{minimize}}\ f_0(\mathbf{x})$$

subject to

$$f_i(\mathbf{x}) \leq 0 \qquad i = 1, \ldots, m$$

As illustrated by the example of Section 9.1, inequality constraints often arise in the formulation of nonlinear programming problems, just as they do in the formulation of linear programs. One basic approach to dealing with inequality constraints analytically is to convert the problem to an equivalent equality-constrained problem. Then we can solve the equality-constrained problem using the Lagrange method.

Active and Inactive Constraints

To see how a problem with inequality constraints can be converted to an equivalent equality-constrained problem, consider the following simple example:

$$\underset{\mathbf{x} \in R^2}{\text{minimize}}\ f_0(\mathbf{x}) = (x_1 - 14)^2 + (x_2 - 11)^2$$

subject to

$$f_1(\mathbf{x}) = (x_1 - 11)^2 + (x_2 - 13)^2 - 7^2 \leq 0$$
$$f_2(\mathbf{x}) = x_1 + x_2 - 19 \leq 0$$

It might seem that the most straightforward way to handle the inequalities would be to change them to equality constraints by introducing slack variables. Some numerical algorithms for nonlinear programming take that approach, but the classical theory makes use instead of the even simpler idea outlined next.

A contour plot for the preceding example is given in Figure 9.10*a*, from which we can easily see that the optimal point is $\mathbf{x}^* = [11, 8]^T$ and lies on the $f_2(\mathbf{x}) = 0$ contour. At the optimal point,

$$f_1(\mathbf{x}^*) = -24 < 0$$

$$f_2(\mathbf{x}^*) = 0$$

so that constraint 1 is **slack** or **inactive,** while constraint 2 is **tight** or **active.**

In Figure 9.10*a* the feasible set for this problem is shaded and labeled S. In this problem the fact that constraint 1 is inactive at optimality means that the point $\mathbf{x}^* = [11, 8]^T$ would be optimal if constraint 1 were missing from the problem altogether. This is analogous to the situation in a linear programming resource allocation problem where a resource constraint that is slack at optimality can be omitted from the problem without changing the optimal solution. In the nonlinear case it is not always true that removing an inactive constraint leaves the optimal point unchanged, but we shall give conditions that ensure that outcome, and from inspection of Figure 9.10*a*, this is clearly true for this problem.

Also, the fact that constraint 2 holds with equality at the optimal point means that it could have been expressed as an equality in the first place if only we had known that it would be active.

Combining these two observations, we can deduce that the equality-con-

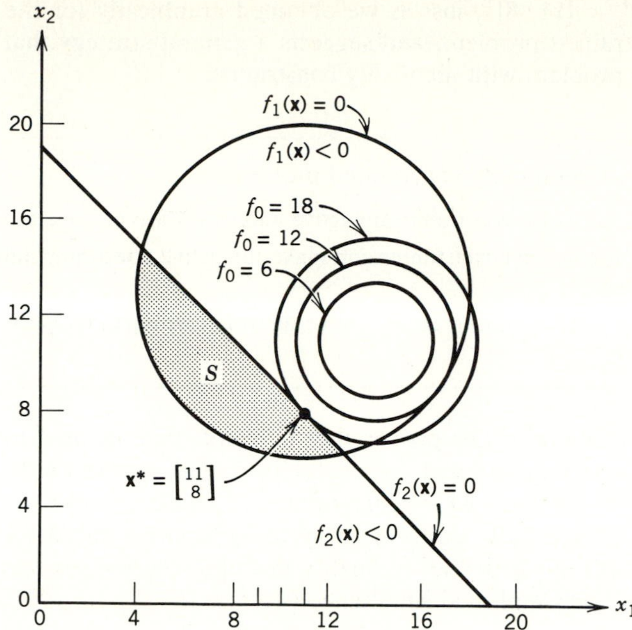

FIGURE 9.10*a*   An inequality-constrained problem.

strained problem

$$\underset{\mathbf{x} \in R^2}{\text{minimize}}\ f_0(\mathbf{x}) = (x_1 - 14)^2 + (x_2 - 11)^2$$

subject to

$$f_2(\mathbf{x}) = x_1 + x_2 - 19 = 0$$

has exactly the same solution as the original inequality-constrained problem. Then we could use the method of Lagrange as follows:

$$\begin{aligned} L(\mathbf{x}, \mathbf{u}) &= f_0(\mathbf{x}) + u_2 f_2(\mathbf{x}) \\ &= (x_1 - 14)^2 + (x_2 - 11)^2 + u_2(x_1 + x_2 - 19) \end{aligned}$$

$$\frac{\partial L}{\partial x_1} = 2(x_1 - 14) + u_2 = 0 \quad \text{(a)}$$

$$\frac{\partial L}{\partial x_2} = 2(x_2 - 11) + u_2 = 0 \quad \text{(b)}$$

$$\frac{\partial L}{\partial u_2} = x_1 + x_2 - 19 = 0 \quad \text{(c)}$$

$$\text{(a) \& (b)} \Rightarrow x_1 - 14 = x_2 - 11 \Rightarrow x_1 = x_2 + 3 \quad \text{(d)}$$

$$\text{(c)} \Rightarrow x_2 = 19 - x_1 \quad \text{(e)}$$

$$\text{(d) \& (e)} \Rightarrow x_2 = 19 - [x_2 + 3] \Rightarrow x_2 = 8 \quad \text{(f)}$$

$$\text{(d) \& (f)} \Rightarrow x_1 = [8] + 3 = 11$$

$$\text{(b)} \Rightarrow u_2 = -2(x_2 - 11) \quad \text{(g)}$$

$$\text{(f) \& (g)} \Rightarrow u_2 = -2([8] - 11) = 6$$

This yields $\mathbf{x}^* = [11, 8]^T$, just as we obtained graphically for the original inequality-constrained problem, and suggests a general strategy that we can use to attack any problem with inequality constraints.

To solve an inequality-constrained problem:

1. Figure out which constraints are active and which are inactive at optimality.
2. Ignore the inactive constraints and make the active constraints into equalities.
3. Use the method of Lagrange to solve the resulting equality-constrained problem.

The only difficulty with this plan is that it is not obvious how to accomplish step 1 for any arbitrary problem. In general, of course, we cannot draw a contour diagram and do not know which constraints will be active until we have solved the problem. Fortunately, there is a way to incorporate the determination of which constraints are active at optimality into the solution process, so that all of the steps are accomplished simultaneously.

The Orthogonality Condition

The scheme that we shall use to automatically identify which constraints are active depends on a simple observation about the Lagrange multipliers. In the preceding example we wrote the Lagrangian function as

$$L(\mathbf{x}, \mathbf{u}) = f_0(\mathbf{x}) + u_2 f_2(\mathbf{x})$$

simply omitting any reference to the inactive constraint function $f_1(\mathbf{x})$. Instead, we could have written

$$\begin{aligned} L(\mathbf{x}, \mathbf{u}) &= f_0(\mathbf{x}) + u_1 f_1(\mathbf{x}) + u_2 f_2(\mathbf{x}) \\ u_1 &= 0 \end{aligned}$$

where the condition that $u_1 = 0$ expresses our knowledge that constraint 1 is slack at optimality. Setting the Lagrange multiplier equal to zero effectively removes the corresponding constraint from the problem because the term for that constraint then adds zero to the Lagrangian function. Thus we can retain our earlier definition of the Lagrangian function as

$$L(\mathbf{x}, \mathbf{u}) = f_0(\mathbf{x}) + \sum_{i=1}^{m} u_i f_i(\mathbf{x})$$

and exclude inactive constraints by simply requiring the Lagrange multiplier corresponding to each slack inequality to be zero. For the preceding example we can summarize the situation at the optimal point as follows:

$$\begin{aligned} f_1(\mathbf{x}^*) &= -24 \qquad u_1^* = 0 \quad \text{constraint 1 inactive} \\ f_2(\mathbf{x}^*) &= 0 \qquad u_2^* = 6 \quad \text{constraint 2 active} \end{aligned}$$

Note that $u_1^* f_1(\mathbf{x}^*) = 0$ and $u_2^* f_2(\mathbf{x}^*) = 0$. This suggests a way to ensure that the Lagrange multipliers corresponding to any inactive constraints will be zero.

Whenever a constraint function has a nonzero value at optimality (the constraint is inactive), we want the corresponding Lagrange multiplier to be zero. This requirement can be enforced by the condition

$$u_i f_i(\mathbf{x}) = 0 \qquad i = 1, \ldots, m$$

These m equations are together called the **orthogonality condition** because they require the vector $\mathbf{u}$ to be orthogonal to the vector

$$\begin{bmatrix} f_1(\mathbf{x}) \\ f_2(\mathbf{x}) \\ \vdots \\ f_m(\mathbf{x}) \end{bmatrix} = \begin{bmatrix} \partial L/\partial u_1 \\ \partial L/\partial u_2 \\ \vdots \\ \partial L/\partial u_m \end{bmatrix}$$

Enforcing the orthogonality condition conveniently ensures that the Lagrangian function will ignore inactive constraints and incorporate active ones as if they were equalities. For the example the orthogonality condition yields

$$u_1^* f_1(\mathbf{x}^*) = 0 \;\Rightarrow\; u_1^*[-24] = 0 \Rightarrow u_1^* = 0$$
$$u_2^* f_2(\mathbf{x}^*) = 0 \;\Rightarrow\; u_2^*[0] = 0 \qquad \Rightarrow u_2^* \text{ may be nonzero}$$

which is equivalent to the condition $u_1 = 0$ that we imposed previously.

The Karush–Kuhn–Tucker Conditions

We have now arranged, by means of the orthogonality condition, for the Lagrangian function to ignore inactive inequalities and to incorporate active ones as though they were equalities. Thus, when accompanied by the orthogonality condition, the Lagrangian function for an inequality-constrained problem is indistinguishable from that for an equivalent equality-constrained problem involving only the active constraints. The gradient condition of Section 9.3,

$$\nabla f_0(\mathbf{x}) + \sum_{i=1}^{m} u_i \nabla f_i(\mathbf{x}) = \mathbf{0}$$

must therefore also hold at a stationary point for the inequality-constrained problem, providing a constraint qualification is satisfied. Because the orthogonality condition excludes inactive constraints from the problem, a satisfactory constraint qualification is to require the gradients of the active constraints to be linearly independent vectors.

In the Lagrange method we used the gradient condition and the original equality constraint equations to find the stationary points of an equality-constrained problem. In a similar way we can now use the gradient condition, the orthogonality condition, and the original constraint inequalities to find the stationary points for an inequality-constrained problem. For the example introduced at the beginning of this section the resulting system is as follows:

$$L(\mathbf{x}, \mathbf{u}) = (x_1 - 14)^2 + (x_2 - 11)^2 + u_1[(x_1 - 11)^2 + (x_2 - 13)^2 - 7^2] + u_2(x_1 + x_2 - 19)$$

$$\frac{\partial L}{\partial x_1} = 2(x_1 - 14) + 2u_1(x_1 - 11) + u_2 = 0$$

$$\frac{\partial L}{\partial x_2} = 2(x_2 - 11) + 2u_1(x_2 - 13) + u_2 = 0$$

$$\frac{\partial L}{\partial u_1} = (x_1 - 11)^2 + (x_2 - 13)^2 - 7^2 \leq 0$$

$$\frac{\partial L}{\partial u_2} = x_1 + x_2 - 19 \leq 0$$

$$u_1 f_1(\mathbf{x}) = 0 \quad \text{and} \quad u_2 f_2(\mathbf{x}) = 0$$

It is easy to show that the solution found earlier, namely, $\mathbf{x}^* = [11, 8]^T$ and $\mathbf{u}^* = [0, 6]^T$, satisfies this system.

In general, combining the gradient condition, the orthogonality condition, and the inequality constraints of a problem yields a system of nonlinear equations and inequalities that can be written in the following form:

$$\frac{\partial L}{\partial x_j} = 0 \qquad j = 1, \ldots, n$$

$$u_i f_i(\mathbf{x}) = 0 \qquad i = 1, \ldots, m$$

$$f_i(\mathbf{x}) \leq 0 \qquad i = 1, \ldots, m$$

As is true of the Lagrange conditions for an equality-constrained problem, this system may have zero, one, or several solutions, and some of them might not correspond to minimizing points of the nonlinear program. In the inequality-constrained case, however, it is possible to say something more about solutions to this system. Whereas the Lagrange multipliers in an equality-constrained problem can be of either sign,

> if a constraint qualification holds for an inequality-constrained problem, the Lagrange multipliers must be nonnegative at a minimizing point.

To see why this is true, consider a problem with the feasible set S illustrated in Figure 9.10*a* but with a different objective function, namely,

$$\underset{\mathbf{x} \in \mathbb{R}^2}{\text{minimize}} \quad f_0(\mathbf{x}) = (x_1 - 20)^2 + (x_2 - 4)^2$$

subject to

$$f_1(\mathbf{x}) = (x_1 - 11)^2 + (x_2 - 13)^2 - 49 = 0$$
$$f_2(\mathbf{x}) = x_1 + x_2 - 19 = 0$$

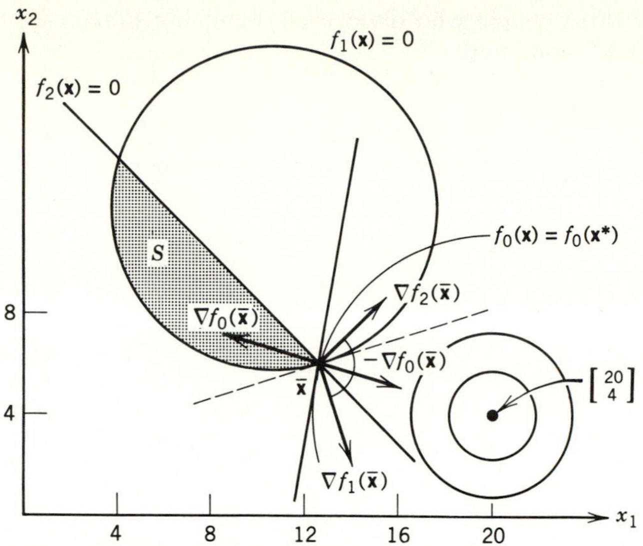

FIGURE 9.10*b* $-\nabla f_0(\bar{\mathbf{x}})$ **is a nonnegative linear combination of** $\nabla f_1(\bar{\mathbf{x}})$ **and** $\nabla f_2(\bar{\mathbf{x}})$**.**

By examining the contours of the objective function as given in Figure 9.10*b*, it is clear that the point $\bar{\mathbf{x}}$ is the optimal point.

Here $-\nabla f_0(\bar{\mathbf{x}})$ lies in the arc between $\nabla f_1(\bar{\mathbf{x}})$ and $\nabla f_2(\bar{\mathbf{x}})$, which means that $-\nabla f_0(\bar{\mathbf{x}})$ can be written as a nonnegative linear combination of $\nabla f_1(\bar{\mathbf{x}})$ and $\nabla f_2(\bar{\mathbf{x}})$. It is important to realize that if $-\nabla f_0(\bar{\mathbf{x}})$ did not lie somewhere in this arc (or on top of one of the constraint gradients), then the contour line $f_0(\mathbf{x}) = f_0(\bar{\mathbf{x}})$ would cut into the interior of the feasible set S as indicated in Figure 9.10*c*. This would be the case, for example, if the objective function f_0 in Figure 9.10*a* is used. In this case $\bar{\mathbf{x}}$ could not be an optimal point because feasible points in S such as $\hat{\mathbf{x}}$, on the same side of the contour line as $-\nabla f_0(\bar{\mathbf{x}})$, would be such that $f_0(\hat{\mathbf{x}}) < f_0(\bar{\mathbf{x}})$.

The fact that the Lagrange multipliers must be nonnegative permits us to exclude some unwanted solutions to the conditions for a minimum by adding the **nonnegativity condition** $u_i \geq 0$. The resulting enlarged system of nonlinear equations and inequalities is known as the Karush–Kuhn–Tucker conditions, in

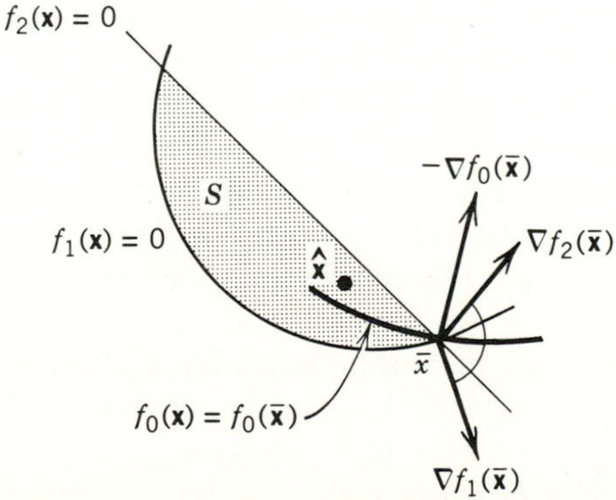

FIGURE 9.10*c* $-\nabla f_0(\bar{\mathbf{x}})$ **is not a nonnegative linear combination of** $\nabla f_1(\bar{\mathbf{x}})$ **and** $\nabla f_2(\bar{\mathbf{x}})$ **so** $\bar{\mathbf{x}}$ **cannot be a minimizing point.**

honor of the three people who discovered them. For brevity we will refer to them as the KKT conditions.

THE KKT CONDITIONS

Given the canonical form nonlinear programming problem

$$\underset{\mathbf{x}\in R^n}{\text{minimize}} \quad f_0(\mathbf{x})$$

subject to

$$f_i(\mathbf{x}) \le 0 \qquad i = 1, \ldots, m$$

if the Lagrangian function is given by

$$L(\mathbf{x}, \mathbf{u}) = f_0(\mathbf{x}) + \sum_{i=1}^{m} u_i f_i(\mathbf{x})$$

then the KKT conditions for the problem are

$$\frac{\partial L}{\partial x_j} = 0 \qquad j = 1, \ldots, n \qquad \text{gradient condition}$$

$$u_i f_i(\mathbf{x}) = 0 \qquad i = 1, \ldots, m \qquad \text{orthogonality}$$

$$f_i(\mathbf{x}) \le 0 \qquad i = 1, \ldots, m \qquad \text{feasibility}$$

$$u_i \ge 0 \qquad i = 1, \ldots, m \qquad \text{nonnegativity}$$

A point $(\bar{\mathbf{x}}, \bar{\mathbf{u}})$ satisfying the KKT conditions is called a **KKT point.**

The KKT conditions play a central role in the theory of nonlinear programming generally, and in particular they are the conditions that must be used in solving problems with inequality constraints.

In Section 9.3 we noted various difficulties in connection with the solution of equality-constrained problems. Some problems do not satisfy the hypotheses of the Lagrange multiplier theorem, and when the hypotheses are satisfied and Lagrange points can be found, it is often difficult to classify them as maxima, minima, or neither. Fortunately, the situation is often somewhat better in the case of inequality-constrained problems.

First, the constraint qualification can be weakened somewhat for inequality-constrained problems. The Lagrange multiplier theorem demands that the gradients of the active constraints be linearly independent, but even if they are not it is sometimes possible to show that the minimizing points of an inequality-constrained problem must satisfy the KKT conditions. The precise circumstances in which this can be done are beyond the scope of this text, but it is encouraging to know that there is a large taxonomy of constraint qualifications that are slightly weaker (and sometimes easier to check) than the linear independence requirement.

Second, for some problems it is unnecessary to check whether a KKT point is a minimizing point. As we shall see in the next section, this is because it may be possible to determine that any point satisfying the KKT conditions is a global minimizing point. It is sometimes even possible to determine that the global minimizing point is unique.

9.5 CONVEXITY

From our study of convex sets in Section 4.2, it might seem that convexity would not have much to do with nonlinear programming. The feasible sets of some inequality-constrained problems are convex sets, but in our study of nonlinear programming up to this point we have made no use of that observation. In fact, nothing that we have done so far in this chapter depends on convexity in any way.

Actually, convexity plays an important role in nonlinear programming. For example, it is a property that can be used to determine whether a KKT point is a minimizing point.

Convex Functions

The usefulness of convexity in nonlinear programming has more to do with **convex functions** than convex sets. Convex functions and convex sets are related in many ways, but the simplest connection involves the **epigraph** of a function. Figure 9.11 shows the graph and epigraph of two functions that we studied in Section 9.2.

The epigraph of a function $f(\mathbf{x})$ where $\mathbf{x} \in \mathrm{R}^n$ is a set in R^{n+1} that is defined as

$$\operatorname{epi}(f) = \left\{ \begin{bmatrix} \mathbf{x} \\ y \end{bmatrix} \in \mathrm{R}^{n+1} \;\middle|\; y \geq f(\mathbf{x}) \right\}$$

In the example of Figure 9.11*a*, epi(f) is clearly not a convex set. Recall that a set is convex if and only if it contains the line segment between any two of its points. In Figure 9.11*a* it is clear that the line segment between the two points $[\bar{\mathbf{x}}, f(\bar{\mathbf{x}})]$ and $[\hat{\mathbf{x}}, f(\hat{\mathbf{x}})]$ contains points below the graph of the function f and thus not in epi(f).

Recall from Section 4.2 that any point between $\bar{\mathbf{x}}$ and $\hat{\mathbf{x}}$ can be written as

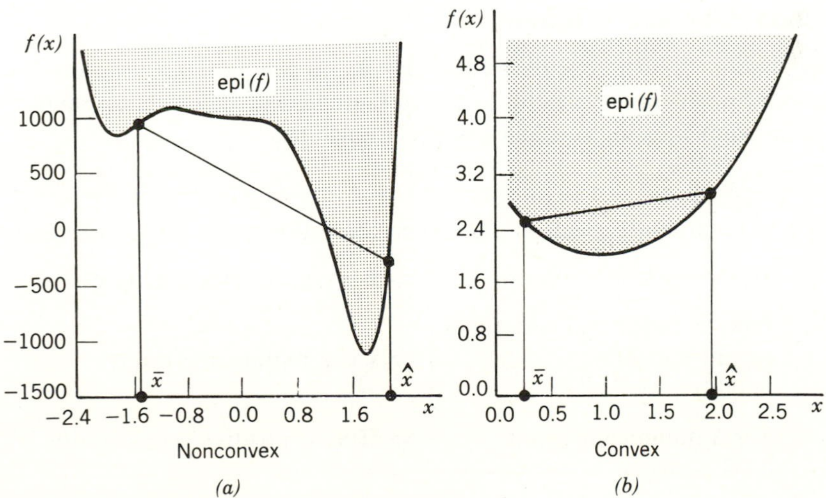

FIGURE 9.11 Nonconvex and convex functions.

a convex combination of $\bar{\mathbf{x}}$ and $\hat{\mathbf{x}}$, namely, as $\lambda\bar{\mathbf{x}} + (1 - \lambda)\hat{\mathbf{x}}$ for some λ between 0 and 1. For example, $\lambda = 0$ corresponds to $\hat{\mathbf{x}}$ and $\lambda = 1$ corresponds to $\bar{\mathbf{x}}$.

In order for all points on the line segment in Figure 9.11*a* to be in epi(f), it would have to be true that for any convex combination of $\bar{\mathbf{x}}$ and $\hat{\mathbf{x}}$,

$$\begin{pmatrix}\text{value of } f \text{ at that}\\ \text{convex combination}\end{pmatrix} \le \begin{pmatrix}\text{height of the line segment}\\ \text{above that convex combination}\end{pmatrix}$$

that is,

$$f(\lambda\bar{\mathbf{x}} + (1 - \lambda)\hat{\mathbf{x}}) \le \lambda f(\bar{\mathbf{x}}) + (1 - \lambda)f(\hat{\mathbf{x}})$$

Note that the height of the line segment above the convex combination $\lambda\bar{\mathbf{x}} + (1 - \lambda)\hat{\mathbf{x}}$ is just the same convex combination of the function values $f(\bar{\mathbf{x}})$ and $f(\hat{\mathbf{x}})$.

A convex function is shown in Figure 9.11*b* from which it is clear that epi(f) is a convex set. It is not hard to show in general that

$$\begin{pmatrix}\text{epi}(f) \text{ is a}\\ \text{convex set}\end{pmatrix} \quad \text{if and only if} \quad \begin{pmatrix}\text{for any two points } \bar{\mathbf{x}} \text{ and } \hat{\mathbf{x}}\\ \text{the preceding inequality holds}\\ \text{for all values } \lambda,\ 0 \le \lambda \le 1\end{pmatrix}$$

We therefore use the inequality to define a **convex function** as follows:

> A function f is convex if and only if, for all choices of points $\bar{\mathbf{x}}$ and $\hat{\mathbf{x}}$
>
> $$f(\lambda\bar{\mathbf{x}} + (1 - \lambda)\hat{\mathbf{x}}) \le \lambda f(\bar{\mathbf{x}}) + (1 - \lambda)f(\hat{\mathbf{x}})$$
>
> for all λ satisfying $0 \le \lambda \le 1$.

A function f is said to be **strictly convex** if each of the inequalities in the preceding definition is a strict inequality. A function f is called **concave** if $-f$ is a convex function.

Convexity and Minima

In Section 9.2 we considered the classifiction of stationary points for unconstrained problems according to the definiteness of the Hessian matrix at a particular point $\bar{\mathbf{x}}$, using the following relationships:

> If $\bar{\mathbf{x}}$ is a stationary point of $f(\mathbf{x})$, then the following implications hold.
>
> $\mathbf{H}(\bar{\mathbf{x}})$ is positive definite $\Rightarrow$ $\bar{\mathbf{x}}$ is a strict minimizing point
>
> $\mathbf{H}(\mathbf{x})$ is positive semidefinite for all $\mathbf{x}$ in some neighborhood of $\bar{\mathbf{x}}$ $\Rightarrow$ $\bar{\mathbf{x}}$ is a minimizing point
>
> $\bar{\mathbf{x}}$ is a minimizing point $\Rightarrow$ $\mathbf{H}(\bar{\mathbf{x}})$ is positive semidefinite

There is also a relationship between the convexity of a function and the definiteness of its Hessian matrix.

$\mathbf{H}(\mathbf{x})$ is positive semidefinite for all values of $\mathbf{x}$.	$\Leftrightarrow$	$f(\mathbf{x})$ is a convex function
$\mathbf{H}(\mathbf{x})$ is positive definite for all values of $\mathbf{x}$.	$\Rightarrow$	$f(\mathbf{x})$ is a strictly convex function

Note that a function can be strictly convex and yet $\mathbf{H}(\mathbf{x})$ need not be positive definite for all $\mathbf{x}$.

If $f(\mathbf{x})$ is a convex function, $\mathbf{H}(\mathbf{x})$ is positive semidefinite for all values of $\mathbf{x}$, so $\mathbf{H}(\mathbf{x})$ is surely positive semidefinite everywhere in a neighborhood of a stationary point $\bar{\mathbf{x}}$, and thus $\bar{\mathbf{x}}$ must be a minimum.

The important thing about this result is that it is global, in the sense that it describes a property of the function that is true everywhere and need not be checked at a particular point. Thus, if a function is convex, like the one pictured in Figure 9.11*b*, there is no need to check whether its stationary point is a minimum; we know from the positive semidefiniteness of the Hessian that it must be a minimum. The function pictured in Figure 9.11*a*, on the other hand, is nonconvex, so its stationary points must be examined individually and classified according to the definiteness of the Hessian matrix (or the sign of the second derivative) in the neighborhood of each stationary point.

Of course, what we really want is a way of classifying the solutions of the KKT conditions for a nonlinear programming problem with constraints. As we shall see in Section 9.6, adding the requirement that the objective and constraint functions be convex ensures that solutions of the KKT conditions correspond to global minimizing points.

Checking Whether a Function is Convex

It is sometimes possible to determine whether a function is convex by using the definition. That is, it may be possible to show that

$$f(\lambda\bar{\mathbf{x}} + [1 - \lambda]\hat{\mathbf{x}}) \leq \lambda f(\bar{\mathbf{x}}) + [1 - \lambda]f(\hat{\mathbf{x}})$$

for all values of $\bar{\mathbf{x}}$ and $\hat{\mathbf{x}}$ and for all $\lambda \in [0, 1]$. However, usually this is quite difficult, particularly if the vector $\mathbf{x}$ is of high dimension. Practical tests for convexity are therefore based on determining directly the definiteness of the Hessian matrix.

In Section 9.2 we checked whether a given matrix was positive definite by using the definition. That is, for the example given there we showed that $\mathbf{z}^T\mathbf{H}\mathbf{z} > \mathbf{0}$ for all vectors $\mathbf{z} \neq \mathbf{0}$. In most cases this is, unfortunately, also quite difficult. For example, consider the question whether the following matrix is positive semidefinite:

$$H = \begin{bmatrix} 4 & 2 & 0 \\ 2 & 2 & -1 \\ 0 & -1 & 4 \end{bmatrix}$$

Using the definition, we would calculate

$$[z_1, z_2, z_3]\begin{bmatrix} 4 & 2 & 0 \\ 2 & 2 & -1 \\ 0 & -1 & 4 \end{bmatrix}\begin{bmatrix} z_1 \\ z_2 \\ z_3 \end{bmatrix} = 4z_1^2 + 4z_1z_2 + 2z_2^2 - 2z_2z_3 + 4z_3^2$$

It is not immediately obvious whether this quantity is always nonnegative. Fortunately, positive definite and positive semidefinite matrices have many other properties that can be used for testing definiteness.

The properties that we shall use for checking definiteness of a matrix concern the determinants of certain submatrices. These determinants are called **minors** of the matrix. A **principal minor** of **H** is the determinant of a square submatrix whose diagonal lies on the diagonal of **H**. The 1×1 principal minors of the preceding matrix are the diagonal elements 4, 2, and 4. The 2×2 principal minors are the determinants

$$\begin{vmatrix} 4 & 2 \\ 2 & 2 \end{vmatrix} = 4 \qquad \begin{vmatrix} 2 & -1 \\ -1 & 4 \end{vmatrix} = 7, \qquad \begin{vmatrix} 4 & 0 \\ 0 & 4 \end{vmatrix} = 16$$

The single 3×3 principal minor is just the determinant of the matrix,

$$\begin{vmatrix} 4 & 2 & 0 \\ 2 & 2 & -1 \\ 0 & -1 & 4 \end{vmatrix} = 12$$

Among the principal minors, some are further distinguished as leading. A **leading principal minor** is the determinant of a square submatrix whose (1, 1) element is the (1, 1) element of **H**. These submatrices are illustrated for the example matrix:

$$\begin{bmatrix} \boxed{4} & 2 & 0 \\ 2 & 2 & -1 \\ 0 & -1 & 4 \end{bmatrix}$$

The determinants of these submatrices are the leading principal minors of the matrix

The leading principal minors of this matrix are

$$4 \quad \text{and} \quad \begin{vmatrix} 4 & 2 \\ 2 & 2 \end{vmatrix} = 4 \quad \text{and} \quad \begin{vmatrix} 4 & 2 & 0 \\ 2 & 2 & -1 \\ 0 & -1 & 4 \end{vmatrix} = 12$$

The following properties of symmetric positive definite and positive semidefinite matrices permit us to check the definiteness of matrix by computing minors.

For symmetric matrices **H**,

H is positive definite $\Leftrightarrow$ all of its leading principal minors are positive

H is positive semidefinite $\Leftrightarrow$ all of its principal minors are nonnegative

The preceding matrix is positive definite (recall that we abbreviate this pd) because all of its leading principal minors are strictly positive. Of course, this means the matrix is also positive semidefinite (psd) because

$$\begin{aligned} \mathbf{H} \text{ pd} &\Rightarrow \mathbf{z}^T\mathbf{H}\mathbf{z} > 0 \quad \text{for all } \mathbf{z} \neq \mathbf{0} \\ &\Rightarrow \mathbf{z}^T\mathbf{H}\mathbf{z} \geq 0 \quad \text{for all } \mathbf{z} \neq \mathbf{0} \Rightarrow \mathbf{H} \text{ psd} \end{aligned}$$

By computing minors, it is possible to check small matrices quite rapidly by hand calculations. There are fewer leading principal minors than there are principal minors, so in checking the definiteness of a matrix it makes sense to compute them first. If they are all positive, so that the matrix is pd, then it is surely also psd. If any of the leading principal minors are negative, then the matrix is surely neither pd nor psd. If some of the leading principal minors are zero, then the matrix is not pd but might be psd, and it is necessary to check the remaining principal minors.

Now consider the problem of deciding whether the following function is convex:

$$f(\mathbf{x}) = x_1^4 + 2x_2^2 + 3x_3^2 - 4x_1 - 4x_2x_3$$

To use the approach already outlined, we must first compute the Hessian matrix for this function:

$$\mathbf{H}(\mathbf{x}) = \begin{bmatrix} 12x_1^2 & 0 & 0 \\ 0 & 4 & -4 \\ 0 & -4 & 6 \end{bmatrix}$$

Note that this Hessian matrix is not constant but is a function of x_1. Next we compute the leading principal minors and check their signs.

$$12x_1^2 \geq 0 \qquad \begin{vmatrix} 12x_1^2 & 0 \\ 0 & 4 \end{vmatrix} = 48x_1^2 \geq 0 \qquad \begin{vmatrix} 12x_1^2 & 0 & 0 \\ 0 & 4 & -4 \\ 0 & -4 & 6 \end{vmatrix} = 96x_1^2 \geq 0$$

The leading principal minors are not strictly positive because x_1 might be zero (in which case the 2×2 and 3×3 leading principal minors would be zero). Thus the function is not strictly convex. None of the leading principal minors are negative, however, so the Hessian might still be psd. To decide, we must calculate the remaining principal minors and check their signs.

$$4 > 0 \qquad 6 > 0 \qquad \begin{vmatrix} 4 & -4 \\ -4 & 6 \end{vmatrix} = 8 > 0 \qquad \begin{vmatrix} 12x_1^2 & 0 \\ 0 & 6 \end{vmatrix} = 72x_1^2 \geq 0$$

None of the nonleading principal minors is negative, so we conclude that the matrix $\mathbf{H}$ is psd (though not pd) and the function $f(\mathbf{x})$ is convex (though not strictly convex).

It is possible for a nonsymmetric matrix to be positive definite or positive semidefinite, but the test is slightly more complicated than that outlined (see Exercise 9.21).

9.6 KARUSH–KUHN–TUCKER THEORY OF NONLINEAR PROGRAMMING

We are now prepared to state two important theorems in nonlinear programming, called the Karush–Kuhn–Tucker theorems. The first theorem states what is necessary to ensure that a local minimizing point must satisfy the KKT conditions. The second theorem states what is sufficient to ensure that a point satisfying the KKT conditions is a global minimizing point.

THE KARUSH–KUHN–TUCKER THEOREMS

Given the canonical form nonlinear programming problem

$$\underset{\mathbf{x}\in R^n}{\text{minimize}} \quad f_0(\mathbf{x})$$

subject to

$$f_i(\mathbf{x}) \leq 0 \qquad i = 1, \ldots, m$$

(necessary conditions)

if f_i is differentiable, $i = 0, \ldots, m$

and $\mathbf{x}$ is a local minimizing point

and a constraint qualification holds

then there exists a vector $\bar{\mathbf{u}}$ such that $(\bar{\mathbf{x}}, \bar{\mathbf{u}})$ satisfies the KKT conditions;

(sufficient conditions)

if $(\bar{\mathbf{x}}, \bar{\mathbf{u}})$ satisfies the KKT conditions

and f_i is a convex function, $i = 0, \ldots, m$

then $\bar{\mathbf{x}}$ is a global minimizing point.

The KKT Method

The Karush–Kuhn–Tucker theorems play a role for inequality-constrained problems similar to that played by the Lagrange multiplier theorem for problems with equality constraints, and suggest a general method for solving inequality-constrained nonlinear programming problems, as follows.

THE KKT METHOD

1. Form the Lagrangian function

$$L(\mathbf{x}, \mathbf{u}) = f_0(\mathbf{x}) + \sum_{i=1}^{m} u_i f_i(\mathbf{x})$$

2. Find all solutions $(\bar{\mathbf{x}}, \bar{\mathbf{u}})$ to the following system of nonlinear algebraic equations and inequalities:

$$\frac{\partial L}{\partial x_j} = 0, \qquad j = 1, \ldots, n \quad \text{gradient condition}$$

$$f_i(x) \leq 0, \qquad i = 1, \ldots, m \quad \text{feasibility}$$

$$u_i f_i(x) = 0, \qquad i = 1, \ldots, m \quad \text{orthogonality}$$

$$u_i \geq 0, \qquad i = 1, \ldots, m \quad \text{nonnegativity}$$

3. If the functions $f_i(\mathbf{x})$, are all convex, the points $\bar{\mathbf{x}}$ are global minimizing points. Otherwise, examine each solution $(\bar{\mathbf{x}}, \bar{\mathbf{u}})$ to see if $\bar{\mathbf{x}}$ is a minimizing point.

To show how the KKT method can be used in practice, we shall use it to solve the following problem.

$$\begin{aligned}
\underset{\mathbf{x}\in R^2}{\text{minimize}}\ & f_0(\mathbf{x}) = -3x_1 + \tfrac{1}{2}x_2^2 \\
\text{subject to } & f_i(\mathbf{x}) = x_1^2 + x_2^2 - 1 \le 0 \\
& f_2(\mathbf{x}) = -x_1 \le 0 \\
& f_3(\mathbf{x}) = -x_2 \le 0
\end{aligned}$$

To apply the KKT method to this problem we proceed as follows:

1. Form the Lagrangian function.

$$\begin{aligned} L(\mathbf{x}, \mathbf{u}) = & (-3x_1 + \tfrac{1}{2}x_2^2) \\ & + u_1[x_1^2 + x_2^2 - 1] + u_2[-x_1] + u_3[-x_2] \end{aligned}$$

2. Write down the KKT conditions and solve them

$$\left.\begin{aligned} \frac{\partial L}{\partial x_1} &= -3 + 2u_1x_1 - u_2 = 0 & \text{(a)} \\ \frac{\partial L}{\partial x_2} &= x_2 + 2u_1x_2 - u_3 = 0 & \text{(b)} \end{aligned}\right\} \quad \text{gradient condition}$$

$$\left.\begin{aligned} f_1(x) &= x_1^2 + x_2^2 - 1 \le 0 & \text{(c)} \\ f_2(x) &= -x_1 \le 0 & \text{(d)} \\ f_3(x) &= -x_2 \le 0 & \text{(e)} \end{aligned}\right\} \quad \text{feasibility}$$

$$\left.\begin{aligned} u_1[x_1^2 + x_2^2 - 1] &= 0 & \text{(f)} \\ u_2[-x_1] &= 0 & \text{(g)} \\ u_3[-x_2] &= 0 & \text{(h)} \end{aligned}\right\} \quad \text{orthogonality}$$

$$u_1 \ge 0,\ u_2 \ge 0,\ u_3 \ge 0 \qquad \text{nonnegativity} \qquad \text{(k)}$$

If $x_1^2 + x_2^2 - 1 \neq 0$, then $(f) \Rightarrow u_1 = 0$
$u_1 = 0$ & (a) $\Rightarrow -3 - u_2 = 0 \Rightarrow u_2 = -3$
but $(k) \Rightarrow u_2 \ge 0$ so $u_1 \neq 0$ and $[x_1^2 + x_2^2 - 1] = 0$ (p)
$u_1 \neq 0$ & $(k) \Rightarrow u_1 > 0$ (q)

If $u_2 \neq 0$, then (g) $\Rightarrow x_1 = 0$
$x_1 = 0$ & (a) $\Rightarrow -3 + 2u_1\,[0] - u_2 = 0 \Rightarrow u_2 = -3$

but $(k) \Rightarrow u_2 \ge 0$ so $x_1 \neq 0$ and $u_2 = 0$ (r)

If $x_2 \neq 0$, then (h) $\Rightarrow u_3 = 0$
$u_3 = 0$ & (b) $\Rightarrow x_2 + 2u_1x_2 - [0] = 0$
$x_2\,(1 + 2u_1) = 0$ & (q) $\Rightarrow x_2 = 0$
but we assumed $x_2 \neq 0$, so $x_2 = 0$ (s)

(p) & (s) $\Rightarrow x_1^2 + [0]^2 - 1 = 0 \Rightarrow x_1 = \pm 1$
$x_1 = \pm 1$ & (d) $\Rightarrow x_1 = 1$ (t)
(a) & (t) & (r) $\Rightarrow -3 + 2u_1\,[1] - [0] = 0 \Rightarrow u_1 = \frac{3}{2}$ (v)
(b) & (v) & (s) $\Rightarrow [0] + 2[\frac{3}{2}][0] - u_3 = 0 \Rightarrow u_3 = 0$
Thus we find that the point $\bar{\mathbf{x}} = [1, 0]^T$ satisfies the KKT conditions, with $\bar{\mathbf{u}} = [\frac{3}{2}, 0, 0]^T$.

3. Check whether the functions are convex.

$$f_0(\mathbf{x}) = -3x_1 + \frac{x_2^2}{2} \qquad \frac{\partial f_0}{\partial x_1} = -3 \qquad \frac{\partial f_0}{\partial x_2} = x_2$$

$$\frac{\partial^2 f_0}{\partial x_1^2} = 0 \qquad \frac{\partial^2 f_0}{\partial x_2^2} = 1 \qquad \frac{\partial^2 f_0}{\partial x_1\, \partial x_2} = \frac{\partial^2 f_0}{\partial x_2\, \partial x_1} = 0 \qquad \mathbf{H} = \begin{bmatrix} 0 & 0 \\ 0 & 1 \end{bmatrix}$$

leading principal minors:

$$|0| \geq 0, \qquad \begin{vmatrix} 0 & 0 \\ 0 & 1 \end{vmatrix} = 0 \geq 0$$

other principal minors:

$$|1| > 0$$

The function $f_0(\mathbf{x})$ is convex.

$$f_1(x) = x_1^2 + x_2^2 - 1 \qquad \frac{\partial f_1}{\partial x_1} = 2x_1 \qquad \frac{\partial f_1}{\partial x_2} = 2x_2$$

$$\frac{\partial^2 f_1}{\partial x_1^2} = 2 \qquad \frac{\partial^2 f_1}{\partial x_2^2} = 2 \qquad \frac{\partial^2 f_1}{\partial x_1\, \partial x_2} = \frac{\partial^2 f_1}{\partial x_2\, \partial x_1} = 0 \qquad \mathbf{H} = \begin{bmatrix} 2 & 0 \\ 0 & 2 \end{bmatrix}$$

leading principal minors:

$$|2| > 0 \qquad \begin{vmatrix} 2 & 0 \\ 0 & 2 \end{vmatrix} = 4 > 0$$

The function $f_1(\mathbf{x})$ is strictly convex.

The functions $f_2(\mathbf{x})$ and $f_3(\mathbf{x})$ are linear and thus convex by the definition of a convex function. All of the functions are thus convex, and the point $\mathbf{x}^* = [1, 0]^T$ is a global minimum.

Note that for this problem there is one inactive constraint, $f_2(\bar{\mathbf{x}}) < 0$, so $\bar{u}_2 = 0$. The gradients of the active constraints,

$$\nabla f_1(\bar{\mathbf{x}}) = \begin{bmatrix} 2\bar{x}_1 \\ 2\bar{x}_2 \end{bmatrix} = \begin{bmatrix} 2 \\ 0 \end{bmatrix}$$

and

$$\nabla f_2(\bar{\mathbf{x}}) = \begin{bmatrix} 0 \\ -1 \end{bmatrix}$$

are linearly independent vectors.

Strategy in Solving the KKT Conditions

At several places in solving the KKT conditions for the preceding example, we used the orthogonality conditions to argue that

$$\text{if } u_i \neq 0 \quad \text{then} \quad f_i(\mathbf{x}) = 0$$

and

$$\text{if } f_i(\mathbf{x}) \neq 0 \quad \text{then} \quad u_i = 0$$

By following one of the conclusions to a contradiction with the remaining conditions, we were able to rule out the corresponding hypothesis and deduce that $u_i = 0$ or that $f_i(\mathbf{x}) = 0$ (or perhaps both). Thinking back on the general approach to inequality constraints outlined in Section 9.4, one can view this method of reasoning about the KKT conditions as simply a way of deciding which constraints are active at the optimal point. As it is discovered that certain constraints

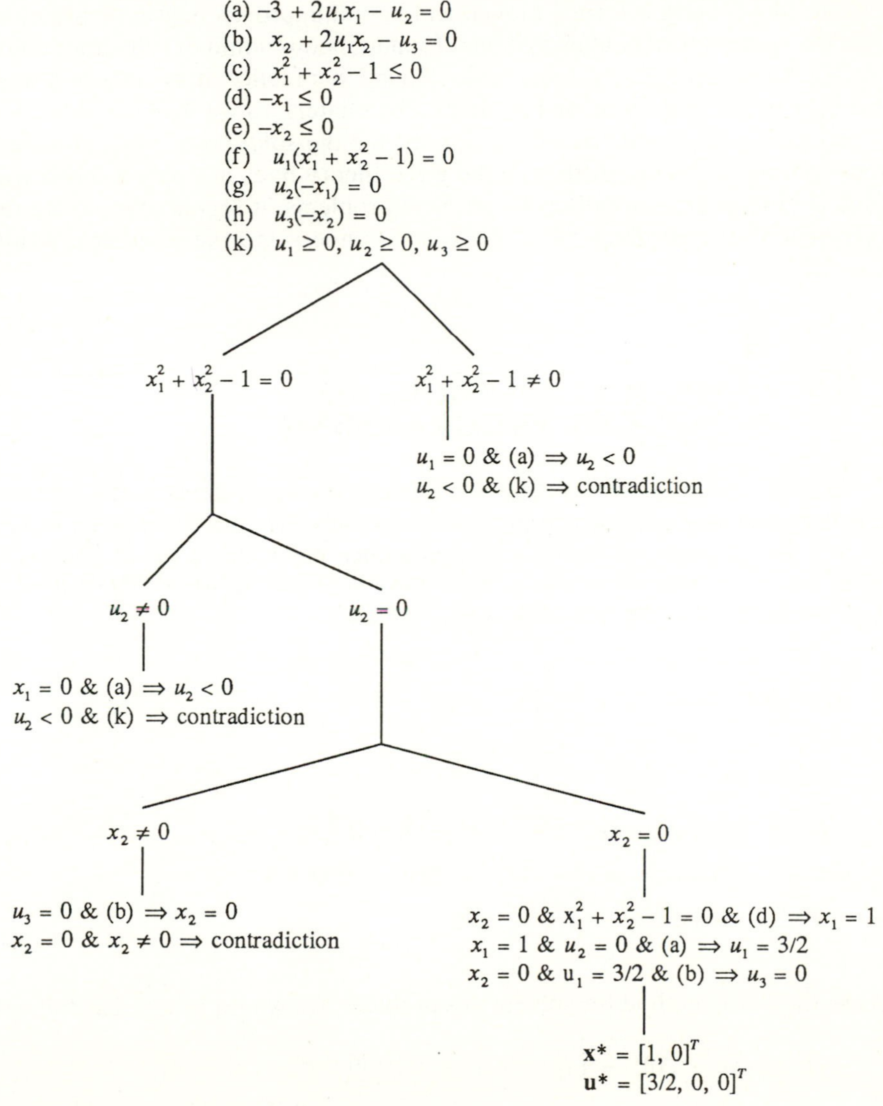

FIGURE 9.12 Branching diagram for the KKT example.

are active and others are inactive, some of the KKT inequalities become equations and some of the Lagrange multipliers can be fixed at 0, so that the complexity of the remaining problem is reduced. This kind of argument based on the orthogonality conditions is thus often quite helpful in dealing with KKT conditions analytically.

Each of the m inequality constraints in a problem can be either active or inactive, so there are 2^m possible combinations of active and inactive constraints. This suggests organizing the calculations by means of a branching diagram similar to those used in solving integer programs. To illustrate this idea, the solution of the KKT conditions for the preceding example is repeated in Figure 9.12 in the form of a branching diagram.

From a practical point of view, the most important observation of all about solving the KKT conditions analytically, and the saddest, is that it usually cannot be done. Even using a formal branch-and-bound approach is usually futile because the subproblems contain systems of simultaneous nonlinear algebraic equations that are analytically intractable. Thus, in practice it is almost always necessary to resort to numerical methods. Fortunately, as we shall see in Section 9.7, effective algorithms have been devised for the solution of most nonlinear programming problems. Although the KKT theory does not play a direct role in the computation of solutions to practical problems, it serves to motivate the development of algorithms and it helps us to understand the numerical results they yield.

9.7 NUMERICAL METHODS OF NONLINEAR PROGRAMMING

In earlier sections of this chapter we referred several times to the practical difficulty of solving nonlinear programs analytically. The problem given in Section 9.1, for example, is harmless in appearance but yields a set of KKT conditions that probably do not have a closed-form analytical solution. The following simple example also illustrates this phenomenon.

$$\underset{\mathbf{x}\in\mathbf{R}^2}{\text{minimize}}\ f_0(\mathbf{x}) = (x_1 - 20)^4 + (x_2 - 12)^4$$

subject to

$$\begin{aligned} f_1(\mathbf{x}) &= 8e^{(x_1-12)/9} - x_2 + 4 \le 0 \\ f_2(\mathbf{x}) &= 6(x_1 - 12)^2 + 25x_2 - 600 \le 0 \\ f_3(\mathbf{x}) &= -x_1 + 12 \le 0 \end{aligned}$$

To use the KKT method for solving this problem, we would proceed as follows:

$$\begin{aligned} L = (x_1 - 20)^4 + (x_2 - 12)^4 &+ u_1[8e^{(x_1-12)/9} - x_2 + 4] \\ &+ u_2[6(x_1 - 12)^2 + 25x_2 - 600] \\ &+ u_3[-x_1 + 12] \end{aligned}$$

KKT Conditions:

$$\frac{\partial L}{\partial x_1} = 4(x_1 - 20)^3 + \left(\frac{8}{9}\right)u_1 e^{(x_1-12)/9} + 12u_2(x_1 - 12) - u_3 = 0$$

$$\frac{\partial L}{\partial x_2} = 4(x_2 - 12)^3 - u_1 + 25u_2 = 0$$

$$\frac{\partial L}{\partial u_1} = 8e^{(x_1-12)/9} - x_2 + 4 \leq 0$$

$$\frac{\partial L}{\partial u_2} = 6(x_1 - 12)^2 + 25x_2 - 600 \leq 0$$

$$\frac{\partial L}{\partial u_3} = -x_1 + 12 \leq 0$$

$$u_1[8e^{(x_1-12)/9} - x_2 + 4] = 0$$

$$u_2[6(x_1 - 12)^2 + 25x_2 - 600] = 0$$

$$u_3[-x_1 + 12] = 0$$

$$u_1, u_2, u_3 \geq 0$$

This KKT system also probably does not have an analytical solution. In view of the fact that simple examples such as this one and the problem of Section 9.1 yield KKT systems that cannot be solved, it seems reasonable to expect that numerical methods will be needed for most nonlinear programming problems that arise in practical applications.

Because of the effectiveness of the KKT method for problems in which the conditions can be solved analytically, one might suppose that a workable approach for more difficult problems would be to solve the KKT conditions numerically. It turns out in practice that it is much simpler to seek the minimizing point for a nonlinear program directly, without ever writing down the KKT conditions for the problem. The direct numerical approach also has the virtue that some algorithms can succeed even when a constraint qualification is not satisfied, and might find a global minimum even when the problem is not convex. This section is therefore devoted to the presentation of algorithms for direct numerical optimization.

Before we begin, we need to introduce some new notation. Throughout this section we shall have occasion to refer to the **iterates,** or trial solution points, generated by numerical algorithms. In the case of a one-dimensional search these points will be scalars, but later we shall consider algorithms that generate iterates that are vectors. To distinguish between vector components and entire vectors generated at different iterations of an algorithm, we will use the following scheme.

$\mathbf{x}$ is a vector of n components

x_3 is the third component of $\mathbf{x}$

x_3^2 means $x_3 \cdot x_3$

$\mathbf{x}^2$ is a vector $\mathbf{x}$ generated at iteration 2

$\mathbf{x}_3^2$ is the third component of iterate $\mathbf{x}^2$

Line Searching

Just as we began the study of nonlinear programming theory with one-dimensional problems in Section 9.1, it is logical to start with one-dimensional problems in the study of numerical optimization. Consider the function of Figure 9.2, which is reproduced for convenience as Figure 9.13.

One of the simplest algorithms for finding the minimum of a function of one variable is the **bisection line search.** The idea of this method is to look for a value of x at which the derivative $df(x)/dx$ is zero. Suppose we want to search the interval $[a^0, b^0]$ for a minimum of the function. We begin by finding the midpoint $x^0 = \frac{1}{2}(a^0 + b^0)$ of the interval, and evaluate the derivative of the function there.

Recall that the slope of a function's graph is negative to the left of a minimum and positive to the right. Thus, if the derivative at x^0 is positive, as shown in Figure 9.13, we can deduce for a convex function that the minimum must be between a^0 and x^0. In that case we can discard the right half of the original interval to obtain a new problem in which the interval to be searched is $[a^0, x^0]$. In the other case, when the slope at the midpoint is negative, the minimum must be between x^0 and b^0 and we can discard the left half of the original interval. Repeating this process reduces the interval of uncertainty until the location of the minimum is known to within any prescribed tolerance. Formally, the algorithm is as follows.

BISECTION LINE SEARCH

0. Initialize
   a^0 = left end of interval containing minimum
   b^0 = right end of interval containing minimum
   T = maximum allowable uncertainty in x
   Set $k = 0$.
1. Compute midpoint
   $x^k = \frac{1}{2}(a^k + b^k)$
2. Reduce interval
   if $df(x^k)/dx < 0$, $a^{k+1} = x^k$ and $b^{k+1} = b^k$
   if $df(x^k)/dx > 0$, $a^{k+1} = a^k$ and $b^{k+1} = x^k$
3. Test
   if $df(x^k)/dx = 0$, x^k is the minimum; STOP
   if $|b^k - a^k| < T$, x^k is close enough; STOP
4. Repeat
   increase k by 1
   GO TO 1

Applying the algorithm to the function $f(x) = x^2 - 2x + 3$ with $T = 0.1$ and the starting interval shown in Figure 9.13 yields the following sequence of intervals and midpoints.

k	a^k	b^k	x^k	$df(x^k)/dx$	$b^k - a^k$
0	0.5	2.0	1.25	+0.5	1.5
1	0.5	1.25	0.875	−0.25	0.75
2	0.875	1.25	1.0625	+0.125	0.375
3	0.875	1.0625	0.96875	−0.0625	0.1875
4	0.96875	1.0625	1.015625	+0.03125	0.09375

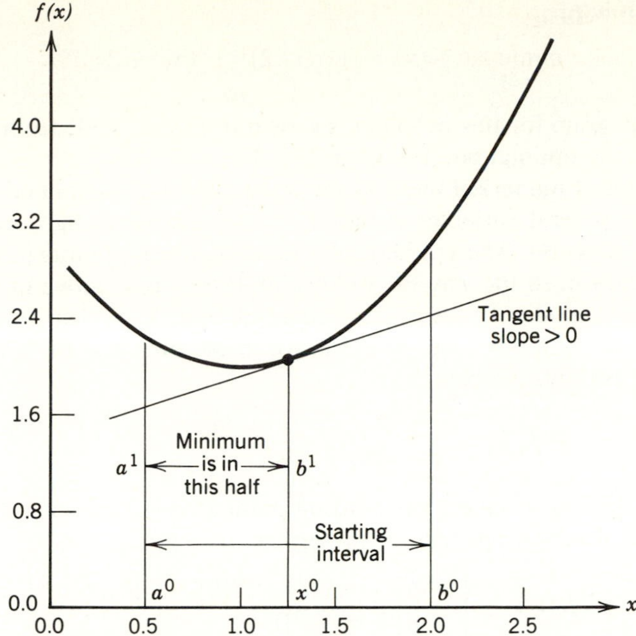

FIGURE 9.13 Bisection line search on the function $f(x) = x^2 - 2x + 3$.

The iterates x^k are converging to the correct value of 1, and the interval $[a^k, b^k]$ brackets the true minimum at each iteration.

Bisection reduces the interval of uncertainty by a factor of 2 for each iteration, so after k iterations the original interval is reduced to

$$b^k - a^k = (b^0 - a^0) \cdot 2^{-k}$$

This behavior is referred to as **linear convergence** because the error after iteration k is a linear function (half, to be precise) of the error after iteration $k - 1$. More sophisticated methods of line searching have been devised in which the convergence is **superlinear,** so that the error after iteration k is proportional to some power $p > 1$ of the error after iteration $k - 1$. Superlinearly convergent methods require fewer iterations to reach a specified convergence tolerance T, when they work.

All line search schemes can converge to local minima when the function has them, instead of to a global minimum, and can fail because of roundoff errors or irregular behavior (such as nondifferentiability) of the function. In fact, all algorithms for nonlinear programming are subject to failure for these reasons. It has been observed empirically that simple methods, such as bisection, are less sensitive to such disturbances than sophisticated methods, so speedy convergence is typically obtained only at the price of decreased robustness. A careful consideration of these complications is important in the design of practical algorithms.

The Method of Steepest Descent

The next step in our exposition of nonlinear programming theory was to consider unconstrained problems with more than one variable, such as the following example:

$$\underset{\mathbf{x}\in R^2}{\text{minimize}}\ f_0(\mathbf{x}) = (x_1 - 2)^4 + (x_1 - 2x_2)^2$$

A contour diagram for this problem is shown in Figure 9.14, from which it can be seen that the optimal point is $\mathbf{x}^* = [2, 1]^T$.

The simplest numerical method for finding the unconstrained minimum of a function of several variables is based on the idea of going downhill on the graph of the function. The gradient of a function always points in the direction of fastest increase, so the way to go downhill fastest is to move in the direction opposite to the gradient vector. This suggests that, if our current estimate of the minimizing point is $\mathbf{x}^k$, we should move in the direction of $-\nabla f_0(\mathbf{x}^k)$.

For the example problem,

$$\nabla f_0(\mathbf{x}) = \begin{bmatrix} 4(x_1 - 2)^3 + 2(x_1 - 2x_2) \\ -4(x_1 - 2x_2) \end{bmatrix}$$

Suppose we arbitrarily select the starting point $\mathbf{x}^0 = [0, 3]^T$ at which $f_0(\mathbf{x}^0) = 52$. At $\mathbf{x}^0$ the direction opposite to the gradient vector is

$$\mathbf{d}^0 = -\nabla f_0(\mathbf{x}^0) = \begin{bmatrix} -4([0] - 2)^3 - 2([0] - 2[3]) \\ 4([0] - 2[3]) \end{bmatrix} = \begin{bmatrix} 44 \\ -24 \end{bmatrix}$$

and moving in the direction $\mathbf{d}^0$ will decrease the function value. For instance, we could move to the point

$$\bar{\mathbf{x}} = \mathbf{x}^0 + 0.1\mathbf{d}^0 = \begin{bmatrix} 0 \\ 3 \end{bmatrix} + \begin{bmatrix} 4.4 \\ -2.4 \end{bmatrix} = \begin{bmatrix} 4.4 \\ 0.6 \end{bmatrix}$$

at which $f_0(\bar{\mathbf{x}}) = 43.4^+$. Ideally, the distance we move in the direction $\mathbf{d}^0$ should be chosen so as to achieve the greatest possible decrease in the objective function value. In other words we should move to a point $\mathbf{x}^1 = \mathbf{x}^0 + \alpha\mathbf{d}^0$, where α is

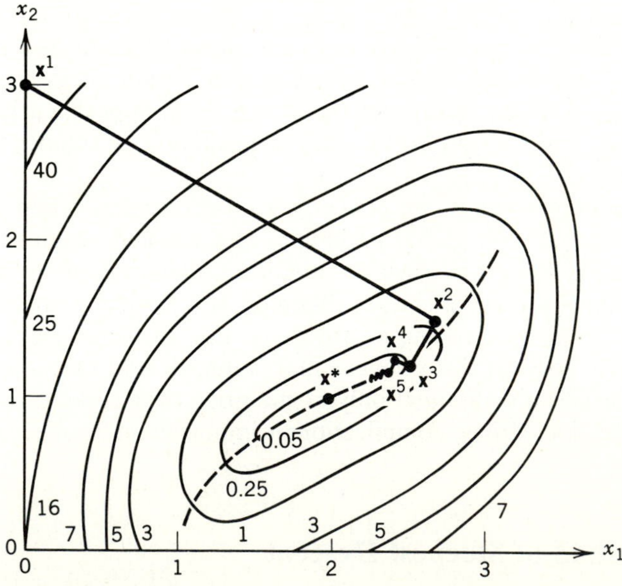

FIGURE 9.14 Contours of $f_0(\mathbf{x}) = (x_1 - 2)^4 + (x_1 - 2x_2)^2$.

chosen so as to minimize the objective value. The optimal value of α can thus be found by solving the following one-dimensional minimization problem:

$$\underset{\alpha \in R^1}{\text{minimize}}\ f_0(\mathbf{x}^0 + \alpha\mathbf{d}^0) = f_0(\alpha)$$

Using the technique of Section 9.1, we would solve this problem by setting the derivative of $f_0(\mathbf{x}^0 + \alpha\mathbf{d}^0)$ with respect to α equal to zero, as follows:

$$\begin{aligned} f_0(\mathbf{x}^0 + \alpha\mathbf{d}^0) &= ([0 + 44\alpha] - 2)^4 + ([0 + 44\alpha] - 2[3 - 24\alpha])^2 \\ &= 3748096\alpha^4 - 681472\alpha^3 + 54928\alpha^2 - 2512\alpha + 52 \\ df_0/d\alpha &= 14992384\alpha^3 - 2044416\alpha^2 + 109856\alpha - 2512 = 0 \end{aligned}$$

This cubic equation is difficult to solve analytically but can be solved numerically to yield $\alpha^* = 0.062^-$.

Alternatively, of course, we could simply perform a line search (using a method such as bisection) in the direction $\mathbf{d}^0$. This is in fact where the term "line search" originates; the locus of points

$$\{\mathbf{x} \in R^n \mid \mathbf{x} = \mathbf{x}^k + \alpha\mathbf{d}\}$$

is a line passing through the point $\mathbf{x}^k$ and having direction $\mathbf{d}$. The scalar parameter α measures distance on the line from the point $\mathbf{x}^k$. In implementing a numerical algorithm such as the method of steepest descent, it is usually both more convenient and computationally faster to use a numerical line search rather than calculating the derivative analytically and then using a numerical method to solve the resulting nonlinear equation $df_0/d\alpha = 0$. A numerical line search also has the advantage that it can be designed to detect and ignore stationary points that are maxima or inflections rather than minima.

The main difficulty with using most numerical line search procedures for the method of steepest descent is that it is necessary to specify an interval over which to conduct the search. In our example the left-hand end of the initial search interval could obviously be chosen as $\alpha = 0$, but what should we select as the right-hand end? In practice it is usually necessary to use some **heuristic procedure,** a rule of thumb that often works but cannot be guaranteed, to determine the starting interval. For example, we might check function values along the line at larger and larger values of α, watching for the function value to go down and then up again. One way of generating the successive values of α would be to start with an increment of $0.01 \cdot \|\mathbf{d}\|$, where $\|\mathbf{d}\|$ is the length $\|\mathbf{d}\| = \sqrt{\mathbf{d}^T\mathbf{d}}$ of the direction vector $\mathbf{d}$, and double the increment in α after each step. This would yield α values of 0, $(0.01)\|\mathbf{d}\|$, $(0.03)\|\mathbf{d}\|$, $(0.07)\|\mathbf{d}\|$, and so on. As soon as the function value goes down and then up, we stop incrementing α and take two values that bracket the minimum as the ends of the initial search interval. Bisection or some other method can then be used to locate the minimum with the required precision.

Having completed the line search in the direction $\mathbf{d}^0$ and obtained $\alpha^* = 0.062^-$ by one means or another, we find

$$\mathbf{x}^1 = \mathbf{x}^0 + \alpha^*\mathbf{d}^0 = \begin{bmatrix} 0 \\ 3 \end{bmatrix} + (0.062^-)\begin{bmatrix} 44 \\ -24 \end{bmatrix} = \begin{bmatrix} 2.70 \\ 1.51 \end{bmatrix}$$

Next we repeat the process starting from this point, by calculating the search direction $\mathbf{d}^1 = -\nabla f_0(\mathbf{x}^1)$, conducting a line search in that direction, and so forth.

The iterations continue until successive values of $\mathbf{x}^k$ are close enough together, say $\|\mathbf{x}^{k+1} - \mathbf{x}^k\| < T$. Formally, the algorithm can be summarized as follows.

STEEPEST DESCENT

0. Initialize
   $\mathbf{x}^0$ = starting point
   T = convergence tolerance
   set $k = 0$
1. Find search direction
   $\mathbf{d}^k = -\nabla f_0(\mathbf{x}^k)$
   if $\mathbf{d}^k = 0$, STOP; $\mathbf{x}^k$ is the minimum
2. Search the line $\mathbf{x} = \mathbf{x}^k + \alpha \mathbf{d}^k$ for a minimum
   use a heuristic to get the starting interval
   return the value α^* at which $f_0(\alpha)$ is minimized
3. Update estimate of minimum
   $\mathbf{x}^{k+1} = \mathbf{x}^k + \alpha^* \mathbf{d}^k$
4. Check for convergence
   if $\|\mathbf{x}^{k+1} - \mathbf{x}^k\| < T$, STOP; $\mathbf{x}^{k+1}$ is the minimum
5. Repeat
   increase k by 1
   GO TO 1

Applying the algorithm to the example problem, the starting point $\mathbf{x}^0 = [0, 3]^T$ yields the following sequence of iterates $\mathbf{x}^k$. Using $T = 0.05$, the algorithm stops after determining $\mathbf{x}^5$ but the table is continued to show the exact values to which the various quantities would converge if the iterates were continued indefinitely.

k	$(\mathbf{x}^k)^T$	$(\mathbf{d}^k)^T$	α^*	$\|\mathbf{x}^{k+1} - \mathbf{x}^k\|$
0	[0.00, 3.00]	[44.00, −24.00]	0.062	3.08
1	[2.70, 1.51]	[−0.73, −1.28]	0.24	0.36
2	[2.52, 1.20]	[−0.80, 0.48]	0.11	0.10
3	[2.43, 1.25]	[−0.18, −0.28]	0.31	0.11
4	[2.37, 1.16]	[−0.30, 0.20]	0.12	0.04
5	[2.33, 1.18]			
⋮	⋮ ⋮	⋮ ⋮	⋮	⋮
∞	[2.00, 1.00]	[0.00, 0.00]	0.00	0.00

In reading this table, it is important to remember that each iteration contains a line search as a subproblem.

These iterates are shown on the contour plot of Figure 9.14, along with some of the subsequent iterates generated by the algorithm. After $\mathbf{x}^3$, the algorithm follows a long, narrow valley in the graph of the function, taking tiny steps and requiring many iterations to get near the optimal point. This phenomenon, known as **zigzagging,** causes slow convergence and is thus a serious drawback to the method of steepest descent.

The problem of zigzagging suggests that faster convergence might be obtained by using more information than just the local gradient to determine the search direction. For example, using $\mathbf{d}^k = -\frac{1}{2}\{\nabla f_0(\mathbf{x}^k) + \nabla f_0(\mathbf{x}^{k-1})\}$ for the example yields search directions that run more nearly parallel to the valley in the graph of the function. Considerable attention has been devoted to the design of algorithms that generate good search directions, and sophisticated methods have resulted from this work.

The Generalized Reduced-Gradient Method

In our study of equality constraints in Section 9.3, we made use of the constraint equations to eliminate variables. Then we could solve the resulting reduced problem, which is unconstrained, by the methods of Section 9.2. In most cases the only way we could eliminate variables analytically was by using a parameterization of the constraints to express the variables of the original problem in terms of a smaller number of new variables, the parameters. In some cases, however, we could eliminate variables directly, by simply solving the constraint equations for some variables in terms of the others. One important circumstance in which it is often possible to do this is when the constraints are linear equations.

Consider the following problem, with a nonlinear objective and linear equality constraints:

$$\underset{\mathbf{x}\in R^4}{\text{minimize}}\ f_0(\mathbf{x}) = x_1^2 + x_2 + x_3^2 + x_4$$

subject to

$$f_1(\mathbf{x}) = x_1 + x_2 + 4x_3 + 4x_4 - 4 = 0$$
$$f_2(\mathbf{x}) = -x_1 + x_2 + 2x_3 - 2x_4 + 2 = 0$$

It is easy to solve the constraint equations for two of the variables in terms of the others. Solving for x_2 and x_3 in terms of x_1 and x_4 gives

$$x_2 = 3x_1 + 8x_4 - 8 \quad \text{and} \quad x_3 = -x_1 - 3x_4 + 3$$

Substituting these expressions into the objective function yields the following reduced problem:

$$\underset{x_1, x_4}{\text{minimize}}\ f_0(x_1, x_4) = x_1^2 + [3x_1 + 8x_4 - 8] + [-x_1 - 3x_4 + 3]^2 + x_4$$

This problem is unconstrained, so we can solve it by setting the partial derivatives equal to zero.

$$\left.\begin{aligned} \frac{\partial f_0}{\partial x_1} &= 2x_1 + 3 + 2[-x_1 - 3x_4 + 3](-1) = 0 \\ \frac{\partial f_0}{\partial x_4} &= 8 + 2[-x_1 - 3x_4 + 3](-3) + 1 = 0 \end{aligned}\right\} \Rightarrow \begin{aligned} x_1 &= 0 \\ x_4 &= \tfrac{1}{2} \end{aligned}$$

Using the formulas for x_2 and x_3, we find $\mathbf{x}^* = [0, -4, \frac{3}{2}, \frac{1}{2}]^T$. What is new about this example, and what distinguishes it from the ones studied in Section 9.3, is the idea that it is easy to solve several constraint equations simultaneously if they are linear. The gradient of the reduced objective function is called the **reduced gradient,** and the method is therefore called the **reduced-gradient method.**

Now consider the possibility of approximating a problem having nonlinear constraints by a problem with linear constraints, which can then be solved like the preceding example. The easiest way to perform this linearization of the constraints is by expanding each constraint function in a Taylor series and neglecting terms beyond the linear one. To see how this works, consider the following example, which resembles the preceding example but has constraints that are slightly different and nonlinear:

$$\begin{aligned}
\underset{\mathbf{x}\in\mathbf{R}^4}{\text{minimize}}\quad & f_0(\mathbf{x}) = x_1^2 + x_2 + x_3^2 + x_4 \\
\text{subject to}\quad & \\
& f_1(\mathbf{x}) = x_1^2 + x_2 + 4x_3 + 4x_4 - 4 = 0 \\
& f_2(\mathbf{x}) = -x_1 + x_2 + 2x_3 - 2x_4^2 + 2 = 0
\end{aligned}$$

The Taylor series expansion for a function of n variables, calculated at a point $\bar{\mathbf{x}}$, is given by the following formula:

$$f(\mathbf{x}) = f(\bar{\mathbf{x}}) + \nabla f(\bar{\mathbf{x}})^T(\mathbf{x} - \bar{\mathbf{x}}) + \frac{(\mathbf{x} - \bar{\mathbf{x}})^T\mathbf{H}(\bar{\mathbf{x}})(\mathbf{x} - \bar{\mathbf{x}})}{2!} + \cdots$$

This is precisely analogous to the formula for a function of one variable, with the first derivative replaced by the gradient, the second derivative replaced by the Hessian, and so forth. If we neglect terms beyond the linear one, we obtain the approximation

$$f(\mathbf{x}) \approx f(\bar{\mathbf{x}}) + \nabla f(\bar{\mathbf{x}})^T(\mathbf{x} - \bar{\mathbf{x}})$$

Using this formula, we can replace the inequality constraints in the preceding problem by equalities that approximate the true constraints in the vicinity of the point $\bar{\mathbf{x}}$ at which the linearization is performed.

$$f_1(\mathbf{x}) \approx [\bar{x}_1^2 + \bar{x}_2 + 4\bar{x}_3 + 4\bar{x}_4 - 4] + [2\bar{x}_1, 1, 4, 4]\begin{bmatrix} x_1 - \bar{x}_1 \\ x_2 - \bar{x}_2 \\ x_3 - \bar{x}_3 \\ x_4 - \bar{x}_4 \end{bmatrix}$$

$$f_1(\mathbf{x}) \approx (2\bar{x}_1)x_1 + x_2 + 4x_3 + 4x_4 - [\bar{x}_1^2 + 4] = 0$$

Making a similar approximation for the second constraint function, we get

$$f_2(\mathbf{x}) \approx -x_1 + x_2 + 2x_3 - (4\bar{x}_4)x_4 + [2\bar{x}_4^2 + 2]$$

The idea of the **generalized reduced-gradient algorithm (GRG)** is to solve a sequence of subproblems, each of which uses a linear approximation of the constraints. In each iteration of the algorithm, the constraint linearization is recalculated at the point found from the previous iteration. Typically, even though the constraints are only approximated, the subproblems yield points that are progressively closer to the optimal point. As the optimal point is approached, the linearized constraints of the subproblems become progressively better approximations to the original nonlinear constraints in the neighborhood of the optimal point. At the optimal point, the linearized problem has the same solution as the original nonlinear problem.

The first step in applying GRG is to pick a starting point at which to perform the first linearization. Suppose that for our example we pick the starting point $\mathbf{x}^0 = [0, -8, 3, 0]^T$, which happens to satisfy the original nonlinear constraints. It is possible to start GRG from an infeasible point, but the details of how to do that need not concern us until later. Next we use the approximation formulas already given to linearize the constraint functions at the point $\mathbf{x}^0$, and form the first approximate problem, as follows.

$$\underset{\mathbf{x}\in\mathbf{R}^4}{\text{minimize}}\ f_0(\mathbf{x}) = x_1^2 + x_2 + x_3^2 + x_4$$

subject to

$$\begin{aligned} x_2 + 4x_3 + 4x_4 &= 4 \\ -x_1 + x_2 + 2x_3 &= -2 \end{aligned}$$

Now we solve the equality constraints of the approximate problem to express two of the variables in terms of the others. Arbitrarily selecting x_2 and x_3 to be **basic variables,** we solve the linear system to write them in terms of x_1 and x_4, which thus become the **nonbasic variables.**

$$x_2 = 2x_1 + 4x_4 - 8$$

$$x_3 = -\tfrac{1}{2}x_1 - 2x_4 + 3$$

Substituting these expressions in the objective function yields the reduced problem

$$\underset{x_1, x_4}{\text{minimize}}\ f_0(x_1, x_4) = x_1^2 + [-8 + 2x_1 + 4x_4] + [-\tfrac{1}{2}x_1 - 2x_4 + 3]^2 + x_4$$

Solving this unconstrained minimization by setting the partial derivatives equal to zero, we get

$$\left.\begin{aligned} \frac{\partial f_0}{\partial x_1} &= 2x_1 + 2 + 2\left[-\frac{x_1}{2} - 2x_4 + 3\right]\left(-\frac{1}{2}\right) = 0 \\ \frac{\partial f_0}{\partial x_4} &= 4 + 2\left[-\frac{x_1}{2} - 2x_4 + 3\right](-2) + 1 = 0 \end{aligned}\right\} \Rightarrow \begin{aligned} x_1 &= -\frac{3}{8} \\ x_4 &= \frac{31}{32} \end{aligned}$$

Substituting in the equations for x_2 and x_3 gives $x_2 = -\frac{39}{8}$ and $x_3 = \frac{5}{4}$. Thus the first iteration of GRG has produced the new point

$$\mathbf{x}^1 = [-0.375, -4.875, 1.25, 0.96875]^T.$$

In a numerical implementation of the GRG algorithm, we would use a method like steepest descent to solve the reduced problems, rather than setting derivatives equal to zero and solving the resulting (possibly nonlinear) equations.

To continue the solution process for the example, we would relinearize the constraint functions at the new point, use the resulting system of linear equations to express two of the variables in terms of the others, substitute into the objective to get a new reduced problem, solve the reduced problem for $\mathbf{x}^2$, and so forth.

Formally, the algorithm is as follows:

GENERALIZED REDUCED-GRADIENT ALGORITHM

0. Initialize
   $\mathbf{x}^0$ = starting point
   T = convergence tolerance
   set $k = 0$.
1. Linearize any nonlinear constraints
   $f_i(\mathbf{x}) \approx f_i(\mathbf{x}^k) + \nabla f_i(\mathbf{x}^k)^T(\mathbf{x} - \mathbf{x}^k)$
2. Solve the m linearized constraints for m (basic) variables in terms of the other (nonbasic) ones.
3. Form the reduced problem
   minimize f_0 (nonbasic variables).
4. Solve the (unconstrained) reduced problem
   find basic variables from linearized constraints.
5. Check for convergence
   if $|\mathbf{x}^{k+1} - \mathbf{x}^k| < T$, STOP; $\mathbf{x}^{k+1}$ is the minimum
6. Repeat
   increase k by 1
   GO TO 1

We can think of the system that results from linearizing the constraints as $\mathbf{Ax} = \mathbf{b}$, where $\mathbf{A}$ will be $m \times n$ if the problem has n variables and m constraints. The rows of $\mathbf{A}$ are just the gradients of the constraint functions, evaluated at the current point $\mathbf{x}^k$, so it will be possible to solve the system $\mathbf{Ax} = \mathbf{b}$ for m of the variables in terms of the others, as required by algorithm step 2, if those gradient vectors are linearly independent. Thus, the same constraint qualification encountered as a hypothesis in the Lagrange multiplier theorem turns out to be necessary for the success of this numerical method as well, but it must hold at every iterate $\mathbf{x}^k$ in addition to the optimal point. One way to identify a set of basic variables and solve the system $\mathbf{Ax} = \mathbf{b}$ is by thinking of the system as the lower part of a linear programming tableau and pivoting to obtain an identity as discussed in Section 3.5 (see Exercise 9.46). It is necessary in a numerical implementation of GRG to allow for the possibility that which of the variables are basic might change from one iteration to the next.

Applying the GRG algorithm to the example problem, using $T = 0.0025$ and the starting point $\mathbf{x}^0 = [0, -8, 3, 0]^T$ yields the sequence of iterates $\mathbf{x}^k$ given in the table on page 315, converging to the values given on the last line. In reading this table, it is important to remember that each iteration contains an unconstrained minimization as a subproblem. The method exhibits a varying rate of convergence, as shown by the last column in the table. The performance of complicated algorithms like GRG on real problems is difficult to predict on theoretical grounds.

The method used with GRG for starting at an infeasible point is first to find a feasible point by solving a **phase 1 problem.** To form the phase 1 problem it is necessary to determine which of the constraints in the original problem are violated at the starting point and which are satisfied. The objective function for the phase 1 problem is the sum of the absolute values of the constraint functions

k	$(\mathbf{x}^k)^T$	$f_0(\mathbf{x}^k)$	$\lVert\mathbf{x}^{k+1} - \mathbf{x}^k\rVert$
0	[0.00000, −8.00000, 3.00000, 0.00000]	1.00000	3.72924
1	[−0.37500, −4.87500, 1.25000, 0.96875]	−2.20313	0.57172
2	[−0.42262, −5.13387, 1.61905, 0.62034]	−1.71361	0.35338
3	[−0.45844, −4.79205, 1.53713, 0.60866]	−1.61045	0.02165
4	[−0.47781, −4.80187, 1.53377, 0.60971]	−1.61139	0.01520
5	[−0.48815, −4.81301, 1.53408, 0.60963]	−1.61170	0.00776
6	[−0.49367, −4.81846, 1.53405, 0.60964]	−1.61178	0.00417
7	[−0.49662, −4.82141, 1.53406, 0.60964]	−1.61181	0.00223
8	[−0.49820, −4.82298, 1.53406, 0.60964]	−1.61182	
⋮	⋮ ⋮ ⋮ ⋮	⋮	⋮
∞	[−0.50000, −4.82479, 1.53406, 0.60964]	−1.61182	0.00000

for the constraints that are violated. The constraints for the phase 1 problem are those from the original problem that are satisfied at the starting point. Suppose that for the example we had started at the point $\mathbf{x}^0 = [1, 1, 0, 1]^T$. This point violates the first constraint but satisfies the second, so the phase 1 problem would be

$$\underset{\mathbf{x}\in\mathbf{R}^4}{\text{minimize}} \quad |x_1^2 + x_2 + 4x_3 + 4x_4 - 4|$$

subject to

$$-x_1 + x_2 + 2x_3 - 2x_4^2 + 2 = 0$$

Once a feasible point has been found by solving the phase 1 problem, the method illustrated above is used to solve the **phase 2 problem,** maintaining feasibility from one iteration to the next.

By introducing slack variables, GRG can be used to solve problems having inequality constraints as well as equalities. Slack inequality constraints do not enter into the system of linear equations used for eliminating variables, so it is necessary to decide at each iteration which of the inequality constraints should be regarded as tight and which as slack. The heuristic procedures that have been developed to manage this process are known as **active set strategies** because they maintain a list of which constraints are active. The process of tightening and releasing constraints causes the number of equations in the linearized system to vary from one iteration to another, and the basic algorithm described above is complicated considerably by the extra logic required to account for this fact and to carry out the active set strategy.

The Ellipsoid Algorithm

A simple algorithm for inequality-constrained nonlinear programming problems is the **ellipsoid algorithm.** To see how it works, consider the example used to introduce this section.

$$\underset{\mathbf{x}\in\mathbf{R}^2}{\text{minimize}} \quad f_0(\mathbf{x}) = (x_1 - 20)^4 + (x_2 - 12)^4$$

subject to

$$\begin{aligned} f_1(\mathbf{x}) &= 8e^{(x_1-12)/9} - x_2 + 4 \le 0 \\ f_2(\mathbf{x}) &= 6(x_1 - 12)^2 + 25x_2 - 600 \le 0 \\ f_3(\mathbf{x}) &= -x_1 + 12 \le 0 \end{aligned}$$

Figure 9.15*a* shows the feasible set S for this problem, bounded by segments of the zero contours of the constraint functions. Neither the optimal point nor the contours of the objective function are shown in Figure 9.15*a*.

Suppose that, without knowing the optimal point, we can nonetheless specify an ellipse E_0 that is known to contain the optimal point. One way to guarantee that E_0 contains the optimal point is to make it large enough to contain the entire feasible set S. Because the optimal point must be feasible, an ellipse that contains all of S is sure to contain $\mathbf{x}^*$. If S is of unknown extent, as will usually be the case, or is unbounded, other methods must be used to obtain a starting ellipsoid, but those need not concern us until later. In Figure 9.15*a*, S is bounded, E_0 contains S, and we can take the center of E_0 as the starting point $\mathbf{x}^0$ of the solution process.

In general, of course, there will be more than two variables, so instead of an ellipse we will need to start with an n-dimensional **ellipsoid** of the form

$$E_0 = \{\mathbf{x} \in \mathrm{R}^n \mid (\mathbf{x} - \mathbf{x}^0)^T \mathbf{Q}_0^{-1} (\mathbf{x} - \mathbf{x}^0) \leq 1\}$$

In order for this definition to describe an ellipsoid rather than some other quadratic form, the matrix $\mathbf{Q}_0$ must be positive definite. It is not hard to show that when $n = 2$ the preceding definition describes an ellipse. The starting ellipsoid E_0 in Figure 9.15*a* is defined by

$$\mathbf{Q}_0 = \begin{bmatrix} 81 & 0 \\ 0 & 169 \end{bmatrix} \quad \text{and} \quad \mathbf{x}^0 = \begin{bmatrix} 18 \\ 21 \end{bmatrix}$$

The basic idea of the ellipsoid algorithm is to generate a sequence of successively smaller ellipsoids each containing the optimal point. The optimal point is localized within regions of diminishing volume as the iterations of the algorithm progress, and under the right conditions the ellipsoid centers $\mathbf{x}^k$ approach $\mathbf{x}^*$. Each ellipsoid (after the first one) is constructed by cutting the previous ellipsoid in half and then enclosing the half that is known to contain the optimal point with the smallest possible new ellipsoid. The details of this process depend on

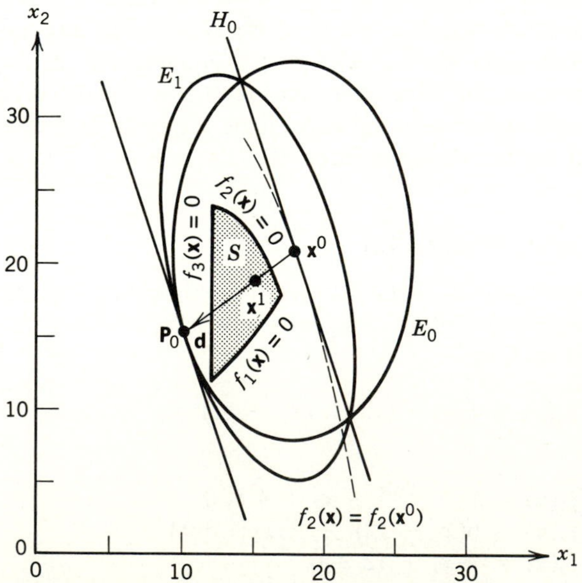

FIGURE 9.15*a* A phase 1 iteration of the ellipsoid algorithm.

whether the current ellipsoid center is feasible or not, so there are two cases to consider.

The first case is illustrated by Figure 9.15*a*. We can see by inspection of the figure that $\mathbf{x}^0 = [18, 21]^T$ is infeasible because it is not in S, but this fact could also be discovered by calculating the values of the constraint functions as follows:

$$\begin{aligned} f_1(\mathbf{x}^0) &= 8e^{([18]-12)/9} - [21] + 4 &&\approx -1.4 < 0 \\ f_2(\mathbf{x}^0) &= 6([18] - 12)^2 + 25[21] - 600 &&= +141 > 0 \\ f_3(\mathbf{x}^0) &= -[18] + 12 &&= -6 < 0 \end{aligned}$$

This analysis shows that the second constraint is the only one that is violated at the point $\mathbf{x}^0$.

Because the optimal point must be a feasible point, it must be in the part of E_0 that contains S. There are many ways to cut the ellipsoid E_0 in half so as to have S contained entirely in one of the halves, but a particularly easy cut is suggested by inspection of the constraint contour $f_2(\mathbf{x}) = f_2(\mathbf{x}^0)$. This contour is shown in Figure 9.15*a* as a dashed line, and it is clear from the picture that points above and to the right of that contour violate the second constraint even more than the point $\mathbf{x}^0$ does. We can therefore safely discard that half of E_0, by cutting the ellipsoid with a line drawn tangent to the $f_2(\mathbf{x}) = f_2(\mathbf{x}^0)$ contour at the point $\mathbf{x}^0$. This **cutting hyperplane** H_0 is also shown in the figure, and is given analytically by

$$H_0 = \{\mathbf{x} \mid -\nabla f_2(\mathbf{x}^0)^T(\mathbf{x} - \mathbf{x}^0) = 0\}$$

The hyperplane H_0 is said to **support** the contour $f_2(\mathbf{x}) = f_2(\mathbf{x}^0)$ at the point $\mathbf{x}^0$, and is therefore sometimes referred to as a **supporting hyperplane.**

To complete the first iteration of the algorithm, it remains only to construct the smallest new ellipsoid E_1 containing the half of E_0 that is below and to the left of the cutting hyperplane. It turns out that, if the cutting hyperplane is translated parallel to itself until it is tangent to E_0 at the point P_0, the smallest ellipsoid containing the correct half of E_0 passes through P_0 and $E_0 \cap H_0$, as shown.

The process of finding a violated constraint, computing the cutting hyperplane that supports the constraint contour at the current ellipsoid center, and constructing the next ellipsoid, as illustrated in Figure 9.15*a*, is called a **phase 1 iteration** of the ellipsoid algorithm. The name derives from the fact that the starting point for a phase 1 iteration is infeasible, and the goal of the iteration is to exclude infeasible points. Frequently more than one phase 1 iteration must be performed before a feasible $\mathbf{x}^k$ is obtained, but in this example $\mathbf{x}^1$ happens to be feasible. Figure 9.15*b* shows the situation after the ellipsoid E_1 has been constructed. It is easy to see from Figure 9.15*b* that the new ellipsoid center $\mathbf{x}^1$ is in S, and we could also discover that fact by evaluating the constraint functions at that point, but $\mathbf{x}^1$ is far from optimal. The unconstrained minimum is at the point $[20, 12]^T$, below and to the right of $\mathbf{x}^1$, and the objective function contour $f_0(\mathbf{x}) = f_0(\mathbf{x}^1)$ passing through the point $\mathbf{x}^1$ shows that all points in E_1 above and to the left have higher objective function values than the point $\mathbf{x}^1$. We can therefore safely discard the upper left half of E_1, by cutting the ellipsoid in half with a hyperplane H_1 supporting the $f_0(\mathbf{x}) = f_0(\mathbf{x}^1)$ contour, as shown. This cutting hyperplane is then translated parallel to itself until it is tangent to E_1 at P_1, and the next ellipsoid E_2 is constructed as the smallest one passing through P_1 and $E_1 \cap H_1$.

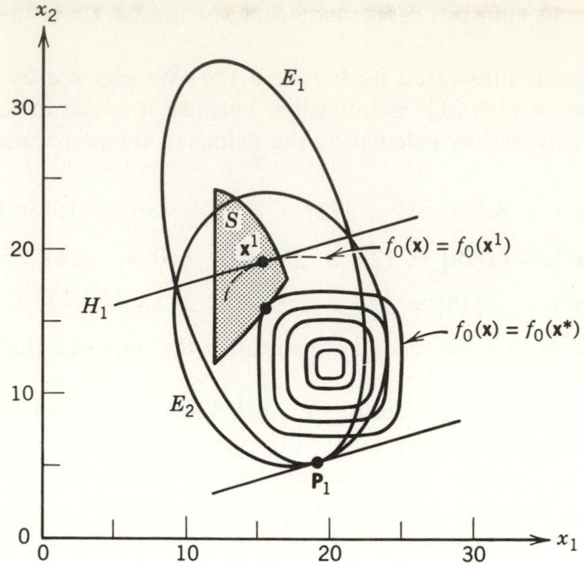

FIGURE 9.15*b* A phase 2 iteration of the ellipsoid algorithm.

The preceding description omits the important detail of just how to construct E_{k+1} from E_k and H_k in such a way that E_{k+1} is the smallest ellipsoid containing the appropriate half of E_k. The following formal statement of the ellipsoid algorithm provides **update formulas** that have this effect.

ELLIPSOID ALGORITHM

0. Initialize
   n = number of variables
   $\mathbf{x}^0$ = starting ellipsoid center
   $\mathbf{Q}_0$ = starting ellipsoid matrix
   T = convergence tolerance
   set $k = 0$
1. Check whether $\mathbf{x}^k$ is feasible
   if $\mathbf{x}^k \in S$, $i = 0$
   if $\mathbf{x}^k \notin S$, i = index of a violated constraint
2. Find the unit normal vector to cutting hyperplane
   $\mathbf{g} = \nabla f_i(\mathbf{x}^k)/\|\nabla f_i(\mathbf{x}^k)\|$
3. Find the direction in which to move $\mathbf{x}$
   $\mathbf{d} = -\mathbf{Q}_k\mathbf{g}/\sqrt{\mathbf{g}^T\mathbf{Q}_k\mathbf{g}}$
4. Update $\mathbf{x}$ and $\mathbf{Q}$

$$\mathbf{x}^{k+1} = \mathbf{x}^k + \frac{\mathbf{d}}{n+1}$$

$$\mathbf{Q}_{k+1} = \frac{n^2}{n^2-1}\left(\mathbf{Q}_k - \frac{2}{n+1}\mathbf{d}\mathbf{d}^T\right)$$

5. Check for convergence
   if $\|\mathbf{x}^{k+1} - \mathbf{x}^k\| < T$, STOP; $\mathbf{x}^{k+1}$ is the minimum
6. Repeat
   increase k by 1
   GO TO 1

Once a violated constraint has been found or $\mathbf{x}^k$ has been shown to be feasible, the process of computing the cutting hyperplane and constructing the new ellipsoid is performed implicitly by updating $\mathbf{x}$ and $\mathbf{Q}$ according to the formulas given in step 4. Thus we never find (or need) $\mathbf{Q}_k^{-1}$, the matrix in the definition of E_k, explicitly. In searching for violated constraints, we can stop after finding the first one because the algorithm only needs one.

Applying the algorithm to the preceding test problem yields the following result for the first iteration, which is pictured in Figure 9.15*a*

$$\mathbf{x}^0 = \begin{bmatrix} 18 \\ 12 \end{bmatrix} \qquad \mathbf{Q}_0 = \begin{bmatrix} 81 & 0 \\ 0 & 169 \end{bmatrix}$$

$$\left.\begin{aligned} f_1(\mathbf{x}^0) &= -1.4 \\ f_2(\mathbf{x}^0) &= +141 \\ f_3(\mathbf{x}^0) &= -6 \end{aligned}\right\} i = 2 \text{ (constraint 2 is violated)}$$

$$\nabla f_2(\mathbf{x}^0) = \begin{bmatrix} 12([18] - 12) \\ 25 \end{bmatrix} = \begin{bmatrix} 72 \\ 25 \end{bmatrix}$$

$$\mathbf{g} = \frac{\nabla f_2(\mathbf{x}^0)}{\|\nabla f_2(\mathbf{x}^0)\|} = \begin{bmatrix} 72 \\ 25 \end{bmatrix} \cdot \frac{1}{\sqrt{72^2 + 25^2}} = \begin{bmatrix} 0.945 \\ 0.328 \end{bmatrix}$$

$$\mathbf{Q}_0\mathbf{g} = \begin{bmatrix} 81 & 0 \\ 0 & 169 \end{bmatrix}\begin{bmatrix} 0.945 \\ 0.328 \end{bmatrix} = \begin{bmatrix} 76.519 \\ 55.434 \end{bmatrix}$$

$$\mathbf{g}^T\mathbf{Q}_0\mathbf{g} = [0.945 \quad 0.328]\begin{bmatrix} 76.519 \\ 55.434 \end{bmatrix} = 90.468$$

$$\mathbf{d} = \frac{-\mathbf{Q}_0\mathbf{g}}{\sqrt{\mathbf{g}^T\mathbf{Q}_0\mathbf{g}}} = \begin{bmatrix} -76.519 \\ -55.434 \end{bmatrix} \cdot \frac{1}{\sqrt{90.468}} = \begin{bmatrix} -8.045 \\ -5.828 \end{bmatrix}$$

$$\mathbf{x}^1 = \mathbf{x}^0 + \frac{1}{n+1} \cdot \mathbf{d} = \begin{bmatrix} 18 \\ 21 \end{bmatrix} + \frac{1}{3}\begin{bmatrix} -8.045 \\ -5.828 \end{bmatrix} = \begin{bmatrix} 15.318 \\ 19.057 \end{bmatrix}$$

$$\frac{2}{n+1}\mathbf{d}\mathbf{d}^T = \frac{2}{3}\begin{bmatrix} -8.045 \\ -5.828 \end{bmatrix}[-8.045 \quad -5.828] = \begin{bmatrix} 43.147 & 31.258 \\ 31.258 & 22.645 \end{bmatrix}$$

$$\mathbf{Q}_1 = \frac{n^2}{n^2 - 1}\left[\mathbf{Q}_0 - \frac{2}{n+1}\mathbf{d}\mathbf{d}^T\right] = \frac{4}{3}\begin{bmatrix} 81 - 43.147 & 0 - 31.258 \\ 0 - 31.258 & 169 - 22.645 \end{bmatrix}$$

$$= \begin{bmatrix} 50.471 & -41.677 \\ -41.677 & 195.140 \end{bmatrix}$$

The direction vector $\mathbf{d}$ is shown in Figure 9.15*a*, and it can be verified by inspection that $\mathbf{x}^1$ is one third of the way between $\mathbf{x}^0$ and T_0.

The second iteration of the algorithm, pictured in Figure 9.15*b*, begins with a check for feasibility of $\mathbf{x}^1$.

$$\left.\begin{aligned} f_1(\mathbf{x}^1) &= -3.5 \\ f_2(\mathbf{x}^1) &= -5.8 \\ f_3(\mathbf{x}^1) &= -3.3 \end{aligned}\right\} i = 0 \quad (\mathbf{x}^1 \text{ is feasible})$$

All the constraints are satisfied at $\mathbf{x}^1$, so the objective function must be used for defining the cutting hyperplane. With this change the calculations for the second iteration are similar to those for the first.

Continuing the solution process yields the following sequence of iterates, converging to the final values given on the last line:

k	*Phase*	$(\mathbf{x}^k)^T$	$f_0(\mathbf{x}^k)$
0	1	[18.00000, 21.00000]	6577.00000
1	2	[15.31838, 19.05729]	2960.95894
2	1	[16.57538, 14.43788]	172.86941
3	2	[14.62410, 16.67265]	1311.93440
4	1	[15.96378, 15.24105]	375.74172
5	2	[15.32857, 16.56945]	912.17786
6	1	[14.53396, 14.18825]	915.60152
7	2	[14.27799, 14.89288]	1142.03542
8	1	[17.38571, 18.49486]	1826.13310
9	2	[15.80618, 17.13177]	1002.87805
10	2	[14.07535, 14.48906]	1270.49325
11	2	[15.50111, 16.15683]	708.22816
12	1	[15.55425, 15.45525]	533.17867
13	2	[15.39878, 15.76771]	649.73504
14	1	[16.51681, 17.05889]	802.16824
15	2	[16.47217, 17.32619]	959.65457
16	2	[15.32958, 15.61843]	647.22587
17	1	[16.05855, 16.54553]	668.24868
18	2	[16.30057, 17.15896]	895.65246
19	2	[15.66238, 16.18184]	659.82602
20	1	[15.43693, 15.70718]	622.41498
⋮	⋮	⋮ ⋮	⋮
30	2	[15.47794, 15.79735]	626.09406
⋮	⋮	⋮ ⋮	⋮
40	1	[15.66190, 16.01582]	614.23491
⋮	⋮	⋮ ⋮	⋮
50	2	[15.64725, 15.99845]	614.56854
⋮	⋮	⋮ ⋮	⋮
∞	2	[15.62949, 15.97377]	614.21210

It is easy to confirm that the solution given on the last line satisfies the KKT conditions for the problem, with the Lagrange multipliers $u_1 = 250.99653$, $u_2 = 0$, and $u_3 = 0$.

The average convergence of the ellipsoid algorithm is linear, so that over the long run the error in the solution decreases as if it were multiplied by a constant factor at each iteration. This is because the volumes of the ellipsoids decrease in geometric progression. The ratio of the volumes of any two successive ellipsoids is a constant $q(n)$ depending only on the number of variables n, and is given by the following formula:

$$q(n) = \frac{\text{volume}(E_{k+1})}{\text{volume}(E_k)} = \sqrt{\frac{n-1}{n+1}}\left(\frac{n}{\sqrt{n^2-1}}\right)^n$$

For this example $n = 2$ so the **volume reduction ratio** is $q(n) = 0.7698^+$. The linear average convergence of the ellipsoid algorithm is reflected in the table by the fact that many iterations are required before the new iterates are all close to the optimal point. The ratio $q(n)$ approaches 1 as n increases, so the convergence of the algorithm is slower for larger values of n.

Over the short run the algorithm makes wild excursions from one iteration to the next, particularly near the beginning of the solution process. Cuts using the objective function frequently yield new points that are infeasible, so phase 1 and phase 2 iterations occur in an irregular sequence. A feasible point with a low objective value may be produced early in the iterations (such as iteration 13 in the table), only to be followed by many iterations in which the feasible points all have higher objective values. This sort of behavior is in contrast to that of an algorithm such as GRG, which maintains feasibility once it has been attained, converges superlinearly when it converges at all, and generates iterates each of which is usually better than the previous one. The fact that the ellipsoid algorithm can generate occasional good iterates very early in the solution process makes it desirable to keep a record of the best point found so far, rather than taking as the solution the first point that satisfies the convergence criterion. The best point found so far, that is, the feasible point having the lowest objective value, is called the **record point,** and the objective function value at the record point is called the **record value.**

In order for the ellipsoid algorithm to work, it is necessary that the intersection of the starting ellipsoid E_0 with the feasible set S have positive volume relative to R^n. Thus, for example, if $n = 3$, the intersection $E_0 \cap S$ must be a three-dimensional region with nonzero volume. For convergence of the algorithm to be guaranteed, the functions $f_i(\mathbf{x})$, $i = 0, \ldots, m$ must be convex, but often the algorithm converges to a KKT point even when the convexity requirement is not satisfied.

The most convenient way to get a starting ellipsoid E_0 is usually from bounds on the variables. If bounds $\mathbf{U}$ and $\mathbf{L}$ on the variables are known, E_0 can be chosen as the smallest ellipsoid containing the right parallelepiped or box B defined by the bounds,

$$B = \{\mathbf{x} \mid \mathbf{L} \le \mathbf{x} \le \mathbf{U}\}$$

The following formulas give $\mathbf{x}^0$ and $\mathbf{Q}_0$ for the smallest ellipsoid containing B, in terms of the bounds.

$$\mathbf{x}^0 = \frac{\mathbf{L} + \mathbf{U}}{2}$$

$$\mathbf{Q}_0 = \frac{n}{4}\begin{bmatrix} (U_1 - L_1)^2 & & & 0 \\ & (U_2 - L_2)^2 & & \\ & & \ddots & \\ 0 & & & (U_n - L_n)^2 \end{bmatrix}$$

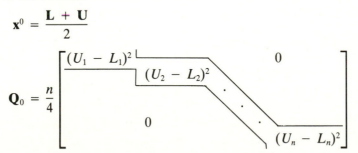

In most practical applications of nonlinear programming, the underlying problem permits the determination of bounds on the variables from physical or economic principles. For example, if the problem concerns the design of a chemical plant and some of the variables are pipe diameters, it will probably be easy to specify the largest and smallest values that could plausibly be used.

The ellipsoid algorithm has several desirable properties that make it the algorithm of choice for many practical applications even though it has only linear convergence. Although the algorithm is not guaranteed to work on nonconvex problems, it is actually one of the most robust algorithms for such problems. It typically yields a rough solution very quickly, and if run for many iterations it is capable of obtaining extremely accurate results. It is much simpler than most other methods for inequality-constrained problems, so it is also comparatively easy to implement in a computer program.

Conclusion

Long books have been written on numerical methods for nonlinear programming, and new research results appear every day, so this section can do little more than give the flavor of the subject. Some considerations that are important in practice but must be omitted from an elementary treatment concern the effects of nonconvexity and nondifferentiability on the behavior of algorithms, the approximation of gradient vectors by finite differencing, the management of numerical roundoff errors, and the design of experimental methods for the evaluation of algorithm performance. The topic of current research activity that is most conspicuous by its unavoidable absence is the design of sequential quadratic programming algorithms and similar sophisticated techniques that are based directly on the KKT theory of nonlinear programming. Hopefully, the introduction provided by this section is sufficient to make more advanced textbooks and some of the research literature accessible to the student.

9.8 NONLINEAR PROGRAMS WITH SPECIAL STRUCTURE

Several important classes of nonlinear programs have special features that can be exploited in developing an algorithm for finding an optimal solution. These special-purpose algorithms are often much more successful than algorithms designed for general nonlinear programs. As an extreme example, a linear program can be viewed as a nonlinear program with special structure, namely, the objective function and all the constraints are linear functions. The simplex algorithm is designed to take advantage of this special structure. In this section we give a brief overview of some other special classes and algorithms.

Quadratic Programming

A **quadratic program** has the form

$$\begin{aligned} \text{minimize} \quad & \tfrac{1}{2}\mathbf{x}^T\mathbf{Q}\mathbf{x} + \mathbf{c}^T\mathbf{x} \\ \text{subject to} \quad & \\ & \mathbf{A}\mathbf{x} \le \mathbf{b} \\ & \mathbf{x} \ge \mathbf{0} \end{aligned}$$

where $\mathbf{Q}$ is a symmetric $n \times n$ matrix and $\mathbf{A}$ is $m \times n$.

Thus, a quadratic program only differs from a linear program in that its objective function is a quadratic function of several variables instead of a linear function. The close similarity of a quadratic program to a linear program has permitted an algorithm to be developed that is similar to the simplex algorithm in that it only requires pivoting on a tableau. This special algorithm for quadratic programming is called **Lemke's method** after the person who discovered it. If the matrix $\mathbf{Q}$ is positive semidefinite, then the method will either yield an optimal solution to the problem or show that no such solution exists.

Lemke's method begins by considering the Karush–Kuhn–Tucker conditions for the preceding quadratic program. With a little work (see Exercise 9.57), the KKT conditions can be calculated as

$$\mathbf{c} = -\mathbf{Q}\mathbf{x} - \mathbf{A}^T\mathbf{u} + \mathbf{y}$$

$$\mathbf{b} = \mathbf{A}\mathbf{x} + \mathbf{v}$$

$$\mathbf{x}^T\mathbf{y} = 0 \qquad \mathbf{u}^T\mathbf{v} = 0$$

$$\mathbf{x} \geq \mathbf{0},\ \mathbf{y} \geq \mathbf{0},\ \mathbf{u} \geq \mathbf{0},\ \mathbf{v} \geq \mathbf{0}$$

The first two linear equations can be written in tableau form as

	x	u	y	v
$\mathbf{c}$	$-\mathbf{Q}$	$-\mathbf{A}^T$	$\mathbf{I}_n$	$\mathbf{0}$
$\mathbf{b}$	$\mathbf{A}$	$\mathbf{0}$	$\mathbf{0}$	$\mathbf{I}_m$

where $\mathbf{I}_n$ is an $n \times n$ identity matrix and $\mathbf{I}_m$ is an $m \times m$ identity matrix. It will be convenient to refer to a variable that has an identity column as a **basic variable** and, otherwise, a variable will be called a **nonbasic variable.** Thus, initially, the variables y_i are basic for $i = 1, \ldots, n$ and so are the variables v_j for $j = 1, \ldots, m$.

Because of the orthogonality (or **complementarity**) condition, $\mathbf{x}^T\mathbf{y} = 0$, we say that, for each i, the variables x_i and y_i are a **complementary pair of variables.** Similarly, for each j the variables u_j and v_j are called a complementary pair because $\mathbf{u}^T\mathbf{v} = 0$.

In the case that $\mathbf{Q}$ is positive semidefinite, the objective function of the quadratic program is convex (because its Hessian matrix is $\mathbf{Q}$), and so, if $\bar{\mathbf{x}}, \bar{\mathbf{u}}, \bar{\mathbf{y}}, \bar{\mathbf{v}}$ solves the KKT conditions, then $\bar{\mathbf{x}}$ must be optimal for the quadratic program. Lemke's method is an ingenious way of finding a solution to the KKT conditions if one exists.

Note that if $\mathbf{c} \geq \mathbf{0}$ and $\mathbf{b} \geq \mathbf{0}$, then

$$\bar{\mathbf{x}} = \mathbf{0} \qquad \bar{\mathbf{u}} = \mathbf{0} \qquad \bar{\mathbf{y}} = \mathbf{c} \qquad \bar{\mathbf{v}} = \mathbf{b}$$

would be a solution to the above KKT conditions, and $\bar{\mathbf{x}} = \mathbf{0}$ would be a solution to the quadratic program. Of course, in this case it is clear from examining the original problem that $\bar{\mathbf{x}} = \mathbf{0}$ is an optimal vector because it is feasible and the objective function is nonnegative (since $\mathbf{Q}$ is positive semidefinite and $\mathbf{c} \geq \mathbf{0}$). The only interesting case is when $\mathbf{c}$ and $\mathbf{b}$ are not both nonnegative.

Lemke's method introduces an additional variable, z_0, and considers the following system

$$\begin{bmatrix} -\mathbf{Q} & -\mathbf{A}^T & \mathbf{I}_n & \mathbf{0} \\ \mathbf{A} & \mathbf{0} & \mathbf{0} & \mathbf{I}_m \end{bmatrix} \begin{bmatrix} \mathbf{x} \\ \mathbf{u} \\ \mathbf{y} \\ \mathbf{v} \end{bmatrix} = \begin{bmatrix} \mathbf{c} \\ \mathbf{b} \end{bmatrix} + z_0\mathbf{e}$$

where $\mathbf{e}$ is a vector having each component equal to 1. If z_0 is large enough, then the right-hand side of the system is nonnegative and a nonnegative solution can be found by setting $\mathbf{x} = \mathbf{0}$ and $\mathbf{u} = \mathbf{0}$ as we did previously. Unfortunately, because of the variable z_0, this solution does not solve the KKT conditions. However, if we had a nonnegative solution $(\bar{\mathbf{x}}, \bar{\mathbf{u}}, \bar{\mathbf{y}}, \bar{\mathbf{v}}, \bar{z}_0)$ to the enlarged problem with $\bar{z}_0 = 0$, then $(\bar{\mathbf{x}}, \bar{\mathbf{u}}, \bar{\mathbf{y}}, \bar{\mathbf{v}})$ would solve the KKT conditions.

The tableau for the enlarged system is

	x	u	y	v	z_0
c	$-\mathbf{Q}$	$-\mathbf{A}^T$	$\mathbf{I}_n$	0	-1 $\vdots$
b	A	0	0	$\mathbf{I}_m$	-1

For example, when the quadratic program has

$$\mathbf{Q} = \begin{bmatrix} 2 & -2 \\ -2 & 3 \end{bmatrix} \quad \mathbf{A} = \begin{bmatrix} -1 & -2 \\ 3/2 & -1 \end{bmatrix} \quad \mathbf{c} = \begin{bmatrix} 2 \\ -8 \end{bmatrix} \quad \mathbf{b} = \begin{bmatrix} -6 \\ -6 \end{bmatrix}$$

then this tableau becomes

	x_1	x_2	u_1	u_2	y_1	y_2	v_1	v_2	z_0
2	-2	2	1	$-3/2$	1	0	0	0	-1
-8	2	-3	2	1	0	1	0	0	(-1)
-6	-1	-2	0	0	0	0	1	0	-1
-6	3/2	-1	0	0	0	0	0	1	-1

Note that pivoting in the z_0 column and in the row with the most negative constant column entry will always yield a tableau with a nonnegative constant column. Doing so for this example yields the following tableau:

	x_1	x_2	u_1	u_2	y_1	y_2	v_1	v_2	z_0
10	-4	5	-1	$-5/2$	1	-1	0	0	0
8	-2	3	-2	-1	0	-1	0	0	1
2	-3	1	-2	-1	0	-1	1	0	0
2	$-1/2$	(2)	-2	-1	0	-1	0	1	0

Setting the nonbasic variables equal to zero and solving for the basic variables gives the nonnegative solution

$$z_0 = 8 \quad \mathbf{x} = \begin{bmatrix} 0 \\ 0 \end{bmatrix} \quad \mathbf{y} = \begin{bmatrix} 10 \\ 0 \end{bmatrix} \quad \mathbf{u} = \begin{bmatrix} 0 \\ 0 \end{bmatrix} \quad \mathbf{v} = \begin{bmatrix} 2 \\ 2 \end{bmatrix}$$

which still satisfies the orthogonality conditions. But $z_0 \neq 0$, and so this solution does not satisfy the KKT conditions.

> However, if we could pivot and maintain nonnegativity of the constant column, and maintain the orthogonality conditions, and eventually get z_0 to be nonbasic, then we would have a solution to the KKT conditions.

We know that pivoting by the minimum ratio rule maintains nonnegativity of the constant column. A key observation in Lemke's method is that

> pivoting in the column associated with the complement of the variable that became nonbasic on the last pivot maintains the orthogonality condition.

For example, y_2 became nonbasic in the preceding tableau, so the next pivot should be in the x_2 column. The minimum ratio pivot in that column is circled in the preceding tableau and results in the following tableau:

	x_1	x_2	u_1	u_2	y_1	y_2	v_1	v_2	z_0
5	−11/4	0	4	0	1	3/2	0	−5/2	0
5	−5/4	0	1	(1/2)	0	1/2	0	−3/2	1
1	−11/4	0	−1	−1/2	0	−1/2	1	−1/2	0
1	−1/4	1	−1	−1/2	0	−1/2	0	1/2	0

Note that the solution obtained by setting the nonbasic variables equal to zero satisfies the orthogonality conditions. In the preceding tableau v_2 became nonbasic, so the next pivot is in the column corresponding to the complement of that variable, namely, the u_2 column. Pivoting on the circled entry, in the minimum ratio row, gives the tableau

	x_1	x_2	u_1	u_2	y_1	y_2	v_1	v_2	z_0
5	−11/4	0	4	0	1	3/2	0	−5/2	0
10	−5/2	0	2	1	0	1	0	−3	2
6	−4	0	0	0	0	0	1	−2	1
6	−3/2	1	0	0	0	0	0	−1	1

On this last pivot z_0 became nonbasic, so the solution obtained by setting the nonbasic variables to zero will satisfy the KKT conditions. This solution is

$$z_0 = 0 \qquad \mathbf{x} = \begin{bmatrix} 0 \\ 6 \end{bmatrix} \qquad \mathbf{y} = \begin{bmatrix} 5 \\ 0 \end{bmatrix} \qquad \mathbf{u} = \begin{bmatrix} 0 \\ 10 \end{bmatrix} \qquad \mathbf{v} = \begin{bmatrix} 6 \\ 0 \end{bmatrix}$$

so $\bar{\mathbf{x}} = [0, 6]^T$ is optimal for the original quadratic program.

A statement of Lemke's method for solving the quadratic program is thus quite simple:

Step 1. Introduce the additional variable z_0 and form the tableau

	$\mathbf{x}$	$\mathbf{u}$	$\mathbf{y}$	$\mathbf{v}$	z_0
$\mathbf{c}$	$-\mathbf{Q}$	$-\mathbf{A}^T$	$\mathbf{I}_n$	$\mathbf{0}$	-1
					$\vdots$
$\mathbf{b}$	$\mathbf{A}$	$\mathbf{0}$	$\mathbf{0}$	$\mathbf{I}_m$	-1

Pivot in the z_0 column and in the row with the most negative constant column entry.

Step 2. Pivot by the minimum ratio rule in the column associated with the complement of the variable that became nonbasic on the last pivot.

Step 3. Continue pivoting as in step 2 until z_0 becomes nonbasic or until no further pivots are possible.

When $\mathbf{Q}$ is positive semidefinite, one can show that Lemke's method will terminate in a finite number of pivots with z_0 becoming nonbasic or with no further pivots being possible. The only way that no further pivots could be possible is that the chosen pivot column has no positive entries. In this case it is possible to show that no optimal solution exists to the quadratic program. We will not include those details here.

Geometric Programming

Geometric programs exhibit a special structure that arises quite naturally in engineering design optimization problems, where the objective and constraint functions often have the form

$$f(\mathbf{t}) = \sum_{i=1}^{N} u_i(\mathbf{t}) \quad \text{with} \quad u_i(\mathbf{t}) = c_i t_1^{a_{i1}} t_2^{a_{i2}} \ldots t_m^{a_{im}}$$

where each component t_j of the variable $\mathbf{t}$ is positive, the exponents a_{ij} are arbitrary real numbers, and the coefficients c_i are positive. A function having the form of $f(\mathbf{t})$ is called a **posynomial** because of the requirement that the components t_j and the coefficients c_i are positive. The terms $u_i(\mathbf{t})$ are called **posynomial terms.** A **geometric program** has the form

$$\begin{aligned} &\text{minimize } f_0(\mathbf{t}) \\ &\text{subject to} \\ &\qquad f_k(\mathbf{t}) \leq 1 \qquad k = 1, \ldots, p \\ &\qquad t_j > 0 \qquad j = 1, \ldots, m \end{aligned}$$

where each $f_k(\mathbf{t})$ is a posynomial, $k = 0, 1, \ldots, p$.

The name "geometric programming" was adopted because of the crucial role that the geometric-arithmetic mean inequality played in the development of a duality theory for this class of problems. Geometric programs are also called **posynomial programs.** To illustrate how this duality theory can be used to help solve geometric programs, we consider the following simple example, due to Duffin, Peterson, and Zener, who discovered this duality theory.

Suppose that we wish to construct a single open-top box to be used in several trips for transporting 400 yd^3 of gravel from one place to another. The sides and bottom of the box must be made from special material, but only 4 yd^2 of this (free) material is available. The ends of the box must be made from a material costing \$20/yd^2. The box has no salvage value, and each trip necessary to transport the gravel costs 10 cents. The problem is to design the box so that the cost of transporting the gravel is minimized.

If we let t_1, t_2, and t_3 denote the length, width, and height, respectively, of the box, then the total cost $f_0(\mathbf{t})$ in dollars is given by

$$f_0(\mathbf{t}) = 40t_1^{-1}t_2^{-1}t_3^{-1} + 40t_2t_3$$

where the first term in the posynomial $f_0(\mathbf{t})$ gives the transportation cost and the second term gives the materials cost for the ends of the box. Because the total area of the sides plus the bottom can be no greater than 4, we have the constraint

$$2t_1t_3 + t_1t_2 \leq 4$$

This results in the geometric program

$$\begin{aligned} &\text{minimize } f_0(\mathbf{t}) = 40t_1^{-1}t_2^{-1}t_3^{-1} + 40t_2t_3 \\ &\text{subject to} \\ &\qquad f_1(\mathbf{t}) = \tfrac{1}{2}t_1t_3 + \tfrac{1}{4}t_1t_2 \leq 1 \\ &\qquad\qquad t_j > 0 \qquad j = 1, 2, 3 \end{aligned}$$

In this geometric program there are four posynomial terms, $u_i(\mathbf{t})$, two in the objective function and two in the constraint. The **exponent matrix, A**, of a geometric program is constructed so that its ith row, $\mathbf{A}_i$, consists of the exponents of the ith posynomial term. Thus, here

$$\mathbf{A} = \begin{bmatrix} -1 & -1 & -1 \\ 0 & 1 & 1 \\ 1 & 0 & 1 \\ 1 & 1 & 0 \end{bmatrix}$$

In general, the exponent matrix of a geometric program is $m \times n$ where m is the number of variables t_j and n is the total number of posynomial terms. The dual of a geometric program has a dual variable δ_i associated with each of the n posynomial terms. Given the exponent matrix $\mathbf{A}$ and the coefficient vector $\mathbf{c}$, the general form of the **dual of a geometric program** is

$$\begin{aligned} &\text{maximize} \quad \prod_{i=1}^{n} \left(\frac{c_i}{\delta_i}\right)^{\delta_i} \prod_{k=1}^{p} \lambda_k(\boldsymbol{\delta}) \\ &\text{subject to} \quad \mathbf{A}^T\boldsymbol{\delta} = 0 \qquad \lambda_0(\boldsymbol{\delta}) = 1 \qquad \boldsymbol{\delta} \geq \mathbf{0} \end{aligned}$$

where $\lambda_k(\boldsymbol{\delta})$ is the sum of the dual variables δ_i associated with the terms in the constraint $f_k(\mathbf{t}) \leq 1$. Thus, the dual of our example geometric program is given by

$$\text{maximize } v(\boldsymbol{\delta}) = \left(\frac{c_1}{\delta_1}\right)^{\delta_1} \left(\frac{c_2}{\delta_2}\right)^{\delta_2} \left(\frac{c_3}{\delta_3}\right)^{\delta_3} \left(\frac{c_4}{\delta_4}\right)^{\delta_4} (\delta_3 + \delta_4)^{(\delta_3 + \delta_4)}$$

$$\begin{bmatrix} -1 & 0 & 1 & 1 \\ -1 & 1 & 0 & 1 \\ -1 & 1 & 1 & 0 \end{bmatrix} \begin{bmatrix} \delta_1 \\ \delta_2 \\ \delta_3 \\ \delta_4 \end{bmatrix} = \begin{bmatrix} 0 \\ 0 \\ 0 \end{bmatrix} \qquad \delta_1 + \delta_2 = 1 \qquad \boldsymbol{\delta} \geq \mathbf{0}$$

In the dual problem, $(c_i/\delta_i)^{\delta_i}$ is defined to be zero when $\delta_i = 0$, and so the objective function $v(\boldsymbol{\delta})$ is defined for all $\boldsymbol{\delta} \geq \mathbf{0}$. One can show that the function $\ln v(\boldsymbol{\delta})$ is a concave function (the negative of a convex function). This means that the dual is equivalent to minimizing a convex function subject to linear constraints. Thus, even though the objective function of the dual is complicated, the constraints are linear, and nonlinear programming algorithms that take advantage of linear constraints can be used to solve the dual problem. For example, the reduced gradient method, discussed earlier, is such a method. Note, however, that the function $v(\boldsymbol{\delta})$ is not differentiable at points where some components δ_i equal zero. Thus, special care has to be taken when using the reduced-gradient method to solve the dual problem.

Given a solution to the dual problem, duality relations exist that allow a solution to the original primal problem to be calculated. These duality relations can be derived by using the Karush–Kuhn–Tucker theory, but we shall not go through that derivation here.

The Duality Relations

If the dual problem has a feasible point $\boldsymbol{\delta} > \mathbf{0}$, and if the primal problem has a feasible point satisfying $f_k(\mathbf{t}) < 1$ for $k = 1, \ldots, p$, then,

(i) the primal and dual problems both have optimal vectors,

(ii) the optimal values of the primal and dual are equal,

(iii) each minimizing point **t** for the primal satisfies

$$u_i(\mathbf{t}) = \delta_i^* v(\boldsymbol{\delta}^*) \quad \text{for each term } i \text{ in } f_0(\mathbf{t})$$

$$u_i(\mathbf{t}) = \frac{\delta_i^*}{\lambda_k(\boldsymbol{\delta}^*)} \quad \begin{array}{l}\text{for each term } i \text{ in } f_k(\mathbf{t}), \text{ and} \\ \text{for each } k \text{ where } \lambda_k(\boldsymbol{\delta}^*) > 0\end{array}$$

We shall use our example to illustrate how relation (iii) can be used to calculate a minimizing point for the primal problem. Before doing so we need to find an optimal vector for the dual.

For this example it is actually quite easy to find an optimal dual vector $\boldsymbol{\delta}^*$ because the linear constraints consist of four equations and four unknowns that have a unique solution, namely,

$$\delta_1^* = \tfrac{2}{3} \qquad \delta_2^* = \tfrac{1}{3} \qquad \delta_3^* = \tfrac{1}{3} \qquad \delta_4^* = \tfrac{1}{3}$$

with $v(\boldsymbol{\delta}^*) = 60$. Of course, it does not always happen that the dual problem has a unique feasible point. In general, however, the dual constraint equations can be used to eliminate $m + 1$ of the dual variables from the problem. In our example $m + 1 = 4$, so we could simply solve the equations to obtain the unique solution $\boldsymbol{\delta}^*$.

Using this optimal dual vector and part (iii) of the duality relations, we see that any minimizing point **t** satisfies

$$40t_1^{-1}t_2^{-1}t_3^{-1} = \frac{2}{3}v(\boldsymbol{\delta}^*) = 40$$

$$40t_2t_3 = \frac{1}{3}v(\boldsymbol{\delta}^*) = 20$$

$$\tfrac{1}{2}t_1t_3 = \frac{1/3}{2/3}$$

$$\tfrac{1}{4}t_1t_2 = \frac{1/3}{2/3}$$

It is easy to solve this system by taking logs of the equations to obtain a linear system in the variables $\ln t_j$, $j = 1, 2, 3$. Solving the resulting linear system yields the minimizing point

$$t_1^* = 2 \qquad t_2^* = 1 \qquad t_3^* = \tfrac{1}{2}$$

for the original geometric program. If the dual problem has to be solved numerically, then the overdetermined system of equations may be slightly inconsistent because an exact solution $\boldsymbol{\delta}^*$ has not been obtained. This system can be solved by using linear programming to find the point that minimizes the sum of the absolute values of the deviations (see Chapter 2).

This example shows how the special structure of a geometric program dual can sometimes be used to rather easily find an optimal solution. It also points out how the general theory of linear and nonlinear programming plays an important role in developing methods for solving problems with special structure.

There are other approaches that exploit the structure of the primal geometric program to find an optimal vector directly, without using the duality relations. One of the most successful approaches is to approximate the posynomial function by linear functions and solve a sequence of linear programs. The survey paper by Ecker, listed in the selected references to this chapter, gives more details on this approach.

Separable Programming

A **separable function** f defined for $\mathbf{x}$ in R^n is one that can be written as the sum of n functions f_i of the single variable x_i, where x_i is the ith component of the vector $\mathbf{x}$, that is,

$$f(\mathbf{x}) = f_1(x_1) + f_2(x_2) + \cdots + f_n(x_n)$$

For example, the function

$$f(\mathbf{x}) = 3x_1^2 - 2x_1 + x_2^3 + 2x_3^2 - x_3$$

is a separable function equal to the sum of the three single variable functions

$$f_1(x_1) = 3x_1^2 - 2x_1 \qquad f_2(x_2) = x_2^3 \qquad f_3(x_3) = 2x_3^2 - x_3$$

A **separable program** is a nonlinear program where the objective and constraint functions are separable. This class of specially structured nonlinear programs is important because many practical problems involve separable functions. We shall see that a separable program can be approximated by using **piecewise linear** functions to approximate the separable functions. Under certain convexity assumptions the simplex method can then be used to solve the approximating problem.

The key ideas used in constructing this piecewise linear approximation are really quite simple and will be illustrated using the following example of a separable program with linear constraints:

$$\begin{aligned} \text{minimize} \quad & 3\,x_1^2 - 2x_1 + x_2^3 + 2x_3^2 - x_3 \\ \text{subject to} \quad & \\ & x_1 + x_2 \le 5 \\ & -x_1 - x_2 + x_3 \le 0 \\ & x_i \ge 0 \qquad i = 1, 2, 3 \end{aligned}$$

First, we show how to construct a piecewise linear function to approximate

$$f_1(x_1) = 3x_1^2 - 2x_1$$

The constraints imply that the values of x_1 are bounded between 0 and 5. In general, we assume that the variables x_i have known upper and lower bounds. Figure 9.16 gives the graph of f_1 in the region of interest and gives the **breakpoints**

$$x_1 = 0 \qquad x_1 = 2 \qquad x_1 = 3 \qquad x_1 = 5$$

used in defining a piecewise linear approximation. The first thing to note is that any point x_1 between 0 and 5 can be represented as

$$x_1 = 0\lambda_1 + 2\lambda_2 + 3\lambda_3 + 5\lambda_4$$

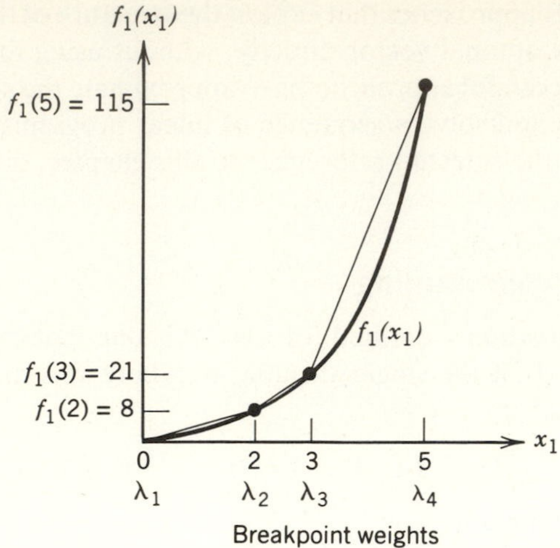

FIGURE 9.16 Approximating $f_1(x_1)$ with a piecewise linear function having four breakpoints.

where λ_1, λ_2, λ_3, and λ_4 are **weights** associated with the four breakpoints and satisfy

$$\lambda_1 + \lambda_2 + \lambda_3 + \lambda_4 = 1 \quad \text{with } \lambda_i \geq 0, i = 1, \ldots, 4$$

and the condition that at most two weights are positive and any two positive weights are adjacent. We shall refer to the condition on the weights as the **adjacency condition.** This representation for x_1 is possible because any point can be written as a convex combination of the two breakpoints between which it lies. For example, if x_1 were between the breakpoints 3 and 5, then λ_3 and λ_4 would be positive and sum to 1, and the other weights would be zero. Note that if x_1 equals one of the breakpoints, then only one weight is positive.

The reason for using such weights is that values of the approximating piecewise linear function at any point between 0 and 5 can then be written as the same weighted sum of the values of f_1 at the breakpoints, that is,

$$\begin{aligned} f_1^a(x_1) &= \lambda_1 f_1(0) + \lambda_2 f_1(2) + \lambda_3 f_1(3) + \lambda_4 f_1(5) \\ &= 0\lambda_1 + 8\lambda_2 + 21\lambda_3 + 115\lambda_4 \end{aligned}$$

where f_1^a is the piecewise linear approximation to f_1. For example, from the values of f_1 given in Figure 9.16, we see that the value of the linear approximation, f_1^a, at a point x_1 between the breakpoints 2 and 3 is given by $8\lambda_2 + 21\lambda_3$ where λ_2 and λ_3 are the positive weights adding to 1 so that $x_1 = 2\lambda_2 + 3\lambda_3$.

These observations show that the variable x_1 can be replaced in the constraints by its representation

$$x_1 = 0\lambda_1 + 2\lambda_2 + 3\lambda_3 + 5\lambda_4$$

and $f_1(x_1)$ can be approximated by

$$f_1^a(x_1) = 0\lambda_1 + 8\lambda_2 + 21\lambda_3 + 115\lambda_4$$

provided we add the constraint that weights sum to 1 and that the adjacency condition holds.

Repeating this process for the other two variables x_2 and x_3 results in an approximating program that would be a linear program if the adjacency condition on each set of weights were ignored. The variables in the approximate problem are the weights used in representing each of the original variables. Unfortunately, the adjacency condition on the weights cannot be reformulated as a linear constraint. However, if we ignore the adjacency condition on each set of weights and solve the resulting linear program, then

> the process of minimization will force the adjacency condition to be automatically satisfied if the functions f_i are convex.

To see why this is true, suppose that a minimizing point for the linear program gives two positive weights that are not adjacent. For example, suppose that the optimal weights associated with x_1 are such that λ_2 and λ_4 are positive and the other weights are 0. The contribution to the linear program's objective function due to this set of weights is $8\lambda_2 + 115\lambda_4$. However, this contribution can be made smaller by choosing positive adjacent weights to represent the same value of x_1 (and thus still satisfy the constraints). For example, if the point x_1 corresponding to this set of weights were between 3 and 5, as indicated in Figure 9.17, then the value $8\lambda_2 + 115\lambda_4$ is the height of the dotted line. However, if x_1 were represented as a convex combination of λ_3 and λ_4, then the contribution to the objective function would be $21\lambda_3 + 115\lambda_4$, namely, the height of the piecewise linear function above x_1 that is smaller because f_1 is a convex function.

Thus, if all the functions f_i are convex, the adjacency condition on each set of weights will be satisfied by any solution to the linear program. This means that the approximating piecewise linear program can be solved by using the simplex algorithm.

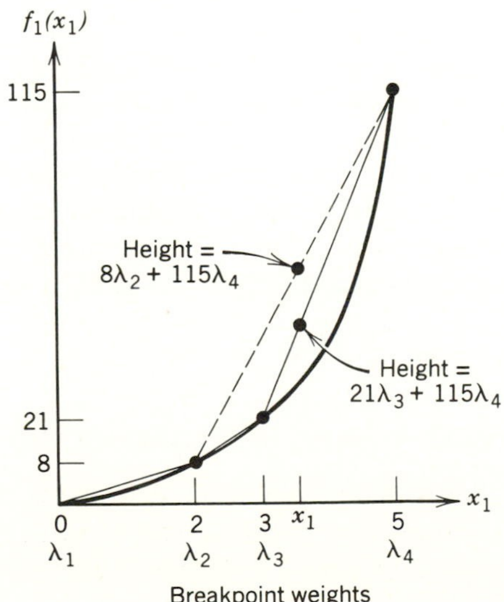

FIGURE 9.17 Convexity guarantees that the adjacency condition holds at optimality.

We conclude this discussion by writing down the linear program that results when breakpoints are chosen for the variables x_2 and x_3 and the adjacency condition is ignored. Recall for this example that each variable is bounded between 0 and 5.

If the breakpoints for x_2 are 0, 3, and 5, then

$$x_2 = 0\delta_1 + 3\delta_2 + 5\delta_3$$

and using $f_2(x_2) = x_2^3$, the piecewise linear approximation is

$$f_2^a(x_2) = 0\delta_1 + 27\delta_2 + 125\delta_3$$

If the breakpoints for x_3 are 0, 1, 3, and 5, then

$$x_3 = 0\beta_1 + 1\beta_2 + 3\beta_3 + 5\beta_4$$

and using $f_3(x_3) = 2x_3^2 - x_3$, the piecewise linear approximation is

$$f_3^a(x_3) = 0\beta_1 + 1\beta_2 + 15\beta_3 + 45\beta_4$$

Thus, from the original separable program, the approximating piecewise linear program without the adjacency conditions is constructed as

$$\begin{aligned}
&\text{minimize} \quad 8\lambda_2 + 21\lambda_3 + 115\lambda_4 + 27\delta_2 + 125\delta_3 + \beta_2 + 15\beta_3 + 45\beta_4 \\
&\text{subject to} \\
&\qquad 2\lambda_2 + 3\lambda_3 + 5\lambda_4 + 3\delta_2 + 5\delta_3 \le 5 \\
&\qquad -2\lambda_2 - 3\lambda_3 - 5\lambda_4 - 3\lambda_2 - 5\delta_3 + \beta_2 + 3\beta_3 + 5\beta_4 \le 0
\end{aligned}$$

$$\begin{aligned}
&\lambda_1 + \lambda_2 + \lambda_3 + \lambda_4 = 1 \qquad && \lambda_i \ge 0,\ i = 1, 2, 3, 4 \\
&\delta_1 + \delta_2 + \delta_3 = 1 && \delta_i \ge 0,\ i = 1, 2, 3 \\
&\beta_1 + \beta_2 + \beta_3 + \beta_4 = 1 && \beta_i \ge 0,\ i = 1, 2, 3, 4
\end{aligned}$$

Solving this linear program gives an approximate solution to the original separable program. Note that the number of variables in this linear program is the total number of breakpoints used for the piecewise linear approximations. To improve the approximation, one need only choose more breakpoints.

If the constraints are nonlinear but separable functions satisfying the convexity assumption, this same type of approximation can be obtained if we use the same breakpoints for the constraints as are used in the objective function for a given variable.

Other Theoretical Aspects of Nonlinear Programming

In our study of linear programming in Part I, we devoted considerable attention to duality and sensitivity analysis. These are important topics in nonlinear optimization as well, but they are at present less well developed than the corresponding theories in linear programming.

The theory of nonlinear programming duality began about 1961 with the discovery of the classical or **Wolfe dual,** which is defined as follows.

$$\text{primal:} \quad \underset{\mathbf{x}\in R^n}{\text{minimize}} \quad f_0(\mathbf{x})$$
$$\text{subject to} \quad f_i(\mathbf{x}) \leq 0 \qquad i = 1, \ldots, m$$

$$\text{Wolfe dual:} \quad \underset{\mathbf{x},\,\mathbf{u}}{\text{maximize}} \quad L(\mathbf{x}, \mathbf{u})$$
$$\text{subject to} \quad \nabla L(\mathbf{x}, \mathbf{u}) = \mathbf{0}$$
$$\mathbf{u} \geq \mathbf{0}$$

Recent work in nonlinear programming duality has focused on the **Lagrangian dual.**

$$\text{Lagrangian dual:} \quad \underset{\mathbf{u}\in R^m}{\text{maximize}} \{\underset{\mathbf{x}\in S}{\text{minimum}}\ L(\mathbf{x}, \mathbf{u})\}$$
$$\text{subject to } \mathbf{u} \geq \mathbf{0}$$

Under certain conditions, both the Wolfe and the Lagrangian duals can be shown to have properties that relate them to the primal problem in ways reminiscent of the ways in which a dual linear program is related to its primal.

Sensitivity analysis in nonlinear programming is often referred to as the study of problems having **data perturbations** because it is assumed that the functions $f_i(\mathbf{x})$, $i = 0, \ldots, m$ do not change in form. As in linear programming, the idea is to find ways of computing a new optimal solution from an old one without completely re-solving the perturbed problem from scratch. Sensitivity analysis is an active area of research in mathematical programming.

SELECTED REFERENCES

Avriel, M., *Nonlinear Programming, Analysis and Methods,* Prentice-Hall, Englewood Cliffs, NJ, 1976.

Bazaraa, M. S., and Shetty, C. M., *Nonlinear Programming, Theory and Algorithms,* Wiley, New York, 1979.

Duffin, R. J., Peterson, E. L., and Zener, C., *Geometric Programming—Theory and Application,* Wiley, New York, 1967.

Ecker, J. G., Geometric Programming: Methods, Computations and Applications, SIAM Review 22:338–362, 1980.

Ecker, J. G., and Kupferschmid, M., A Computational Comparison of the Ellipsoid Algorithm with Several Nonlinear Programming Algorithms, *SIAM Journal on Control and Optimization* **23**:657–674 (1985).

Fiacco, A. V., *Introduction to Sensitivity and Stability Analysis in Nonlinear Programming,* Academic Press, New York, 1983.

Gill, P. E., Murray, W., and Wright, M. H., *Practical Optimization,* Academic Press, New York, 1981.

Himmelblau, D. M., *Applied Nonlinear Programming,* McGraw-Hill, New York, 1972.

Lasdon, L. S., Waren, A. D., Jain, A., and Ratner, M., Design and Testing of a Generalized Reduced Gradient Code for Nonlinear Programming, *ACM Transactions on Mathematical Software* **4**:34–50 (1978).

Lemke, C. E., Bimatrix Equilibrium Points and Mathematical Programming, *Management Science* **11**:681–689 (1965).

Liebman, J., Schrage, L., Lasdon, L., and Waren, A., Applications of Modeling and Optimization with GINO, Graduate School of Business, University of Chicago, Chicago, 1984.

Mangasarian, O. L., *Nonlinear Programming,* McGraw-Hill, New York, 1969.

McCormick, G. P., *Nonlinear Programming: Theory, Algorithms, and Applications,* Wiley, New York, 1983.

Tapia, R. A., On the Role of Slack Variables in Quasi-Newton Methods for Constrained Optimization, Department of Mathematical Sciences, Rice University, Houston, 1976.

Wolfe, P., A Duality Theorem for Nonlinear Programming, *Quarterly of Applied Mathematics* **19**:239–244 (1961).

Zangwill, W. I., *Nonlinear Programming, A Unified Approach,* Prentice-Hall, Englewood Cliffs, 1969.

EXERCISES

9.1 Consider the nonlinear Oakwood Furniture problem of Section 9.1

(a) Reformulate the problem as a linear program (LP) by neglecting the effects that make the nonlinear program (NLP) nonlinear. That is, simplify the NLP by assuming that numbers near 1 are exactly 1 and that numbers much smaller than 1 are zero to obtain an approximate LP.

(b) Solve the approximate LP and compare the solution to that reported for the NLP.

(c) Make a contour plot for the LP and compare it to Figure 9.1.

(d) For what purposes might the LP approximation be satisfactory? Would the fidelity of the LP approximation improve if Oakwood's inventory were reduced from 700 to 70?

9.2 Some nonlinear programming formulation problems.

(a) It is desired to construct a rectangular tank of volume at least V, while using the least possible area of material for the bottom, sides, and top. Formulate a nonlinear program to find the height h, width w, and breadth b of the tank.

(b) A farmer wishes to fence in a rectangular pen for his animals. He only has 30 ft of fencing to use, but the back wall of his barn is 40 ft long, and he can use it on one side of the pen, thereby increasing the enclosed area somewhat. If the sides of the pen perpendicular to the back of the barn are of length s, and the side of the pen parallel to the back of the barn is of length w, formulate a nonlinear program to find s and w so that the enclosed area will be as big as possible.

(c) For a certain industrial process it is necessary to build a tank that has the shape of a circular cylinder of radius r and height h. Within the cylinder is a conical funnel equal in radius at its top to the radius of the cylinder, and having straight sides ending in a point of negligible radius in the center of the bottom face of the tank. The cylindrical tank, its circular bottom, and the cone are all to be fabricated from the same material, which weighs 3 lb/ft^2. The assembly is open on top and must not weigh more than 3000 lb when completed. Formulate a nonlinear program to find r and h so as to maximize the volume contained between the cylinder walls and the cone.

(d) It is required to find the point on the curve described by

$$7x_1 - 3x_2^2 = 0$$

that is closest to the point $[1, 0]^T$. Formulate a nonlinear programming problem that will yield the desired point as its optimal point.

9.3 In the following schematic diagram v is the open-circuit voltage of a battery, r is its internal resistance, and R is the resistance of a load.

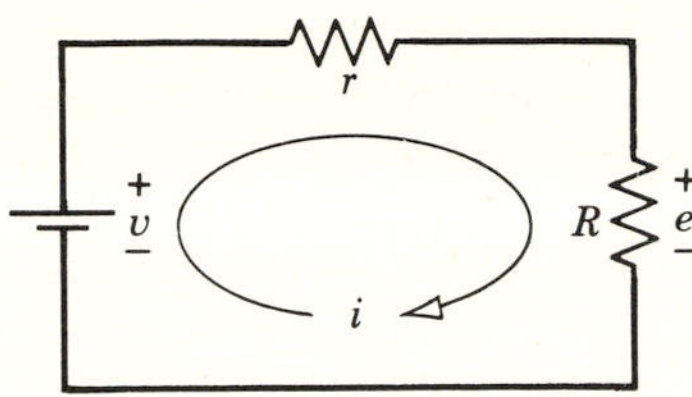

From Ohm's law the current flowing in this circuit is $i = v/(r + R)$ and the voltage across the load resistance is $e = iR$. The power dissipated in the load resistance is $P = i^2R$.

(a) Formulate a nonlinear program to find the value of R that maximizes the power dissipation in the load resistance.

(b) How does the formulation change if the allowable voltage across the load resistance is E?

9.4 In Section 9.2 the following problem is solved:

$$\underset{\mathbf{x}\in\mathbf{R}^2}{\text{minimize}}\ f(\mathbf{x}) = (x_1 - 1)^2 + x_2^2 + 1$$

Solve the following related problems by setting the partial derivatives equal to zero and solving the resulting systems. Should all of these problems have the same optimal point? If not, explain why not. If so, explain any differences between the answers you obtain.

(a) $\underset{\mathbf{x}\in\mathbf{R}^2}{\text{minimize}}\ f(\mathbf{x}) = \sqrt{(x_1 - 1)^2 + x_2^2 + 1}$

(b) $\underset{\mathbf{x}\in\mathbf{R}^2}{\text{minimize}}\ f(\mathbf{x}) = (x_1 - 1)^2 + x_2^2$

(c) $\underset{\mathbf{x}\in\mathbf{R}^2}{\text{minimize}}\ f(\mathbf{x}) = \sqrt{(x_1 - 1)^2 + x_2^2}$

9.5 Consider the following function of two variables.

$$f(\mathbf{x}) = \frac{x_1 + x_2}{3 + x_1^2 + x_2^2 + x_1x_2}$$

(a) Find all the stationary points of this function.

(b) Classify each stationary point as to whether it is a strict local minimizing point, a local minimizing point that is not strict, a strict local maximizing point, a local maximizing point that is not strict, or neither a minimizing nor a maximizing point.

9.6 Consider the following function of two variables:

$$f(\mathbf{x}) = (x_2^2 - x_1)^2$$

(a) Find all the stationary points of this function.

(b) From among the stationary points found in part (a), identify those that are minima.

9.7 Consider the following function:

$$f(\mathbf{x}) = (x_2 - x_1^2)^2 + (1 - x_1)^2$$

(a) Find all the stationary points of this function.

(b) Classify the points found in (a) according to whether they are strict minimizing points, minimizing points, strict maximizing points, maximizing points, or points that are neither minimizing nor maximizing points.

9.8 Consider the following function of three variables:

$$f(\mathbf{x}) = ax_1^2e^{x_2} + x_2^2e^{x_3} + x_3^2e^{x_1}$$

(a) For what values of a is the point $\mathbf{x} = [0, 0, 0]^T$ a stationary point?
(b) Classify the point as a strict minimum, minimum, maximum, or strict maximum for the values of a found in part (a).

9.9 The following questions concern the example of Figure 9.8*a*
(a) Show that the parameterization

$$x_1 = g_1(t) = \frac{\ln(t)}{1 + \sqrt{1 - t \cdot \ln(t)}}$$

$$x_2 = g_2(t) = 1 + \sqrt{1 - t \cdot \ln(t)},$$

satisfies the constraint equation

$$f_1(x) = x_1e^{x_1x_2} + x_2 - 2 = 0$$

for all values of t such that $t \cdot \ln(t) \leq 1$.
(b) Find the numerical range of t over which this parameterization is equivalent to the original constraint.
(c) Find $[d\mathbf{x}/dt]$ as a function of t. Then find the gradient $\nabla f_0(\mathbf{x})$ of the objective function

$$f_0 = -x_2(x_1 + 2)$$

and use the parameterization to write it as a function of t. Finally, use the condition

$$\nabla f_0(t)^T[d\mathbf{x}/dt] = 0$$

to show that $t^* = 1$ at the optimal point.
(d) Show that the values $\mathbf{x}^* = [0, 2]^T$ and $u^* = 2$ satisfy the equations

$$\nabla f_0(\mathbf{x}) = -u\nabla f_1(\mathbf{x})$$

$$f_1(\mathbf{x}) = 0$$

as claimed in Section 9.3.
(e) Find a parameterization that is equivalent to the original constraint for values of t larger than that for which $t \cdot \ln(t) = 1$.

9.10 Consider the following equation in two variables:

$$f_1(\mathbf{x}) = x_1^3x_2 + x_2^3x_1 - a = 0$$

(a) Find a parametric representation for this equation so that part or all of the locus of points that it describes is also given by two equations of the form

$$x_1 = g_1(t) \quad \text{and} \quad x_2 = g_2(t)$$

(b) Over what range of values for t does this parameterization hold? What range of values for $\mathbf{x}$ does that correspond to?
(c) Is your parameterization unique?

9.11 The following questions concern the example of Figure 9.9*a*, which is solved in Section 9.3
(a) Find $\nabla f_0(\mathbf{x}^*)$, $\nabla f_1(\mathbf{x}^*)$, and $\nabla f_2(\mathbf{x}^*)$ numerically.
(b) Show that the three vectors found in part (a) lie in the same plane.
(c) Find u_1 and u_2 such that $\nabla f_0(\mathbf{x}^*) = -[u_1 \nabla f_1(\mathbf{x}^*) + u_2 \nabla f_2(\mathbf{x}^*)]$.

9.12 Consider the following equality-constrained problem:

$$\underset{\mathbf{x} \in R^2}{\text{minimize}} \ f_0(\mathbf{x}) = (x_1 - 1)^2 + x_2^2 + 1$$

subject to

$$f_1(\mathbf{x}) = x_1 - \tfrac{1}{4}x_2^2 = 0$$

This problem was solved in Section 9.3 in several ways, including by use of the substitution $x_1 = x_2^2/4$.
(a) Solve the problem by substituting $x_2 = \sqrt{4x_1}$ instead.
(b) Solve the problem by the Lagrange method. Does the same answer result in all cases? If not, explain why not.

9.13 Solve the following nonlinear programming problems by the Lagrange method.

(a) $\underset{\mathbf{x} \in R^3}{\text{maximize}} \quad x_1 x_2 x_3$

subject to

$$x_1 + x_2 + x_3 = 9$$

(b) $\underset{\mathbf{x} \in R^3}{\text{maximize}} \quad x_1 x_2 + x_2 x_3 + x_1 x_3$

subject to

$$x_1 + x_2 + x_3 = 9$$

(c) $\underset{\mathbf{x} \in R^2}{\text{minimize}} \quad x_1 - x_2$

subject to

$$x_1^2 + x_2^2 - 1 = 0$$

(d) $\underset{\mathbf{x} \in R^3}{\text{maximize}} \quad 3x_1 x_3 + 4x_2 x_3$

subject to

$$x_2^2 + x_3^2 = 4$$
$$x_1 x_3 = 3$$
$$x_j \geq 0, j = 1, \ldots, 3$$

9.14 Consider the equality-constrained nonlinear program

$$\text{minimize } f_0(\mathbf{x}) \quad \text{subject to} \quad f_i(\mathbf{x}) = 0, i = 1, \ldots, m$$

Suppose that $(\bar{\mathbf{x}}, \bar{\mathbf{u}})$ is a solution to the Lagrange conditions, and consider the function of $\mathbf{x}$ defined by letting $\mathbf{u} = \bar{\mathbf{u}}$ in the Lagrangian to obtain

$$L(\mathbf{x}, \bar{\mathbf{u}}) = f_0(\mathbf{x}) + \sum_{i=1}^{m} \bar{u}_i f_i(\mathbf{x})$$

Let $\mathbf{H}_L(\mathbf{x})$ be the Hessian of this function. The following is an important sufficient condition for $\bar{\mathbf{x}}$ to be a local minimizing point.

If $\mathbf{x}^T \mathbf{H}_L(\bar{\mathbf{x}})\mathbf{x} > 0$ for all nonzero vectors $\mathbf{x}$ such that
$$\mathbf{x}^T \nabla f_i(\bar{\mathbf{x}}) = 0, \qquad i = 1, \ldots, m$$
then $\bar{\mathbf{x}}$ is a strict local minimizing point.

Formulate a nonlinear programming problem that could be used to check whether this sufficient condition holds.

9.15 Consider the example problem of Section 9.4, pictured in Figure 9.10
(a) Show that $\mathbf{x}^* = [11, 8]^T$ and $\mathbf{u}^* = [0, 6]^T$ satisfies the KKT conditions for the problem.
(b) Solve the KKT conditions to obtain $\mathbf{x}^* = [11, 8]^T$ and $\mathbf{u}^* = [0, 6]^T$. Is this the only solution to the KKT conditions?

9.16 Let $f(\mathbf{x})$ be a scalar function of $\mathbf{x} \in R^n$. Use the definition of convexity to show that the epigraph of f is a convex set if and only if $f(\mathbf{x})$ is a convex function.

9.17 Consider a function $f(\mathbf{x})$, with level sets given by $S(a) = \{\mathbf{x} \mid f(\mathbf{x}) \leq a\}$.
(a) Prove that if $f(\mathbf{x})$ is a convex function, the level sets $S(a)$ are convex sets.
(b) If the level set $S(a)$ is a convex set for all values of a, is the function $f(\mathbf{x})$ necessarily convex? Explain.

9.18 Suppose $f(\mathbf{x})$ is a convex function and $g(\mathbf{x})$ is a strictly convex function. Prove that $f(\mathbf{x}) + g(\mathbf{x})$ is a strictly convex function.

9.19 A function $f(\mathbf{x})$ is concave if $-f(\mathbf{x})$ is convex.
(a) Use the definition of convexity to show that any linear function is both convex and concave.
(b) If $f(\mathbf{x})$ is concave and h is an increasing concave function, show that the function g defined by $g(\mathbf{x}) = h(f(\mathbf{x}))$ is concave.

9.20 Consider the function $f(x) = x^2$, where x is a scalar.
(a) Use the definition of convexity to prove that $f(x)$ is a convex function.
(b) Use the Hessian matrix definiteness test to prove that $f(x)$ is a convex function.

9.21 Consider the problem of determining whether a square nonsymmetric matrix is positive semidefinite.
(a) Show that a matrix $\mathbf{A}$ is psd if and only if $\mathbf{A} + \mathbf{A}^T$ is psd.
(b) Show that $\mathbf{A} + \mathbf{A}^T$ is a symmetric matrix even if $\mathbf{A}$ is not.
(c) Use the results of (a) and (b) to formulate a method of testing whether a nonsymmetric matrix is psd.
(d) Determine whether the following matrix is psd.

$$\begin{bmatrix} 1 & 3 \\ 0 & 2 \end{bmatrix}$$

9.22 If $\mathbf{H}$ is an $n \times n$ matrix, give formulas for
(a) the number of principal minors;
(b) the number of leading principal minors.

9.23 Consider the following function:

$$f(\mathbf{x}) = 2x_1^2 + 2x_1x_2 + x_2^2 - x_2x_3 + 2x_3^2$$

(a) Is $f(\mathbf{x})$ convex?
(b) Is $f(\mathbf{x})$ strictly convex?

9.24 Show analytically that the function of Figure 9.3 is nonconvex.

9.25 Determine whether the function

$$f(x_1, x_2) = 2x_1^2x_2^{-1}$$

is convex over the positive orthant, $\{\mathbf{x} \in \mathrm{R}^2 \mid x_1 > 0, x_2 > 0\}$.

9.26 The Gerschgorin circle theorem concerns the locations in the complex plane of the eigenvalues of a matrix: "Every eigenvalue of $\mathbf{H}$ lies in at least one of the circles $C_1, C_2, \ldots, C_n$, where C_i is centered at the diagonal entry $\mathbf{H}(i, i)$ and has radius equal to the sum of the absolute values of the entries along the rest of the ith row." A matrix is positive definite if and only if all of its eigenvalues are positive, and positive semidefinite if and only if all of its eigenvalues are nonnegative.
(a) Explain how the Gerschgorin circle theorem can sometimes be used to quickly identify a matrix as pd or psd.

(b) Determine whether the following matrix is psd.

$$\begin{bmatrix} 9 & 2 & 1 & 4 & -1 \\ 2 & 7 & -3 & 1 & 0 \\ 1 & -3 & 9 & 2 & 2 \\ 4 & 1 & 2 & 9 & 1 \\ -1 & 0 & 2 & 1 & 4 \end{bmatrix}$$

9.27 Consider the canonical form nonlinear programming problem

$$\underset{\mathbf{x} \in R^n}{\text{minimize}} \; f_0(\mathbf{x})$$

subject to

$$f_i(\mathbf{x}) \le 0, \qquad i = 1, \ldots, m$$

with feasible set $S = \{\mathbf{x} \mid f_i(\mathbf{x}) \le 0, i = 1, \ldots, m\}$.

(a) If the $f_i(\mathbf{x})$ are convex functions, is S necessarily a convex set? Explain.

(b) If S is a convex set, are the functions $f_i(\mathbf{x})$ necessarily convex functions? Explain.

9.28 Consider the following nonlinear programming problem:

$$\underset{\mathbf{x} \in R^3}{\text{minimize}} \; x_1 - x_2 + x_3$$

subject to

$$x_1^2 + x_2^2 + x_3^2 - 1 \le 0$$

(a) Use the KKT conditions to find a solution.

(b) Is the solution unique? Give a geometric interpretation of your answer.

9.29 Consider the following nonlinear programming problem:

$$\underset{\mathbf{x} \in R^2}{\text{minimize}} \; -x_1 - 2x_2 + x_2^2$$

subject to

$$x_1 + x_2 \le 1$$

$$x_1 \ge 0, x_2 \ge 0$$

(a) Is this a convex programming problem? Explain.

(b) Write down the KKT conditions for this problem.

(c) Use the KKT conditions to find a solution. Is it unique?

(d) Does a constraint qualification hold for this problem?

9.30 Consider the following nonlinear programming problem:

$$\underset{\mathbf{x} \in R^2}{\text{maximize}} \; -x_1 - 2x_2 + x_2^3$$

subject to

$$x_1 + x_2 \le 1$$

$$x_1 \ge 0, x_2 \ge 0$$

(a) Write down the KKT conditions for this problem.

(b) Find all the solutions $(\mathbf{x}^*, \mathbf{u}^*)$ of the KKT conditions found in part (a) above.

(c) Determine whether the objective function is convex over the feasible set.

(d) If $(\mathbf{x}^*, \mathbf{u}^*)$ is a solution to the KKT conditions for this problem, can we conclude that $\mathbf{x}^*$ is a global minimizing point? Explain.

9.31 Use the KKT conditions to solve the following problem:

$$\underset{\mathbf{x}\in\mathbf{R}^2}{\text{minimize}} \quad x_1^2 + 4x_2^2 - 4x_1x_2 - x_1 + 12x_2$$

subject to

$$x_1 + x_2 \geq 4$$
$$x_1 \geq 0,\ x_2 \geq 0$$

9.32 Consider the following nonlinear programming problem:

$$\underset{\mathbf{x}\in\mathbf{R}^2}{\text{minimize}} \quad x_1^2 - x_2^2$$

subject to

$$-(x_1 - 2)^2 - x_2^2 + 4 \leq 0$$

(a) Write down the KKT conditions for the problem.
(b) Find all solutions of the KKT conditions.
(c) Find all local minimizing points.

9.33 Consider the following nonlinear programming problem:

$$\underset{\mathbf{x}\in\mathbf{R}^3}{\text{minimize}} \quad (x_1 - 10)^2 + (x_2 - 10)^2 + (x_3 - 10)^2$$

subject to

$$x_1^2 + x_2^2 + x_3^2 \leq 12$$
$$-x_1 - x_2 + 2x_3 \leq 0$$

(a) Write down the KKT conditions for the problem.
(b) Solve the KKT conditions analytically to find the optimal point $\mathbf{x}^*$ and the corresponding Lagrange multipliers $\mathbf{u}^*$.
(c) Which of the constraints (if either) are active at $\mathbf{x}^*$?
(d) Is the point $\mathbf{x}^*$ solving the KKT conditions a global minimum? Is the global minimum unique for this problem?

9.34 Consider the following inequality-constrained problem:

$$\underset{\mathbf{x}\in\mathbf{R}^2}{\text{minimize}} \quad x_1 + x_2$$

subject to

$$x_1^2 + x_2^2 \leq +4$$
$$-x_1^2 - x_2^2 \leq -4$$

(a) Draw a contour diagram for the problem, showing the feasible set and some contours of the objective function.
(b) Solve the problem graphically. Which constraint is active?
(c) Solve the problem using the KKT conditions. Are the Lagrange multipliers uniquely determined?

9.35 Consider the following problem, which was discussed in Section 9.6:

$$\underset{\mathbf{x}\in\mathbf{R}^2}{\text{minimize}} \quad f_0(\mathbf{x}) = -3x_1 + \tfrac{1}{2}x_2^2$$

subject to

$$f_1(\mathbf{x}) = x_1^2 + x_2^2 - 1 \leq 0$$
$$f_2(\mathbf{x}) = -x_1 \leq 0$$
$$f_3(\mathbf{x}) = -x_2 \leq 0$$

The KKT point for this problem is $\mathbf{x} = [1, 0]^T$, with $\mathbf{u} = [\frac{3}{2}, 0, 0]^T$. Thus, $u_3^* = 0$ even though the third constraint is active. With the help of a contour diagram for this problem, explain the significance of a zero Lagrange multiplier for an active constraint.

9.36 Consider the following nonlinear programming problem:

$$\underset{\mathbf{x}\in R^2}{\text{minimize}} \quad x_1^2 - \tfrac{1}{2}x_2^2$$

subject to

$$x_1 - x_2 \leq 0$$
$$x_1 + x_2 \leq 0$$
$$x_1 \leq 0$$

(a) Show that the KKT conditions hold at $\mathbf{x}^* = [0, 0]^T$, with $\mathbf{u}^* = [0, 0, 0]^T$.
(b) Are the gradients of the active constraints linearly independent at the optimal point? Explain.

9.37 Consider the following nonlinear programming problem:

$$\underset{\mathbf{x}\in R^2}{\text{minimize}} \quad -x_1$$

subject to

$$x_2 - (1 - x_1)^3 \leq 0$$
$$-x_2 \leq 0$$

(a) Solve the problem graphically.
(b) Write down the KKT conditions.
(c) Does the $\mathbf{x}^*$ found in part (a) satisfy the KKT conditions for any Lagrange multipliers? Explain.
(d) Change the second constraint to read $-x_2 \leq 1$ and solve the resulting KKT conditions. Does a constraint qualification hold at the optimal point for this problem?

9.38 Consider the following nonlinear programming problem:

$$\underset{\mathbf{x}\in R^2}{\text{minimize}} \quad x_1^2 + x_2^2$$

subject to

$$(x_1 - 1)^3 - x_2^2 \geq 0$$

(a) Solve the problem graphically
(b) Write down the KKT conditions for the problem.
(c) Do the KKT conditions yield the optimal solution? Explain.

9.39 Consider the linear program

$$\text{minimize} \quad \mathbf{c}^T\mathbf{x}$$

subject to

$$\mathbf{Ax} \leq \mathbf{b}$$

(a) Write down the KKT conditions for this problem.
(b) If $(\bar{\mathbf{x}}, \bar{\mathbf{u}})$ solves the KKT conditions, show that $\bar{\mathbf{x}}$ is optimal for the linear program and $\bar{\mathbf{u}}$ is optimal for its dual.

9.40 Is it ever possible for the bisection line search algorithm to locate a minimum exactly, that is, with precisely zero error, in a finite number of iterations? If not, explain why not. If so, construct an example illustrating this behavior.

9.41 Will the bisection line search algorithm ever converge to a maximum rather than a minimum? Explain.

9.42 Apply the bisection line search algorithm to the function of Figure 9.3, starting with the interval $[-3, 2]$ and performing 10 iterations. Does the algorithm converge to the global minimum? Is it guaranteed to do so?

9.43 Apply the bisection line search algorithm to find the minimum of the function $y = |x|$,
(a) starting with the interval $[-1, 2]$;
(b) starting with the interval $[-1, 1]$.
In each case, perform five iterations of the algorithm. Is bisection doomed to fail on all nondifferentiable functions?

9.44 Carry out the calculations to find $\mathbf{x}^6$, starting from $\mathbf{x}^5$, for the steepest-descent example of Section 9.7,
(a) using the search direction $\mathbf{d} = -\nabla f(\mathbf{x}^5)$;
(b) using the search direction $\mathbf{d} = -0.5\{\nabla f(\mathbf{x}^4) + \nabla f(\mathbf{x}^5)\}$.
Does the search direction in (b) follow the valley better than that in (a)? How far is $\mathbf{x}^6$ from $\mathbf{x}^*$ in each case?

9.45 Consider the following unconstrained problem:

$$\underset{\mathbf{x}\in R^2}{\text{minimize}}\ f(x) = 4x_1^2 - 9x_1 + x_2^2$$

Let $\mathbf{x}^0 = [1, 1]^T$ be the starting point for the method of steepest descent. Use the algorithm to find $\mathbf{x}^1$.

9.46 Consider the following nonlinear programming problem, which is discussed in Section 9.7 as an example for the GRG algorithm:

$$\underset{\mathbf{x}\in R^4}{\text{minimize}} \quad x_1^2 + x_2 + x_3^2 + x_4$$

subject to

$$x_1 + x_2 + 4x_3 + 4x_4 - 4 = 0$$

$$-x_1 + x_2 + 2x_3 - 2x_4 + 2 = 0$$

In Section 9.7, the following solution of the constraints is given, expressing x_2 and x_3 in terms of x_1 and x_4.

$$x_2 = 3x_1 + 8x_4 - 8 \quad \text{and} \quad x_3 = -x_1 - 3x_4 + 3$$

(a) Confirm that this result is correct.
(b) Show how pivoting in a linear programming tableau can be used to obtain this result.
(c) Solve the constraint equations for x_1 and x_2 in terms of x_3 and x_4.
(d) Substitute the expressions found in (c) into the objective function to obtain an unconstrained problem in x_3 and x_4.
(e) Solve the reduced problem found in (d) and back substitute to get the optimal values of x_1 and x_2. Does your answer agree with the one in Section 9.7?

9.47 Consider the following function of three variables:

$$f(\mathbf{x}) = x_1^2x_2x_3 + x_3^2$$

(a) Find the first-order Taylor series expansion of this function about the point $\mathbf{x} = [1, 1, 1]^T$.
(b) Find the second-order Taylor series expansion of the function about the point $\mathbf{x} = [1, 1, 1]^T$.

9.48 Consider the following nonlinear programming problem, which is discussed in Section 9.7 as an example for the GRG algorithm.

$$\underset{\mathbf{x}\in R^4}{\text{minimize}}\ f_0(\mathbf{x}) = x_1^2 + x_2 + x_3^2 + x_4$$

subject to

$$f_1(\mathbf{x}) = x_1^2 + x_2 + 4x_3 + 4x_4 - 4 = 0$$

$$f_2(\mathbf{x}) = -x_1 + x_2 + 2x_3 - 2x_4^2 + 2 = 0$$

(a) Continue the solution process outlined in the text to show how the algorithm yields the iterate $\mathbf{x}^2$.
(b) Use the nonlinear constraints to express two of the variables in terms of the others, substitute to obtain a reduced objective function, and solve the resulting unconstrained problem. Confirm that your solution agrees with that given on the last line of the table of GRG results in Section 9.7

9.49 The following equation describes an ellipse in R^2.

$$\frac{(x_1 - 2)^2}{4} + \frac{(x_2 - 1)^2}{9} = 1$$

(a) Graph this ellipse.
(b) Find the positive definite matrix $\mathbf{Q}$ and the vector $\mathbf{x}^0$ so that the equation for this ellipse is expressed in the following form:

$$(\mathbf{x} - \mathbf{x}^0)^T\mathbf{Q}^{-1}(\mathbf{x} - \mathbf{x}^0) = 1$$

9.50 Consider the following nonlinear programming problem:

$$\underset{\mathbf{x}\in R^2}{\text{minimize}}\ f_0(\mathbf{x}) = 2x_1^2 - x_1 + x_2^2$$

subject to

$$f_1(\mathbf{x}) = 8x_1 + 8x_2 \leq 1$$

The minimizing point for this problem is inside the circle

$$x_1^2 + x_2^2 \leq 1$$

so we can use this circle as the initial ellipsoid E_0 in the ellipsoid algorithm.
(a) In R^2 draw a picture of E_0. Indicate the intersection of E_0 with the feasible set, and the starting point $\mathbf{x}^0$ for the ellipsoid algorithm.
(b) On the picture draw and label with an **H** the hyperplane used to cut the circle in half in performing the first iteration of the ellipsoid algorithm.
(c) Use the update formulas to calculate $\mathbf{x}^1$.

9.51 Consider the following nonlinear programming problem:

$$\underset{\mathbf{x}\in R^2}{\text{minimize}}\ (x_1 - 2)^2 + x_2^2$$

subject to

$$x_1^2 + x_2^2 \leq 1$$

(a) Show that the objective and constraint functions are convex.
(b) Suppose the circle defined by the constraint is used as the initial ellipsoid for the ellipsoid algorithm, so that the starting point is $\mathbf{x}^0 = [0, 0]^T$. What is $\mathbf{x}^1$?
(c) With the starting conditions described in part (b), what is the value of the first component of $\mathbf{x}^4$?

9.52 In each iteration of the ellipsoid algorithm, the cutting hyperplane is translated parallel to itself until it is tangent to the current ellipsoid at a point $\mathbf{y}$.
(a) Show that the point $\mathbf{y}$ can be found analytically by solving the following optimization problem.

$$\underset{\mathbf{x}\in \mathbf{R}^n}{\text{minimize}} \quad \nabla f(\mathbf{x}^k)^T\mathbf{y}$$

subject to

$$(\mathbf{y} - \mathbf{x}^k)^T\mathbf{Q}_k^{-1}(\mathbf{y} - \mathbf{x}^k) \le 1$$

(b) Use the Lagrange method to solve the problem, obtaining a formula for $\mathbf{y}$ in terms of $\mathbf{x}^k$, $\nabla f(\mathbf{x}^k)$, and $\mathbf{Q}_k$.

9.53 The ellipsoid algorithm volume reduction factor is given by the following formula:

$$q(n) = \frac{\text{volume}(E_{k+1})}{\text{volume}(E_k)} = \sqrt{\frac{n-1}{n+1}}\left(\frac{n}{\sqrt{n^2-1}}\right)^n$$

(a) Evaluate the expression and tabulate $q(n)$ for values of $n = 1, 2, 3, 4, 5, 10$, and 100.

(b) Show analytically that for large n, $q(n) \approx 1 - \dfrac{1}{2n}$.

9.54 In the ellipsoid algorithm example of Section 9.7, the following starting values were used:

$$\mathbf{x}^0 = \begin{bmatrix} 18 \\ 12 \end{bmatrix} \qquad \mathbf{Q}_0 = \begin{bmatrix} 81 & 0 \\ 0 & 169 \end{bmatrix}$$

Suppose these values were obtained from bounds on the variables. What must the bounds have been?

9.55 When the ellipsoid algorithm is used on a nonconvex problem, it is possible for it to converge to a point that is not a minimizing point. As an illustration of this phenomenon, consider the following problem:

$$\underset{\mathbf{x}\in \mathbf{R}^2}{\text{minimize}} \quad (x_1 - 15)^2 + x_2^2$$

subject to

$$x_1^2 + x_2^2 \ge 25$$

$$(x_1 - 3)^2 + x_2^2 \le 25$$

The ellipsoid E_0 defined by

$$\frac{(x_1 + 1)^2}{100} + \frac{x_2^2}{25} \le 1$$

contains the optimal point $\mathbf{x}^*$ and can be used as a starting ellipsoid.

(a) Find the exact minimizing point $\mathbf{x}^*$ graphically. Indicate the feasible set and $\mathbf{x}^*$ on your graph.

(b) Confirm analytically that $\mathbf{x}^* \in E_0$.

(c) Draw the starting ellipsoid E_0 on your graph, and perform the first iteration of the algorithm. Show the starting point $\mathbf{x}^0$, the cutting hyperplane $\mathbf{H}_0$, the new ellipsoid center $\mathbf{x}^1$, and the new ellipsoid E_1. Is $\mathbf{x}^*$ contained in E_1?

9.56 Answer whether each of the following statements is true or false:

(a) The Lagrange multiplier corresponding to an inactive constraint is always zero.

(b) The Lagrange multiplier corresponding to an active constraint is never zero.

(c) A positive semidefinite matrix is never singular.

(d) A positive definite matrix is never singular.

(e) If f_i, $i = 1, \ldots, m$, are differentiable and $\bar{\mathbf{x}}$ is a local minimum, there exists a vector $\mathbf{u}$ such that $(\bar{\mathbf{x}}, \bar{\mathbf{u}})$ satisfies the KKT conditions.

(f) If there exists a vector $\bar{\mathbf{u}}$ such that $(\bar{\mathbf{x}}, \bar{\mathbf{u}})$ satisfies the KKT conditions and f_i, $i = 0, \ldots, m$, are differentiable convex functions, then $\bar{\mathbf{x}}$ is a global minimum.

9.57 Consider the quadratic programming problem

minimize $\frac{1}{2}\mathbf{x}^T\mathbf{Q}\mathbf{x} + \mathbf{c}^T\mathbf{x}$

subject to

$$\mathbf{A}\mathbf{x} \leq \mathbf{b}$$
$$\mathbf{x} \geq \mathbf{0},$$

where $\mathbf{Q}$ is a symmetric matrix.
(a) Show that the gradient of $\mathbf{x}^T\mathbf{Q}\mathbf{x}$ is given by $2\mathbf{Q}\mathbf{x}$.
(b) Using the result in (a), show that the KKT conditions for the quadratic programming problem can be written in the form

$$\mathbf{Q}\mathbf{x} + \mathbf{c} + \mathbf{A}^T\mathbf{u} = \mathbf{y} \qquad \mathbf{A}\mathbf{x} - \mathbf{b} = \mathbf{v}$$
$$\mathbf{x}^T\mathbf{y} = 0 \qquad \mathbf{u}^T\mathbf{v} = 0 \qquad \mathbf{x} \geq \mathbf{0},\ \mathbf{y} \geq \mathbf{0},\ \mathbf{u} \geq \mathbf{0},\ \mathbf{v} \geq \mathbf{0}$$

9.58 Consider the following quadratic program:

minimize $2x_1^2 + 2x_1x_2 + 4x_2^2 - x_1 + x_2$

subject to

$$x_1 + x_2 \leq 2$$
$$x_1 - x_2 \leq 1$$
$$x_1 \geq 0,\ x_2 \geq 0$$

(a) Find matrices $\mathbf{Q}$ and $\mathbf{A}$ so that this problem can be expressed in the form of the general quadratic program in Exercise 9.57.
(b) Show that the resulting matrix $\mathbf{Q}$ is positive semidefinite.
(c) Solve this problem using Lemke's method.

9.59 Consider the geometric program

minimize $f_0(\mathbf{t}) = 40t_1t_2 + 20t_2t_3$

subject to

$$f_1(\mathbf{t}) = \tfrac{1}{5}t_1^{-1}t_2^{-1/2} + \tfrac{3}{5}t_2^{-1}t_3^{-2/3}$$
$$t_j > 0,\ j = 1, 2, 3$$

(a) Write down the dual of this problem.
(b) Find an optimal vector, $\boldsymbol{\delta}^*$, for the dual by finding the unique feasible point for the dual.
(c) Use the duality relations to find an optimal vector, $\mathbf{t}^*$, for the geometric program.

9.60 Consider the geometric program

minimize $40t_1^{-1}t_2^{-1/2}t_3^{-1} + 20t_1t_3 + 20t_1t_2t_3$

subject to

$$\tfrac{1}{3}t_1^{-2}t_2^{-2} + \tfrac{4}{3}t_2^{1/2}t_3^{-1} \leq 1$$
$$t_j > 0,\ j = 1, 2, 3$$

(a) Write down the dual of this geometric program.
(b) Use the fact that there are four dual constraint equations in five variables to reformulate the dual problem as a minimization problem in a single variable r and having the form

minimize $g(\mathrm{r})$

subject to

$$\tfrac{1}{4} \leq r \leq \tfrac{1}{2}$$

9.61 The function

$$f(\mathbf{t}) = (t_1^2 + t_2^2)^{1/2} + t_1^{-2}t_2^{-1}$$

is not a posynomial because of the first term.
(a) Show that the problem of minimizing $f(\mathbf{t})$ can be reformulated as a geometric program by substituting the variable

$$t_3 = t_1^2 + t_2^2$$

into the objective function $f(\mathbf{t})$ and adding the constraint

$$t_3 \geq t_1^2 + t_2^2$$

(b) Find an optimal vector for the resulting geometric program.

9.62 Consider the convex separable problem

minimize $x_1^3 + 2x_2^2$

subject to

$$x_1 + x_2 \leq 2 \qquad x_1 \geq 0, x_2 \geq 0$$

(a) Use breakpoints 0, 1, and 2 for each variable and write down the piecewise linear program that approximates this separable program.
(b) Solve the linear program that results when the adjacency conditions are ignored.
(c) Compare the approximate solution obtained from the linear program with the true optimal vector.

9.63 Suppose that the constraint $x_1 + x_2 \leq 2$ in Exercise 9.62 is replaced by the nonlinear constraint

$$x_1^2 + x_2^2 \leq 4$$

and the variables are still constrained to be nonnegative.
(a) Construct an approximating piecewise linear program using the breakpoints given in Exercise 9.62 for the objective and the constraint.
(b) Why does using the same breakpoints for the objective and the constraint guarantee that the adjacency property holds?

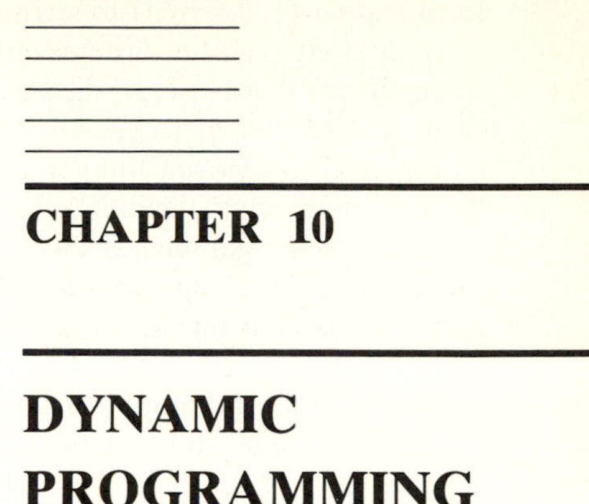

CHAPTER 10

DYNAMIC PROGRAMMING

Dynamic programming is an optimization method that can be used for solving certain problems requiring sequential decisions that must be made at various **stages.** A decision made at one stage of the problem affects the **state** of the problem and the possible decisions that can be made at the next stage. Unlike linear programming, there is no "standard form" for a dynamic programming problem that allows one algorithm to be used in solving all such problems. It is better to think of dynamic programming as a general computational approach. After solving a small, one-stage, subproblem, larger and larger subproblems are recursively solved with the final subproblem yielding a solution to the original problem. The recursive relations used in one application can be quite different from those used in another application even though the underlying approach is essentially the same.

10.1 AN INTRODUCTORY EXAMPLE

To illustrate the general dynamic programming solution approach and to introduce some useful notation, we consider the following simple problem.

The Stagecoach Problem

Suppose that, during the days of stagecoach travel, a passenger is faced with traveling from the state of California to the state of New York. The following directed graph (Figure 10.1) represents all the possible routes that can be taken from state 1 (California) to state 13 (New York). States 2 through 12 are intermediate states, and a directed arc from state i to state j indicates that there is a stagecoach link between the two states. The number on each arc is the number of days required to travel by stage along that link. No matter what route is taken, five stagecoach links must be traveled. On the first stage (stagecoach) the passenger must decide to go to state 2 or state 3 or state 4. On the second stage the passenger must decide which of the states, 5, 6, or 7, to travel to (from the current state) and so on for the remaining three stages. Of course, the decision made at one stage affects the possible state that the passenger is in at the next stage. The problem is to find the route from state 1 to state 13 that minimizes the total number of travel days.

For example, the "northernmost" route (from state 1 to state 2 to state 5 to state 8 to state 10 to state 13) has a total travel time of 15 days and the "southern most" route takes 21 days.

This particular problem is simple enough to be solved by inspection because there are only a few routes to consider. It is not hard to see that there are only $54 = (3 \cdot 3 \cdot 2 \cdot 3 \cdot 1)$ possible routes from state 1 to state 13. Each of those routes requires 4 additions to give the associated total cost, and so only 216 additions are required to solve the problem by completely enumerating all of the routes.

However, imagine a similarly structured directed graph involving 20 stages with 10 possible states to travel to on each stage except the last. In this case there would be 10^{19} possible routes and $19 \cdot 10^{19}$ additions would be required to solve the problem by complete enumeration (as well as $10^{19} - 1$ comparisons). On a good mainframe computer that takes 52 nsec to do one fixed-point (integer) addition, it would take about 300,000 yr just to do the $19 \cdot 10^{19}$ additions required in a solution by complete enumeration. (One nanosecond equals 10^{-9} sec.) We shall see below, however, that a dynamic programming approach to this problem would yield a solution in only a few seconds of mainframe computer time.

To see how the dynamic programming approach works, consider our example problem represented in Figure 10.1. When the passenger is in state 1, it is not immediately clear which route to state 13 is best. At this stage the problem

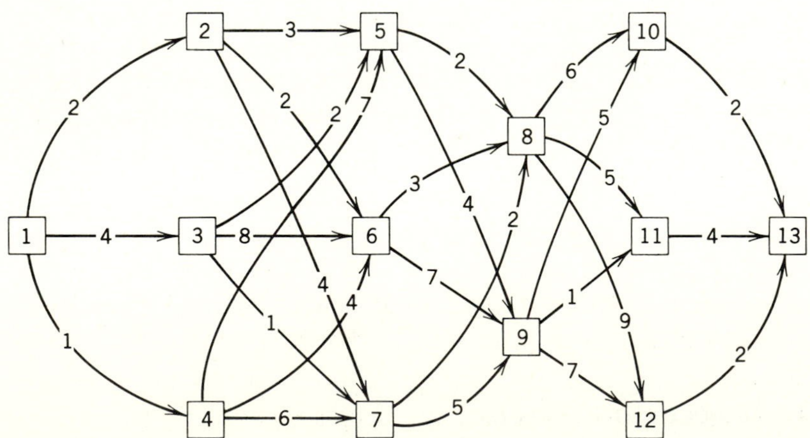

FIGURE 10.1 A directed graph representing all possible routes between state 1 and state 13.

is complicated because there are too many stages *remaining* in the problem. However, if there were only one stage remaining, then the best route from the current state to the final destination would be easy to find. For example, if the passenger is in state 10, the cost of the best route to state 13 is clearly 2 days. If the passenger is in state 11 or 12, then the cost of the best route to state 13 is, respectively, 4 days or 2 days.

We let N denote the total number of stages in the problem, so $N = 5$ in our example. Thinking of cost as the number of travel days, we let

$f_n(s)$ = cost of the best route for stages n through N given that the passenger is in state s

We saw above that

$$f_5(10) = 2 \qquad f_5(11) = 4 \qquad f_5(12) = 2$$

Of course, to solve the original problem, we wish to find

$f_1(1)$ = cost of the best route for stages 1 through N given that the passenger is in state 1

Dynamic programming finds $f_1(1)$ recursively once the easily computed values $f_5(s)$ are known.

The Backward Recursive Relations

To see how $f_1(1)$ can be recursively calculated in our example, note that if the passenger is in a certain state s at stage n, then there are only a limited number of states x_n that can be traveled to on the nth stage. Suppose that we let

$f_n(s, x_n)$ = cost of the best route for stages $n, \ldots, N$ given that the passenger is in state s, and given that the passenger goes to state x_n on stage n.

If we have already calculated $f_{n+1}(x_n)$ as the cost of the best route from state x_n to the final destination on stages $n + 1$ through N, then the values $f_n(s, x_n)$ can be calculated using the recursive relations

$$f_n(s, x_n) = c(s, x_n) + f_{n+1}(x_n)$$

where $c(s, x_n)$ is the cost of going from state s to state x_n on stage n. In words, this relation simply states that

$$f_n(s, x_n) = \begin{pmatrix}\text{cost of going} \\ \text{from } s \text{ to } x_n \text{ on} \\ \text{the } n\text{th stage}\end{pmatrix} + \begin{pmatrix}\text{cost of the \textit{best}} \\ \text{route for stages } n + 1 \\ \text{through } N \text{ given that} \\ \text{the passenger is in} \\ \text{state } x_n\end{pmatrix}$$

Finding recursive relations that allow us to calculate $f_n(s, x_n)$ in terms of already calculated values $f_{n+1}(s)$ is the key to solving a problem by dynamic programming. We can then calculate $f_n(s)$ using

$f_n(s) = \text{minimum}\{f_n(s, x_n)\}$ over all possible x_n's

The above recursive relations are called **backward recursive relations** because the computations start with the final stage and work backwards through the stages to compute the optimal value $f_1(1)$. After showing how this procedure works in our example problem, we shall derive **forward recursive relations** that can also be used in solving this problem by dynamic programming.

In our example, we have already determined the values $f_5(s)$ for all s. These values are given in the following table along with the state, x_5^*, to travel to on the best route from state s.

s	$f_5(s)$	x_5^*
10	2	13
11	4	13
12	2	13

For the stage $n = 4$, the passenger could be in state $s = 8$ or 9 and from either of these states could travel to state $x_4 = 10$, 11, or 12. To calculate $f_4(s)$ for each possible s, we use the recursive relations

$$f_4(s, x_4) = c(s, x_4) + f_5(x_4)$$

The details of these calculations are summarized in the following table.

s	$f_4(s, x_4) = c(s, x_4) + f_5(x_4)$			$f_4(s)$	x_4^*
	$x_4 = 10$	$x_4 = 11$	$x_4 = 12$		
8	$6 + f_5(10)$	$5 + f_5(11)$	$9 + f_5(12)$	8	10
9	$5 + f_5(10)$	$1 + f_5(11)$	$7 + f_5(12)$	5	11

Stepping back one more stage to $n = 3$, we calculate $f_3(s)$ using the values $f_4(s)$ already calculated. The details are summarized in the following table:

s	$f_3(s, x_3) = c(s, x_3) + f_4(x_3)$		$f_3(s)$	x_3^*
	$x_3 = 8$	$x_3 = 9$		
5	$2 + f_4(8)$	$4 + f_4(9)$	9	9
6	$3 + f_4(8)$	$7 + f_4(9)$	11	8
7	$2 + f_4(8)$	$5 + f_4(9)$	10	8 or 9

Enlarging the problem even further to include stages 2 through 5, we now calculate $f_2(s)$ using the values $f_3(s)$. The following table summarizes these calculations:

s	$f_2(s, x_2) = c(s, x_2) + f_3(x_2)$			$f_2(s)$	x_2^*
	$x_2 = 5$	$x_2 = 6$	$x_2 = 7$		
2	$3 + f_3(5)$	$2 + f_3(6)$	$4 + f_3(7)$	12	5
3	$2 + f_3(5)$	$8 + f_3(6)$	$1 + f_3(7)$	11	5 or 7
4	$7 + f_3(5)$	$4 + f_3(6)$	$6 + f_3(7)$	15	6

Finally, given the values $f_2(s)$, we calculate $f_1(1)$ as summarized in the following table:

s	$f_1(s, x_1) = c(s, x_1) + f_2(x_1)$			$f_1(s)$	x_1^*
	$x_1 = 2$	$x_1 = 3$	$x_1 = 4$		
1	$2 + f_2(2)$	$4 + f_2(3)$	$1 + f_2(4)$	14	2

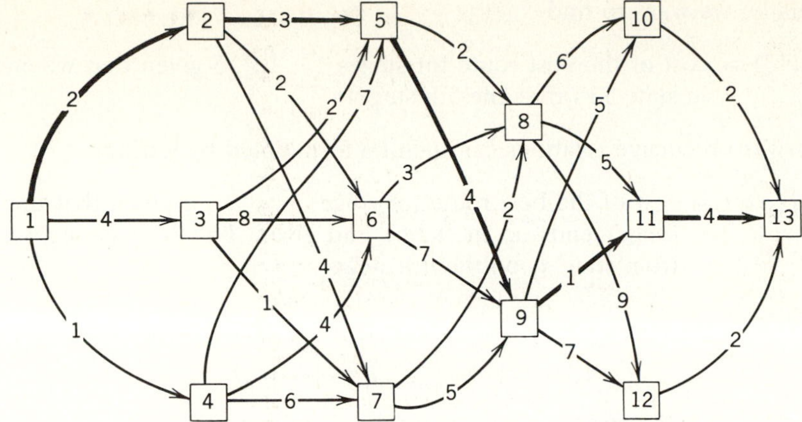

FIGURE 10.2 The minimum travel time route between state 1 and state 13.

Thus, the best route from state 1 to state 13 takes $f_1(1) = 14$ travel days.

To find the actual route that gives this minimum value of 14 days, we use the values x_n^* recorded in each table. On stage $n = 1$ the best state to go to is $x_1^* = 2$. Looking back at the table for stage 2, we see that if we are in state $s = 2$, the optimal decision is to go to state $x_2^* = 5$. If we are in state $s = 5$ at stage 3, we select $x_3^* = 9$. From the table for stage $n = 4$, if we are in state $s = 9$, then we go to $x_4^* = 11$. Finally, if we are in state 11 on the last stage, then we go to the final destination $x_5^* = 13$. Thus, the optimal policy is

$$x_1^* = 2 \qquad x_2^* = 5 \qquad x_3^* = 9 \qquad x_4^* = 11 \qquad x_5^* = 13$$

as indicated in Figure 10.2 by the bold arcs.

The optimal route in Figure 10.2 provides a good illustration of the **principle of optimality** stated by Richard Bellman when he first developed the dynamic programming approach, namely,

> an optimal policy has the property that whatever the current state and decision are, the remaining decisions must constitute an optimal policy with regard to the state resulting from the current decision.

The Forward Recursive Relations

There is another set of recursive relations that can be used to solve this same problem by dynamic programming, namely, the **forward recursive relations.** These relations are developed using a slightly different point of view. Ultimately, we wish to find the best route from state 1 to state 13. Starting with state 1, it is easy to find the best route on the first stage given that we end up in one of the states 2, 3, or 4. If we let

$g_n(s)$ = cost of the best route for stages 1, . . . , n given that we end up in state s (after the nth stage)

then it is easy to calculate

$$g_1(2) = 2 \qquad g_1(3) = 4 \qquad g_1(4) = 1$$

Eventually, we want to find

$g_5(13)$ = cost of the best route for stages 1, . . . , 5 given that we end up in state 13 (after the 5th stage)

The forward recursive relations can then be formulated by letting

$g_n(s, x_n)$ = cost of the best route for stages 1, . . . , n given that the passenger ends up in state s and given that the passenger came from state x_n on the nth stage

With this notation,

$$g_n(s, x_n) = c(x_n, s) + g_{n-1}(x_n)$$

where $c(x_n, s)$ is the cost of going from state x_n to state s. Thus,

$g_n(s)$ = minimum$\{g_n(s, x_n)\}$ over all possible x_n's

Given that we know the values $g_1(2)$, $g_1(3)$, and $g_1(4)$, we can use the above recursive relations to calculate $g_2(s)$ for all s. By considering larger and larger subproblems and progressing in a forward manner through the stages, we recursively calculate the values $g_1(s)$, $g_2(s)$, $g_3(s)$, $g_4(s)$, and finally $g_5(13)$. Of course, the two approaches give the same optimal route (see Exercise 10.2).

Some problems can be solved using either the forward or the backward recursive relations. However, when the stages correspond to actual time periods, then it is more natural to use backward recursive relations. In fact, for some problems the solution procedure must use backward recursive relations.

Basic Features of a Dynamic Programming Formulation

The stagecoach example illustrates the basic features of problems that can be solved by dynamic programming, namely:

The problem can be *divided into stages* with a decision x_n that needs to be made at each stage.

There are a certain number of *states s associated with each stage,* and the decision made at one stage affects the state of the system at the next stage.

Beginning with the last stage, a one-stage subproblem can be solved giving the optimal decisions for each state in the final stage.

Finally, **recursive relations** can be found that allow the solution of the one-stage subproblem to be used to find solutions to larger and larger subproblems with the final subproblem being the original problem.

Thus, to solve a problem by dynamic programming, the crucial step is first to

determine the **stages** and the **states**

In many problems this is not as easy as it was in the example problem of this section. With a little experience, however, it will become easier to see how certain problems can be formulated as dynamic programming problems. In the remaining sections of this chapter, we show how a wide variety of problems can be solved by dynamic programming.

10.2 A LOADING PROBLEM

Suppose a ship is to be loaded with crates of four different types of products. The following table gives the weight w_k (in tons) and the value v_k (in thousands of dollars) for a single crate of product k.

k	*Weight* w_k	*Value* v_k
1	2	4
2	8	10
3	3	6
4	4	8

Suppose the ship has a total weight capacity of 10 tons. The problem is to determine the number of crates of each product that should be selected to maximize the total value of the cargo.

It is not hard to see how this problem can be formulated as an integer programming problem. Letting

x_k = number of crates of product k selected

the linear integer programming formulation becomes

$$\begin{aligned}&\text{maximize } v_1x_1 + \cdots + v_4x_4\\&\text{subject to}\\&\qquad w_1x_1 + \cdots + w_4x_4 \leq 10\\&\qquad x_k \geq 0, \quad x_k \text{ integer } k = 1, \ldots, 4\end{aligned}$$

Thus, the branch-and-bound method of Chapter 8 could be used to solve this problem. An alternate approach is to use dynamic programming.

To determine the stages for this problem, it is helpful to think of the actual process of loading the ship. First, we have to pick a type of product and decide how many crates of that product should be loaded. We would then pick one of the remaining product types and decide on how many crates of that product should be chosen, and so on for each of the products. A natural sequence in which to make these decisions is to start with product 1, then go to product 2, then product 3, and finally product 4. Thus,

at stage n, decide on the number of crates, x_n of product n to be selected

At stage n the one thing that affects the possible decisions, x_n, that can be made is the amount of weight capacity left on the ship. Thus, for any stage n,

the state s of the problem is the amount of remaining capacity

A loading policy is then a sequence of values (x_1, x_2, x_3, x_4) that gives the number of crates of each product type to be selected. If we have already made the decisions x_1, x_2, and x_3, then deciding on the optimal value for the variable x_4 at the last stage is easy, namely, select as many crates of that product as will fit in the remaining capacity s. Thinking of the problem this way, we let

$f_n(s, x_n)$ = value of the best loading policy for products $n, \ldots, 4$ given that s units of capacity remain, and given that x_n units of product n are selected

$f_n(s)$ = value of the best loading policy for products $n, \ldots, 4$ given that s units of capacity remain

In this notation we wish to find

$f_1(10)$ = value of the best loading policy for products $1, \ldots, 4$ given that 10 units of capacity remain

If s units of capacity remain at stage n, then selecting x_n crates of product n contributes $v_n x_n$ to the value of the cargo and leaves $s - w_n x_n$ units of capacity for the remaining products. Therefore, the recursive relations are

$$f_n(s, x_n) = v_n x_n + f_{n+1}(s - w_n x_n)$$

and

$$f_n(s) = \text{maximum}\{f_n(s, x_n)\} \text{ over all possible } x_n\text{'s}$$

Note here that we need to maximize $f_n(s, x_n)$ over all possible decisions x_n. For $n = 4$ we could be in any of the states $s = 0, 1, \ldots, 10$, but if we are in states 0, 1, 2, or 3, we cannot select any crates of product 4 because one crate weighs 4 tons. If we are in states 4, 5, 6, or 7, we can select $x_4^* = 1$ crate of product 4. Finally, if we are in states 8, 9, or 10, we can select $x_4^* = 2$ crates of product 4. Thus, $f_4(s)$ is easily determined as summarized in the following table:

s	$f_4(s)$	x_4^*
0	0	0
1	0	0
2	0	0
3	0	0
4	8	1
5	8	1
6	8	1
7	8	1
8	16	2
9	16	2
10	16	2

Given the values $f_4(s)$, we can calculate the values $f_3(s)$ by using the recursive relations

$$f_3(s, x_3) = v_3 x_3 + f_4(s - w_3 x_3)$$

and

$$f_3(s) = \text{maximum}\{f_3(s, x_3)\} \text{ over all possible } x_3\text{'s}$$

At stage 3 we could be in any one of the states $s = 0, 1, \ldots, 10$, but the possible values for x_3 depend on the state. In the following table, if a value of x_3 is not possible for a particular value of s, then there is an asterisk (*) for that entry. Recall that $v_3 = 6$ and $w_3 = 3$ so that $f_3(s, x_3) = 6x_3 + f_4(s - 3x_3)$.

s	$f_3(s, x_3) = 6x_3 + f_4(s - 3x_3)$ $x_3 = 0$	$x_3 = 1$	$x_3 = 2$	$x_3 = 3$	$f_3(s)$	x_3^*
0	$0 + f_4(0)$	*	*	*	0	0
1	$0 + f_4(1)$	*	*	*	0	0
2	$0 + f_4(2)$	*	*	*	0	0
3	$0 + f_4(3)$	$6 + f_4(0)$	*	*	6	1
4	$0 + f_4(4)$	$6 + f_4(1)$	*	*	8	0
5	$0 + f_4(5)$	$6 + f_4(2)$	*	*	8	0
6	$0 + f_4(6)$	$6 + f_4(3)$	$12 + f_4(0)$	*	12	2
7	$0 + f_4(7)$	$6 + f_4(4)$	$12 + f_4(1)$	*	14	1
8	$0 + f_4(8)$	$6 + f_4(5)$	$12 + f_4(2)$	*	16	0
9	$0 + f_4(9)$	$6 + f_4(6)$	$12 + f_4(3)$	$18 + f_4(0)$	18	3
10	$0 + f_4(10)$	$6 + f_4(7)$	$12 + f_4(4)$	$18 + f_4(1)$	20	2

Enlarging the problem to include stages 2, 3, and 4, we again could be in any of the states $s = 0, 1, \ldots, 10$, but because one crate of product 2 weighs 8 tons, we can select at most $x_2 = 1$ unit of product 2 provided we have $s \geq 8$ units of capacity left. The following table summarizes the calculations for the stage $n = 2$.

s	$f_2(s, x_2) = 10x_2 + f_3(s - 8x_2)$ $x_2 = 0$	$x_2 = 1$	$f_2(s)$	x_2^*
0	$0 + f_3(0)$	*	0	0
1	$0 + f_3(1)$	*	0	0
2	$0 + f_3(2)$	*	0	0
3	$0 + f_3(3)$	*	6	0
4	$0 + f_3(4)$	*	8	0
5	$0 + f_3(5)$	*	8	0
6	$0 + f_3(6)$	*	12	0
7	$0 + f_3(7)$	*	14	0
8	$0 + f_3(8)$	$10 + f_3(0)$	16	0
9	$0 + f_3(9)$	$10 + f_3(1)$	18	0
10	$0 + f_3(10)$	$10 + f_3(2)$	20	0

Finally, to calculate $f_1(10)$, we use the values $f_2(s)$ and the recursive relations

$$f_1(10) = \text{maximum}\{v_1x_1 + f_2(10 - w_1x_1)\} \text{ over all } x_1\text{'s}$$

Given that we are in state 10 at stage 1 and that a crate of product 1 weighs 2 tons, we know that the only possible values for x_1 are 0, 1, 2, 3, 4, or 5. The calculations for this final stage are given in the following table:

s	$f_1(s, x_1) = 4x_1 + f_2(s - 2x_1)$ $x_1 = 0$	$x_1 = 1$	$x_1 = 2$	$x_1 = 3$	$x_1 = 4$	$x_1 = 5$	$f_1(s)$	x_1^*
10	$0 + f_2(10)$	$4 + f_2(8)$	$8 + f_2(6)$	$12 + f_2(4)$	$16 + f_2(2)$	$20 + f_2(0)$	20	0, 1, 2, 3, 5

Thus, using the values $f_2(s)$, we see that the maximum value of the cargo is 20 and occurs for 5 different loading policies, namely,

$$f_1(10) = 20 \quad \text{for } x_1^* = 0, 1, 2, 3, \text{ or } 5$$

If $x_1^* = 0$, then at stage 2, $s = 10$ because there are still 10 units of capacity left. Thus from the table for stage 2 we see that $x_2^* = 0$, which in turn means that $s = 10$ at stage 3. Thus, $x_3^* = 2$ and so $s = 4$ at stage 4, which means that $x_4^* = 1$. Therefore one optimal loading policy is

$$x_1^* = 0 \qquad x_2^* = 0 \qquad x_3^* = 2 \qquad x_4^* = 1$$

The other four optimal policies are similarly determined from the tables and are given by

$$x_1^* = 1 \qquad x_2^* = 0 \qquad x_3^* = 0 \qquad x_4^* = 2$$

$$x_1^* = 2 \qquad x_2^* = 0 \qquad x_3^* = 2 \qquad x_4^* = 0$$

$$x_1^* = 3 \qquad x_2^* = 0 \qquad x_3^* = 0 \qquad x_4^* = 1$$

$$x_1^* = 5 \qquad x_2^* = 0 \qquad x_3^* = 0 \qquad x_4^* = 0$$

10.3 THE BOXES PROBLEM

In both of the preceding examples the recursive relations allowed the values $f_n(s)$ for stage n to be calculated from the values $f_{n+1}(s)$ calculated at stage $n + 1$. Sometimes the recursive relations are such that the values $f_n(s)$ depend on values $f_k(s)$ where k is greater than $n + 1$. To illustrate how this can occur, consider the following problem.

Suppose that 10 different items must be packed in wooden boxes. Assume that the cost of building a box of size i for item i is c_i dollars where $c_1 > c_2 > c_3 > \cdots > c_{10}$, reflecting the fact that the items have been ordered so that item 1 is the largest, item 2 is second largest, and so on with item 10 being the smallest. For purposes of this example, we shall use the values

$$(c_1, c_2, \ldots, c_{10}) = (32, 30, 20, 17, 15, 10, 9, 5, 3, 2)$$

If we could build one box of each size, then the cost of packing all the items is simply $c_1 + c_2 + \cdots + c_{10}$. Suppose, however, that we can only build boxes having five different sizes, but a box of size i can be used to pack item i and any other item j smaller than item i, namely, any item j with $i \leq j$. For example, we could build two boxes of size 1 (for items 1 and 2), two boxes of size 3 (for items 3 and 4), three boxes of size 5 (for items 5, 6, and 7), one box of size 8 (for item 8), and two boxes of size 9 (for items 9 and 10). The cost of this packing policy is $2c_1 + 2c_3 + 3c_5 + c_8 + 2c_9 = 160$. The problem is to determine which five sizes should be built and how many of each of those five sizes should be built to minimize the cost of packing all the items.

To determine the stages and the states for this problem, consider the actual process of packing the items. We know that at least one box of size 1 must be built because item 1 must be packed. The question is how many boxes of size 1 should be made. Similarly, if items 1 through $n - 1$ have already been packed and some of the five sizes have been used up, then we know that at least one box of size n must be made. Again, we need to determine how many boxes of size n should be made. Thus,

at stage n, we need to decide on the number of boxes of size n to build

If we have only one size left that can be used, then clearly we must pack all the remaining items in boxes of size n. Note that we do not need to consider the case when all five sizes have been used up because then we could not pack all the items. In general, we may have several sizes left that can be built, and so for any stage n,

the state s is the number of sizes left that can be built

To determine the recursive relations, we let

$f_n(s, x_n)$ = cost of the best packing policy for items n through 10, given s sizes left that can be built, and given that x_n boxes of size n are built

$f_n(s)$ = cost of the best packing policy for items n through 10, given s sizes left that can be built

In this notation we wish to find $f_1(5)$. If we build x_n boxes of size n at stage n, then we pack x_n of the items, and we use up one of the remaining sizes. Thus, the recursive relations have the form

$$f_n(s, x_n) = c_n x_n + f_{n+x_n}(s - 1)$$

or, in words,

$$f_n(s, x_n) = \begin{pmatrix}\text{cost of building} \\ x_n \text{ boxes of size } n\end{pmatrix} + \begin{pmatrix}\text{best way to pack} \\ \text{items } n + x_n \text{ through } 10 \\ \text{using } s - 1 \text{ sizes}\end{pmatrix}$$

Starting with stage 10, the only state that need be considered in an optimal policy is $s = 1$ because if two or more sizes remained at stage 10, then clearly that packing policy could be improved. This means that at stage 10 we would build $x_{10}^* = 1$ box of size 10 for a cost of 2. Thus, for $n = 10$, the values $f_{10}(s)$ needed for subsequent stages are those given in the following table:

s	$f_{10}(s)$	x_{10}^*
1	2	1

At stage $n = 9$, when items 9 and 10 remain to be packed, the only states that need be considered are states $s = 1$ or $s = 2$. If $s = 1$, then clearly the optimal decision is to make $x_9^* = 2$ boxes of size 9 to pack items 9 and 10. If $s = 2$, then we should make one box for item 9 and one box for item 10. Thus, at stage 9, the values $f_9(s)$ are given as follows:

s	$f_9(s)$	x_9^*
1	6	2
2	5	1

The calculations for stage 8 are summarized in the following table. Note that this is the first table where we begin to use the recursive relations

$$f_n(s, x_n) = c_n x_n + f_{n+x_n}(s - 1)$$

s	$f_8(s, x_8) = 5x_8 + f_{8+x_8}(s-1)$			$f_8(s)$	x_8^*
	$x_8 = 1$	$x_8 = 2$	$x_8 = 3$		
1	*	*	15	15	3
2	$5 + f_9(1)$	$10 + f_{10}(1)$	*	11	1
3	$5 + f_9(2)$	*	*	10	1

An asterisk in the table indicates that the corresponding value of x_n need not be considered in an optimal policy. For example, when $s = 3$, it would not be optimal to make $x_8 = 2$ boxes of size 8 because this would result in not using 5 sizes to cover all 10 items (which is clearly not optimal).

Stepping back one stage, the values $f_7(s)$ can now be calculated using the values $f_8(s)$, $f_9(s)$, and $f_{10}(s)$.

s	$f_7(s, x_7) = 9x_7 + f_{7+x_7}(s-1)$				$f_7(s)$	x_7^*
	$x_7 = 1$	$x_7 = 2$	$x_7 = 3$	$x_7 = 4$		
1	*	*	*	36	36	4
2	$9 + f_8(1)$	$18 + f_9(1)$	$27 + f_{10}(1)$	*	24	1, 2
3	$9 + f_8(2)$	$18 + f_9(2)$	*	*	20	1
4	$9 + f_8(3)$	*	*	*	19	1

The table for stage $n = 6$ can now be calculated, and the details of these calculations follow:

s	$f_6(s, x_6) = 10x_6 + f_{6+x_6}(s-1)$					$f_6(s)$	x_6^*
	$x_6 = 1$	$x_6 = 2$	$x_6 = 3$	$x_6 = 4$	$x_6 = 5$		
1	*	*	*	*	50	50	5
2	$10 + f_7(1)$	$20 + f_8(1)$	$30 + f_9(1)$	$40 + f_{10}(1)$	*	35	2
3	$10 + f_7(2)$	$20 + f_8(2)$	$30 + f_9(2)$	*	*	31	2
4	$10 + f_7(3)$	$20 + f_8(3)$	*	*	*	30	1, 2

The calculations for stages 5, 4, 3, 2, and 1 are summarized in the following tables:

s	$f_5(s, x_5) = 15x_5 + f_{5+x_5}(s-1)$						$f_5(s)$	x_5^*
	$x_5 = 1$	$x_5 = 2$	$x_5 = 3$	$x_5 = 4$	$x_5 = 5$	$x_5 = 6$		
1	*	*	*	*	*	90	90	6
2	$15 + f_6(1)$	$30 + f_7(1)$	$45 + f_8(1)$	$60 + f_9(1)$	$75 + f_{10}(1)$	*	60	3
3	$15 + f_6(2)$	$30 + f_7(2)$	$45 + f_8(2)$	$60 + f_9(2)$	*	*	50	1
4	$15 + f_6(3)$	$30 + f_7(3)$	$45 + f_8(3)$	*	*	*	46	1

s	$f_4(s, x_4) = 17x_4 + f_{4+x_4}(s-1)$						$f_4(s)$	x_4^*
	$x_4 = 1$	$x_4 = 2$	$x_4 = 3$	$x_4 = 4$	$x_4 = 5$	$x_4 = 6$		
2	$17 + f_5(1)$	$34 + f_6(1)$	$51 + f_7(1)$	$68 + f_8(1)$	$85 + f_9(1)$	$102 + f_{10}(1)$	83	4
3	$17 + f_5(2)$	$34 + f_6(2)$	$51 + f_7(2)$	$68 + f_8(2)$	$85 + f_9(2)$	*	69	2
4	$17 + f_5(3)$	$34 + f_6(3)$	$51 + f_7(3)$	$68 + f_8(3)$	*	*	65	2

s	$f_3(s, x_3) = 20x_3 + f_{3+x_3}(s-1)$						$f_3(s)$	x_3^*
	$x_3 = 1$	$x_3 = 2$	$x_3 = 3$	$x_3 = 4$	$x_3 = 5$	$x_3 = 6$		
3	$20 + f_4(2)$	$40 + f_5(2)$	$60 + f_6(2)$	$80 + f_7(2)$	$100 + f_8(2)$	$120 + f_9(2)$	95	3
4	$20 + f_4(3)$	$40 + f_5(3)$	$60 + f_6(3)$	$80 + f_7(3)$	$100 + f_8(3)$	*	89	1

s	$f_2(s, x_2) = 30x_2 + f_{2+x_2}(s-1)$ $x_2 = 1$	$x_2 = 2$	$x_2 = 3$	$x_2 = 4$	$x_2 = 5$	$x_2 = 6$	$f_2(s)$	x_2^*
4	$30 + f_3(3)$	$60 + f_4(3)$	$90 + f_5(3)$	$120 + f_6(3)$	$150 + f_7(3)$	$180 + f_8(3)$	125	1

s	$f_1(s, x_1) = 32x_1 + f_{1+x_1}(s-1)$ $x_1 = 1$	$x_1 = 2$	$x_1 = 3$	$x_1 = 4$	$x_1 = 5$	$x_1 = 6$	$f_1(s)$	x_1^*
5	$32 + f_2(4)$	$64 + f_3(4)$	$96 + f_4(4)$	$128 + f_5(4)$	$160 + f_6(4)$	$192 + f_7(4)$	153	2

Note that the number of states to be considered decreases as we work backward to stage 1. For example, at stage 4, only states $s = 2, 3$, and 4 need be considered because it is not possible to have used up more than 3 sizes in packing the first three items.

From the table for stage 1 we see that the optimal packing policy has

$$\text{minimum cost} = f_1(5) = 153$$

and that at stage 1, we should build $x_1^* = 2$ boxes of size 1. Given that the first two items are packed and four sizes can still be used, we see from the table corresponding to stage 3 that the optimal decision is to make $x_3^* = 1$ box of size 3. Continuing to read the tables backward, we see that the optimal packing policy is to

make $x_1^* = 2$ boxes of size 1.
make $x_3^* = 1$ box of size 3
make $x_4^* = 2$ boxes of size 4
make $x_6^* = 2$ boxes of size 6
make $x_8^* = 3$ boxes of size 8

It is important to realize that solving a problem by dynamic programming provides much more information than just the optimal solution to the original problem. For example, suppose that items 1, 2, 3, and 4 do not need to be packed but items 5 through 10 need to be packed using four different sizes. The minimum packing cost for this new problem is just

$f_5(4)$ = cost of the best packing policy for items 5 through 10 given that four sizes can be used

Of course, we have already calculated $f_5(4) = 46$ in the table corresponding to stage 5. Reading backward from that table, we see that, with regard to this new problem, the optimal policy is

make $x_5^* = 1$ box of size 5
make $x_6^* = 2$ boxes of size 6
make $x_8^* = 1$ box of size 8
make $x_9^* = 2$ boxes of size 9

10.4 AN EQUIPMENT REPLACEMENT PROBLEM

Suppose a house painter needs to have a paint sprayer in each of the next 7 yr. Currently, the painter has a sprayer that is 2 yr old. At the beginning of each year the painter must either trade in the old sprayer and buy a new one or continue to use the old sprayer. The painter wishes to minimize total costs over the next 7 yr and has the following data to work with in making decisions:

p = price of a new sprayer = \$50
$c(i)$ = cost of operating a sprayer that is i years old at the beginning of the year
$t(i)$ = trade-in value of a sprayer that is i years old at the beginning of the year
$v(i)$ = salvage value of a sprayer that is i years old when it is sold at the end of year 7

The following table gives these values in dollars for each appropriate value of i:

i	$c(i)$	$t(i)$	$v(i)$
0	6	*	*
1	9	27	22
2	12	21	19
3	20	10	7
4	32	3	0
5	40	0	0
6	55	0	0
7	70	0	0
8	80	0	0
9	90	0	0

If we were at the beginning of year 7, so that only one year remained, it would be easy to decide what to do. The crucial factor governing this decision would be the current age of the sprayer. Given that current age, we could use the above table to determine whether it was best to replace the sprayer with a new one or to continue using it for another year. Thus,

at stage n, we need to decide to replace or not replace the sprayer at the beginning of year n

and for any stage,

the state s is current age of the sprayer

The possible decisions x_n that can be made at any stage are to *replace* or *not replace*. Starting out with a 2-yr-old sprayer, we need to determine the best replacement policy for the next 7 yr. To determine the recursive relations, we let

$f_n(s, x_n)$ = cost of the best replacement policy for years n through 7, given that the sprayer is s years old at the beginning of year n, and given that we make decision x_n for year n
$f_n(s)$ = cost of the best replacement policy for years n through 7, given that the sprayer is s years old at the beginning of year n

Using the above notation, it is not hard to see that the recursive relations become

$$f_n(s, x_n) = \begin{cases} p - t(s) + c(0) + f_{n+1}(1) & \text{if } x_n = \text{replace} \\ c(s) + f_{n+1}(s + 1) & \text{if } x_n = \text{not replace} \end{cases}$$

and

$$f_n(s) = \text{minimum } \{f_n(s, x_n)\} \text{ over all possible } x_n\text{'s}.$$

For the final stage, $n = 7$, we could be in state 1, 2, 3, 4, 5, 6, or 8. Note that it is not possible to be in state 7 at the beginning of year 7. The calculations for stage 7 are summarized in the following table:

s	*Replace* $x_7 = r$	*Not Replace* $x_7 = \bar{r}$	$f_7(s)$	x_7^*
1	$50 - 27 + 6 - 22$	$9 - 19$	-10	$\bar{r}$
2	$50 - 21 + 6 - 22$	$12 - 7$	5	$\bar{r}$
3	$50 - 10 + 6 - 22$	$20 - 0$	20	$\bar{r}$
4	$50 - 3 + 6 - 22$	$32 - 0$	31	r
5	$50 - 0 + 6 - 22$	$40 - 0$	34	r
6	$50 - 0 + 6 - 22$	$55 - 0$	34	r
8	$50 - 0 + 6 - 22$	$80 - 0$	34	r

For example, when the system is in state 2 at the beginning of year 7 and the decision is to replace (r) the sprayer, then the total cost for the final year is given by

$$\begin{pmatrix}\text{purchase price}\\ \text{of a new sprayer}\end{pmatrix} - \begin{pmatrix}\text{trade-in value of}\\ \text{a 2-yr-old sprayer}\end{pmatrix}$$
$$+ \begin{pmatrix}\text{cost of operating}\\ \text{a new sprayer for a year}\end{pmatrix} - \begin{pmatrix}\text{salvage value of}\\ \text{a 1-yr-old sprayer}\end{pmatrix}$$
$$= 50 - 21 + 6 - 22$$

If the decision is to not replace ($\bar{r}$), then the total cost is given by

$$\begin{pmatrix}\text{cost of operating}\\ \text{a 2-yr-old sprayer}\end{pmatrix} + \begin{pmatrix}\text{salvage value of}\\ \text{a 3-yr-old sprayer}\end{pmatrix}$$
$$= 12 - 7$$

Given the values $f_7(s)$, we can now consider the last two stages and determine the values $f_6(s)$. These calculations are summarized in the following table. Note that at this stage it is not possible to be in state 6.

s	*Replace* $x_6 = r$	*Not Replace* $x_6 = \bar{r}$	$f_6(s)$	x_6^*
1	$50 - 27 + 6 + f_7(1)$	$9 + f_7(2)$	14	$\bar{r}$
2	$50 - 21 + 6 + f_7(1)$	$12 + f_7(3)$	25	r
3	$50 - 10 + 6 + f_7(1)$	$20 + f_7(4)$	36	r
4	$50 - 3 + 6 + f_7(1)$	$32 + f_7(5)$	43	r
5	$50 - 0 + 6 + f_7(1)$	$40 + f_7(6)$	46	r
7	$50 - 0 + 6 - f_7(1)$	$55 + f_7(8)$	46	r

Stepping back one more stage to the beginning of the fifth year, we use the recursive relations to calculate the values $f_5(s)$ in terms of the values $f_6(s)$. These calculations are given in the following table:

s	*Replace* $x_5 = r$	*Not Replace* $x_5 = \bar{r}$	$f_5(s)$	x_5^*
1	$50 - 27 + 6 + f_6(1)$	$9 + f_6(2)$	34	$\bar{r}$
2	$50 - 21 + 6 + f_6(1)$	$12 + f_6(3)$	48	$\bar{r}$
3	$50 - 10 + 6 + f_6(1)$	$20 + f_6(4)$	60	r
4	$50 - 3 + 6 + f_6(1)$	$32 + f_6(5)$	67	r
6	$50 - 0 + 6 + f_6(1)$	$55 + f_6(7)$	70	r

The calculations for the remaining four stages proceed in the same manner, and the details are given in the following four tables:

s	*Replace* $x_4 = r$	*Not Replace* $x_4 = \bar{r}$	$f_4(s)$	x_4^*
1	$50 - 27 + 6 + f_5(1)$	$9 + f_5(2)$	57	$\bar{r}$
2	$50 - 21 + 6 + f_5(1)$	$12 + f_5(3)$	60	r
3	$50 - 10 + 6 + f_5(1)$	$20 + f_5(4)$	80	r
5	$50 - 0 + 6 + f_5(1)$	$40 + f_5(6)$	90	r

s	*Replace* $x_3 = r$	*Not Replace* $x_3 = \bar{r}$	$f_3(s)$	x_3^*
1	$50 - 27 + 6 + f_4(1)$	$9 + f_4(2)$	78	$\bar{r}$
2	$50 - 21 + 6 + f_4(1)$	$12 + f_4(3)$	92	r or $\bar{r}$
4	$50 - 3 + 6 + f_4(1)$	$40 + f_4(4)$	110	r

s	*Replace* $x_2 = r$	*Not Replace* $x_2 = \bar{r}$	$f_2(s)$	x_2^*
1	$50 - 27 + 6 + f_3(1)$	$9 + f_3(2)$	101	$\bar{r}$
3	$50 - 21 + 6 + f_3(1)$	$12 + f_3(4)$	124	r

s	*Replace* $x_1 = r$	*Not Replace* $x_1 = \bar{r}$	$f_1(s)$	x_1^*
2	$50 - 21 + 6 + f_2(1)$	$9 + f_2(2)$	136	r or $\bar{r}$

Thus, the best replacement policy for years 1 through 7 has a total cost of $f_1(2) = 136$, and from the last table we see that the optimal policy is not unique. Reading backwards through the tables, we obtain the following three optimal replacement policies all with the cost of 136.

$$x_1, x_2, x_3, x_4, x_5, x_6, x_7 = r, \bar{r}, r, \bar{r}, \bar{r}, r, \bar{r}$$

$$\text{or} \quad r, \bar{r}, \bar{r}, r, \bar{r}, r, \bar{r}$$

$$\text{or} \quad \bar{r}, r, \bar{r}, r, \bar{r}, r, \bar{r}$$

10.5 PROBLEMS WITH SEVERAL STATE VARIABLES

In all the examples considered so far, the recursive relations involved only a single state variable, s. Some dynamic programming problems have more than one state variable. For example, suppose that S crates of apples and R crates of oranges are to be distributed to N stores where $P_n(x_n, y_n)$ is the profit returned if store n gets x_n crates of apples and y_n crates of oranges. The problem is to determine how many crates of apples and oranges should be distributed to each of the N stores to maximize profit. We assume that all crates must be distributed.

Of course, this problem can be formulated as an integer programming problem, namely,

$$\text{maximize } \sum_{n=1}^{N} P_n(x_n, y_n)$$

subject to

$$\sum_{n=1}^{N} x_n = S \quad \text{and} \quad \sum_{n=1}^{N} y_n = R$$

$$x_n \geq 0 \text{ and integer}$$

$$y_n \geq 0 \text{ and integer}$$

The profit function, $P_n(x_n, y_n)$, might not be a linear function of x_n and y_n and so this integer programming problem might not be linear. The branch-and-bound algorithm of Chapter 8 can be used to solve nonlinear integer programs, but the subproblems would be nonlinear programming problems. Another approach is to use dynamic programming.

To formulate this problem as a dynamic programming problem, we first identify the stages and the states. Clearly,

at stage n, we decide on how many crates x_n of apples and how many crates y_n of oranges to give to store n

The crucial factor in making a decision at stage n is the number, s, of crates of apples and the number, r, of crates of oranges that are left to be distributed. Thus, the state of the problem is measured by the two state variables s and r and is given by the pair (s, r). At the final stage (when only one store remains), it is easy to determine x_n and y_n, namely, give all the remaining crates to the final store so that $x_n = s$ and $y_n = r$.

When there are two (or several) state variables, the general dynamic programming approach is essentially the same as when there is only one state variable. To obtain the recursive relations, we let

$f_n(s, r, x_n, y_n)$ = cost of the best allocation policy for stores n through N, given s crates of apples and r crates of oranges left to be allocated, and given that store n gets x_n crates of apples and y_n crates of oranges

$f_n(s, r)$ = cost of the best allocation policy for stores n through N, given s crates of apples and r crates of oranges left to be allocated

Using this notation, the recursive relations are

$$f_n(s, r, x_n, y_n) = P_n(x_n, y_n) + f_{n+1}(s - x_n, r - y_n)$$

and

$$f_n(s, r) = \text{maximum } \{f_n(s, r, x_n, y_n)\} \text{ over all possible } x_n, y_n$$

The Curse of Dimensionality

Conceptually, there is very little difference between the formulation for this two-state-variable problem and the one-state-variable problems considered earlier. However, from a computational point of view there is a considerable difference because of the large number of possible states and the large number of possible decisions that need to be considered.

To illustrate this point, suppose that we only had to distribute $S = 9$ crates of apples to the stores. In this case the formulation would involve only a single state variable, namely, the number s of remaining crates of apples. Thus, at each stage (except the first), we could be in any of the 10 possible states, $s = 0, 1, \ldots, 9$. However, if there were also $R = 9$ crates of oranges to be distributed, then the number of states increases multiplicatively, and there would be $(S + 1)\cdot(R + 1) = 10\cdot 10 = 100$ states that need to be considered. In addition, for each of these 100 states, we would have to calculate values $f_n(s, r, x_n, y_n)$ for all possible pairs (x_n, y_n) where $x_n \leq s$ and $y_n \leq r$.

It is not difficult to imagine a similar problem with 3 or 4 or 5 or $\cdots$ or L types of products to be distributed, and it is easy to see that such problems could (at least conceptually) be formulated as dynamic programming problems. However, as the number of state variables increases, the amount of calculation required to solve the problems becomes enormous. This is often referred to as the **curse of dimensionality** that makes dynamic programming formulations computationally impractical when there are too many state variables.

10.6 CONTINUOUS-STATE DYNAMIC PROGRAMMING PROBLEMS

In all the examples considered so far, the state variables have taken on discrete values. Some dynamic programming formulations involve state variables that are continuous and can take on any value over some specified interval. In this section we illustrate how the dynamic programming approach works for such problems.

A Simple Example

Given a real number $Q > 0$, suppose that we wish to find real values $x_i \geq 0$, $i = 1, 2, 3$ such that the product $x_1x_2x_3$ is maximized subject to $x_1 + x_2 + x_3 = Q$. That is, suppose we wish to solve the nonlinear programming problem

maximize $x_1x_2x_3$
subject to

$$x_1 + x_2 + x_3 = Q$$
$$\text{and } x_i \geq 0 \qquad i = 1, 2, 3$$

Of course, this example problem could involve any number of variables, but for simplicity we consider only three variables. It is not hard to see that the solution to the problem is $x_i^* = Q/3$ for $i = 1, 2, 3$. Recall a cube is the rectangular parallelepiped of maximum volume where the sum of its height, width, and length is constant.

To see how this solution can be obtained by dynamic programming, think of sequentially allocating the amount Q to the three variables. Thus,

at stage n, decide on the amount x_n to be allocated to the nth variable

Of course, at any stage, we cannot allocate more than the amount remaining so,

the state s is the amount of Q that has not yet been allocated

Note that this state variable can be any real value s satisfying

$$0 < s < Q$$

The recursive relations are then obtained by letting

$f_n(s, x_n)$ = best (largest) value for the product of the variables n through 3, given that s is the amount that remains to be allocated, and given that the nth variable is allocated x_n

$f_n(s)$ = best value for the product of variables n through 3, given that s is the amount that remains to be allocated

We wish to determine the optimal value for the problem, namely, $f_1(Q)$. In this notation the recursive relations are

$$f_n(s, x_n) = x_n \cdot f_{n+1}(s - x_n)$$

and

$$f_n(s) = \text{maximum } \{f_n(s, x_n)\} \text{ over all } x_n \text{ such that } 0 \le x_n \le s$$

For the last stage, when only the third variable remains, it is clearly best to allocate all that remains to that variable so

$$f_3(s) = s \quad \text{with} \quad x_3^* = s$$

At stage 2, when the second and third variable have not yet been allocated values, $f_2(s)$ is calculated using its definition, namely,

$$\begin{aligned} f_2(s) &= \underset{0 \le x_2 \le s}{\text{maximum}} \, [f_2(s, x_2)] \\ &= \underset{0 \le x_2 \le s}{\text{maximum}} \, [x_2 f_3(s - x_2)] \\ &= \underset{0 \le x_2 \le s}{\text{maximum}} \, [x_2(s - x_2)] \\ &= \underset{0 \le x_2 \le s}{\text{maximum}} \, [s x_2 - (x_2)^2] \end{aligned}$$

Note here that the value of x_2 can be any real number between 0 and s. To maximize the function $h(x_2) = sx_2 - (x_2)^2$ over the interval $[0, s]$, set its derivative to zero to obtain $s - 2x_2 = 0$, which gives the stationary point $x_2^* = s/2$. The second derivative $h''(x_2) = -2 < 0$ so x_2^* is a maximizing point.

Evaluating $f_2(s, x_2)$ at $x_2^* = s/2$ gives

$$f_2(s) = \frac{s^2}{4} \quad \text{with} \quad x_2^* = \frac{s}{2}$$

Stepping back one more stage, we use $f_2(s)$ to calculate $f_1(s)$. Of course, at the first stage the amount s remaining to be allocated is $s = Q$, so we calculate

$$\begin{aligned} f_1(Q) &= \underset{0 \le x_1 \le Q}{\text{maximum}} \, [f_1(Q, x_1)] \\ &= \underset{0 \le x_1 \le Q}{\text{maximum}} \, [x_1 f_2(Q - x_1)] \\ &= \underset{0 \le x_1 \le Q}{\text{maximum}} \left[\frac{x_1(Q - x_1)^2}{4} \right] \\ &= \underset{0 \le x_1 \le Q}{\text{maximum}} \left[\frac{x_1^3 - 2Qx_1^2 + Q^2 x_1}{4} \right] \end{aligned}$$

To maximize the function of x_1 in the last set of brackets, simply set its derivative to zero to obtain

$$3x_1^2 - 4Qx_1 + Q^2 = 0$$

which implies (using the quadratic formula) that

$$x_1 = Q \quad \text{or} \quad x_1 = \frac{Q}{3}$$

The second derivative is positive at $x_1 = Q$ and negative at $x_1 = Q/3$ so $x_1^* = Q/3$ is the maximizing value over the interval $[0, Q]$. Evaluating $f_1(Q, x_1)$ at $Q/3$ gives

$$f_1(Q) = \frac{Q^3}{27} \quad \text{with} \quad x_1^* = \frac{Q}{3}$$

If the first variable is allocated $Q/3$, then $s = 2Q/3$ is the amount that remains to be allocated to the remaining variables at the second stage. But $x_2^* = s/2$, so the second variable is allocated $x_2^* = Q/3$. From the stage 3 calculations it then follows that $x_3^* = Q/3$.

A More Complicated Example

Continuous-state dynamic programming problems can be quite complicated even when there is only one state variable. To see how complications can arise, consider the following nonlinear program:

maximize $-2x_1 + x_1^2 + 3x_2 - 2x_2^2 + 5x_3 - x_3^2$
subject to

$$x_1 + x_2 + x_3 \le 2, \qquad x_i \ge 0, i = 1, 2, 3$$

We can think of sequentially allocating a nonnegative amount to each of the variables, so

at stage n, we decide on the amount x_n to be allocated to the nth variable

the state s at any stage is the amount left that can be allocated, namely, the slack in the constraint

Thinking of the problem this way,

$f_n(s, x_n)$ = maximum contribution to the objective function due to variables n through 3, given that s units of slack remain to be allocated, and given that the nth variable is allocated the amount x_n

so that

$f_n(s)$ = maximum contribution to the objective function due to variables n through 3, given that s units of slack remain to be allocated

= maximum $\{f_n(s, x_n)\}$ over all possible x_n

In this notation we wish to find $f_1(2)$. At the final stage, when only the third variable has not yet received an allocation, the problem could be in any state s in the interval $[0, 2]$. We calculate $f_3(s)$ using

$$\begin{aligned} f_3(s) &= \underset{0 \le x_3 \le s}{\text{maximum}}\ f_3(s, x_3) \\ &= \underset{0 \le x_3 \le s}{\text{maximum}}\ [5x_3 - x_3^2] \end{aligned}$$

The function in brackets is a concave function of x_3 and has an unconstrained maximum at $\frac{5}{2}$ and increases from zero as x_3 increases from zero. The largest value for the state variable s is $s = 2 < \frac{5}{2}$. Thus, given that $s \ge 0$ remains to be allocated to the third variable, the optimal choice is $x_3^* = s$, and so

$$f_3(s) = 5s - s^2 \quad \text{with} \quad x_3^* = s$$

At stage 2, when the second and third variable remain to be allocated $f_3(s)$ is used to determine $f_2(s)$. Note that

$$f_2(s, x_2) = 3x_2 - 2x_2^2 + f_3(s - x_2)$$

and so

$$\begin{aligned} f_2(s) &= \underset{0 \le x_2 \le s}{\text{maximum}}\ f_2(s, x_2) \\ &= \underset{0 \le x_2 \le s}{\text{maximum}}\ [3x_2 - 2x_2^2 + f_3(s - x_2)] \\ &= \underset{0 \le x_2 \le s}{\text{maximum}}\ [3x_2 - 2x_2^2 + 5(s - x_2) - (s - x_2)^2] \\ &= \underset{0 \le x_2 \le s}{\text{maximum}}\ [5s - s^2 + (2s - 2)x_2 - 3x_2^2] \end{aligned}$$

The last set of brackets contains a concave function of x_2 when s is fixed and setting its derivative to zero gives

$$x_2 = \frac{2s - 2}{6}$$

Note that

$$\frac{2s - 2}{6} \le 0 \quad \text{when} \quad 0 \le s \le 1$$

and

$$0 \le \frac{2s - 2}{6} \le \frac{1}{3} \quad \text{when} \quad 1 \le s \le 2$$

Thus, when s is in the interval [0, 1], the function in brackets has an unconstrained maximum at a negative point. Because the function is concave, this means that the maximum over $[0, s]$ for $0 \leq s \leq 1$ occurs for $x_2^* = 0$. Evaluating $f_2(s, x_2)$ gives

$$f_2(s) = 5s - s^2 \quad \text{at} \quad x_2^* = 0 \text{ when } s \in [0, 1]$$

However, for $1 \leq s \leq 2$, the maximum occurs at $x_2^* = (2s - 2)/6 \geq 0$ and evaluating $f_2(s, x_2)$ at this value of x_2^* gives

$$f_2(s) = \frac{1 + 13s - 2s^2}{3} \quad \text{at} \quad x_2^* = \frac{2s - 2}{6} \text{ when } s \in [1, 2]$$

These results are summarized in the following table. Note the difference between this table and the ones obtained earlier for problems having discrete state variables.

s	$f_2(s)$	x_2^*
$s \in [0, 1]$	$5s - s^2$	0
$s \in [1, 2]$	$\frac{1 + 13s - 2s^2}{3}$	$\frac{2s - 2}{6}$

Using these expressions for $f_2(s)$, we can now step back one more stage to stage 1 and recursively calculate

$$f_1(2, x_1) = -2x_1 + x_1^2 + f_2(2 - x_1)$$

Thus,

$$\begin{aligned} f_1(2) &= \underset{0 \leq x_1 \leq 2}{\text{maximum}}\; f_1(2, x_1) \\ &= \underset{0 \leq x_1 \leq 2}{\text{maximum}}\; -2x_1 + x_1^2 + \begin{cases} 5(2 - x_1) - (2 - x_1)^2 & \text{if } 0 \leq 2 - x_1 \leq 1 \\ \dfrac{1 + 13(2 - x_1) - 2(2 - x_1)^2}{3} & \text{if } 1 \leq 2 - x_1 \leq 2 \end{cases} \end{aligned}$$

Simplifying the preceding expression for $f_1(2)$ gives

$$f_1(2) = \underset{0 \leq x_1 \leq 2}{\text{maximum}} \begin{cases} 6 - 3x_1 & \text{if } 1 \leq x_1 \leq 2 \\ \frac{19}{3} - \frac{11}{3}x_1 + \frac{1}{3}x_1^2 & \text{if } 0 \leq x_1 \leq 1 \end{cases}$$

Over [1, 2], the maximum of this function is attained at $x_1 = 1$ and has a value of 3. Over [0, 1], the maximum is attained at $x_1 = 0$ and has a value of $\frac{19}{3}$. Therefore,

$$f_1(2) = \tfrac{19}{3} \quad \text{with} \quad x_1^* = 0$$

This means that $s = 2$ at stage 2 and so

$$x_2^* = \frac{2s - 2}{6} = \frac{2}{3}$$

and from the stage 3 calculations we see, when $s = \frac{5}{3}$ remains to be allocated, that

$$x_3^* = \tfrac{5}{3}$$

SELECTED REFERENCES

Bellman, R., and Dreyfus, S., *Applied Dynamic Programming,* Princeton University Press, Princeton, 1962.

Denardo, E. V., *Dynamic Programming Models and Applications,* Prentice-Hall, Englewood Cliffs, NJ, 1982.

Dreyfus, S. E. and Law, A. M., *The Art and Theory of Dynamic Programming,* Academic Press, New York, 1977.

Hadley, G., *Nonlinear and Dynamic Programming,* Addison-Wesley, Reading, MA, 1964.

Nemhauser, G. L., *Introduction to Dynamic Programming,* Wiley, New York, 1966.

Wagner, H. M., *Principles of Operations Research,* Prentice-Hall, Englewood Cliffs, NJ, 1969.

EXERCISES

10.1 Consider the stagecoach problem discussed in Section 10.1.
(a) Use the tables calculated in Section 10.1 to find all the optimal routes and their associated costs from state 3 to state 13.
(b) Suppose you wish to find the optimal route from state 1 to state 8. Note that the tables given in Section 10.1 cannot be used to find such a route. Find this route by dynamic programming.

10.2 (a) Solve the stagecoach problem of Section 10.1 using the forward recursive relations.
(b) Use the tables calculated in part (a) to find the optimal route from state 1 to state 8.

10.3 For the boxes problem discussed in Section 10.3, what is the best way to pack items 4 through 10 using only three different sizes of boxes?

10.4 Consider the following directed graph with 13 nodes where the number on an arc between two nodes is the cost of using that arc in a path. Use dynamic programming to find all minimum cost paths from node 1 to node 13.

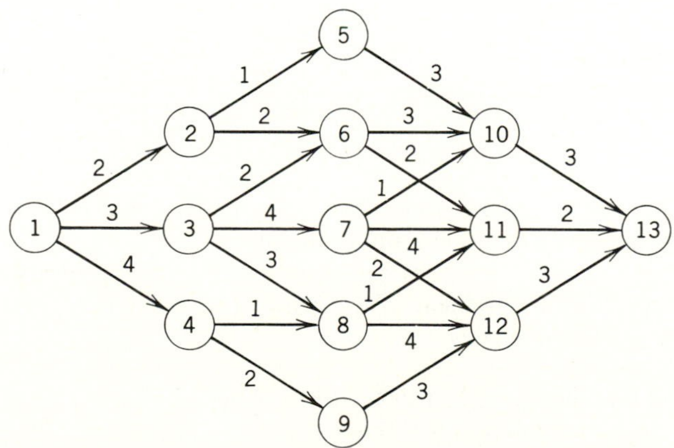

In Exercises 10.5 through 10.9, dynamic programming formulations are required. In each such formulation, use the notation developed in the chapter and be sure to indicate what the stages and states are. Also, carefully define $f_n(s, x_n)$ and $f_n(s)$ in words before stating the recursive relations.

10.5 Charlie Dolittle has 5 days to prepare for final exams in three courses. His method of studying requires that he study one and only one course each day. The entries in the following table give his estimate of the final exam score for each of the courses if he studies that course for the number of days in the left-hand column.

	Final Exam Score in		
Days of Study	*Course 1*	*Course 2*	*Course 3*
0	60	70	40
1	65	75	45
2	70	75	45
3	80	80	60
4	85	90	80
5	85	95	90

Charlie wishes to allocate his five study days to the three courses so as to maximize the sum of his estimated final exam scores.

(a) Formulate this problem as a dynamic programming problem. Use the notation developed in the chapter and be sure to indicate what the stages and states are as well as carefully define $f_n(s, x_n)$ in words.
(b) Solve this problem using your formulation in part (a).

10.6 An owner of five grocery stores has just bought six bushels of apples that he wants to sell in the produce departments of the stores. From the past history of apple sales, the owner has constructed the following table that gives the amount of profit returned (in dollars) by each store for each integer number of bushels it receives. Only integer numbers of bushels are to be distributed to the stores.

		Store				
		1	2	3	4	5
Bushels Allocated	0	0	0	0	0	0
	1	2	2	3	5	3
	2	3	4	4	6	5
	3	5	6	7	6	7
	4	7	8	7	6	9
	5	7	9	7	6	9
	6	7	9	8	7	9

The owner wishes to determine how many bushels should be allocated to each store to maximize the total profit returned.

(a) Formulate this problem as a dynamic programming problem and find all optimal solutions.

Suppose that the owner had 1000 bushels of apples to allocate to the 5 stores and that a table similar to the preceding was known giving the various profits returned.
(b) How many ways can 1000 bushels be allocated to the 5 stores assuming that the bushels are indistinguishable from one another?

(c) How many additions would be required to find the total profit returned for each of the possible allocations?
(d) How many additions would be required in solving this larger problem by using dynamic programming?

10.7 A company has agreed to build at least one bridge across a raging river in the Amazon basin. Three sites have been chosen, but for each site there is a positive probability that the company will fail in its attempt to build a bridge at that site. These probabilities of failure for sites 1, 2, and 3 are .60, .80, and .70, respectively. However, the probability of failure can be reduced by assigning additional teams of workers to the site. Three additional teams are available, and the following table gives the probability that a site will fail in its attempt when 0, 1, 2, or 3 additional teams are assigned to the site.

		Probability of Failure		
		Site 1	Site 2	Site 3
Number of Additional Teams	0	.60	.80	.70
	1	.40	.40	.50
	2	.25	.30	.30
	3	.20	.20	.20

The company wants to minimize the probability that it will fail to build a bridge across the river, so it wants to minimize the product of the probabilities that each site will fail. Formulate this problem as a dynamic programming problem and find all optimal solutions.

10.8 Suppose that you need to plan purchases of oranges for the next N days. Your purchase plan must provide D_i oranges for day i. Oranges can be bought at the beginning of day i at a cost of c_i cents each. There is a fixed charge of F cents each time oranges are purchased. Oranges purchased at the beginning of the day can be used to meet the demand for that day or for a later day. However, oranges left over at the end of each day must be stored and incur a storage cost of p cents each. The problem is to determine how many oranges to purchase at the beginning of each day to meet the required demands at minimum cost.
(a) Formulate this problem as a dynamic programming problem.
(b) Solve this problem using your formulation for the case when $N = 5$, $F = 20$, and $p = 2$, with D_i and c_i given by

i	1	2	3	4	5
D_i	3	1	2	1	1
c_i	9	5	5	7	6

10.9 A local hardware store must carry bags of salt for the upcoming winter months, and currently there is no salt in stock. The store buys salt from a wholesaler who ships the salt to the store. Bags of salt must be bought in boxes and each box contains five bags. The store has predicted the demands for each of the five winter months. The maximum number of bags the store can buy at the beginning of each month are given in the table along with the number of bags needed for each month. The store can buy more than it needs in a given month and use it later, which incurs a storage cost of $1 per bag left over at the end of the month. The price for a bag of salt and the shipping cost per bag fluctuate from month to month as indicated:

Month	*Demand in Bags*	*Maximum Number of Bags that Can Be Ordered*	*Per Bag Shipping Cost*	*Wholesale Price per Bag*
1	10	20	$2	$2
2	20	25	$1	$3
3	15	20	$2	$3
4	20	25	$2	$4
5	10	15	$1	$3

Use dynamic programming to find all optimal buying schedules that minimize the total cost while meeting the demands.

10.10 An automobile manufacturer must schedule its production for the next 4 months. To meet its deadlines, the company must produce cars in the quantities indicated in the second column of the table. Assume that production quantities must be integer multiples of 5. The third and fourth columns of the table give the maximum number of cars that can be produced and the per-unit production costs for each month. Cars produced in one month and stored to meet demand in a later month incur a storage cost of $20 per car left in stock at the end of the month.

Month	*Cars Required*	*Maximum Production*	*Per-unit Production Cost*
1	15	30	$1400
2	20	40	$1300
3	30	35	$1100
4	25	15	$1400

The company wishes to schedule production to meet demand and minimize total costs. Formulate and solve this problem as a dynamic programming problem.

10.11 A painting company needs to decide on the number of painters it employs over the next 5 weeks. The company has estimated that the minimum number of painters it needs over the next 5 weeks is 4, 6, 7, 3, and 5, respectively. Painters can be hired or fired at the end of each week. However, whenever the work force size at the beginning of the current week exceeds that at the beginning of the previous week, a supplemental cost of $4 times the number of extra painters hired is incurred. On the other hand, if the work force at the beginning of the current week exceeds the minimum requirement, another supplemental cost is incurred, namely, $3 times the number of painters over the minimum required for that week.

Use dynamic programming to determine the size of the work force at the end of each week that minimizes the total supplemental costs.

10.12 Consider the nonlinear integer programming problem:

maximize $(x_1 + 5)(x_2 + 1)(x_3 + 2)$

subject to

$$3x_1 + 2x_2 + x_3 \leq 6$$

$$x_i \geq 0 \text{ and integer}, \qquad i = 1, 2, 3$$

Use dynamic programming to find all optimal solutions to this problem.

10.13 Solve the following nonlinear integer programming using dynamic programming. Be sure to give the recursive relations for $f_1(s, x_1)$ and $f_2(s, x_2)$.

$$\text{minimize} \quad x_1^2 - 20x_1 + x_2^2 - 14x_2$$
subject to
$$2x_1 + x_2 \leq 10$$
$$x_i \geq 0 \text{ and integer}, \quad i = 1, 2$$

10.14 Consider the following nonlinear integer programming:

$$\text{maximize } x_1x_2 + x_2 + 2x_1$$
subject to
$$x_1 + 2x_2 \leq 6$$
$$x_i \geq 0 \text{ and integer}, \quad i = 1, 2$$

(a) How does the objective function for this problem differ from those in Exercises 10.12 and 10.13?
(b) Find an equivalent objective function that allows the problem to be solved using dynamic programming.
(c) Solve the equivalent problem using dynamic programming.

10.15 Solve the following nonlinear integer program using a dynamic programming formulation. Be sure to give the recursive relations for $f_n(s, x_n)$, $n = 1, 2, 3$.

$$\text{maximize } x_1x_2^2x_3^3$$
subject to
$$x_1 + 4x_2 + 3x_3 \leq 11$$
$$x_i \geq 1 \text{ and integer}, \quad i = 1, 2, 3$$

10.16 An industrial firm must distribute M gallons of waste oil among N different sites where it is to be burned. If x gallons of oil are burned at site n, then the environmental damage number is assessed to be $nx^2 - 4x$. The amount, x, sent to a site need not be an integer number of gallons. The firm wishes to distribute its oil among the sites so that the sum of the environmental damage numbers is minimized.
(a) Formulate this problem as a dynamic programming problem. Be sure to carefully define all notation.
(b) When $M = 12$ and $N = 3$, use your dynamic programming formulation to solve the problem.
(c) In the case that $M = 15$ and $N = 2$ what is the optimal solution?

10.17 Solve the following nonlinear programming problem using dynamic programming:

$$\text{minimize} \quad x_1^2 - 8x_1 + x_2^2 - 10x_2$$
subject to
$$2x_1 + x_2 \leq 6 \quad \text{and} \quad x_i \geq 0, i = 1, 2$$

10.18 Solve the following nonlinear programming problem using dynamic programming:

$$\text{maximize } 3x_1 - x_1^2 + 2x_2 - 2x_2^2 - x_3 + x_3^2$$
subject to
$$x_1 + 2x_2 + x_3 \leq 2$$
$$x_i \geq 0, \quad i = 1, 2, 3$$

10.19 An electronic system consists of N stages, and each stage has at least one component. Redundant components can be added to a stage to increase its reliability. If x_n components of reliability r_n are used in stage n, the overall reliability of the stage is

$$R_n = 1 - (1 - r_n)^{x_n}$$

The reliability of the entire system is the product

$$R = R_1R_2R_3 \cdots R_N$$

Suppose that each component used in stage n costs c_n dollars and that the total amount of money available to spend on components is C dollars. The problem is to determine the number of components to be used at each stage to maximize the system reliability R subject to the limited budget for buying components.
(a) Formulate this problem as a dynamic programming problem.
(b) Use the formulation in part (a) to solve the problem in the case when $N = 3$, $c_1 = 11$, $c_2 = 7$, $c_3 = 12$, $C = 63$, $r_1 = 0.99$, $r_2 = 0.81$, and $r_3 = 0.95$.

10.20 A political party has two groups of workers who raise money for its campaign fund in three different political districts. Two yuppies make up the first group, and two senior citizens make up the second group. After a careful analysis of the three districts the party has determined how much money can be raised in each district whenever i yuppies and j senior citizens are assigned to that district. The following three tables summarize this information with the rows corresponding to yuppies and the columns corresponding to senior citizens. The entries inside the tables are in hundreds of dollars. For example, if 1 yuppie and 0 senior citizens are assigned to district 2, then the amount raised will be $400.

	0	1	2
0	1	3	1
1	2	4	4
2	3	1	2

District 1

	0	1	2
0	2	1	2
1	4	3	1
2	1	5	3

District 2

	0	1	2
0	0	2	4
1	5	1	3
2	2	4	2

District 3

The party wants to know how many yuppies and how many senior citizens should be assigned to each district to maximize the total number of dollars raised for the campaign fund. Use dynamic programming to solve this problem.

PART
THREE

PROBABILISTIC MODELS

CHAPTER 11

QUEUEING MODELS

Queueing or waiting line problems arise in many different settings. Airplanes queueing up for takeoff on a runway, telephone calls arriving at a switching center, and customers arriving at a bank are just a few of the many examples of **calling units** arriving at a **queueing system** for service. We shall usually refer to the calling units as **customers.** In general, a queueing system is described by specifying

the **input process** (how customers arrive),

the **queueing discipline** (how customers are selected from the waiting line for service), and

the **service mechanism** (how service is performed).

The queueing discipline used for models developed in this chapter is the **first-come first-served** rule whereby arriving customers stand in a waiting line and the customer at the front of the line is the first to be served by the next free server. The basic components of such a queueing system are shown in Figure 11.1.

11.1 A SIMPLE EXAMPLE

As a simple example of a queueing system, consider a large bank with several tellers. Arriving customers wait in a single waiting line (or queue), and

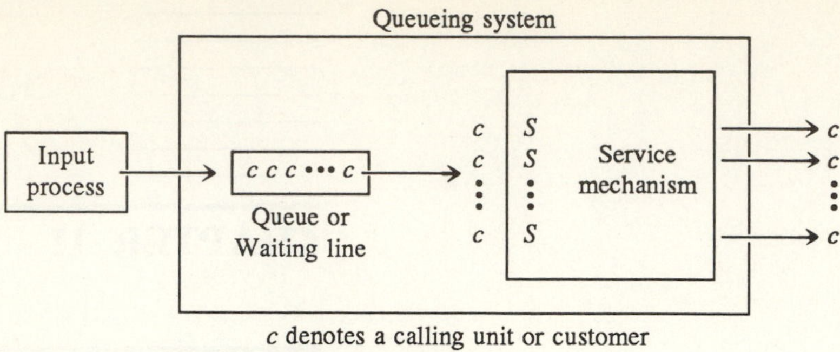

FIGURE 11.1 Components of a queueing system.

when a teller is free, the first customer in line goes to that teller. Several customers have recently been lost to competing banks because once the day gets started, the waiting line is usually quite long. Suppose that we have been asked to do something to shorten the long waiting lines. We could, of course, simply recommend that more tellers be added to serve customers, but each additional teller is quite costly to the bank. It would be extremely helpful if we had an analytical model for studying exactly how the number of tellers affects the length of the waiting line.

The problem is complicated by the random (probabilistic) nature of customer arrivals and by the random variations in the service time required for different customers. For example, if a customer arrived *exactly* every 3 min and if service time took *exactly* 3 min (or less), then with one teller there would never be anyone waiting for service. On the other hand, if the **average arrival rate** is 20 customers per hour (1 customer every 3 min) but arrivals are random so that several customers could arrive at nearly the same time, we might occasionally see a lot of people waiting in line even if service took exactly 3 min.

The **state** of a queueing system at time t is

$N(t)$ = number of customers in the system at time t including those waiting as well as those being served

Throughout this chapter we shall use the following notation:

$P_n(t)$ = probability that n customers are in the queueing system at time t

It will be convenient to let

$P(A)$ = probability of the event A

Thus,

$P_n(t) = P(n$ customers are in the queueing system at time $t)$

that is,

$$P_n(t) = P[N(t) = n]$$

In our example we want to know an important **operating characteristic** of the queueing system, namely, the expected number of customers in the waiting line. The operating characteristics of a queueing system include:

the time that an arriving customer should expect to spend in the system, waiting and being served,

the time that an arriving customer should expect to wait prior to service,

the expected number of customers in the waiting line,

the expected number of customers in the queueing system (waiting or being served).

The main focus of this chapter is to show how these operating characteristics can be calculated for a variety of queueing models. The key to these calculations is determining the probabilities $P_n(t)$ for $n \geq 1$. However, we first need to know more about the probabilistic nature of the queueing system.

11.2 A SIMPLE MODEL SUGGESTED BY OBSERVING INTERARRIVAL TIMES

One way to gain some insight into the random nature of the arrivals and departures is to simply record the times of arrivals and departures. Suppose that the bank opens at 9:00 AM. Often there is a group of customers waiting at the door for the bank to open. Thus, at the start of the day, all the tellers are busy with these customers. Other customers arrive during this start-up time, but after a while the effects of the initial group of customers is no longer noticed. The queueing system is then said to be in the **steady-state condition.**

We give a precise definition of the steady-state condition in Section 11.4, and we shall see later that most of the results in this chapter depend on the queueing system being in the steady-state condition. We assume for this example that the steady-state condition has been reached by 10:00 AM.

By observing arrivals and departures after 10:00, suppose that the list of events shown in Table 11.1 has been tabulated. This list gives the history of arriving and departing customers after 10:00 AM, and hopefully from this history we shall be able to draw some conclusions regarding the random nature of arrivals and departures.

If such a list were kept for several hours each day, we would accumulate a lot of information about the nature of the arrivals and departures. In particular, given the **history of events,** we can compute the **interarrival times** (the times between consecutive arrivals). For example, from Table 11.1, we see that the time between the arrival of customer 1 and the arrival of customer 2 is 3 min. From Table 11.1 the interarrival times listed in Table 11.2 can be calculated.

Imagine that thousands of interarrival times are calculated over several days of observations. One way to summarize the data collected is to plot a **histogram** showing the fraction of the total interarrival times that are t minutes in length, for $t = 1, 2, 3, \ldots$. We assume here that interarrival times have been rounded off to the nearest minute. Suppose that for all the calculated interarrival times, Figure 11.2 gives the associated histogram.

To draw quantitative conclusions from these observations, a mathematical model is needed that describes the random nature of the interarrival times. From the data in Figure 11.2, it appears that the fraction of interarrival times of length

TABLE 11.1 History of Events: Arrivals and Departures

Customer 1 arrives at 10:01
Customer 2 arrives at 10:04
Customer 1 departs at 10:05
Customer 3 arrives at 10:06
Customer 4 arrives at 10:07
Customer 5 arrives at 10:10
Customer 3 departs at 10:11
Customer 6 arrives at 10:14
Customer 2 departs at 10:15
Customer 4 departs at 10:16
Customer 7 arrives at 10:17
Customer 5 departs at 10:20
Customer 7 departs at 10:21
Customer 8 arrives at 10:22
etc.

t decreases exponentially as a function of t. One function displaying exponential decay of this sort has the form

$$f(t) = \lambda e^{-\lambda t} \quad \text{for } \lambda > 0$$

In fact, after some trial and error, we see that with the parameter $\lambda = \frac{1}{3}$, the curve $f(t)$ fits the data summarized in Figure 11.2 rather well. In particular, for $\lambda = \frac{1}{3}$ we obtain the following values for $f(t)$:

t	$f(t) = \lambda e^{-\lambda t}$ with $\lambda = \frac{1}{3}$
1	0.24
2	0.17
3	0.12
4	0.09
5	0.06
6	0.05
7	0.03
8	0.02

Comparing these values with the data as summarized in the histogram in Figure 12.2, we see that for $\lambda = \frac{1}{3}$,

TABLE 11.2 Interarrival Times

10:01 to 10:04	3 min between 1st and 2nd customers
10:04 to 10:06	2 min between 2nd and 3rd customers
10:06 to 10:07	1 min between 3rd and 4th customers
10:07 to 10:10	3 min between 4th and 5th customers
10:10 to 10:14	4 min between 5th and 6th customers
10:14 to 10:17	3 min between 6th and 7th customers
10:17 to 10:22	5 min between 7th and 8th customers

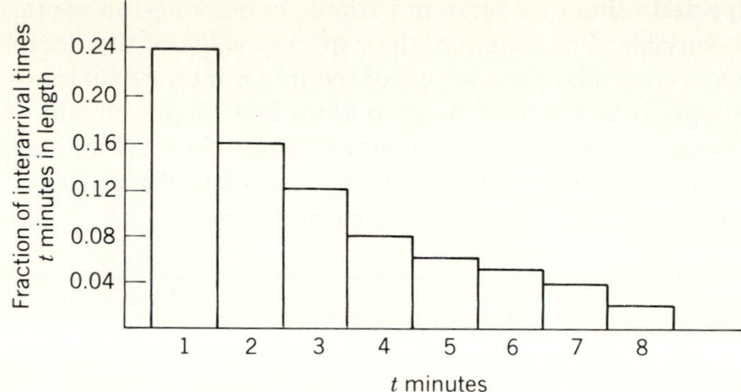

FIGURE 11.2 Histogram of interarrival times.

> $f(t) = \lambda e^{-\lambda t}$ is a very good approximation to the fraction of interarrival times having length t

The function $f(t) = \lambda e^{-\lambda t}$ for $t \geq 0$, is the **density function** for the **exponential distribution.** It will be convenient to let I be the **random variable** denoting the interarrival time. Recall from probability that

> If $f(x)$ is the density function for a nonnegative random variable I, then
>
> $$P(I \leq t) = \int_0^t f(x)\, dx$$
>
> where $P(I \leq t)$ denotes the probability that I takes on a value less than or equal to t.

Integrating the exponential density function, gives

$$P(I \leq t) = \int_0^t \lambda e^{-\lambda x}\, dx = 1 - e^{-\lambda t}$$

The function $F(t) = P(I \leq t)$ is called the **distribution function** for the random variable I. Note, from elementary calculus, that the density function is the derivative of the distribution function. In this case

$$\frac{d}{dt}(1 - e^{-\lambda t}) = \lambda e^{-\lambda t}$$

Thus, from the observations summarized by the histogram of Figure 11.2, we see for this example that

> it is reasonable to assume that the interarrival times are exponentially distributed, that is,
>
> $$P(I \leq t) = 1 - e^{-\lambda t}$$

The **expected value** of a random variable is the long-run average value of that random variable. For example, the expected value of the interarrival time could be approximated by the average of the first n interarrival times, provided n is large enough. It is usually difficult to know how large n should be to ensure that this average is a good approximation to the expected interarrival time. Fortunately, the following result allows us to calculate the expected value of a random variable provided we know its density function.

If $f(x)$ is the density function for a nonnegative continuous-valued random variable I, and if $E(I)$ denotes the expected value of I, then

$$E(I) = \int_0^\infty x f(x)\, dx$$

Thus, we can calculate the expected interarrival time as

$$E(I) = \int_0^\infty t\lambda e^{-\lambda t}\, dt = \frac{1}{\lambda}$$

In the bank example $\lambda = \frac{1}{3}$ so the expected interarrival time is 3 min.

Consequences of Exponentially Distributed Interarrival Times

If interarrival times are exponentially distributed, we can gain some important insights into the nature of the arrivals process. In particular, suppose an arrival has just occurred and we wish to determine the probability that no arrival occurs in a subsequent interval of time Δt. Note that

$$\begin{aligned} P(\text{no arrival in } \Delta t) &= P(I > \Delta t) \\ &= 1 - P(I \le \Delta t) \\ &= e^{-\lambda \Delta t} \end{aligned}$$

But using the fact that, for any x,

$$e^x = 1 + x + \frac{x^2}{2!} + \frac{x^3}{3!} + \cdots$$

we obtain

$$P(\text{no arrival in } \Delta t) = 1 - \lambda\, \Delta t + \frac{(-\lambda\, \Delta t)^2}{2!} + \frac{(-\lambda\, \Delta t)^3}{3!} + \cdots$$

The terms with power 2 or more are negligible compared to $1 - \lambda\, \Delta t$ when Δt is near zero so,

when Δt is small,

$$P(\text{no arrival in } \Delta t) \approx 1 - \lambda\, \Delta t.$$

Here and elsewhere in this chapter,

> $a \approx b$ means that $a = b + o(\Delta t)$ where $o(\Delta t)$ is some function of Δt such that
>
> $$\frac{o(\Delta t)}{\Delta t} \longrightarrow 0 \text{ as } \Delta t \longrightarrow 0$$

A function satisfying this limiting property is called a **function of order Δt.** It will be convenient to let $o(\Delta t)$ denote any function of order Δt. For example,

$$\frac{(-\lambda\ \Delta t)^2}{2!} + \frac{(-\lambda\ \Delta t)^3}{3!} + \cdots$$

is a function of order Δt because dividing each term by Δt and taking the limit as $\Delta t \rightarrow 0$ gives a limit of 0. Thus, we see that

$$P(\text{no arrival in } \Delta t) = 1 - \lambda\ \Delta t + o(\Delta t)$$

For many queueing systems, arrivals are **independent** in the sense that the arrival of one customer is independent of what the other customers are doing. Recall from probability that if events A and B are independent, then

$$P(A \text{ and } B) = P(A)P(B)$$

Assuming that arrivals are independent, we can now calculate the probability of *exactly one arrival* in time Δt. Suppose that

$$p = P(\text{one arrival in } \Delta t)$$

Thus, if arrivals are independent, it follows that

$$P(k \text{ arrivals in } \Delta t) = p^k$$

Using the fact that

$$P(\text{no arrival in } \Delta t) + \sum_{k=1}^{\infty} P(k \text{ arrivals in } \Delta t) = 1$$

we calculate

$$P(\text{no arrival in } \Delta t) = 1 - \sum_{k=1}^{\infty} p^k$$

Using the geometric series expansion, $\Sigma_{k=0}^{\infty} x^k = 1/(1 - x)$, for $-1 < x < 1$ we obtain

$$\sum_{k=1}^{\infty} p^k = \frac{1}{1-p} - 1$$

But we have already seen that $P(\text{no arrival in } \Delta t) \approx 1 - \lambda\ \Delta t$, so it follows that

$$(1 - \lambda\ \Delta t) + o(\Delta t) = 1 - \left(\frac{1}{1-p} - 1\right)$$

Solving this last expression for p gives

$$p \approx \frac{\lambda\ \Delta t}{1 + \lambda\ \Delta t}$$

Again using the expansion $1/(1 - x) = 1 + x + x^2 + \cdots$, we see that for $-1 < \lambda\,\Delta t < 1$,

$$\begin{aligned} p &= \frac{\lambda\,\Delta t}{1 + \lambda\,\Delta t} + o(\Delta t) \\ &= \lambda\,\Delta t \frac{1}{1 - (-\lambda\,\Delta t)} + o(\Delta t) \\ &= \lambda\,\Delta t - (\lambda\,\Delta t)^2 + (\lambda\,\Delta t)^3 - (\lambda\,\Delta t)^4 + \cdots + o(\Delta t) \\ &= \lambda\,\Delta t + o(\Delta t) \end{aligned}$$

But recalling that $p = P(\text{one arrival in } \Delta t)$, we see that

> when Δt is small,
>
> $$P(\text{one arrival in } \Delta t) \approx \lambda\,\Delta t.$$

For $k \geq 2$, p^k is negligible compared to $p \approx \lambda\,\Delta t$ when Δt is small, so another consequence is that

> when Δt is small,
>
> $$P(k \text{ arrivals in } \Delta t) \approx 0, \qquad \text{for } k \geq 2.$$

It is important to realize that the above derivations regarding the probabilistic nature of arrivals are a direct consequence of the assumption that the interarrival times are exponentially distributed.

By observing the **interdeparture times,** as calculated from the history of events in Table 11.1, we can also make a histogram similar to that in Figure 11.2. For many queueing systems the interdeparture times are also exponentially distributed with some parameter μ. Actually, it is helpful to think of "departures from" the queueing system as "arrivals out of" the queueing system. In any case, if interdeparture times are exponentially distributed with parameter μ, it follows that

> when Δt is small,
>
> $$P(\text{one departure in } \Delta t) \approx \mu\,\Delta t$$
>
> and for $k \geq 2$,
>
> $$P(k \text{ departures in } \Delta t) \approx 0$$

Actually, if we think of an arrival or a departure as an **event** in the queueing system, we can show that the probability of having two or more events in a small time interval Δt is essentially zero when arrivals and departures are independent. For example, if the two events were one arrival and one departure, then

$$\begin{aligned}
&P(\text{one arrival and one departure in } \Delta t)\\
&\quad = P(\text{one arrival in } \Delta t)P(\text{one departure in } \Delta t)\\
&\quad = [\lambda\, \Delta t + o(\Delta t)][\mu\, \Delta t + o(\Delta t)]\\
&\quad = \lambda\mu(\Delta t)^2 + \lambda\, \Delta t[o(\Delta t)] + \mu\, \Delta t[o(\Delta t)] + o(\Delta t)\\
&\quad = o(\Delta t)
\end{aligned}$$

Here we have used the fact that the product and the sum of two functions of order Δt are also a function of order Δt. Also, note that $\lambda\Delta t[o(\Delta t)]$ and $\mu\Delta t[o(\Delta t)]$ are functions of order Δt. Using similar arguments, it is not hard to derive the following result.

$$P\begin{pmatrix}\text{number of arrivals plus number}\\ \text{of departures in } \Delta t \text{ is two or greater}\end{pmatrix} = o(\Delta t)$$

In summary, if the interarrival and interdeparture times are exponentially distributed, we have shown that arrivals and departures must satisfy certain probability rules. Of course, it is important to realize that *not all queueing systems have exponentially distributed interarrival and interdeparture times.* We shall consider some such systems in Section 11.11, but the analysis of these systems is very difficult and very few analytical results are available.

For queueing systems where the random arrivals and departures are exponentially distributed, we can derive some very useful results that allow us to say much more regarding the probabilistic nature of the system. To this end consider the following queueing model.

A Simple Model

For $\Delta t > 0$ and sufficiently small, arrivals and departures satisfy the following conditions:

1. $P(\text{one arrival in } \Delta t) \approx \lambda\, \Delta t$
2. $P(\text{one departure in } \Delta t) \approx \mu\, \Delta t$
3. $P\begin{pmatrix}\text{total number of arrivals and}\\ \text{departures in } \Delta t \text{ is two or greater}\end{pmatrix} \approx 0$
4. arrivals and departures are independent

Conditions 1 through 3 are simply the ones that we derived based on the assumption that interarrival and interdeparture times are exponentially distributed. Condition 4 is a reasonable assumption for many queueing systems because the input process governing arrivals is usually independent of the service (or output) mechanism governing departures.

Consider a queueing system satisfying the four conditions stated in the simple model. The event that there are n customers in the system at time $t + \Delta t$ can be expressed as the union of four mutually exclusive events.

In particular, for $n \geq 1$

if the system is in state n at time $t + \Delta t$

then

$$\left.\begin{array}{l}\text{the system is in state } n \text{ at time } t\\ \text{and no arrivals occur in time } \Delta t\\ \text{and no departures occur in time } \Delta t\end{array}\right\}\text{event 1}$$

or

$$\left.\begin{array}{l}\text{the system is in state } n+1 \text{ at time } t\\ \text{and no arrivals occur in time } \Delta t\\ \text{and one departure occurs in time } \Delta t\end{array}\right\}\text{event 2}$$

or

$$\left.\begin{array}{l}\text{the system is in state } n-1 \text{ at time } t\\ \text{and one arrival occurs in time } \Delta t\\ \text{and no departure occurs in time } \Delta t\end{array}\right\}\text{event 3}$$

or

$$\left.\begin{array}{l}\text{the system is in some state at time } t\\ \text{and the total number of arrivals and}\\ \text{departures in } \Delta t \text{ is 2 or greater}\end{array}\right\}\text{event 4}$$

Note that event 4 includes occurrences such as having the system in state $n - 3$ at time t with four arrivals and three departures so that the system is in state n at time $t + \Delta t$.

In our notation,

$$P(\text{system is in state } n \text{ at time } t + \Delta t) = P_n(t + \Delta t)$$

so

$$P_n(t + \Delta t) = P(\text{event 1}) + P(\text{event 2}) + P(\text{event 3}) + P(\text{event 4})$$

To calculate P(event 1), we first note that because arrivals and departures are independent events,

$$\begin{aligned}
&P(\text{no arrivals in } \Delta t \text{ and no departures in } \Delta t)\\
&\quad = P(\text{no arrivals in } \Delta t)P(\text{no departures in } \Delta t)\\
&\quad = [1 - \lambda\,\Delta t + o(\Delta t)][1 - \mu\,\Delta t + o(\Delta t)]\\
&\quad = 1 - (\lambda + \mu)\Delta t + o(\Delta t)\\
&\quad \approx 1 - (\lambda + \mu)\Delta t
\end{aligned}$$

Therefore,

$$P(\text{event 1}) \approx P_n(t)[1 - (\lambda + \mu)\Delta t]$$

To calculate P(event 2), we similarly note that

$$\begin{aligned}
&P(\text{no arrival in } \Delta t \text{ and one departure in } \Delta t)\\
&\quad = P(\text{no arrival in } \Delta t)P(\text{one departure in } \Delta t)\\
&\quad = [1 - \lambda\,\Delta t + o(\Delta t)][(\mu\,\Delta t) + o(\Delta t)]\\
&\quad = \mu\,\Delta t - \lambda\mu(\Delta t)^2 + o(\Delta t)\\
&\quad \approx \mu\,\Delta t
\end{aligned}$$

Thus,

$$P(\text{event } 2) \approx P_{n+1}(t)\mu\,\Delta t$$

Similarly, it is not hard to show that

$$P(\text{event } 3) \approx P_{n-1}(t)\lambda\,\Delta t$$

$$P(\text{event } 4) \approx 0$$

Summing the probabilities of all four events, we obtain

$$P_n(t + \Delta t) = P_n(t)[1 - (\lambda + \mu)\,\Delta t] + P_{n+1}(t)\mu\,\Delta t + P_{n-1}(t)\lambda\,\Delta t + o(\Delta t)$$

After a little algebra, we obtain

$$\frac{P_n(t + \Delta t) - P_n(t)}{\Delta t} = \frac{[\mu\,\Delta t P_{n+1}(t) + \lambda\,\Delta t P_{n-1}(t) - (\lambda + \mu)\,\Delta t P_n(t)]}{\Delta t} + \frac{o(\Delta t)}{\Delta t}$$

Taking the limit of both sides of the preceding equation as $\Delta t \to 0$, and assuming that $P_n(t)$ is a differentiable function of t, we obtain

$$P_n'(t) = \mu P_{n+1}(t) + \lambda P_{n-1}(t) - (\lambda + \mu)P_n(t) \quad \text{for } n \geq 1$$

where $P_n'(t) = dP_n(t)/dt$. Note that if $n = 0$, then event 3 need not be considered in the above analysis and, in event 1, the probability of no departure would be 1 because there are no customers in the system. Thus, for $n = 0$, we obtain the equation

$$P_0'(t) = \mu P_1(t) - \lambda P_0(t)$$

In summary,

> For our simple model the probabilities $P_n(t)$ satisfy the following differential equations:
>
> $$P_n'(t) = \mu P_{n+1}(t) + \lambda P_{n-1}(t) - (\lambda + \mu)P_n(t) \qquad n \geq 1$$
>
> $$P_0'(t) = \mu P_1(t) - \lambda P_0(t)$$

These two **lag-differential equations** are difficult to solve in general, but there are two special cases that will provide us with some important insights into the nature of arrivals and departures.

Special Case 1: Arrivals But No Departures

Consider a queueing system satisfying the conditions of our simple model and suppose it is in state 0 at time $t = 0$. In addition suppose only arrivals are possible. Thus we assume that $\mu = 0$ so that

$$P(\text{one departure in } \Delta t) \approx 0$$

In this case the differential equations reduce to

$$P_0'(t) = -\lambda P_0(t)$$

and

$$P_n'(t) = \lambda P_{n-1}(t) - \lambda P_n(t) \quad \text{for } n \geq 1$$

Using the fact that $P_0(t) + P_1(t) + P_2(t) + \cdots = 1$, these equations can be solved to obtain the unique solution

$$P_0(t) = e^{-\lambda t}$$

$$P_n(t) = \frac{(\lambda t)^n e^{-\lambda t}}{n!} \quad \text{for } n = 1, 2, 3, \ldots$$

This distribution for the probabilities of being in the various states is called the **Poisson distribution** with parameter λt.

Knowing the probabilities $P_n(t)$ of having n customers in the queueing system at time t, we can now derive some results that provide further insight into the nature of arrivals.

The Mean Arrival Rate

The number of arrivals in the time interval from 0 up to and including t is a random variable, and when there are no departures from the system, it equals the state, $N(t)$, of the system at time t. We shall use $E(X)$ to denote the expected value of a random variable X. Thus

$$E(\text{number of arrivals in } (0, t]) = E[N(t)]$$

From the definition of expected value,

$$E[N(t)] = 0P_0(t) + 1P_1(t) + 2P_2(t) + \cdots$$

But we have already seen that the probabilities $P_n(t)$ are given by the Poisson distribution, so

$$E[N(t)] = \sum_{n=0}^{\infty} nP_n(t) = \sum_{n=0}^{\infty} \frac{n(\lambda t)^n e^{-\lambda t}}{n!}$$

$$= (\lambda t)e^{-\lambda t} \sum_{n=1}^{\infty} \frac{(\lambda t)^{n-1}}{(n-1)!}$$

Thus, using the fact that

$$e^x = 1 + x + \frac{x^2}{2!} + \frac{x^3}{3!} + \cdots$$

we obtain

$$E[N(t)] = (\lambda t)e^{-\lambda t}e^{\lambda t} = \lambda t$$

Therefore, the following important result holds

> $E(\text{number of arrivals in } (0, t]) = \lambda t$, and so $\lambda =$ expected number of arrivals per unit time

The expected number of arrivals per unit time is often referred to as the **mean arrival rate.**

The Time of First Arrival

Suppose that we denote the time of first arrival by the random variable T having a distribution function $F(t)$. If there are 0 units in the system at time $t = 0$, then from the definition of a distribution function

$$\begin{aligned} F(t) &= P(T \leq t) \\ &= 1 - P(T > t) \\ &= 1 - P[N(t) = 0] \\ &= 1 - P_0(t) \\ &= 1 - e^{-\lambda t} \quad \text{for } t \geq 0 \end{aligned}$$

Thus, the random variable T has an exponential distribution with parameter λ. Therefore, the density function for T is

$$f(t) = F'(t) = \lambda e^{-\lambda t}$$

and the **expected time of first arrival** is

$$E(T) = \int_0^\infty t f(t)\, dt = \int_0^\infty t\lambda e^{-\lambda t}\, dt = \frac{1}{\lambda}$$

In summary,

> the time T of first arrival is exponentially distributed with expected time of first arrival given by
>
> $$E(T) = \frac{1}{\lambda} \text{ where } \lambda \text{ is the mean arrival rate.}$$

Another Look at Interarrival Times

It is interesting to note that because the states are Poisson distributed, we can show that the interarrival times are exponentially distributed. Of course, it was the observation that interarrival times are exponentially distributed that allowed us to gain so much insight into the queueing system and to show in particular that the states have a Poisson distribution. We give this derivation only to show that there is consistency in the model.

Suppose then that an arrival occurs at time h, and I denotes the time between h and the next arrival (the interarrival time). To find the distribution function of I, we calculate

$$\begin{aligned} P(I \leq t) &= 1 - P(I > t) \\ &= 1 - P(\text{no arrival in } (h, h + t]) \end{aligned}$$

To calculate $P(\text{no arrival occurs in } (h, h + t])$, we need the definition of **conditional probability,** namely,

$$P(A \mid B) = \frac{P(A \text{ and } B)}{P(B)} \quad \text{provided } P(B) \neq 0$$

Here, $P(A \mid B)$ denotes the probability of event A given that event B has occurred. If A and B are **independent events,** then note that $P(A \text{ and } B) = P(A)P(B)$ and so $P(A \mid B) = P(A)$. If we assume that arrivals are independent, then

$$\begin{aligned} &P(\text{no arrival in } (h, h + t]) \\ &\quad = P(\text{no arrival in } (h, h + t] \mid \text{no arrival in } (0, h]) \\ &\quad = \frac{P(\text{no arrival in } (0, h + t])}{P(\text{no arrival in } 0, h])} \\ &\quad = \frac{P_0(h + t)}{P_0(h)} = \frac{e^{-\lambda(h+t)}}{e^{-\lambda h}} = e^{-\lambda t} \end{aligned}$$

Therefore,

$$P(I \leq t) = 1 - e^{-\lambda t}$$

and so interarrival times are exponentially distributed.

The Memoryless Property of Arrivals.
In the preceding calculation we showed that

$$\frac{P(\text{no arrival in } (0, h + t])}{P(\text{no arrival in } 0, h])} = e^{-\lambda t}$$

But $e^{-\lambda t} = P_0(t) = P(\text{no arrival in } (0, t])$. Therefore, given no arrivals in $(0, h]$, the probability of no arrivals in $(0, h + t]$ depends only on the length of the time interval t and not on h. It is as if the process starts over again at time h and "forgets" that no arrivals occurred in the time $(0, h]$.

Special Case 2: Departures But No Arrivals

Another special case that provides useful information about more general queueing systems is when there are departures but no arrivals. Suppose that as long as there are customers in the queueing system,

$$P(\text{one departure in } \Delta t) \approx \mu \, \Delta t$$

$$P(\text{one arrival in } \Delta t) = 0$$

If departures from the system are considered as arrivals to the outside of the system, then we can use the results of the arrivals-only case to obtain results for the departures-only case. Of course, the parameter μ is used in place of the parameter λ.

In particular, if there are customers in the queueing system, then the time of first departure (time of first arrival to the outside) has an exponential distribution with parameter μ. Also, as long as there are customers in the system, then the interdeparture time has the same distribution as the interarrival time to the outside, namely, exponential with parameter μ. Finally, as long as there are customers in the system, the expected number of departures in the time interval $(0, t]$ must be the same as the expected number of arrivals to the outside of the system, namely, μt. In summary:

> As long as there are customers in the queueing system:
>
> the time of first departure is exponentially distributed with parameter μ and mean $1/\mu$,
>
> the expected number of departures in time $(0, t]$ equals μt, so μ is the expected number of departures per unit time. The parameter μ is called the **mean departure rate,**
>
> the service times are exponentially distributed with parameter μ and mean $1/\mu$.

When there is only a single server, interdeparture times correspond to service times as long as there are customers on the queueing system. Thus, if service times are exponentially distributed then so are interdeparture times.

When there are several servers, each with exponentially distributed service times, then it is also true that interdeparture times are exponentially distributed. To see why this is true, suppose that there are n busy servers, each having exponentially distributed service times with parameter μ. Let T denote the interdeparture time for customers leaving the queueing system. If a customer has just left, then

$$T = \text{minimum}\{T_1, T_2, \ldots, T_n\}$$

where T_i denotes the remaining service time for server i. The random variables T_i have the same distribution as the service times, namely exponential with parameter μ, because of the memoryless property of the exponential distribution. Using the fact that the random variables T_i are independent, we can calculate $P(T > t)$ as

$$\begin{aligned} P(T > t) &= P(\text{minimum}\{T_1, T_2, \ldots, T_n\} > t) \\ &= P(T_1 > t, T_2 > t, \ldots T_n > t) \\ &= P(T_1 > t)P(T_2 > t) \cdots P(T_n > t) \\ &= e^{-\mu t}e^{-\mu t} \cdots e^{-\mu t} \\ &= e^{-n\mu t} \end{aligned}$$

Thus, the interdeparture times are exponentially distributed with parameter $n\mu$ and, therefore, the mean departure rate is $n\mu$. This means that as long as all servers are busy, the queueing system can be viewed as a single-server queue having service times that are exponentially distributed with parameter $n\mu$.

11.3 A MORE GENERAL ARRIVALS AND DEPARTURES MODEL

In the simple model considered in the last section, we saw that the parameter λ can be interpreted as the mean arrival rate, that is, the expected number of arrivals per unit time. Similarly, the parameter μ can be interpreted as the mean departure rate. In many queueing systems these mean rates depend on the number of customers in the system. For example, in the bank model considered earlier, the service time might depend on the number of customers waiting, because the tellers might work faster when the line is long.

In this section we generalize the simple model of the previous section to account for this dependence on the state of the system.

A More General Model

For $\Delta t > 0$ and sufficiently small, arrivals and departures satisfy the following conditions:

1. $P(\text{one arrival in } \Delta t) \approx a_n \Delta t$
2. $P(\text{one departure in } \Delta t) \approx d_n \Delta t$
3. $P\begin{pmatrix}\text{total number of arrivals and} \\ \text{departures in } \Delta t \text{ is two or more}\end{pmatrix} \approx 0$
4. arrivals and departures are independent.

Note that the only difference between this model and the simpler model considered in Section 11.2 is that the parameters λ and μ have been replaced by the parameters a_n and d_n that depend on the state of the system at time t. The parameters a_n and d_n can be interpreted as the mean arrival rate and mean departure rate, respectively, when there are n customers in the system.

As in the simple model of Section 11.2, the probability, $P_n(t + \Delta t)$, of being in state n at time $t + \Delta t$ can be calculated as

$$P_n(t + \Delta t) = P(\text{event } 1) + P(\text{event } 2) + P(\text{event } 3) + P(\text{event } 4)$$

where events 1 through 4 are defined exactly the same as in the simple model. For convenience, we repeat the definition of these events, and to the right of each line we give its associated probability.

Event 1	*Probability*
the system is in state n at time t	$= P_n(t)$
and no arrivals occur in time Δt	$\approx 1 - a_n \Delta t$
and no departures occur in time Δt	$\approx 1 - d_n \Delta t$
Event 2	
the system is in state $n + 1$ at time t	$= P_{n+1}(t)$
and no arrivals occur in time Δt	$\approx 1 - a_{n+1} \Delta t$
and one departure occurs in time Δt	$\approx d_{n+1} \Delta t$
Event 3	
the system is in state $n - 1$ at time t	$= P_{n-1}(t)$
and one arrival occurs in time Δt	$\approx a_{n-1} \Delta t$
and no departure occurs in time Δt	$\approx 1 - d_{n-1} \Delta t$
Event 4	
the system is in some state at time t and two or more arrivals and departures occur in time Δt	≈ 0

You should check that each of the probabilities given on the right agrees with the assumptions of the general model. For example,

$$P(\text{no arrival in } \Delta t) = 1 - P(\text{one or more arrivals in } \Delta t)$$
$$= 1 - P(\text{one arrival in } \Delta t) - P(k \geq 2 \text{ arrivals in } \Delta t)$$

But $P(k \geq 2 \text{ arrivals in } \Delta t) \approx 0$, and so

$$P(\text{no arrival in } \Delta t) \approx 1 - P(\text{one arrival in } \Delta t)$$
$$\approx 1 - a_n \Delta t$$

Similarly,

$$P(\text{no departure in } \Delta t) \approx 1 - d_n \Delta t$$

Thus,

$$P(\text{event } 1) \approx P_n(t)(1 - a_n\Delta t)(1 - d_n\Delta t)$$
$$\approx P_n(t)[1 - (a_n + d_n)\,\Delta t] + o(\Delta t)$$
$$\approx P_n(t)[1 - (a_n + d_n)\,\Delta t]$$

Similar calculations show that

$$P(\text{event } 2) \approx P_{n+1}(t)d_{n+1}\Delta t$$

$$P(\text{event } 3) \approx P_{n-1}(t)a_{n-1}\Delta t$$

$$P(\text{event } 4) \approx 0$$

Therefore,

$$P_n(t + \Delta t) \approx P_n(t)[1 - (a_n + d_n)\,\Delta t] + P_{n+1}(t)d_{n+1}\Delta t$$
$$+ P_{n-1}(t)a_{n-1}\Delta t$$

Then, after a little algebra, we obtain

$$\frac{P_n(t + \Delta t) - P_n(t)}{\Delta t} \approx d_{n+1}P_{n+1}(t) + a_{n-1}P_{n-1}(t) - (a_n + d_n)P_n(t)$$

Taking the limit of both sides, as Δt approaches 0, yields

$$P_n'(t) = d_{n+1}P_{n+1}(t) + a_{n-1}P_{n-1}(t) - (a_n + d_n)P_n(t) \quad \text{for } n \geq 1$$

For the case $n = 0$, there are no customers in the system so event 3 need not be considered, and the probability of no departure in event 1 is then equal to one. Thus,

$$P_0(t + \Delta t) = P_1(t)[d_1\Delta t] + P_0(t)[1 - a_0\Delta t]$$

so

$$\frac{P_0(t + \Delta t) - P_0(t)}{\Delta t} \approx d_1P_1(t) - a_0P_0(t)$$

Taking the limit of both sides, as Δt approaches 0, yields

$$P_0'(t) = d_1P_1(t) - a_0P_0(t)$$

In summary:

> For arrivals and departures occurring according to the assumptions of the general model, the probabilities $P_n(t)$ satisfy the following differential equations:
>
> $$P_n'(t) = d_{n+1}P_{n+1}(t) + a_{n-1}P_{n-1}(t) - (a_n + d_n)P_n(t) \qquad n \geq 1$$
>
> $$P_0'(t) = d_1P_1(t) - a_0P_0(t)$$

Solving this system of infinitely many lag-differential equations is extremely difficult, and the analytic expressions obtained are complicated and not of much practical use because of the way in which the probabilities $P_n(t)$ obtained depend on t. However, in many queueing problems, after a certain time has passed (the **transient period**), the system reaches a **steady-state condition** where the probabilities $P_n(t)$ do not depend on the time t. For queueing systems that have reached the steady state, the preceding equations can be solved quite easily, as we shall see in the next section.

11.4 STEADY-STATE QUEUEING SYSTEMS

After an initial transient period, many queueing systems reach a steady-state condition where the probabilities, $P_n(t)$, of having n customers in the system do not depend on the time t. For the remainder of this chapter we shall make the following assumption:

THE STEADY-STATE ASSUMPTION

> For each n, $P_n(t)$ is a constant P_n independent of t.

The most important consequence of the steady-state assumption is that the lag-differential equations become:

THE STEADY-STATE EQUATIONS

> $$0 = d_{n+1}P_{n+1} + a_{n-1}P_{n-1} - (a_n + d_n)P_n \qquad n \geq 1$$
>
> $$0 = d_1P_1 - a_0P_0.$$

These equations follow simply by using the fact that

$$P_n(t) = P_n \text{ implies that } P_n'(t) = 0$$

where $P_n'(t)$ denotes the derivative of P_n with respect to t.

Using these equations, it is easy to derive analytic expressions for the probabilities P_n for each n. From the equation for $n = 0$, we obtain

$$P_1 = \frac{a_0}{d_1} P_0$$

For $n = 1$ the steady-state equations yield

$$d_2P_2 = a_1P_1 + d_1P_1 - a_0P_0$$

but since $d_1P_1 = a_0P_0$ we obtain

$$P_2 = \frac{a_1}{d_2} P_1 = \frac{a_0a_1}{d_1d_2} P_0$$

For $n = 2$ we obtain

$$d_3P_3 = a_2P_2 + d_2P_2 - a_1P_1 = a_2P_2$$

so

$$P_3 = \frac{a_2}{d_3} P_2 = \frac{a_0a_1a_2}{d_1d_2d_3} P_0$$

In general, we see that

$$\boxed{P_n = \frac{a_0a_1 \cdots a_{n-1}}{d_1d_2 \cdots d_n} P_0 \quad \text{for } n \geq 1}$$

Thus to calculate P_n for $n \geq 1$, we need only determine P_0 and use the above result. To determine P_0, we note that the P_n's are probabilities that must sum to one, and thus

$$\sum_{n=0}^{\infty} P_n = 1 \quad \text{implies that} \quad P_0 + \sum_{n=1}^{\infty} \frac{a_0a_1 \cdots a_{n-1}}{d_1d_2 \cdots d_n} P_0 = 1$$

Therefore,

$$P_0 \left(1 + \sum_{n=1}^{\infty} \frac{a_0a_1 \cdots a_{n-1}}{d_1d_2 \cdots d_n} \right) = 1$$

which implies that

$$\boxed{P_0 = \frac{1}{1 + \sum_{n=1}^{\infty} \frac{a_0a_1 \cdots a_{n-1}}{d_1d_2 \cdots d_n}}}$$

Thus, given the mean arrival rates, a_n, and the mean departure rates, d_n, the preceding results allow us to easily calculate the steady-state probabilities P_n for $n \geq 0$. We shall now show how these probabilities can be used to determine the operating characteristics of a queueing system that is in the steady state.

It will be convenient to use the following notation:

L = expected number of customers in the queueing system that are waiting or being served
L_q = expected number of customers that are waiting for service
W = expected time that a customer spends in the queueing system waiting and being served
W_q = expected time that a customer waits for service

We can calculate L and L_q directly from the definition of expected value; namely,

$$L = \sum_{n=0}^{\infty} nP_n$$

$$L_q = \sum_{n=0}^{\infty} nP_{n+s} \quad \text{where } s = \text{the number of servers}$$

Note that with s servers, there will be n customers in the waiting line when there are $n + s$ customers in the system. For example, with two servers, if there are five customers in the system, then three are in the waiting line and two are being served.

Calculating W is not quite as straightforward, but there are some important relations between the operating characteristics that can be used in many cases to determine W, W_q, L, and L_q given only one of these values.

STEADY-STATE RELATIONS

For a queueing system in the steady state, the following relationships hold:

If $a_n = \lambda$ for $n \geq 0$, then

$$L = \lambda W \quad \text{and} \quad L_q = \lambda W_q$$

If $d_n = \mu$ for $n \geq 1$, then

$$W = W_q + \frac{1}{\mu}$$

If $\bar{\lambda} = \sum_{n=0}^{\infty} a_n P_n$ is the average arrival rate over all the states, then

$$L = \bar{\lambda} W \quad \text{and} \quad L_q = \bar{\lambda} W_q$$

It is important to note that if any one of the operating characteristics is known, then the preceding relationships can be used to determine the other three.

We will not formally prove each of the relations, but we can give a "heuristic argument" that makes the relation $L = \lambda W$ seem plausible. An arriving customer in a steady-state queueing system should expect to find L customers in the system. From our arrivals-only model we know that the interarrival times are exponentially distributed with mean $1/\lambda$. Therefore, on average, an arriving customer should expect to wait until L more customers arrive and so should expect to wait L times $1/\lambda$ units of time. Thus, it seems reasonable that

$$W = \frac{1}{\lambda} L \quad \text{so that } L = \lambda W$$

For some steady-state queueing models, it is possible to derive formulas for L or L_q in terms of the parameters of the model such as the number of servers, the arrival rates a_n, and the departure rates d_n. One can then simply "plug in" the parameters to determine L or L_q and then use the steady-state relations to determine the other operating characteristics. Unfortunately, these formulas are usually of little use in practice because the problem being studied is often slightly different from any of the models for which formulas were derived.

The only derivation of a formula for L that we shall give is in the simple model (of Section 11.2) where the arrival and departure rates do not depend on the state of the system. This derivation is given to illustrate some of the algebra involved and to illustrate some calculations that will be useful for more complicated queueing models.

Our approach to calculating the operating characteristics will be to calculate first the probability P_0 and then calculate the probabilities P_n for $n \geq 1$. Given the P_n's, we shall then calculate L or L_q by simply using the definition of expected value. Given L or L_q, the other operating characteristics can be found by using the steady-state relations.

11.5 A SINGLE-SERVER SYSTEM WITH CONSTANT ARRIVAL AND DEPARTURE RATES

Consider a steady-state queueing system with a single server and arrival and departure rates that do not depend on the state of the system. Recall that this was the case for the simple model introduced in Section 11.2. In particular, suppose that

$$a_n = \lambda \quad \text{for } n \geq 0 \quad \text{and} \quad d_n = \mu \quad \text{for } n \geq 1$$

That is, we are considering a model where the interarrival times are exponentially distributed with parameter λ and service times are exponentially distributed with parameter μ.

In this case we use the results of Section 11.4 to calculate

$$P_0 = \frac{1}{1 + \sum_{n=1}^{\infty} \frac{a_0 a_1 \cdots a_{n-1}}{d_1 d_2 \cdots d_n}} = \frac{1}{1 + \sum_{n=1}^{\infty} \frac{\lambda^n}{\mu^n}}$$

and

$$P_n = \frac{a_0 a_1 \cdots a_{n-1}}{d_1 d_2 \cdots d_n} P_0 = \frac{\lambda^n}{\mu^n} P_0 \quad \text{for } n \geq 1$$

Note that the infinite series used in this calculation of P_0 is a geometric series with ratio $\rho = \lambda/\mu$ and therefore converges if $\rho < 1$. For this one-server model, if the mean arrival rate λ is not less than the mean service rate μ, the number of customers in the system grows without bound. Thus, it is reasonable to consider only the case where $\lambda/\mu < 1$.

Recall that the geometric series

$$1 + \sum_{n=1}^{\infty} \rho^n \quad \text{converges to} \quad \frac{1}{1 - \rho} \qquad \text{when } \rho < 1$$

Consequently, from the preceding expressions for P_0 and P_n, we see that

$$P_0 = 1 - \rho$$

and

$$P_n = \rho^n P_0 = \rho^n(1 - \rho) \quad \text{for } n \geq 1$$

Using these probabilities, we can now calculate the expected number of customers, L, in the system as a function of ρ. From the definition of expected value,

$$L = \sum_{n=1}^{\infty} nP_n = \sum_{n=1}^{\infty} n\rho^n(1 - \rho) = (1 - \rho) \sum_{n=1}^{\infty} n\rho^n$$

To calculate a closed-form expression for the series on the right, we illustrate a technique that will be useful in other queueing problems. The trick is to make the series on the right look like a sum of derivatives by factoring ρ out of the summation, so that

$$\sum_{n=1}^{\infty} n\rho^n = \rho \sum_{n=1}^{\infty} n\rho^{n-1}$$

Now recall from calculus that if a series converges, then

> the derivative of the sum equals the sum of the derivatives.

This means that

$$\frac{d}{d\rho}\left(\sum_{n=1}^{\infty} \rho^n\right) = \sum_{n=1}^{\infty} \frac{d}{d\rho}[\rho^n] = \sum_{n=1}^{\infty} n\rho^{n-1}$$

Thus

$$\begin{aligned} L &= \rho(1 - \rho) \sum_{n=1}^{\infty} n\rho^{n-1} = \rho(1 - \rho) \sum_{n=1}^{\infty} \frac{d}{d\rho}[\rho^n] \\ &= \rho(1 - \rho) \frac{d}{d\rho}\left(\sum_{n=1}^{\infty} \rho^n\right) \\ &= \rho(1 - \rho) \frac{d}{d\rho}\left(\frac{1}{1 - \rho} - 1\right) \\ &= \rho(1 - \rho) \frac{1}{(1 - \rho)^2} = \frac{\rho}{1 - \rho} \end{aligned}$$

Thus

> For a single-server queueing system with $a_n = \lambda$ for $n \geq 0$ and $d_n = \mu$ for $n \geq 1$,
>
> $$L = \frac{\rho}{1 - \rho} = \frac{\lambda}{\mu - \lambda}$$
>
> provided
>
> $$\rho = \frac{\lambda}{\mu} < 1$$

Therefore, the steady-state relations give

$$W = \frac{1}{\lambda} L = \frac{1}{\mu - \lambda}$$

$$W_q = W - \frac{1}{\mu} = \frac{\lambda}{\mu(\mu - \lambda)}$$

$$L_q = \lambda W_q = \frac{\lambda^2}{\mu(\mu - \lambda)}$$

11.6 CALCULATING OPERATING CHARACTERISTICS FROM BASIC PRINCIPLES

The formulas for L and L_q derived in the preceding section are for one particular model, namely, a single-server model with constant mean arrival and mean departure rates. These formulas are usually of little use in practice because of some slight variation in the model that makes the formulas invalid. In this section we illustrate an approach to calculating L or L_q using only the steady-state probabilities P_n derived in Section 11.4 and the basic definition of expected value. A summary of the approach follows.

> Determine the mean arrival and departure rates, a_n and d_n, for each state n.
>
> Calculate the steady-state probabilities
>
> $$P_0 = \frac{1}{1 + \sum_{n=1}^{\infty} \frac{a_0 a_1 \cdots a_{n-1}}{d_1 d_2 \cdots d_n}}$$
>
> $$P_n = \frac{a_0 a_1 \cdots a_{n-1}}{d_1 d_2 \cdots d_n} P_0 \quad \text{for } n \geq 1$$
>
> Calculate L or L_q using
>
> $$L = \sum_{n=1}^{\infty} n P_n \quad \text{or} \quad L_q = \sum_{n=1}^{\infty} n P_{n+s}$$
>
> where s is the number of servers.

This approach is illustrated using a model with a single server and constant mean arrival rates. However, the mean departure rates are not constant, so the formula $L = \lambda/(\mu - \lambda)$, derived in Section 11.5, is not valid.

Consider a small gas station with a single pump and a full-time attendant. Arriving cars have exponentially distributed interarrival times with an expected interarrival time of one sixth of an hour. This means that cars are arriving at the mean rate of $\lambda = 6$ per hour. Whenever there is a single car at the pump, the attendant provides all the service with a mean service time of 5 min (so that on average 12 cars per hour are serviced). However, if there is more than one car at the pump, the mechanic in the garage comes out and helps service the cars, which reduces the mean service time to 4 min. In both cases the service times are exponentially distributed. We shall assume that steady-state conditions hold.

In this example the mean arrival rate is constant at $\lambda = 6$ per hour, but the mean departure rate depends on the state of the system. When there is one customer in the system, the mean departure rate is 12 customers per hour, but with two or more cars at the pump, the mean departure rate is 15 per hour. Thus, in terms of the a_n's and the d_n's, we have

$a_n = 6$ per hour for $n \geq 1$
$d_1 = 12$ per hour, $d_n = 15$ per hour for $n \geq 2$

For bookkeeping purposes it will be convenient to use a **rate diagram** to record the arrival and departure rates for the various states of the system. The rate diagram for this example is shown in Figure 11.3.

We can now calculate P_0 as

$$P_0 = \frac{1}{1 + \sum_{n=1}^{\infty} \frac{a_0 a_1 \cdots a_{n-1}}{d_1 d_2 \cdots d_n}}$$

$$= \left[1 + \frac{a_0}{d_1} + \frac{a_0 a_2}{d_1 d_2} + \frac{a_0 a_1 a_2}{d_1 d_2 d_3} + \cdots\right]^{-1}$$

$$= \left[1 + \frac{6}{12} + \frac{6}{12}\left(\frac{6}{15}\right) + \frac{6}{12}\left(\frac{6}{15}\right)^2 + \frac{6}{12}\left(\frac{6}{15}\right)^3 + \cdots\right]^{-1}$$

$$= \left[1 + \frac{6}{12}\left[1 + \frac{6}{15} + \left(\frac{6}{15}\right)^2 + \left(\frac{6}{15}\right)^3 + \cdots\right]\right]^{-1}$$

$$= \left[1 + \frac{1}{2}\left[1 + \frac{2}{5} + \left(\frac{2}{5}\right)^2 + \left(\frac{2}{5}\right)^3 + \cdots\right]\right]^{-1}$$

$$= \left[1 + \frac{1}{2}\left(\frac{1}{1 - (2/5)}\right)\right]^{-1} = \frac{6}{11}$$

Thus $P_0 = \frac{6}{11}$, and we can now calculate the remaining P_n's using

$$P_n = \frac{a_0 a_1 \cdots a_{n-1}}{d_1 d_2 \cdots d_n} P_0 \quad \text{for } n \geq 1$$

Therefore,

$$P_1 = \frac{6}{12} P_0 = \frac{1}{2}\left(\frac{6}{11}\right)$$

$$P_2 = \frac{6}{12}\left(\frac{6}{15}\right) P_0 = \frac{1}{2}\left(\frac{2}{5}\right)\left(\frac{6}{11}\right)$$

$$P_3 = \frac{6}{12}\left(\frac{6}{15}\right)^2 P_0 = \frac{1}{2}\left(\frac{2}{5}\right)^2\left(\frac{6}{11}\right)$$

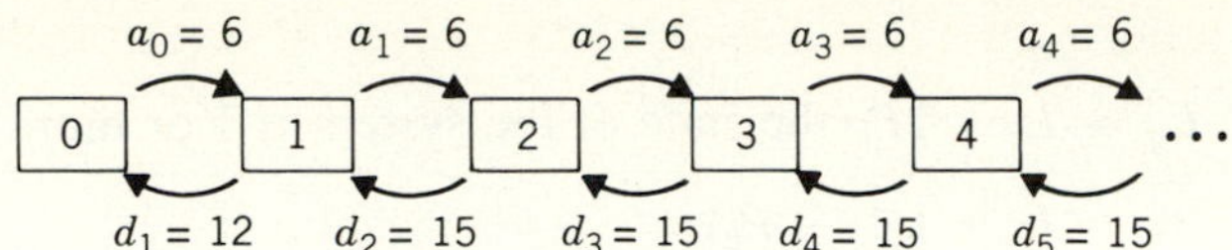

FIGURE 11.3 The rate diagram for the gas station problem.

In general, we see that

$$P_n = \frac{1}{2}\left(\frac{2}{5}\right)^{n-1}\left(\frac{6}{11}\right) \quad \text{for } n \geq 1$$

and so from the definition of expected value

$$\begin{aligned} L &= \sum_{n=1}^{\infty} nP_n = \frac{1}{2}\left(\frac{6}{11}\right)\sum_{n=1}^{\infty} n\left(\frac{2}{5}\right)^{n-1} \\ &= \frac{3}{11}\sum_{n=1}^{\infty} n\rho^{n-1} \quad \text{for } \rho = \frac{2}{5} \\ &= \frac{3}{11}\sum_{n=1}^{\infty}\frac{d}{d\rho}\rho^n \\ &= \frac{3}{11}\frac{d}{d\rho}\sum_{n=1}^{\infty}\rho^n = \frac{3}{11}\frac{d}{d\rho}\left(\frac{1}{1-\rho} - 1\right) \\ &= \frac{3}{11}\frac{1}{(1-\rho)^2} = \frac{25}{33} \quad \text{for } \rho = \frac{2}{5} \end{aligned}$$

Thus, at any given time, the expected number of cars at the pump is $L = \frac{25}{33}$ or about 0.76 cars. Of course, on any given observation of the pump this expected number of cars would never be observed, but the long-run average number of cars observed at the pump is $L = 0.76$.

It is important to realize that the probabilities P_n give the fraction of time the system is in state n. For example, the fraction of time there are no cars at the pump is $P_0 = \frac{6}{11}$ and the fraction of time exactly one car is at the pump is $P_1 = \frac{3}{11}$. Note, therefore, that $\frac{9}{11}$ of the time no cars are waiting for service and so the mechanic can stay in the garage.

Knowing L, we can now determine the other operating characteristics using the steady-state relations. In particular,

$$W = \frac{1}{\lambda}L \quad \text{implies } W = \frac{1}{6}\left(\frac{25}{33}\right) = 0.13 \text{ hr}$$

Thus an arriving car should expect to spend about 0.13 hr or 7.8 min at the station (waiting for and receiving service). The relation $W = W_q + (1/\mu)$ does not hold because the mean departure rates are not constant. However, $L_q = \lambda W_q$ does hold and can be used to calculate W_q once L_q is known. Note that

$$L_q = L - (\text{the expected number being served})$$

But one car will be served as long as there is one or more cars at the station.

Thus,

$$\begin{aligned} L_q &= L - 1P \text{ (the state of the system is 1 or more)} \\ &= L - 1(1 - P_0) \\ &= \frac{25}{33} - \frac{5}{11} = \frac{10}{33} \end{aligned}$$

Therefore, the average waiting time prior to service is

$$W_q = \frac{1}{\lambda} L_q = \frac{1}{6}\left(\frac{10}{33}\right) = 0.05 \text{ hr or } 3 \text{ min}$$

11.7 MULTIPLE-SERVER QUEUES

The general approach to calculating operating characteristics illustrated in Section 11.6 can be used for a wide variety of queueing models. In this section we illustrate the approach using a model with three servers (although any number of servers could have been chosen for this example).

Consider a bank in a steady-state condition having three tellers and a single waiting line. Each teller has a service time that is exponentially distributed with a mean service time of 4 min. Arriving customers have an exponential interarrival time distribution with the mean arrival rate being $\lambda = 30$ customers per hour.

It is important that the mean arrival and departure rates be measured in the same units, and so we note that each teller can service customers at the mean rate of 15 per hour. It is usually helpful to construct a schematic diagram of the queueing system. In this example, the queueing system has the structure indicated in Figure 11.4.

When there are three or more customers in the bank, the mean departure rate is 45 per hour because all the servers are busy. When only two customers are in the bank, the mean departure rate is 30 per hour, and it drops to 15 per hour when only one customer is in the system. The rate diagram for this example is given in Figure 11.5.

Given the rate diagram, the next step in our general approach is to calculate the steady-state probabilities, P_n, starting with P_0. In fact, from this point on, our calculations proceed exactly as they did for the single-server example in

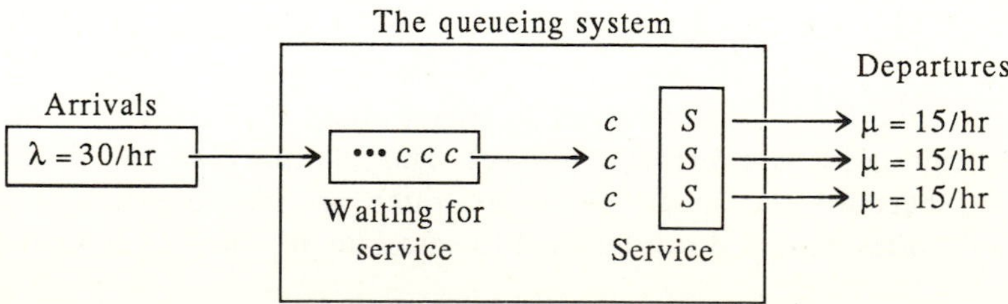

FIGURE 11.4 A schematic diagram of a multiple server queueing system with three servers.

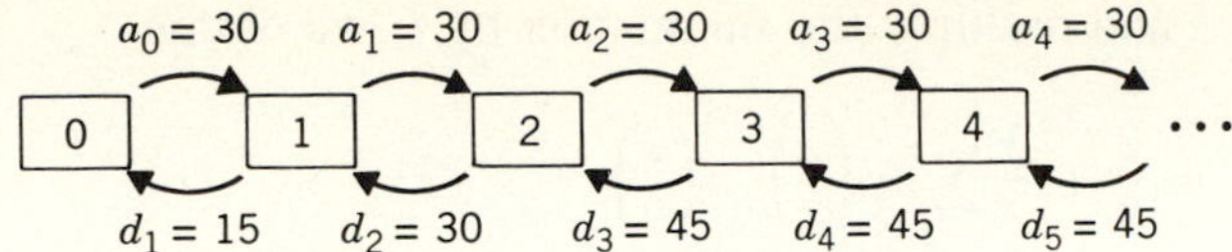

FIGURE 11.5 The rate diagram for the banking problem.

Section 11.6. Calculating P_0 gives

$$P_0 = \frac{1}{1 + \sum_{n=1}^{\infty} \frac{a_0 a_1 \cdots a_{n+1}}{d_1 d_2 \cdots d_n}}$$

$$= \left[1 + \frac{a_0}{d_1} + \frac{a_0 a_1}{d_1 d_2} + \frac{a_0 a_1 a_2}{d_1 d_2 d_3} + \cdots\right]^{-1}$$

$$= \left[1 + \frac{30}{15} + \frac{30}{15}\left(\frac{30}{30}\right) + \frac{30}{15}\left(\frac{30}{30}\right)\frac{30}{45} + \frac{30}{15}\left(\frac{30}{30}\right)\left(\frac{30}{45}\right)^2 + \cdots\right]^{-1}$$

$$= \left[1 + 2 + 2\left(1 + \frac{2}{3} + \left(\frac{2}{3}\right)^2 + \left(\frac{2}{3}\right)^3 + \cdots\right)\right]^{-1}$$

$$= \left[1 + 2 + 2\left[\frac{1}{1 - (2/3)}\right]\right]^{-1} = \frac{1}{9}$$

Thus $P_0 = \frac{1}{9}$, and we can then calculate the other steady-state probabilities using the general relations derived in Section 11.4, namely,

$$P_n = \frac{a_0 a_1 \cdots a_{n-1}}{d_1 d_2 \cdots d_n} P_0 \quad \text{for } n \geq 1$$

Therefore

$$P_1 = 2P_0$$

$$P_2 = 2P_0$$

$$P_3 = 2(\tfrac{2}{3})P_0$$

$$P_4 = 2(\tfrac{2}{3})^2 P_0$$

so in general we see that

$$P_n = 2(\tfrac{2}{3})^{n-2} P_0 \quad \text{for } n \geq 2$$

From the definition of expected value,

$$L = \sum_{n=1}^{\infty} nP_n = \frac{2}{9} + \sum_{n=2}^{\infty} n\left(\frac{2}{3}\right)^{n-2}\left(\frac{2}{9}\right)$$

This last summation does not look like a sum of derivatives, but if we multiply each term by $\frac{2}{3}$ and multiply the summation by $\frac{3}{2}$, we obtain

$$L = \frac{2}{9} + \frac{3}{2}\sum_{n=2}^{\infty} n\left(\frac{2}{3}\right)^{n-1}\left(\frac{2}{9}\right)$$

$$= \frac{2}{9} + \frac{1}{3}\sum_{n=2}^{\infty} n\left(\frac{2}{3}\right)^{n-1} = \frac{2}{9} + \frac{1}{3}\sum_{n=2}^{\infty} n\rho^{n-1} \quad \text{for } \rho = \frac{2}{3}$$

$$= \frac{2}{9} + \frac{1}{3}\sum_{n=2}^{\infty} \frac{d}{d\rho}\rho^n$$

$$= \frac{2}{9} + \frac{1}{3}\frac{d}{d\rho}\sum_{n=2}^{\infty} \rho^n = \frac{2}{9} + \frac{1}{3}\frac{d}{d\rho}\left(\frac{1}{1-\rho} - 1 - \rho\right)$$

$$= \frac{2}{9} + \frac{1}{3}\left(\frac{1}{(1-\rho)^2} - 1\right) \quad \text{for } \rho = \frac{2}{3}$$

So,

$$L = \frac{2}{9} + \left(\frac{1}{3}\right)8 = \frac{26}{9} = 2\frac{8}{9}$$

In this example the arrival rates do not depend on the state of the system, so the steady-state relation $L = \lambda W$ holds. Thus, the expected time spent in the bank is given by

$$W = \frac{1}{\lambda}L = \frac{1}{30}\left(\frac{26}{9}\right) = 0.096 \text{ hr} \quad \text{or } 5.8 \text{ min}$$

To find the expected time, W_q, spent waiting for service, we cannot use the relation $W = W_q + (1/\mu)$ because the departure rate depends on the state of the system. However, the relation

$$W = W_q + (\text{expected time spent in service})$$

always holds. In this case the expected service time is 4 min so

$$W_q = 5.8 \text{ min} - 4 \text{ min} = 1.8 \text{ min}$$

To calculate L_q, the expected number of customers waiting for service, we can use the relation $L_q = \lambda W_q$. Because $\lambda = 30$ customers per hour or 2 customers per minute, we see that

$$L_q = 2(1.8) = 3.6 \text{ customers}$$

In this example there was no particular reason that we calculated L first, instead of L_q. We could just as easily have calculated L_q using the definition of the expected number of customers waiting for service. When a queueing system has s servers, there will be n customers waiting for service when there are $n + s$ customers in the queueing system. Thus, from the definition of expected value,

$$L_q = \sum_{n=1}^{\infty} nP_{n+s}$$

For the preceding example we could have calculated

$$L_q = 1P_4 + 2P_5 + 3P_6 + \cdots$$

and obtained the same answer, namely, $L_q = 3.6$ customers.

11.8 FINITE QUEUES

In the examples considered so far, we assumed the queueing system could be in any state n, for $n \geq 0$. However, there are many practical cases when only a finite number of states are possible. In terms of the general arrivals and departures model of Section 11.3, this means there is some integer N such that

$$a_n = 0 \quad \text{for } n \geq N$$

and

$$d_n = 0 \quad \text{for } n \geq N + 1$$

In this section we consider an example of a finite queue that illustrates how the operating characteristics can be obtained using the same approach followed in Sections 11.6 and 11.7. In fact, for finite queues we shall see that the calculations are actually easier because no infinite series need be evaluated.

Again, our approach will be to use the arrival and departure rates, a_n and d_n, to first calculate the steady-state probabilities, P_n. We shall then be able to calculate L or L_q from the definition of expected value.

Consider a small self-serve car wash, in the steady-state condition, that has cars arriving with exponentially distributed interarrival times at a mean rate of 12 per hour. The car wash has two washing bays, each equipped with a pressure spraying hose that the customer uses to wash off the car. Washing time, in either bay, is exponentially distributed with a mean washing time of 4 min. The car wash has very little space and can accommodate only one waiting car in addition to those being washed. If all three spots are occupied, arriving cars leave and go elsewhere to be washed.

The diagram in Figure 11.6 shows the basic structure of the queueing system for this example. Note that each server has a mean service rate of 15 per hour because the mean service time is 4 min. In this example of a finite queue, there are only four possible states for the queueing system, $n = 0, 1, 2$, and 3. Note that when the system is in state $n = 3$, the arrival rate is $a_3 = 0$. In fact, $a_n = 0$ for all $n \geq 3$ and $d_n = 0$ for $n \geq 4$. The rate diagram for this example is given in Figure 11.7. Note that when only one washing bay is occupied, the mean departure rate is 15 per hour, but, when both bays are being used, the mean departure rate is 30 per hour.

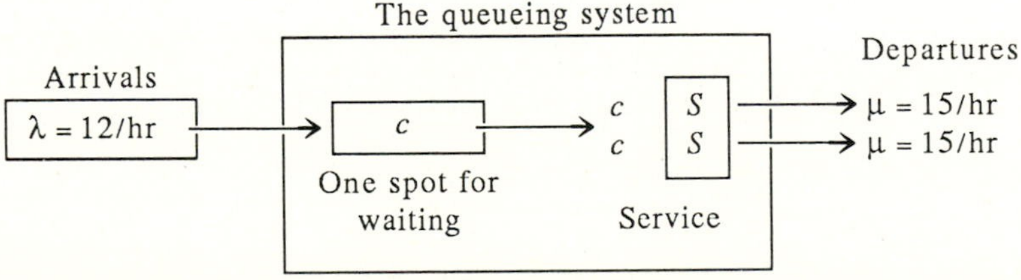

FIGURE 11.6 A schematic diagram of a finite queueing system with two servers.

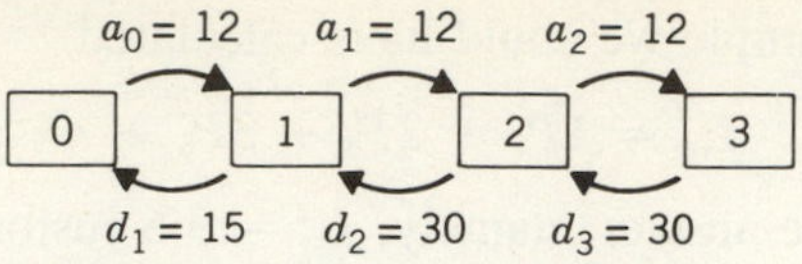

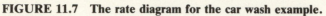

FIGURE 11.7 The rate diagram for the car wash example.

As in all the preceding examples, P_0 is calculated using

$$P_0 = \frac{1}{1 + \sum_{n=1}^{\infty} \frac{a_0 a_1 \cdots a_{n-1}}{d_1 d_2 \cdots d_n}}$$

In this example, where $a_n = 0$ for $n \geq 3$, this expression simplifies to

$$P_0 = \left[1 + \frac{a_0}{d_1} + \frac{a_0 a_1}{d_1 d_2} + \frac{a_0 a_1 a_2}{d_1 d_2 d_3}\right]^{-1}$$

Thus

$$P_0 = \left[1 + \frac{12}{15} + \frac{12}{15}\left(\frac{12}{30}\right) + \frac{12}{15}\left(\frac{12}{30}\right)^2\right]^{-1}$$

$$= \left[1 + \frac{4}{5} + \frac{8}{25} + \frac{16}{125}\right]^{-1} = \frac{125}{281}$$

Therefore

$$P_1 = \frac{4}{5} P_0 = \frac{100}{281}$$

$$P_2 = \frac{8}{25} P_0 = \frac{40}{281}$$

$$P_3 = \frac{16}{125} P_0 = \frac{16}{281}$$

The expected number of cars at the car wash, L, can then be calculated as

$$L = 1P_1 + 2P_2 + 3P_3 = \frac{4}{5} P_0 + \frac{16}{25} P_0 + \frac{48}{125} P_0$$

$$= \frac{228}{281} \approx 0.81 \text{ cars}$$

Because the arrival rates depend on the state of the system, we cannot use the relation $W = (1/\lambda)L$ to calculate W. However, recall from the steady-state relations of Section 11.4 that

$$L = \bar{\lambda} W$$

where $\bar{\lambda}$ is the average arrival rate over all the states. Using

$$\bar{\lambda} = \sum_{n=0}^{\infty} a_n P_n = 12(P_0 + P_1 + P_2)$$

$$= 12\left(\frac{265}{281}\right) = \frac{3180}{281} \approx 11.3$$

we calculate

$$W = \frac{1}{\bar{\lambda}} L = \frac{281}{3180}\left(\frac{228}{281}\right) = \frac{228}{3180} \approx 0.072 \text{ hr}$$

Note that 0.072 hr is about 4.3 min, and so a car entering the car wash should expect to spend a total of 4.3 min waiting for and receiving service. Using the fact that

$$W = W_q + (\text{expected time for service})$$

and that expected service time is 4 min, we see that

$$W_q \approx 0.3 \text{ min}$$

Another way of calculating W_q is to use the steady-state relation

$$L_q = \bar{\lambda} W_q$$

In this example there is one car waiting for service only when there are three cars in the system. Thus

$$L_q = 1P_3 = \frac{16}{281}$$

and so

$$W_q = \frac{1}{\bar{\lambda}} L_q = \frac{281}{3180}\left(\frac{16}{281}\right) = \frac{16}{3180} \text{ hr} \quad \text{or } 0.3 \text{ min}$$

11.9 LIMITED-SOURCE QUEUES

A special type of finite queue occurs when there is only a limited number of arriving customers. Such a queueing system is referred to as a limited-source queue. For example, suppose a mechanic is assigned to two machines that must be repaired whenever they break down. In this case the customers are the two machines. Suppose for each machine the running time between breakdowns is known to be exponentially distributed with a mean running time of 3 hr. The service time required by the mechanic to get either of the machines running again is also exponentially distributed with the mean service time being 30 min. Suppose we wish to know, under steady-state conditions, what fraction of the time all the machines are running.

A schematic diagram for the queueing system is given in Figure 11.8. Note that, in terms of hours, each machine enters the queueing system at a mean rate of $\frac{1}{3}$ of a machine per hour.

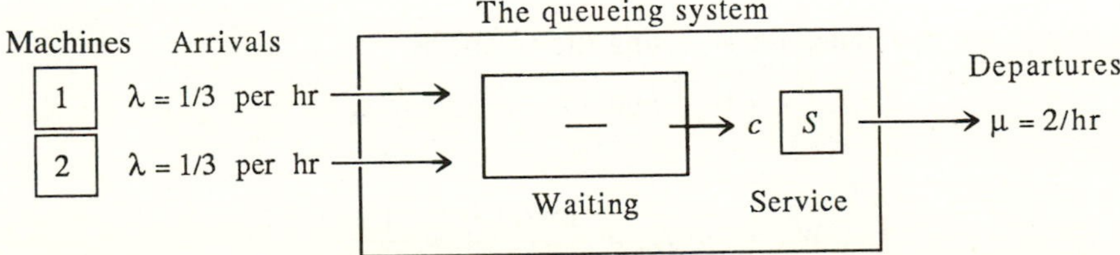

FIGURE 11.8 A schematic diagram of a limited-source queue with one server.

The diagram in Figure 11.8 shows that when all the machines are running, the mean arrival rate to the queueing system is $a_0 = \frac{2}{3}$ of a machine per hour, but when one machine is already in for repair, the mean arrival rate is $a_1 = \frac{1}{3}$ of a machine per hour. Of course, if both of the machines are in for repair, the mean arrival rate is zero. Thus the system has only three states, 0, 1, or 2. In state 1 or state 2 the mean departure rate is 2 machines per hour because the expected service time is one-half hour. The rate diagram for this limited-source queue is as given in Figure 11.9.

In this case the expression for P_0 simplifies to

$$P_0 = \frac{1}{\left(1 + \dfrac{a_0}{d_1} + \dfrac{a_0 a_1}{d_1 d_2}\right)}$$

Thus

$$P_0 = \left[1 + \frac{1}{3} + \frac{1}{3}\left(\frac{1}{6}\right)\right]^{-1} = \frac{12}{85}$$

and this is the fraction of time all the machines are running, that is, the fraction of time no machines are in for repair.

If we wish to calculate all the operating characteristics of the queueing system, we first need to calculate P_1 and P_2. Here

$$P_1 = \frac{1}{3} P_0 = \frac{6}{25} \quad \text{and} \quad P_2 = \frac{1}{18} P_0 = \frac{1}{25}$$

In this example

$$L = 1P_1 + 2P_2 = \frac{8}{25} \text{ machines}$$

We cannot use the relation $L = \lambda W$ to calculate W because the mean arrival rates depend on the states of the system. However, the average mean arrival rate over all the states is

$$\bar{\lambda} = P_0 a_0 + P_1 a_1 = \frac{14}{25}$$

so

$$W = \frac{1}{\bar{\lambda}} L = \frac{4}{7} \quad \text{of an hour}$$

This means the expected time a machine waits for repair once it has stopped running is

$$W_q = W - (\text{expected service time}) = \frac{4}{7} - \frac{1}{2} = \frac{1}{14} \text{ of an hour}$$

Alternately, we can calculate W_q using the relations

$$L_q = \bar{\lambda} W_q \quad \text{and} \quad L_q = 1P_2 = \tfrac{1}{25}$$

so that

$$W_q = \frac{1}{\bar{\lambda}} L_q = \frac{25}{14}\left(\frac{1}{25}\right) = \frac{1}{14}$$

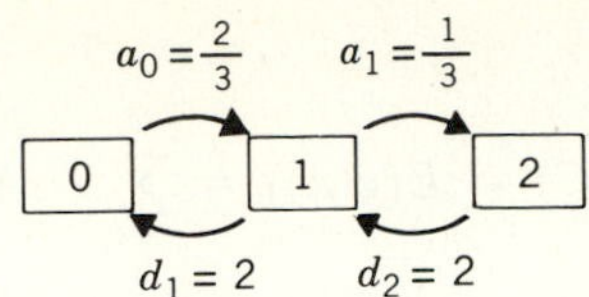

FIGURE 11.9 The rate diagram for the machine repair example.

11.10 OPTIMIZATION IN QUEUES

So far we have been concerned mainly with determining the operating characteristics of various queueing models. In practice, however, it is important to understand how the operating characteristics impact on the costs of the queueing system. We assume throughout this section that interarrival times and service times are exponentially distributed and that steady-state conditions hold. In most queueing situations there are costs associated with the time customers spend in the system and costs associated with providing service. We let

$E(WC)$ = expected waiting cost per unit time
$E(SC)$ = expected service cost per unit time

Typically, the waiting times and $E(WC)$ can be decreased by providing better (and more costly) service. For example, increasing the number of servers or hiring more efficient servers will decrease waiting times, but $E(SC)$ will be increased. The trade-offs between expected waiting costs and expected service costs give rise to optimization problems in queues where the objective is to

$$\text{minimize } E(TC) = E(WC) + E(SC)$$

where $E(TC)$ denotes expected total cost per unit time.

Determining the expected service cost per unit time is usually quite straightforward. In many applications, the service cost is not even random. For example, the service cost per unit time may simply be the hourly wage paid to the servers whether they are busy or not. However, the expected waiting cost per unit time usually is random and depends on the expected number of customers in the queueing system or on the expected time customers spend in the system. To determine $E(WC)$, we consider two cases, namely, when customers are internal to the organization providing service and when customers are external to the organization.

Case 1. Customers are internal so the cost of waiting is lost profit due to lost productivity.

Typical examples of customers who are internal to the organization providing service are workers in a factory arriving at a parts center to get parts for the machines they are building or trucks belonging to a company fleet that arrive at the company garage to be serviced. When customers are internal and the cost of waiting is due to lost productivity, the waiting cost is usually a function of the number of customers in the queueing system. Suppose

the waiting cost per unit time $WC = g(n)$ where n is the number of customers in the system

In this case

$$E(WC) = E(g(n)) = \sum_{n=1}^{\infty} g(n)P_n$$

When $g(n)$ is a linear function of n, say $g(n) = kn$ where k is some constant,

$$E(WC) = \sum_{n=1}^{\infty} knP_n = k\sum_{n=1}^{\infty} nP_n = kL$$

where L is the expected number of customers in the queueing system.

Case 2. Customers are external to the organization providing service so the cost of waiting is lost profit from future business.

If waiting time is too long, individual customers may decide to go to another organization that provides service with less waiting time. Typical examples of customers who are external would be customers arriving at a bank or cars arriving at a car wash. In these cases the waiting time for an individual customer has an important impact on future business. Suppose that

w = waiting time for an individual customer

and

$h(w)$ = cost of waiting w units of time

where h is a continuous function of the random variable w. If $f(w)$ is the density function for w, then

$$E(h(w)) = \int_0^{\infty} h(w)f(w)\,dw$$

To calculate $E(WC)$, the expected waiting cost per unit time, we use the fact that

expected waiting cost per unit time equals
expected waiting cost per customer times the
expected number of customers arriving per unit time.

Thus

$$E(WC) = \lambda E(h(w))$$

In the linear case, when $h(w) = kw$ where k is some constant,

$$E(h(w)) = E(kw) = kE(w) = kW$$

recalling that W is the expected time an arriving customer spends in the queueing system. Therefore,

$$E(WC) = \lambda(kW) = kL$$

With this general discussion as background, we now consider some specific applications.

Decision variables that arise in a large number of queueing problems are

- the number of servers, s
- the efficiency of the servers in terms of the mean departure rate, μ
- the number of service facilities and the mean arrival rate λ to each facility

Note that these decision variables affect L and thus directly affect $E(WC)$ whether the customers are internal or external to the organization.

We consider three optimization models using these decision variables.

Optimizing over the Number of Servers, s

Consider a large garage with several workers repairing cars. The workers get parts from a centrally located parts center that is staffed by two servers. Each server is paid \$15 per hour and has an exponentially distributed service time with a mean service rate of 4 customers per hour. Workers arrive at the parts center at a mean rate of 6 per hour, and the workers are paid at the rate of \$20 per hour. On several occasions, the owner of the garage has noticed many workers waiting for parts. The owner wonders if increasing the number of servers at the parts center would be profitable.

In this situation the owner is paying \$20 per hour for each worker standing idle by the parts center, so in the steady-state condition

$$E(WC) = \$20L$$

$$E(SC) = \$15s$$

where s is the number of servers. Of course, as the number of servers increases, the expected number of workers, L, at the parts center decreases. The question is whether the increase in service costs due to another server is more than offset by the decrease in waiting costs.

To answer this question, we first need to calculate L and see how it varies with the number of servers s. This queueing system is a multiple-server queue with constant mean arrival rates but the mean departure rates depend on the number of servers. The rate diagram depends on the number of servers, but for each fixed value of s we can calculate the steady-state probabilities P_n for $n \geq 0$. Calculating L for each of the systems with $s = 2$, 3, 4, and 5, we find the following results:

s	L	$\$20L + \$15s$
2	3.43	\$98.57
3	1.74	\$79.74
4	1.54	\$90.90
5	1.51	\$105.20

Note that for values of $s \geq 6$, the service costs alone are at least \$90. Thus, the optimal number of servers in this case is $s = 3$, giving an expected total cost per hour of \$79.74.

Optimizing over the Mean Service Rate, μ

Often there is some control over the efficiency of service as reflected through the mean service rate μ for a busy server. For example, suppose that a single server queue has a mean arrival rate fixed at $\lambda = 5$ per hour but the mean departure rate μ depends on the efficiency of the server. Suppose that for $5 \le \mu \le 10$,

$f(\mu) = 5\mu$ gives the cost per hour charged by the server when the service rate is μ

Suppose also that the cost of a waiting customer is estimated to be \$2 for each hour the customer waits. In this case the expected waiting cost per customer is $\$2W$ and $\lambda = 5$ customers arrive per unit time so $E(WC) = \$2\lambda W = \$10L$. Thus

$$\begin{aligned} E(TC) &= E(WC) + E(SC) \\ &= 10L + 5\mu \end{aligned}$$

We know from Section 11.5 that

$$L = \frac{\lambda}{\mu - \lambda}$$

and using $\lambda = 5$, we can give $E(TC)$ as a function of the mean departure rate μ, namely,

$$E(TC) = 10\left(\frac{5}{\mu - 5}\right) + 5\mu$$

Setting the derivative of this function to zero gives

$$\mu^2 - 10\mu + 15 = 0$$

Solving for μ, we obtain

$$\mu = 5 + \sqrt{10} \quad \text{and} \quad \mu = 5 - \sqrt{10}$$

Thus, the minimizing value in the range $5 \le \mu \le 10$ is

$$\mu = 5 + \sqrt{10}$$

Optimizing over the Mean Arrival Rate, λ

In some queueing situations we can make decisions that affect the mean arrival rate, λ. For example, consider a large factory where workers need to get tools from one or more tool centers. Suppose the cost of maintaining a tool center is \$20 per hour. Also suppose that such a tool center has a mean service rate of 30 workers per hour. The mean arrival rate of workers who need tools is 20 workers per hour, and these workers are paid \$25 per hour even when they are at the tool center. The problem is to determine how many tool centers should be distributed throughout the factory to minimize the expected total cost of getting tools to the workers.

Note in this situation, when only one tool center is used the arrival rate is 20 workers per hour, but, if there are T tool centers, the arrival rate to each center is $20/T$, assuming the worker population is divided evenly among the T centers. Thus, each tool center becomes its own queueing system, and the situation is far different from just assuming additional servers have been added to a single tool center.

For each tool center the expected cost per hour is

$$E(WC) = \$25L$$

and the expected service cost per hour is \$20. When there is only one tool center, $L = 2$ so

$$\begin{aligned} E(TC) &= E(WC) + E(SC) \\ &= \$50 + \$20 = \$70 \end{aligned}$$

When there are two tool centers, the arrival rate to each center is 10 workers per hour, so for each center we calculate $L = \frac{1}{2}$. Thus for each tool center

$$E(WC) = \$12.50 \quad \text{and} \quad E(SC) = \$20$$

This means that when two tool centers are used,

$$E(TC) = \$2(12.50) + \$2(20) = \$65$$

The reader should check that when three or more tool centers are used, the $E(TC)$ is greater than \$65, so expected total cost per hour is minimized when two centers are used.

11.11 QUEUEING MODELS INVOLVING NONEXPONENTIAL DISTRIBUTIONS

All the queueing models considered so far in this chapter assume the interarrival and service times are exponentially distributed. There are many queueing problems where this assumption is not realistic, and the general approach used in the previous sections for calculating the operating characteristics cannot be used. Unfortunately, only a few analytic results are available for queueing problems involving nonexponential distributions, and these results are typically very complicated to derive and to use in practice.

If *neither* the interarrival time *nor* the service time is exponentially distributed, very little can be said analytically about the queueing system. In this case one solution approach is to construct a computer simulation of the system and observe its behavior over time. By collecting statistics as the simulation progresses, approximations to the operating characteristics can be obtained. In Chapter 13 the method of simulation is discussed in some detail, and an example is presented that shows how it can be used to provide insight into nonexponential queueing models.

Queueing problems having exponentially distributed interarrival times but nonexponentially distributed service times are frequently encountered in practice. The random nature of arrivals can often be modeled quite accurately using the exponential distribution, but this is often not the case for service times. For example, if the service performed is essentially the same for each customer, the variation in service times will not be as large as that permitted by the exponential distribution. In some models the service time has negligible variation and is best modeled as a fixed constant.

In the case of a single-server queue where the interarrival times are exponentially distributed but the service times have some arbitrary distribution, there are some analytic results that allow the operating characteristics to be calculated. In particular

if there is a single server for a queueing system in the steady-state condition, and

the interarrival times are exponentially distributed with a constant mean arrival rate λ, and

the service time distribution has mean $1/\mu$ with variance σ^2, and

$$\rho = \frac{\lambda}{\mu} < 1$$

then

$$P_0 = 1 - \rho$$

$$L_q = \frac{\lambda^2\sigma^2 + \rho^2}{2(1 - \rho)}$$

$$L = \rho + L_q$$

$$W_q = \frac{1}{\lambda} L_q$$

$$W = W_q + \frac{1}{\mu}$$

These results are called the **Pollaczek–Khintchine equations** in honor of the two researchers who discovered them over 50 years ago. Of course, when the service time distribution is exponential with mean service rate of μ (so that its mean is $1/\mu$ and its variance is $\sigma^2 = 1/\mu^2$), the operating characteristics obtained by using the Pollaczek–Khintchine equations are the same as those obtained in Section 11.5.

The Kendall Notation

There is a notation that is widely used in the queueing literature for conveniently specifying the assumptions being made about a particular queueing model. This notation is called the **Kendall notation** for the person who first introduced it. The notation has the general form

$$a \,/\, b \,/\, c \,/\, d \,/\, e$$

where a, b, c, d, and e, respectively, identify

the interarrival time distribution,

the service time distribution,

the number of servers,

the capacity of the queueing system, and

the size of the calling population.

In specifying the type of distribution being used, it is standard to let

M = exponential distribution
G = general distribution
D = degenerate (constant time) distribution

For example, a queueing model with exponentially distributed interarrival times and service times, three servers, and no limits on the capacity of the system or on the calling population would be identified by the notation

$$M / M / 3 / \infty / \infty$$

As another example, the queueing system for the car wash model used in Section 11.8 would be described using the notation

$$M / M / 2 / 3 / \infty$$

Sometimes a sixth entry is used in the Kendall notation to describe the queueing discipline, namely, how customers are selected for service. In all the queueing models discussed in this chapter, the queueing discipline is the first-come first-served (FCFS) model. In some queueing models customers may be selected on a last-come first-served (LCFS) basis. In other situations customers may be selected randomly from the group that is waiting. It is important to realize that none of the results obtained in this chapter apply to systems in which the queueing discipline is other than FCFS. The general approach introduced here provides a foundation for the analysis of these more complicated queueing systems. However, as mentioned earlier, few useful analytical results are available for such systems, and it is usually necessary to resort to simulation in order to study their behavior.

SELECTED REFERENCES

Cox, D. R., and Smith, W. L., *Queues,* Methuen, London, 1961.

Gross, D., and Harris, C. M., *Fundamentals of Queueing Theory* (2d ed.), Wiley, New York, 1985.

Kleinrock, L., *Queueing Systems, Vol. I, Theory,* Wiley, New York, 1975.

Little, J. D. C., A Proof for the Queueing Formula $L = \lambda W$, *Operations Research* 9:383–387 (1961).

Newell, G. F., *Applications of Queueing Theory* (2d ed.), Chapman and Hall, London, 1982.

EXERCISES

11.1 Consider a queueing system having interarrival times that are exponentially distributed with a mean of 3 min. Suppose an arrival occurs at 8:00 AM.

(a) What is the probability that the next arrival occurs before 8:06 AM?

(b) What is the probability that the next arrival occurs between 8:06 AM and 8:09 AM?

(c) Given that no arrival occurs between 8:00 AM and 8:06 AM, what is the probability that the next arrival occurs between 8:06 AM and 8:09 AM?

(d) What is the probability that the number of arrivals between 8:06 AM and 8:09 AM is at least 1 but no more than 3?
(e) What is the expected number of arrivals between 8:06 AM and 8:09 AM?

11.2 Consider a queueing system satisfying the conditions of the simple model in Section 11.3. Suppose, in addition, that the maximum number of customers allowed in the queueing system is one. For $n = 0$ and 1, let $P_n(t)$ denote the probability of having n customers in the system at time $t > 0$, assuming no customers are in the system at time $t = 0$.
(a) List all the possible ways the system could have 1 customer at time $t + \Delta t$, and list all the possible ways the system could have 0 customers at time $t + \Delta t$.
(b) For $n = 0$ and $n = 1$, use the probabilities of part (a) to calculate $P_n(t + \Delta t)$.
(c) Use the results of part (b) to derive a pair of differential equations that must be satisfied by $P_0(t)$, $P_1(t)$, and their derivatives $P_0'(t)$ and $P_1'(t)$.
(d) Assuming steady-state conditions, use the differential equations in part (c) to derive expressions for P_0 and P_1 in terms of the parameters λ and μ given in the simple model.

11.3 Suppose a jumbo jet has just landed with several hundred passengers who must be processed by a single customs agent. The processing times (service times) are exponentially distributed with a mean of 5 min. Assume the customs agent is continually busy processing passengers.
(a) What is the probability that the processing time for the first customer is greater than 10 min?
(b) What is the expected number of passengers to be processed in the first hour?
(c) What is the probability that the customs agent processes exactly 3 passengers within the first 10 min?

11.4 For a queueing system where arrivals and departures satisfy the conditions of the simple model in Section 11.2, show that

$$P(\text{event } 3) = P_{n-1}(t)\lambda\,\Delta t + 0(\Delta t)$$

and

$$P(\text{event } 4) = 0(\Delta t)$$

11.5 Trucks arrive at a safety inspection station so that interarrival times are exponentially distributed with a mean of $\frac{1}{3}$ hour. The times required for inspection are also exponentially distributed with a mean of $\frac{1}{5}$ hour. Assume the associated queueing system is in steady state.
(a) Construct the rate diagram for this queueing system.
(b) Calculate P_0 and then calculate P_n for each $n \geq 1$.
(c) Calculate L, the expected number of trucks at the inspection station, using its definition $L = 1P_1 + 2P_2 + 3P_3 + \cdots$.
(d) Given L, calculate the other operating characteristics W, L_q, and W_q.

11.6 Cars arrive at a gas station with exponentially distributed interarrival times at a mean arrival rate of 2 per hour. There is only one gas pump and service time is exponentially distributed with a mean service time of 15 min. The station has unlimited waiting space but only 3 cars including the one being served can wait under a roof covering the pump area. Assume steady-state conditions prevail.
(a) Construct the rate diagram for this queueing system.
(b) What is the probability that an arriving car will not have to wait for service?
(c) What is the expected number of cars at the station?
(d) How long should an arriving car expect to wait for service?

(e) What fraction of time will the area covered by the roof be able to accommodate all waiting cars?

11.7 For the queueing system described in Exercise 11.6, suppose arriving cars not able to wait under the roof go elsewhere for service. In addition, suppose that when there are two cars at the station, an arriving car will go elsewhere for service with probability $\frac{1}{2}$.
(a) Construct the rate diagram for this queueing system.
(b) What is the probability an arriving car will not have to wait for service?
(c) Does the steady-state relation $L = \lambda W$ hold for this system? Explain.
(d) Calculate the expected time, W, that an arriving car should spend at the station.

11.8 Telephone calls arrive at a department store catalog ordering center with two operators. If both operators are busy, the calls are automatically put on hold and queued on a first-come first-served basis to be serviced by the next available operator. The time spent by each operator in taking the order is exponentially distributed with a mean service time of 6 min. The calls have exponentially distributed interarrival times and occur at a mean rate of 15 per hour.
(a) Construct the rate diagram for this queueing system.
(b) What is the expected number of calls on hold waiting for service?
(c) How much time should an arriving call expect to spend in the queueing system (waiting for and receiving service)?
(d) What fraction of time are both operators busy?

11.9 For the queueing system described in Exercise 11.8, suppose that $\frac{1}{3}$ of the callers who are put on hold hang up and take their business to another department store.
(a) Construct the rate diagram for this queueing system.
(b) Calculate L, L_q, W, and W_q.
(c) What fraction of the callers take their business to another store?

11.10 Consider a steady-state queueing system that allows at most three customers so that $P_0 + P_1 + P_2 + P_3 = 1$. Assume there are two servers.
(a) Calculate the expected number of customers being served in terms of the P_n's.
(b) Calculate the expected number of customers waiting for service in terms of the P_n's.
(c) What fraction of time can the queueing system accommodate all arriving customers?

11.11 A bank has three tellers and customers have exponentially distributed interarrival times with a mean of $\frac{1}{4}$ minute. There is one waiting line, and the service time for each teller is exponentially distributed with a mean of $\frac{1}{2}$ minute. Assume steady-state conditions hold.
(a) Construct the rate diagram for this queueing system.
(b) What is the expected number of customers in the bank?
(c) How long should an arriving customer expect to wait for service?

11.12 For the queueing system described in Exercise 11.8, suppose that, instead of a single waiting line for the calls put on hold, an arriving call is assigned with equal probability to one of the operators. Thus, there are two waiting lines, one for each operator.
(a) Should the expected time an arriving call spends in the system be larger or

smaller than in the single waiting line case? Explain without actually calculating any values.
(b) Calculate the expected time an arriving call spends in the system and compare this answer with the one obtained in Exercise 11.8 part (c).

11.13 Consider a steady-state queueing system with exponentially distributed interarrival times with a mean arrival rate of 10 per hour. There are three servers each with exponentially distributed service times with a mean of $\frac{1}{5}$ hour.
(a) Show that $L = \frac{26}{9}$.
(b) What is wrong with the following argument. Since $L = \frac{26}{9}$, we should on average expect to see less than three people in the queueing system. So on average we should expect to see no one waiting for service because there are three servers. Therefore, L_q must be zero.

11.14 Consider a steady-state queueing system with s servers each having exponentially distributed service times with mean $1/\mu$. Suppose interarrival times are exponentially distributed with a mean arrival rate of λ. Derive the formula

$$L_q = \frac{P_0(\lambda/\mu)^s\rho}{s!(1-\rho)^2} \quad \text{where } \rho = \lambda/\mu s$$

11.15 A barber shop has two barbers and, in addition to the two barber chairs, there are three chairs for customers to wait for a haircut. When five customers are in the shop and there are no chairs for arriving customers to sit in, all arriving customers go elsewhere for haircuts. Arrivals occur with exponentially distributed interarrival times at a mean arrival rate of 3 per hour, and each barber has an exponentially distributed service time with a mean of $\frac{1}{2}$ hour. Assume steady state.
(a) Construct the rate diagram for this queueing system.
(b) Calculate L, L_q, W, and W_q.

11.16 A plumber has been assigned the responsibility for repairing three washing machines in a laundry whenever they breakdown. Running time between breakdowns is exponentially distributed with a mean of 3 days. The time to repair a machine is exponentially distributed with a mean of 5 days. Assume steady-state conditions hold.
(a) Construct the rate diagram for this queueing system.
(b) What is the fraction of time that all machines are running?
(c) What is the expected number of machines that are running?
(d) What is the expected fraction of machines that are running?

11.17 A mechanic has been assigned to repair two cars when they break down. The running time between breakdowns is exponentially distributed with a mean of $\frac{1}{2}$ month. The repair times are exponentially distributed with a mean of $1/\mu$ months. For what values of μ will both cars be running at least half the time?

11.18 A parts center in a large factory has two servers each having an exponentially distributed service time with a mean of $\frac{1}{4}$ hour. Workers arrive to get parts with exponentially distributed interarrival times at a mean arrival rate of 5 per hour. The plant manager has determined that because of lost time spent at the parts center getting parts, the factory loses \$15 per hour per worker. Each server costs the factory \$10 per hour.
(a) Calculate the expected total cost per hour for this parts center.
(b) Suppose the two servers can be replaced with a single "super server" who has an exponential service time distribution with a mean of $\frac{1}{10}$ hour. How much per hour should management be willing to pay this super server?

11.19 Two repairmen, Ralph and Jack, are being considered for repairing two machines. Ralph is asking \$5 per hour when he is repairing a machine. Ralph's repair time distribution is exponentially distributed with a mean of $\frac{1}{6}$ hour. Jack wants \$6 per hour when he is repairing a machine, and his service time distribution is exponentially distributed with a mean of $\frac{1}{8}$ hour. Each machine has exponentially distributed running times between breakdowns with a mean of $\frac{1}{2}$ hour.
(a) If Ralph is hired, what is the fraction of time that he will be busy?
(b) If Jack is hired, what is the fraction of time that he will be busy?
(c) What is the expected service cost per hour for each of the hiring alternatives?
(d) If the cost of down time is \$2 per hour for each machine, who should be hired to minimize total expected cost per hour?

11.20 For the queueing system described in Exercise 11.18, the plant manager is considering adding a second parts center identical to the first one. The fixed cost of building the new center costs the company \$10 per hour when amortized over the next year. If the manager is interested in minimizing total expected cost per hour over the next year, does it make sense to add a second parts center? Explain.

11.21 Consider a single-server queueing model in the steady state. Suppose the interarrival times are exponentially distributed with mean arrival rate λ, and the service times are exponentially distributed with mean service rate μ where $\lambda/\mu < 1$. Use the Pollaczek–Khintchine equations to show that

$$L = \frac{\lambda}{\mu - \lambda}$$

as was derived in Section 11.5.

11.22 Consider the queueing problem described in Exercise 11.5. Suppose that the time required to inspect a truck is no longer exponentially distributed but is a fixed constant time of $\frac{1}{5}$ hour.
(a) Calculate L_q for this problem.
(b) How does the value for L_q compare with that obtained in Exercise 11.5?
(c) Show in general that as the variance σ^2 decreases, then the value of L_q decreases.

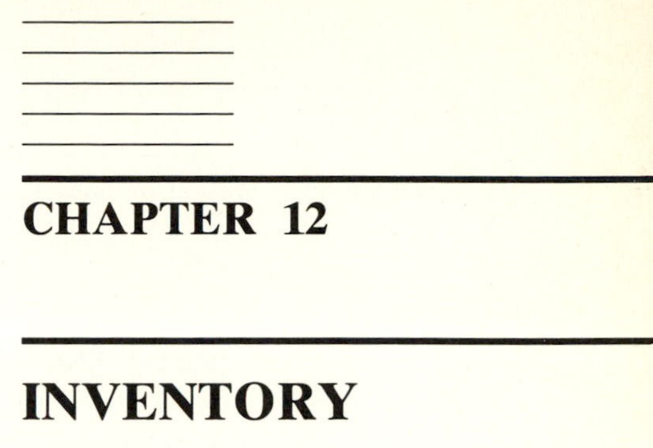

CHAPTER 12

INVENTORY MODELS

Maintaining inventories is a major cost for many organizations. For example, an automobile factory needs to stock several different kinds of parts used in assembling cars. Many of these parts may come from small independent businesses while others may come from production units within the company itself. For each of the many inventories of parts, the fundamental question is

> How much new stock should be ordered and
> when should the orders be placed?

If too many parts are ordered, the company pays excessive storage costs, but if not enough parts are available, the production of cars could be severely hampered and result in economic losses to the company.

In some situations the demand for items from an inventory is deterministic and thus very predictable over time. In other situations the demand is a random variable and probabilistic inventory models need to be developed. In this chapter we develop some deterministic and stochastic inventory models and show how solutions can be obtained by using a variety of techniques. In particular, we shall see that constrained and unconstrained optimization, dynamic programming, and integer programming all have important applications to inventory problems.

12.1 ECONOMIC ORDER-QUANTITY MODELS

Some of the most common inventory models are **economic order-quantity** (or **economic lot-size**) models. These models are applied in situations where the

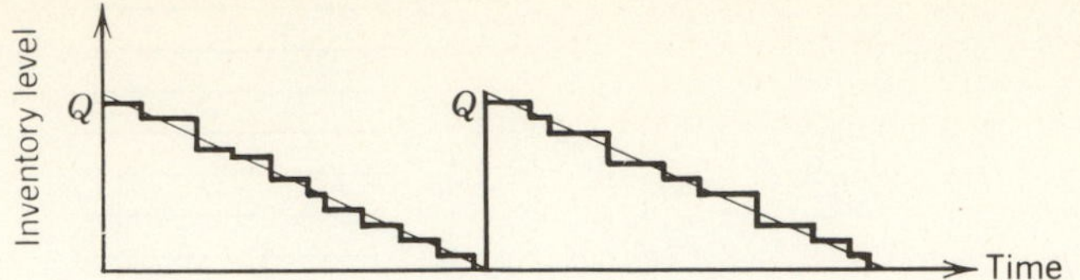

FIGURE 12.1 Inventory level over time approximated by a linear function.

inventory level is reduced over time, and each time the inventory level reaches zero the same quantity Q is ordered. We assume that the inventory level is continuously reduced at a constant rate even though in many applications this is just a close approximation to what actually happens. For example, the staircase graph in Figure 12.1 shows how an inventory level might be reduced over time in a stepwise manner with the height of each step corresponding to the number of items depleted that day. The inventory level can be approximated by the linear function superimposed over the staircase graph.

The basic assumptions for these models are as follows.

1. The inventory is continuously depleted at the constant rate of a items per unit time.
2. Each time the inventory level reaches zero, the same quantity Q is ordered, and there is no lag time for delivery.
3. The per-item ordering cost is c.
4. There is a fixed cost (or setup cost) of K each time the quantity Q is ordered.
5. The holding cost is h per item per unit time.

We wish to determine what quantity Q should be ordered to minimize total cost per unit time. Initially, we shall consider the case when no shortages are allowed; that is, when the inventory level is not allowed to fall below zero. Once the model for this case is developed, it is easy to see how it can be extended to handle the case when shortages are allowed and to the case when there are "price breaks" depending on the quantity Q ordered.

Economic Order-Quantity Models with No Shortages Allowed

Given the preceding basic assumptions 1 through 5, the inventory level over time is given by the graph in Figure 12.2. Note that the inventory level for any

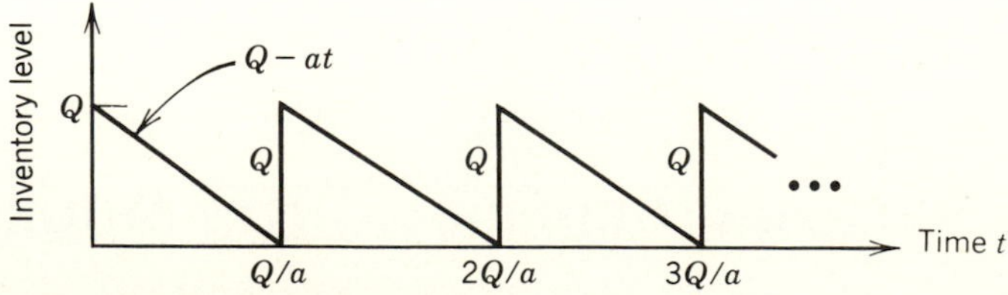

FIGURE 12.2 Inventory level over time.

time t in the time period $[0, Q/a]$ is given by the function $Q - at$ in view of our assumption that for each time period of length Q/a,

$$\begin{aligned}\text{total cost} &= \text{fixed cost} + \text{ordering cost} + \text{holding cost}\\ &= K + cQ + \text{holding cost}\end{aligned}$$

To determine the holding cost, we partition the time interval $[0, Q/a]$ into n subintervals each of length Δt. Then, as illustrated in Figure 12.3, we see that if t_i is some point in the ith subinterval,

$$h(Q - at_i)\,\Delta t = \text{approximate holding cost for the } i\text{th subinterval}$$

This means that the holding cost for the interval $[0, Q/a]$ can be approximated by the sum of holding costs for each of the n subintervals, so

$$\text{holding cost over } [0, Q/a] \approx \sum_{i=1}^{n} h(Q - at_i)\,\Delta t$$

As $\Delta t \to 0$ and the number n of subintervals approaches ∞, this sum more closely approximates the holding costs and we obtain

$$(\text{holding cost over } [0, Q/a]) = \lim_{\substack{\Delta t \to 0\\ n\to\infty}} \sum_{i=1}^{n} h(Q - at_i)\,\Delta t$$

From elementary calculus we recognize this limit as the definition of the definite integral of the function $h(Q - at)$ over the interval $[0, Q/a]$. Thus the holding cost is just h times the area under the curve $Q - at$ from 0 to Q/a, that is,

$$(\text{holding cost over } [0, Q/a]) = \int_0^{Q/a} h(Q - at)\,dt = \frac{hQ^2}{2a}$$

Therefore, for the time period $[0, Q/a]$

$$\text{total cost} = K + cQ + \frac{hQ^2}{2a}$$

To obtain the total cost per unit time, $T(Q)$, we divide the total cost by the length, Q/a, of the time period. This gives

$$T(Q) = \frac{K + cQ + (hQ^2/2a)}{Q/a} = aKQ^{-1} + \frac{hQ}{2} + ac$$

We minimize $T(Q)$ by finding where its derivative equals zero. Differentiating

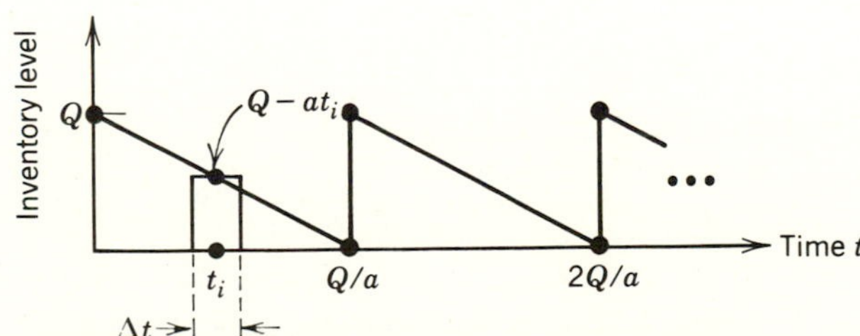

FIGURE 12.3 Approximating the holding cost over the ith subinterval of length Δt.

we obtain

$$\frac{dT(Q)}{dQ} = -aKQ^{-2} + \frac{h}{2}$$

It then follows that

$$\frac{dT(Q)}{dQ} = 0 \Longrightarrow Q^* = \sqrt{\frac{2aK}{h}}$$

is the optimal amount to order. Q^* is usually called the **economic** (or **optimal**) **order-quantity.** To see that Q^* is in fact a minimizing point we calculate

$$\frac{d^2T(Q)}{dQ^2} = 2aKQ^{-3} > 0 \quad \text{when} \quad Q > 0$$

Since the second derivative is positive, we know that $T(Q)$ is a strictly convex function and that Q^* is the unique minimizing point.

Economic Order-Quantity Models with Shortages Allowed

Now consider the same model satisfying assumptions 1 through 5 but suppose in addition that

> for each period a fixed amount of shortage is allowed and there is a penalty cost p per item of unsatisfied demand per unit time.

The graph in Figure 12.4 shows how the inventory level changes over time in this situation. Note that if R is the initial inventory level at time $t = 0$, then the amount of shortage allowed for each period is $S = Q - R$.

Note in particular from Figure 12.4 that when the inventory level reaches zero at time $t = R/a$ we do not order. Shortages are allowed to occur so that the fixed cost, K, of ordering can be spread out over a longer time period. Of course, there is a penalty cost incurred over the time period $[R/a, Q/a]$. In this case for the period of time $[0, Q/a]$ we see that

total cost = fixed cost + ordering cost + holding cost
+ penalty cost

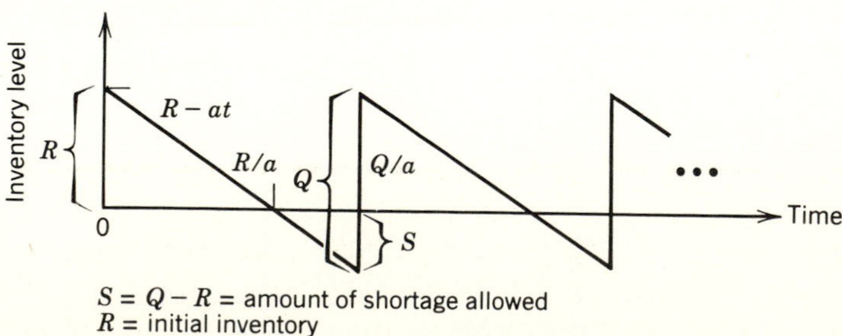

FIGURE 12.4 Inventory level over time with shortages allowed.

Using the same analysis as in the case with no shortages, we see that

$$\text{(holding cost for } [0, R/a]) = \int_0^{R/a} h(R - at)\, dt = \frac{hR^2}{2a}$$

and

$$\text{(penalty cost for } [R/a, Q/a]) = (-1)\int_{R/a}^{Q/a} p(R - at)\, dt = \frac{p(Q - R)^2}{2a}$$

Note that the last integral is multiplied by (-1) because from R/a to Q/a the integral of the function $p(R - at)$ is negative. Thus the total cost per unit time, calculated as a function of Q and R is given by

$$T(Q, R) = \frac{K + cQ + (hR^2/2a) + (p(Q - R)^2/2a)}{Q/a}$$

$$= aKQ^{-1} + ac + \frac{hR^2Q^{-1}}{2} + \frac{p(Q - R)^2Q^{-1}}{2}$$

To find values Q^* and R^* that minimize $T(Q, R)$, we solve the nonlinear program

$$\text{minimize } T(Q, R) \text{ subject to } Q \geq R$$

Note that $Q \geq R$ is a constraint that must be satisfied by Q^* and R^*. The unconstrained minimum of $T(Q, R)$ is found by setting the partial derivatives of $T(Q, R)$ equal to zero. With a little algebraic manipulation, we can solve the simultaneous equations

$$\frac{\partial T}{\partial Q} = 0 \quad \text{and} \quad \frac{\partial T}{\partial R} = 0$$

to obtain

$$Q^* = \sqrt{\left(\frac{2aK}{h}\right)\frac{(p + h)}{p}} \quad \text{and} \quad R^* = \sqrt{\left(\frac{2aK}{h}\right)\frac{p}{(p + h)}}$$

When $h > 0$, we have $p + h > p$ so $Q^* \geq R^*$. Thus the constraint $Q \geq R$ is automatically satisfied by the solution to the simultaneous equations. To see that Q^*, R^* is actually a minimizing point, we need to look more closely at the function $T(Q, R)$. Note that

$$T(Q, R) = aKQ^{-1} + ac + \frac{hR^2Q^{-1}}{2} + \frac{p(Q - R)^2Q^{-1}}{2}$$

$$= aKQ^{-1} + ac + \frac{hR^2Q^{-1}}{2} + \frac{pQ}{2} - pR + \frac{pR^2Q^{-1}}{2}$$

The functions aKQ^{-1}, $pQ/2$, and $-pR$ are convex functions. The sum of convex functions is convex (see Exercise 12.18) so if the function R^2Q^{-1} is a convex function of R and Q, then the function $T(Q, R)$ is a convex function.

Recall from Chapter 9 that a differentiable function is convex if and only if its Hessian matrix has nonnegative principal minors. The Hessian matrix, $H(R, Q)$, for the function $f(R, Q) = R^2Q^{-1}$ is the matrix of second partial derivatives, so

$$H(R, Q) = \begin{bmatrix} \dfrac{\partial^2 f}{\partial R^2} & \dfrac{\partial^2 f}{\partial R\,\partial Q} \\ \dfrac{\partial^2 f}{\partial Q\,\partial R} & \dfrac{\partial^2 f}{\partial Q^2} \end{bmatrix} = \begin{bmatrix} 2Q^{-1} & -2RQ^{-2} \\ -2RQ^{-2} & 2R^2Q^{-3} \end{bmatrix}$$

Recall that a principal minor is the determinant of a submatrix whose diagonal lies on the diagonal of the original matrix. So in this case we see that the 1×1 principal minors are

$$\det(2Q^{-1}) = 2Q^{-1} \quad \text{and} \quad \det(2R^2Q^{-3}) = 2R^2Q^{-3}$$

and the 2×2 principal minor is $\det H = 4R^2Q^{-4} - 4R^2Q^{-4} = 0$. Thus all of the principal minors are nonnegative provided $Q > 0$, and so the function $f(R, Q)$ is a convex function over the region where $Q > 0$. Then because $T(R, Q)$ is a sum of convex functions, it must be convex. This, of course, means that the point Q^*, R^* is a minimizing point.

Economic Order-Quantity Models with Price Breaks

In some cases the cost of ordering depends on the amount Q that is ordered, that is, there are price breaks or quantity discounts that will affect the optimal amount Q^* to be ordered. To illustrate the general situation, consider the case where there are three price break amounts, q_1, q_2, and q_3, so that the ordering cost, c, is given as follows:

$$\text{per-unit ordering cost } c = \begin{cases} c_1 & \text{if } 0 \le Q < q_1 \\ c_2 & \text{if } q_1 \le Q < q_2 \\ c_3 & \text{if } q_2 \le Q < q_3 \\ c_4 & \text{if } q_3 \le Q < \infty \end{cases}$$

with $c_1 > c_2 > c_3 > c_4$. If the ordering cost were c_i, independent of the amount ordered, then (assuming that shortages are not allowed) we would calculate the total cost per unit time as

$$T_i(Q) = aKQ^{-1} + \frac{hQ}{2} + ac_i \qquad i = 1, 2, 3, 4$$

These functions $T_i(Q)$ differ by only a constant, and for any value of Q, $T_1(Q) > T_2(Q) > T_3(Q) > T_4(Q)$. Also, as we saw earlier, each $T_i(Q)$ is a convex function and the point

$$Q^0 = \sqrt{\frac{2aK}{h}}$$

is the unconstrained minimizing point for each T_i. Note that Q^0 does not depend on c_i. Figure 12.5 gives typical graphs of the functions T_i when there are three price breaks, q_1, q_2, and q_3. The actual cost per unit time as a function of the amount Q ordered is given by the piecewise bold curve in Figure 12.5.

For the particular case considered in Figure 12.5, the value Q^0 satisfies $q_1 < Q^0 < q_2$ and so the minimizing point Q^* to the piecewise solid curve cannot lie to the left of Q^0 because each of the functions T_i is convex. The minimizing

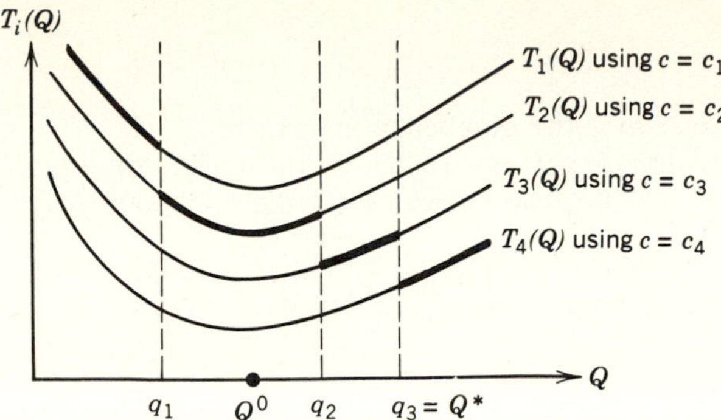

FIGURE 12.5 A typical graph of the cost per unit time with three price breaks.

point Q^* could lie to the right of Q^0, but if it does, it must occur at one of the price break points because the functions T_i are convex. Thus, to find Q^* we simply consider

$$\text{minimum } \{T_2(Q^0),\ T_3(q_2),\ T_4(q_3)\}$$

In the case illustrated in Figure 12.5, $T_4(q_3)$ is the minimum value of the piecewise solid curve so $Q^* = q_3$. Of course, the minimum depends on where the break points are. For example, q_2 can be made to be the optimal point simply by moving q_3 far enough to the right.

Economic Order-Quantity Models with Several Inventories

In many inventory situations, more than one type of item is being ordered, and there are constraints that restrict the amounts that can be ordered. For example, consider the case where items from the various inventories are kept in the same warehouse and are thus competing for the limited storage space. Suppose in particular that

> one item of inventory i uses s_i units of warehouse space and the total space available is B units

Suppose that Q_i is the amount of inventory i ordered and that the basic economic lot size assumptions hold for each of the inventories, namely:

Inventory i is continuously depleted at the constant rate of a_i items per unit time.

Each time the ith inventory level reaches zero, the same quantity Q_i is ordered.

The per-item ordering cost for items in inventory i is c_i.

There is a fixed cost (or setup cost) of K_i each time the quantity Q_i is ordered.

The holding cost is h_i per item per unit time for items in inventory i.

The ordering for each of the inventories typically takes place at different times, but to be sure that the space limitation is never violated, we require that the constraint

$$s_1Q_1 + s_2Q_2 + \cdots + s_mQ_m \le B$$

be satisfied, where m is the number of inventories.

In this model there are m variables and so to find the optimal order quantities, $Q_1^*, Q_2^*, \ldots, Q_m^*$, we need to solve the nonlinear program

$$\text{minimize} \sum_{i=1}^{m} \left(a_iK_iQ_i^{-1} + \frac{h_iQ_i}{2} + a_ic_i \right)$$

subject to

$$s_1Q_1 + s_2Q_2 + \cdots + s_mQ_m \le B$$
$$Q_i \ge 0 \qquad i = 1, 2, \ldots, m$$

This nonlinear program is actually quite easy to solve. To see why, note that ignoring the space constraint, the optimal order quantities are given by

$$Q_i^0 = \sqrt{\frac{2a_iK_i}{h_i}} \quad \text{for } i = 1, 2, \ldots, m$$

If these values satisfy the space constraint, they are the optimal values for the nonlinear program. If the space constraint is violated at the values Q_i^0, then we know that

$$s_1Q_1 + s_2Q_2 + \cdots + s_mQ_m = B$$

must hold at the optimal point. This means that the optimal point can be found by using the Lagrange method for equality-constrained nonliner programs. (The nonnegativity constraints on the Q_i will automatically be satisfied.) Recall that the Lagrange method was discussed in Chapter 9.

In this case the Lagrangian, $L(Q, u)$, is given by

$$L(Q, u) = \sum_{i=1}^{m} \left(a_iK_iQ_i^{-1} + \frac{h_iQ_i}{2} + a_ic_i \right) + u(s_1Q_1 + \cdots + s_mQ_m - B)$$

We then consider the system

$$\frac{\partial L}{\partial Q_i} = 0 \qquad i = 1, 2, \ldots, m \quad \text{and} \quad \frac{\partial L}{\partial u} = 0$$

that is,

$$-a_iK_iQ_i^{-2} + \frac{h_i}{2} + us_i = 0 \qquad i = 1, 2, \ldots, m$$

$$s_1Q_1 + \cdots + s_mQ_m = B$$

Solving the first set of equations for the Q_i gives

$$Q_i = \sqrt{\frac{2a_iK_i}{h_i + 2us_i}} \quad \text{for } i = 1, 2, \ldots, m$$

Note that these values of Q_i are identical to the Q_i^0 values when $u = 0$. This means that the space constraint is not satisfied at these values of Q_i so when

$u = 0$,

$$s_1 \sqrt{\frac{2a_1 K_1}{h_1 + 2us_1}} + \cdots + s_m \sqrt{\frac{2a_m K_m}{h_m + 2us_m}} > B$$

The left-hand side of the above inequality strictly decreases when u increases from 0. Thus there is a unique value $u^* > 0$ such that the left-hand side evaluated at u^* equals B. Given numerical values for the a_i, K_i, h_i, and s_i, it is fairly easy to numerically find a close approximation to this value of u^*.

Thus the optimal order quantities Q_i^* for this model are given by

$$Q_i^* = \sqrt{\frac{2a_i K_i}{h_i + 2u^* s_i}} \quad \text{for } i = 1, 2, \ldots, m$$

where u^* is chosen so that $s_1 Q_1^* + \cdots + s_m Q_m^* = B$.

12.2 DYNAMIC INVENTORY MODELS WITH PERIODIC REVIEW

In the economic order-quantity models of Section 12.1, the inventory level is continuously reduced at a constant rate independent of the time period, so the time periods are all of the same length. In addition, the same amount is ordered for each time period. In **periodic review models** orders are placed at the beginning of fixed time periods, but the demands are not independent of the time periods nor are the time periods necessarily of the same length. These models assume that the inventory is "periodically reviewed" at certain fixed times when varying orders for additional items can be placed.

The basic assumptions for the periodic review models are as follows:

There are N time periods indexed by $t = 1, 2, \ldots, N$, and

D_t = demand in period t that must be satisfied at the beginning of the time period

x_t = amount ordered at the beginning of period t

I_t = amount of inventory at the end of period t, and I_0 is the initial inventory (at the start of period 1)

$c_t(x_t)$ = cost of ordering x_t units in period t

$h_t(I_t)$ = holding cost charged on I_t units left in stock at the end of period t

D_t, x_t, and I_t are required to be integers

Initially, no special assumptions are made regarding the functions c_t and h_t. The problem is to determine values for the variables x_t so that the sum of ordering and holding costs is minimized and the demands in each period are satisfied.

Figure 12.6 contains a schematic of a four-period model, and the graph of the inventory level indicates an example ordering policy, namely,

$$x_1 = 6 \qquad x_2 = 2 \qquad x_3 = 3 \qquad x_4 = 5$$

Note that it is easy to find a feasible ordering policy because such a policy only needs to satisfy the demands at the beginning of each period. Thus, there

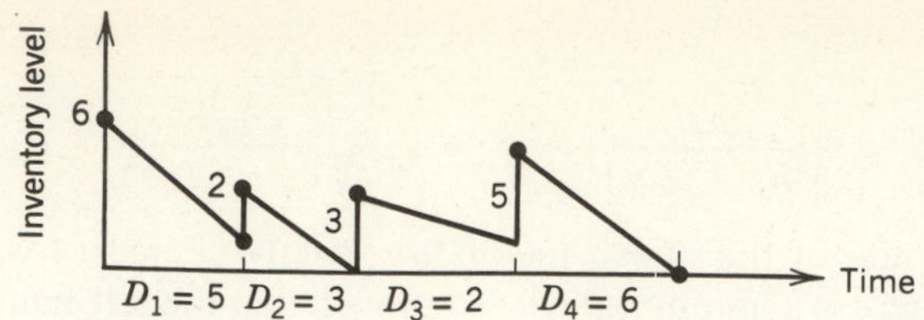

FIGURE 12.6 An example ordering policy for a periodic review model having four periods.

are many feasible policies. However, in order to find an optimal policy, the functions c_t and h_t need to be taken into account. Suppose, for example, that c_t and h_t are defined as follows:

For periods $t = 1$ and 2,

$$c_t(x_t) = \begin{cases} 4x_t & \text{if } 0 \leq x_t \leq 5 \\ 20 + 2(x_t - 5) & \text{if } 5 < x_t \end{cases}$$

$$h_t(I_t) = I_t,$$

and for periods $t = 3$ and 4,

$$c_t(x_t) = 30x_t \quad \text{if} \quad 0 \leq x_t$$

$$h_t(I_t) = 15I_t$$

With these cost functions, for example, the total of the ordering costs and holding costs for the policy, $x_1 = 6$, $x_2 = 2$, $x_3 = 3$, and $x_4 = 5$, given in Figure 12.6, is 286. This policy is surely not optimal because ordering in periods 3 and 4 is so expensive relative to periods 1 and 2, and the holding costs charged in periods 1 and 2 are not excessive. Thus, it is better to meet the demands in periods 3 and 4 with inventory ordered in periods 1 and 2. For example, the total cost of the policy $x_1 = 16$, $x_2 = 0$, $x_3 = 0$, and $x_4 = 0$ is only 151.

It is not difficult to formulate this problem as a nonlinear program. Because D_t, x_t, and I_t are required to be integer valued for each t, the general periodic review inventory problem can be formulated as the following nonlinear integer programming problem with linear constraints:

$$\text{minimize} \sum_{t=1}^{N} [c_t(x_t) + h_t(I_t)]$$

subject to

$$\left.\begin{aligned} I_t &= I_{t-1} + x_t - D_t \\ x_t &\geq 0 \quad \text{and integer} \\ I_t &\geq 0 \quad \text{and integer} \end{aligned}\right\} \quad \text{for } t = 1, 2, \ldots, N$$

Note that the first set of constraints simply requires that the inventory level at the end of period t equals the inventory level at the end of period $t - 1$ plus what is ordered at the beginning of period t minus the demand in period t. These constraints also show that the variables I_t could be eliminated and rewritten in terms of the variables x_t, as

$$I_t = I_0 + x_1 + \cdots + x_t - (D_1 + \cdots + D_t)$$

Solving a nonlinear integer program is not an easy matter. A branch-and-bound algorithm similar to the one studied in Chapter 8 could be used. However, obtaining a lower bound on the objective function over each new subset generated in the branch step would require more than simply solving a linear programming relaxation. One way to find a lower bound would be to solve a linearly constrained nonlinear program obtained by relaxing the integer requirements for each new subset. However, this approach requires considerable computational effort.

Another approach to solving this inventory problem is to use a dynamic programming formulation. Recall from Chapter 10 that dynamic programming can sometimes be a practical method for solving nonlinear integer programs. For this inventory problem the stages and states necessary for a dynamic programming formulation are easily identified as follows

> *at stage n,* decide how much should be ordered at the beginning of period n (that is, at stage n select x_n)
>
> *the state s* of the system at any stage n is the amount of inventory on hand at the beginning of stage (period) n prior to ordering

Using the notation introduced in Chapter 10, the dynamic programming formulation for this N period inventory model (and the preceding nonlinear program) is as follows. Consider the stages (or periods) 1, 2, . . . , N, and let

$f_n(s, x_n)$ = cost of the best inventory policy for periods $n, \ldots, N$ given s units of inventory on hand at the beginning of period n, and given that x_n additional units are ordered at the beginning of period n.

Let

$f_n(s)$ = cost of the best inventory policy for periods $n, \ldots, N$ given s units of inventory on hand at the beginning of period n

If I_0 is the initial inventory at the start of period 1, we wish to find $f_1(I_0)$, the cost of the best policy for periods 1 through N given I_0 units of initial inventory.

The backward recursive relations for this dynamic programming formulation are

$$f_n(s, x_n) = c_n(x_n) + h_n(s + x_n - D_n) + f_{n+1}(s + x_n - D_n)$$

We then calculate $f_n(s)$ by using

$$f_n(s) = \text{minimum } \{f_n(s, x_n) \mid x_n \text{ is a possible decision}\}$$

Unfortunately, the set of all possible decisions x_n can be so large that the dynamic programming formulation is computationally impractical. For example, in the four-period model corresponding to Figure 12.6 with $D_1 = 5$, $D_2 = 3$, $D_3 = 3$, and $D_4 = 5$,

> all possible x_1's at the beginning of period 1 are given by

$$x_1 = D_1 \quad \text{or} \quad D_1 + 1 \quad \text{or} \quad D_1 + 2 \quad \text{or} \cdots \text{or} \quad D_1 + D_2 + D_3 + D_4$$

that is,

$$x_1 = 5 \quad \text{or} \quad 6 \quad \text{or} \quad 7 \quad \text{or} \quad 8 \quad \text{or} \cdots \text{or} \quad 16$$

Of course, a large number of possible decisions at stage n means that there are a large number of states that need to be considered at stage $n + 1$. In the example the possible states s at the beginning of stage 2 are

$$s = 0 \quad \text{or} \quad 1 \quad \text{or} \quad 2 \quad \text{or} \cdots \text{or} \quad D_2 + D_3 + D_4$$

Calculating $f_2(s)$ for these various states would require using the following table:

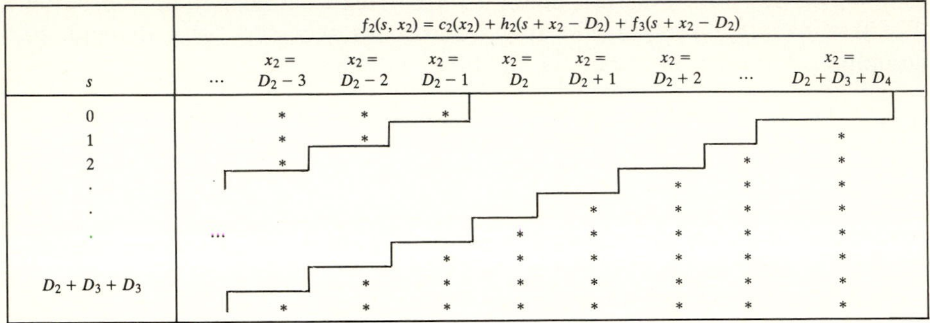

s	$f_2(s, x_2) = c_2(x_2) + h_2(s + x_2 - D_2) + f_3(s + x_2 - D_2)$								
	$\cdots$	$x_2 = D_2 - 3$	$x_2 = D_2 - 2$	$x_2 = D_2 - 1$	$x_2 = D_2$	$x_2 = D_2 + 1$	$x_2 = D_2 + 2$	$\cdots$	$x_2 = D_2 + D_3 + D_4$
0		*	*	*					
1		*	*						*
2		*						*	*
.							*	*	*
.						*	*	*	*
.	$\cdots$				*	*	*	*	*
				*	*	*	*	*	*
$D_2 + D_3 + D_3$			*	*	*	*	*	*	*
		*	*	*	*	*	*	*	*

For each row of this table there are $D_3 + D_4 + 1$ entries, $f_2(s, x_2)$, to be calculated, and there are $D_2 + D_3 + D_4 + 1$ rows in the table. The entries denoted by $*$ correspond to values of x_2 that need not be considered. Thus to calculate the values of $f_2(s)$, we need to calculate

$$(D_3 + D_4 + 1)(D_2 + D_3 + D_4 + 1) \quad \text{entries of the form } f_2(s, x_2)$$

For the small example in Figure 12.6, this only amounts to 108 entries, but imagine a larger problem with $D_2 = 3000$, $D_3 = 2000$, and $D_4 = 6000$. We would then need to calculate over 88 million entries $f_2(s, x_2)$ just to determine $f_2(s)$ for each s.

Fortunately, by looking more closely at the structure of the cost functions c_t and h_t, we can sometimes eliminate many of the possible decisions x_n that need to be considered at each stage n. The computational savings can often be dramatic. For example, instead of calculating more than 88 million values $f_2(s, x_2)$ to determine $f_2(s)$ as in the preceding table when $D_2 = 3000$, $D_3 = 2000$, and $D_4 = 6000$, if the functions c_t and h_t are concave functions, then only *three* values, $f_2(s, x_2)$, need be calculated to determine $f_2(s)$.

Concave Cost Functions

When c_t and h_t are concave functions, the total cost g defined by

$$g(\mathbf{x}) = \sum_{t=1}^{N} [c_t(x_t) + h_t(I_t)]$$

is a concave function of the ordering policy $\mathbf{x} = [x_1, x_2, \ldots, x_N]^T$. To see why g is concave, use the fact that the sum of concave functions is a concave function (see Exercise 12.18). Also, if h_t is a concave function, then using the fact that

$I_t(\mathbf{x})$ is a linear function of the variables x_t, it is not difficult to show that $h_t(I_t(\mathbf{x}))$ is a concave function of $\mathbf{x}$ (see Exercise 12.19).

We will now show that

> when the functions c_t and h_t are concave,
> an optimal policy has the property that
>
> $$I_{t-1}x_t = 0 \quad \text{for all } t \geq 1$$

This property implies that

> when the ordering and holding costs are concave,
> an optimal policy has the property that ordering
> occurs only when the inventory level is zero

This means that in the dynamic programming formulation, the only values of x_n that need to be considered as possible decisions at stage n are

$$x_n = D_n \quad \text{or} \quad D_n + D_{n+1} \quad \text{or} \cdots \text{or} \quad D_n + D_{n+1} + \cdots + D_N$$

Thus, in the dynamic programming formulation for the preceding four-period example, the only values of x_1 that need be considered for an optimal policy are

$$x_1 = 5 \quad \text{or} \quad 8 \quad \text{or} \quad 10 \quad \text{or} \quad 16$$

This drastically reduces the calculations needed for a dynamic programming solution.

To see why $I_{t-1}x_t = 0$ for any optimal policy when the ordering costs and holding costs are concave, consider a policy where this does not hold. Figure 12.7 contains a graphical representation of such a policy. Consider the first instance where $I_{t-1}x_t \neq 0$ in such an ordering policy. In Figure 12.7, this occurs when $I_2x_3 \neq 0$. Let the ordering policy illustrated in Figure 12.7 be denoted by

$$\mathbf{x} = [x_1, x_2, x_3, x_4, \ldots, x_N]^T$$

Given the policy $\mathbf{x}$ with $I_2x_3 \neq 0$, consider the two policies $\mathbf{x}^+$ and $\mathbf{x}^*$ defined from this policy by

$$\mathbf{x}^+ = [x_1, x_2 - I_2, x_3 + I_2, x_4, \ldots, x_N]^T$$

$$\mathbf{x}^* = [x_1, x_2 + x_3, 0, x_4, \ldots, x_N]^T$$

Note that $\mathbf{x}^+$ and $\mathbf{x}^*$ differ from $\mathbf{x}$ only in the second and third components. In $\mathbf{x}^+$ we simply subtracted I_2 from the amount ordered in period 2 and added I_2 to the amount ordered in period 3. In $\mathbf{x}^*$ nothing is ordered in period 3, and the amount ordered in period 2 is the total amount ordered in periods 2 and 3

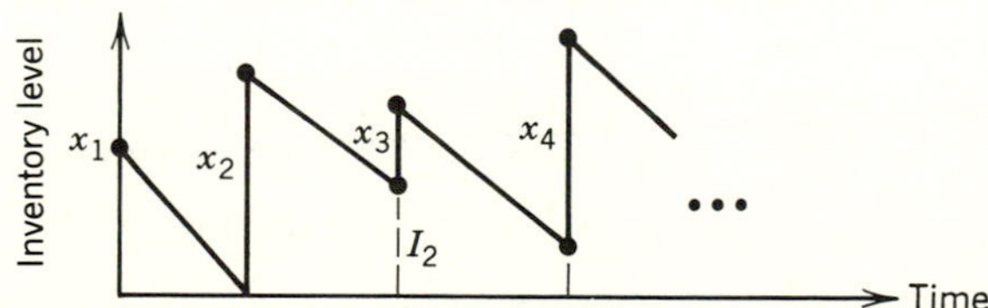

FIGURE 12.7 An ordering policy where the property $I_{t-1}x_t = 0$ for all t does *not* hold.

under the policy $\mathbf{x}$. Both of these new policies, $\mathbf{x}^+$ and $\mathbf{x}^*$, have the property that

> the inventory level at the end of period 2 times the amount ordered at the beginning of period 3 equals zero

We shall now show that either $\mathbf{x}^+$ or $\mathbf{x}^*$ will have a smaller total cost than the original policy $\mathbf{x}$ and so $\mathbf{x}$ could not be optimal.

Comparing the components of $\mathbf{x}$, $\mathbf{x}^+$, and $\mathbf{x}^*$, it is not hard to see that

$$\mathbf{x} = \lambda\mathbf{x}^+ + (1 - \lambda)\mathbf{x}^* \quad \text{where} \quad \lambda = \frac{x_3}{x_3 + I_2}$$

Here $x_3 + I_2 \neq 0$ because $x_3 I_2 \neq 0$ and $x_3 \geq 0$ and $I_2 \geq 0$. This means that $\mathbf{x}$ is a convex combination of $\mathbf{x}^+$ and $\mathbf{x}^*$, so $\mathbf{x}$ is on the line segment $[\mathbf{x}^+, \mathbf{x}^*]$ between $\mathbf{x}^+$ and $\mathbf{x}^*$.

Now consider minimizing the nonlinear concave objective function g over the line segment $[\mathbf{x}^+, \mathbf{x}^*]$.

> Because g is concave, the minimizing point over the line segment $[\mathbf{x}^+, \mathbf{x}^*]$ must be one of the endpoints $\mathbf{x}^+$ or $\mathbf{x}^*$.

This means that either $\mathbf{x}^+$ or $\mathbf{x}^*$ is an ordering policy with a total cost lower than that of the original policy $\mathbf{x}$. Thus $\mathbf{x}$ could not be an optimal policy. If there are any other periods t for which $I_{t-1}x_t \neq 0$, this procedure can be repeated to obtain a new policy with a lower cost. Therefore, no optimal policy can have a period t with $I_{t-1}x_t \neq 0$, provided the ordering costs and the holding costs are concave functions.

We will now show how the computations required for the dynamic programming formulation simplify using the fact that ordering occurs only when the inventory level is zero. The general approach will be illustrated by a simple example.

An Example with Concave Holding and Ordering Costs

Consider a five-period model with demands given by

$$D_1 = 7 \qquad D_2 = 5 \qquad D_3 = 3 \qquad D_4 = 8 \qquad D_5 = 2$$

Suppose that a fixed cost, $K = 10$, is charged each time an order is placed and that items cost 3 per unit for the first 10 units ordered and cost 2 per unit for each unit ordered in excess of 10.

Thus, the ordering cost c_t for each period t is given by

$$c_t(x_t) = \begin{cases} 0 & \text{if } x_t = 0 \\ 10 + 3x_t & \text{if } 0 < x_t \leq 10 \\ 10 + 30 + 2(x_t - 10) = 20 + 2x_t & \text{if } 10 < x_t \end{cases}$$

From the graph of c_t in Figure 12.8, it is clear that c_t is a concave function. Suppose in addition that the holding costs are linear with

$$h_t(I_t) = I_t \quad \text{for each period } t$$

and that the initial inventory is zero. Thus, here c_t and h_t are concave functions and so the total cost is a concave function of the ordering policy $\mathbf{x}$.

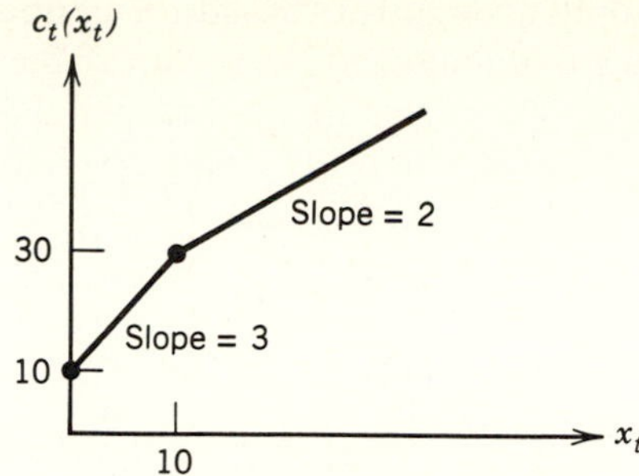

FIGURE 12.8 Graph of a concave ordering cost function c_t.

Recall from our dynamic programming formulation that

$f_n(s, x_n)$ = cost of the best inventory policy for periods $n, \ldots, N$ given s units of inventory on hand at the beginning of period n, and given that x_n additional units are ordered at the beginning of period n

and

$f_n(s)$ = cost of the best policy for periods $n, \ldots, N$ given s units of inventory on hand at the beginning of period n.

The initial inventory $I_0 = 0$, so we wish to find $f_1(0)$, the cost of the best policy for periods 1 through $N = 5$ when the initial inventory level is zero. Of course, in the process of calculating this optimal value, the optimal policy will also be found.

The recursive relations for this problem are

$$f_n(s, x_n) = c_n(x_n) + h_n(I_n) + f_{n+1}(I_n)$$

where $s = I_{n-1}$, and

$$f_n(s) = \text{minimum } \{f_n(s, x_n) \mid x_n \text{ is a possible decision}\}$$

Using the fact that $I_{n-1}x_n = 0$ in an optimal policy,

the only possible state s for any period n
when $x_n \neq 0$, is the state $s = 0$

This follows because if $s \neq 0$ at some stage n, we must order $x_n = 0$. Of course, if $s = 0$, then we must order at least $x_n = D_n$ in order to satisfy the demand for period n.

Thus, the values $f_n(0)$ for $n = 5, 4, 3, 2, 1$ are calculated recursively as follows. At the start of period 5 with $s = 0$, we must order $D_5 = 2$ units and so

$$f_5(0) = 10 + 3\cdot 2 = 16$$

At the start of period 4, x_4 can be either $D_4 + D_5 = 10$ or $D_4 = 8$. If $x_4 = 10$, the ordering cost is $10 + 3\cdot 10$, and there is a holding cost of 1 on each of the 2 units left over at the end of period 4 yielding a total ordering and holding cost of 42. Note that when an order covers the demand of several periods, the holding costs need to be calculated for inventory left in stock at the end of the periods.

If $x_4 = 8$, so that only enough is ordered for period 4, then the total cost is 50. These calculations are summarized along with all the remaining calculations required for the solution. The terms in the brackets $[\cdot]$ are the holding costs incurred at the ends of periods where the inventory level is positive:

$$f_4(0) = \min \begin{Bmatrix} 10 + 3\cdot 10 + [2] & = 42 & x_4 = 10 \\ 10 + 3\cdot 8 \ + f_5(0) & = 50 & x_4 = 8 \end{Bmatrix} = 42$$

$$f_3(0) = \min \begin{Bmatrix} 20 + 2\cdot 13 + [12] & & = 58 & x_3 = 13 \\ 20 + 2\cdot 11 + [8] & + f_5(0) & = 66 & x_3 = 11 \\ 10 + 3\cdot 3 & + f_4(0) & = 61 & x_3 = 3 \end{Bmatrix} = 58$$

$$f_2(0) = \min \begin{Bmatrix} 20 + 2\cdot 18 + [25] & & = 81 & x_2 = 18 \\ 20 + 2\cdot 16 + [19] & + f_5(0) & = 87 & x_2 = 16 \\ 10 + 3\cdot 8 \ + [3] & + f_4(0) & = 79 & x_2 = 8 \\ 10 + 3\cdot 5 & + f_3(0) & = 83 & x_2 = 5 \end{Bmatrix} = 79$$

$$f_1(0) = \min \begin{Bmatrix} 20 + 2\cdot 25 + [43] & & = 113 & x_1 = 25 \\ 20 + 2\cdot 23 + [35] & + f_5(0) & = 117 & x_1 = 23 \\ 20 + 2\cdot 15 + [11] & + f_4(0) & = 103 & x_1 = 15 \\ 20 + 2\cdot 12 + [5] & + f_3(0) & = 107 & x_1 = 12 \\ 10 + 3\cdot 7 & + f_2(0) & = 110 & x_1 = 3 \end{Bmatrix} = 103$$

Thus, the minimum cost for the optimal policy is $f_1(0) = 103$, and the optimal ordering policy is

$$x_1 = 15 \qquad x_2 = 0 \qquad x_3 = 0 \qquad x_4 = 10 \qquad x_5 = 0$$

A graphical representation of this optimal policy is given in Figure 12.9. From this example, it is clear how the dynamic programming approach works in general. The details are summarized as follows:

Summary of the Dynamic Programming Approach

Calculate $f_N(0) = c_N(D_N)$.

Given $f_N(0), f_{N-1}(0), \ldots, f_{n+1}(0)$, calculate $f_n(0)$ using the fact that at stage n, the only decisions x_n that need to be considered are

$$x_n = D_n, D_n + D_{n+1}, \ldots, D_n + \cdots + D_N$$

and the only state that needs to be considered is $s = 0$.

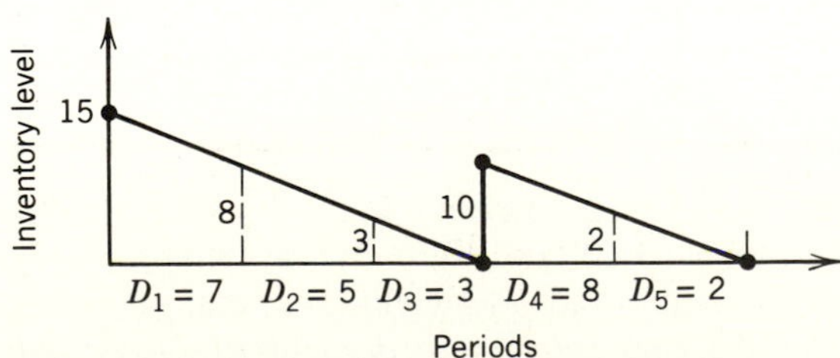

FIGURE 12.9 A graphical representation of the optimal policy $x_1 = 15, x_2 = 0, x_3 = 0, x_4 = 10$, and $x_5 = 0$.

For each possible $x_n = D_n + D_{n+1} + \cdots + D_{n+k}$, $k \le N - n$, recursively calculate each of the values

$$f_n(0, x_n) = c_n(x_n) + h_n(I_n) + h_{n+1}(I_{n+1}) + \cdots + h_{n+k-1}(I_{n+k-1}) + f_{n+k+1}(0)$$

Then,

$$f_n(0) = \min\{f_n(0, x_n) \mid x_n = D_n, D_n + D_{n+1}, \ldots, D_n + \cdots + D_N\}$$

Thus, when the ordering and holding costs are concave functions, the computations required for a dynamic programming solution greatly simplify.

Convex Cost Functions

In some periodic review models the ordering costs c_t and the holding costs h_t are convex functions, and the preceding simplified dynamic programming approach will not necessarily yield an optimal policy. For example, situations where inventory can be ordered from several sources with different per-unit costs give rise to convex ordering costs. To illustrate, suppose that for each period t, orders can be placed with source 1 that has five units available at a cost of one per unit or with source 2 that has an unlimited supply available at a cost of three per unit. Thus, if five or less units are ordered at the start of a period, they would be ordered solely from source 1. If more than five are ordered, the first five would come from source 1 and the remainder would come from source 2. This means that the ordering cost, c_t, for each period t is given by

$$c_t(x_t) = \begin{cases} x_t & \text{if } 0 \le x_t \le 5 \\ 5 + 3(x_t - 5) & \text{if } 5 < x_t \end{cases}$$

This function c_t is a convex function as can be seen from its graph in Figure 12.10. Note that if there were a fixed cost K of ordering, the function c_t would not be convex.

Suppose, in addition, that the holding cost charged on inventory, I_t, in stock at the end of each period is given by the linear function

$$h_t(I_t) = I_t$$

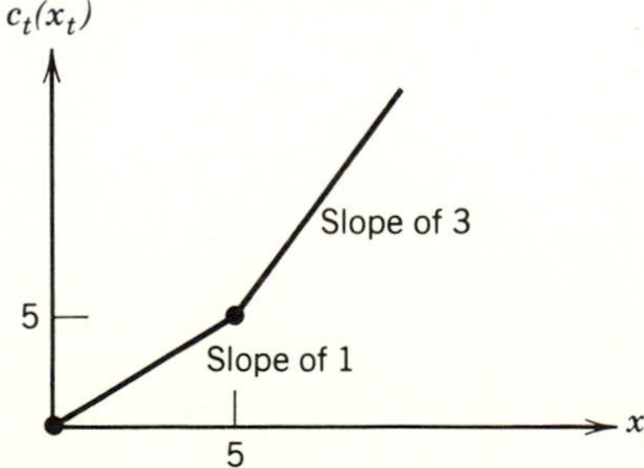

FIGURE 12.10 Graph of a convex ordering cost function c_t resulting from two supply sources.

In this case the total cost is a convex function of an ordering policy $\mathbf{x}$. Thus, it is not necessarily true that $I_{t-1}x_t = 0$ for each period t as is the case when the total cost is a concave function. Therefore, an optimal policy need not have the property that the amount, x_n, ordered in period n must be one of the values $x_n = 0, D_n, D_n + D_{n+1}, \ldots, D_n + D_{n+1} + \cdots + D_N$.

To show why this is true, consider a four-period example with demands

$$D_1 = 5 \qquad D_2 = 3 \qquad D_3 = 2 \qquad D_4 = 6$$

and with the preceding convex ordering and holding costs. Using the *incorrect* assumption that $I_{n-1}x_n = 0$ for each period n, the simplified dynamic programming approach for concave costs yields the following policy:

$$x_1 = 5 \qquad x_2 = 3 \qquad x_3 = 2 \qquad x_4 = 6$$

To see that this policy is not optimal, note that for the 6 units ordered in period 4, the first 5 cost 1 per unit, but a single additional unit (the sixth unit) increases the total cost by 3. The holding cost is only 1 per unit, so it is clearly better to order that 1 extra unit in period 3 (at a cost of 1) and store it (at a cost of 1) for use in period 4. Satisfying the sixth unit of period 4 demand in this manner incrementally increases the total cost by 2 instead of by 3. With this adjustment, we obtain the following policy:

$$x_1^* = 5 \qquad x_2^* = 3 \qquad x_3^* = 3 \qquad x_4^* = 5$$

This is, in fact, the correct optimal policy in this example as we shall now show. This policy is indicated in Figure 12.11.

In this example, it was better to satisfy the sixth unit of the demand in period 4 with one additional unit ordered in period 3. It is not hard to see that if the sixth unit of demand in period 4 were to be satisfied by ordering in periods 1 or 2, then the total cost would have increased by more than 2. Of course, that additional unit of demand in period 4 had to be satisfied by ordering in periods 1, 2, or 3.

The idea of checking on the best way to satisfy a particular unit of demand suggests the following simple algorithm for finding a feasible policy that is optimal when the ordering and holding costs are convex functions.

A Simple Algorithm When Cost Functions Are Convex

Step 0. Order $x_1 = D_1$ to satisfy the demand in period 1.

Step 1. Let k be the earliest period having at least one unit of unsatisfied demand. Consider satisfying one unit of period k's unsatisfied demand by ordering one extra unit in period j for $j = 1, 2, \ldots, k$.

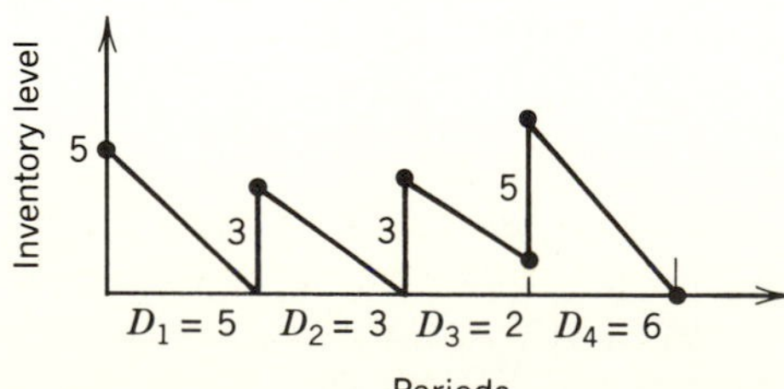

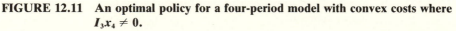

FIGURE 12.11 An optimal policy for a four-period model with convex costs where $I_3x_4 \neq 0$.

Step 2. For each of the k alternatives in step 1, calculate the resulting incremental ordering and holding costs. Select the alternative with minimum incremental cost in as late a period as possible. Revise the current policy by ordering according to the selected alternative.

Step 3. If any units of unsatisfied demand remain, go to step 1. Otherwise, the current policy meets all the demands and is optimal.

The fact that this algorithm yields an optimal policy is not obvious and depends heavily on the convexity of the ordering and holding costs. After illustrating the algorithm on the example, we shall show why convexity of the cost functions plays such an important role in the algorithm.

To illustrate the algorithm, we use the four-period example with convex ordering costs and linear holding costs. Recall that $D_1 = 5$, $D_2 = 3$, $D_3 = 2$, $D_4 = 6$, that the holding cost is 1 per unit left at the end of the period, and that the ordering cost for each period is 1 per unit for the first five units but 3 per unit for each unit greater than five.

In step 0, we initially set $x_1 = 5$ and $x_2 = x_3 = x_4 = 0$ as the current (infeasible) policy.

In step 1, we consider meeting one unit of period 2's unsatisfied demand by ordering one more unit in period 1 (and storing it for use in period 2) or by ordering one more unit in period 2.

In step 2, we calculate the incremental costs of these two alternatives to be 4 (if the unit is produced in period 1) or 1 (if the unit is produced in period 2). Thus, we revise the current policy to be

$$x_1 = 5 \qquad x_2 = 1 \qquad x_3 = 0 \qquad x_4 = 0$$

In step 3, we return to step 1 with this revised policy and continue.

Because of the particular ordering and holding costs used in this example, it is clear that this algorithm successively meets the demands in periods 1, 2, and 3 by ordering only what is needed in those periods so that no holding costs are incurred. The same is true for the first five units of demand in period 4. Continuing the algorithm, we reach the stage where the current (infeasible) policy is

$$x_1 = 5 \qquad x_2 = 3 \qquad x_3 = 2 \qquad x_4 = 5$$

Upon returning to step 1, we then consider meeting the final unit of period 4's demand by ordering one unit in periods 1, 2, 3, or 4. The incremental costs for these four alternatives are 6, 3, 2, and 3, respectively. Thus, the minimum incremental cost results from ordering that one extra unit in period 3, which yields the optimal policy $\mathbf{x}^*$ given in Figure 12.11.

It is important to realize that this algorithm is valid for arbitrary convex ordering and holding costs. In terms of computations this algorithm requires more work than the simplified dynamic programming approach for concave costs. However, it requires far less computation than using a dynamic programming formulation where no special assumptions hold regarding the cost functions.

To see why convexity plays such an important role in assuring that the algorithm yields an optimal policy when the cost functions are convex, consider a simple three-period problem with demands $D_i = 1$ for each period

$i = 1, 2, 3$. Suppose the ordering costs are as follows:

$$c_1(x_1) = \begin{cases} 1 & \text{if } x_1 = 1 \\ 8 & \text{if } x_1 = 2 \\ 9 & \text{if } x_1 = 3 \end{cases} \qquad c_2(x_2) = 6x_2 \qquad c_3(x_3) = 10x_3$$

and that $h_t(I_t) = \frac{1}{5}I_t$ for $t = 1, 2, 3$. Note that, as indicated in Figure 12.12, c_1 is not a convex function, but c_2 and c_3 are convex because they are linear. The holding cost is also linear (convex), but the total cost function is not convex because $c_1(x_1)$ is not convex.

Applying the three-step algorithm, we start in step 0 by satisfying period 1's demand by ordering one unit in period 1 and set

$$x_1 = 1 \qquad x_2 = 0 \qquad x_3 = 0$$

In step 1 we consider satisfying one unit of period 2's demand by ordering one more unit in period 1 at a total incremental cost equal to 7.2 or ordering one more unit in period 2 at a total incremental cost of 6. The latter has the minimum incremental cost, so we revise the current policy to

$$x_1 = 1 \qquad x_2 = 1 \qquad x_3 = 0$$

Returning to step 1, we consider three alternatives for satisfying one unit of period 3 demand, namely,

	Incremental Cost
Order one more unit in period 1 and store for use in period 3	$7\frac{2}{5}$
Order one more unit in period 2 and store for use in period 3	$6\frac{1}{5}$
Order one more unit in period 3	10

The second alternative has minimum incremental cost, so we revise the current policy to become

$$x_1 = 1 \qquad x_2 = 2 \qquad x_3 = 0 \quad \text{with total cost } = 13\tfrac{1}{5}$$

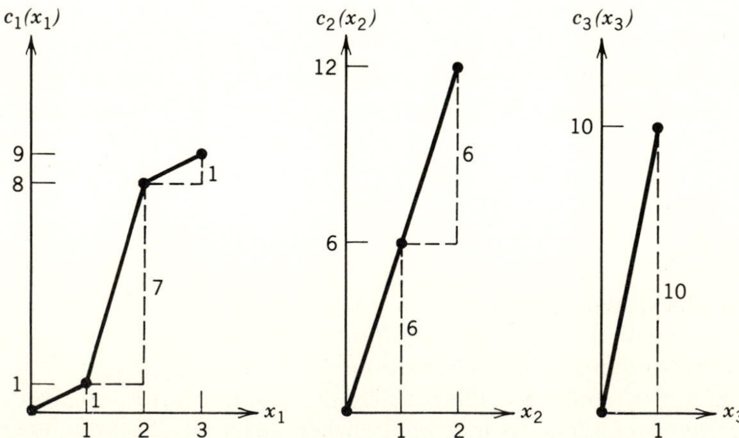

FIGURE 12.12 Nonconvex ordering cost function c_1.

However, this policy is clearly not optimal, and the optimal policy is easily seen to be

$$x_1^* = 3 \qquad x_2^* = 0 \qquad x_3^* = 0 \quad \text{with total cost} = 9\tfrac{3}{5}$$

The reason that $\mathbf{x}^*$ was never encountered in the algorithm is because each extra unit considered in step 1 was cheaper to order in period 2 than in period 1. We were never able to "get over the hump" of the cost curve $c_1(x_1)$ to take advantage of the extremely low marginal cost for the third unit ordered in period 1. Of course, if $c_1(x_1)$ had been a convex function, the marginal cost of that third unit would have been at least as large as 7 (the marginal cost for the second unit ordered in period 1). This is the point of the example:

> When the marginal costs of each successive unit ordered are nondecreasing, as is the case for convex functions, then the policy obtained by the algorithm is guaranteed to be optimal.

12.3 STOCHASTIC INVENTORY MODELS

In the inventory models studied so far, the demand for items was assumed to be known (or reasonably well estimated) in advance. In this section we consider the case where the demand D is a random variable that can take on nonnegative real values. Analysis of inventory problems with random demands is more complicated than the models considered earlier. For this reason we start by developing a model for situations where inventory must be made available at the beginning of a single period of time during which there is a random demand for items.

For example, consider a roadside stand that sells fresh cider only during the month of October. At the beginning of October, the owner can buy cider for \$3.00 per gallon. The stand sells cider for \$6.50 per gallon, and cider left over at the end of October can be sold to a local pig farmer for \$1.00 per gallon. Based on a long history of selling fresh cider each October, the owner knows that the demand is uniformly distributed between 100 and 200 gallons. If the stand runs out of cider during October, the manager has a policy of giving, at no charge, a small sack of apples to each customer who requests a gallon of fresh cider but cannot be accommodated. Each sack of apples costs the owner 50 cents. The problem is to determine how many gallons of cider to make available at the beginning of October in order to maximize expected profit.

We shall return to a solution of this problem after developing a single-period model for problems of this type.

A Single-Period Model

The basic assumptions for a single-period model are as follows:

At the beginning of the period, a single order is placed, bringing the inventory level up to an amount y. This means that if the initial inventory level is x, then $y - x$ is the amount ordered.

D = nonnegative continuous random variable for demand with density function f and distribution function F
p = selling price per unit of inventory
s = shortage price per unit of demand not satisfied
c = ordering cost per unit of inventory
h = holding cost per unit of inventory left at the end of the period

Initially, we assume the starting inventory level, x, is zero so the amount, y, made available is the amount ordered. We also assume there is no fixed cost of placing an order. Both of these assumptions will be eliminated after developing the basic single-period model.

Suppose that

P = profit realized by making y units available at the beginning of the period

The profit, P, depends on the random demand D:

$$P(D) = \begin{cases} pD - cy - h(y - D) & \text{when } y > D \\ py - cy - s(D - y) & \text{when } y \le D \end{cases}$$

Note that P is a continuous function of D because when $D = y$ the top and bottom expressions in $P(D)$ are identical.

The following basic result from probability will allow us to calculate the expected profit:

> If X is a random variable with density function f and if g is a continuous function of X, then
>
> $$E[g(X)] = \int_{-\infty}^{+\infty} g(z)f(z)\,dz$$
>
> where $E[g(X)]$ is the expected value of $g(X)$.

Using this result and the definition of $P(D)$, we obtain

$$\begin{aligned} E[P(D)] &= \int_0^y [pz - cy - h(y - z)]f(z)\,dz \\ &\quad + \int_y^\infty [py - cy - s(z - y)]f(z)\,dz \\ &= -\int_0^y h(y - z)f(z)\,dz - \int_y^\infty s(z - y)f(z)\,dz \\ &\quad - \int_0^\infty cyf(z)\,dz + \int_0^y pzf(z)\,dz + \int_y^\infty pyf(z)\,dz \end{aligned}$$

From the definition of a density function for a nonnegative random variable we know that

$$\int_0^\infty f(z)\,dz = 1$$

and so

$$-\int_0^\infty cyf(z)\,dz = -cy\int_0^\infty f(z)\,dz = -cy$$

Also, note that

$$\int_0^y pzf(z)\,dz = \int_0^\infty pzf(z)\,dz - \int_y^\infty pzf(z)\,dz$$

From the definition of the expected value of D, we know that

$$\int_0^\infty pzf(z)\,dz = p\int_0^\infty zf(z)\,dz = pE[D]$$

and so

$$\int_0^y pzf(z)\,dz = pE[D] - \int_y^\infty pzf(z)\,dz$$

Using all these observations, we obtain the following expression for the expected profit as a function of D, namely,

$$E[P(D)] = pE[D] - [cy + L(y)]$$

where

$$L(y) = \underbrace{\int_y^\infty p(z-y)f(z)\,dz}_{\substack{\text{expected loss}\\ \text{of revenue}}} + \underbrace{\int_y^\infty s(z-y)f(z)\,dz}_{\substack{\text{expected shortage}\\ \text{cost}}} + \underbrace{\int_0^y h(y-z)f(z)\,dz}_{\substack{\text{expected holding}\\ \text{cost}}}$$

It is important to keep in mind that the three integrals used in the expression for $L(y)$ are

the expected loss of revenue when the demand D is greater than the amount y made available,

the expected shortage cost when the demand D is greater than the amount y made available, and

the expected holding cost when the demand D is less than the amount y made available.

Because the first two integrals in $L(y)$ have the same limits of integration, it will be convenient to combine them together into a single integral as

$$\int_y^\infty p(z-y)f(z)\,dz + \int_y^\infty s(z-y)f(z)\,dz = \int_y^\infty r(z-y)f(z)\,dz$$

where $r = p + s$. Therefore,

$$cy + L(y) = cy + \int_y^\infty r(z-y)f(z)\,dz + \int_0^y h(y-z)f(z)\,dz$$

Finally, to maximize expected profit $E[P(D)]$, where

$$E[P(D)] = pE[D] - [cy + L(y)]$$

it suffices to minimize $cy + L(y)$ over all $y \geq 0$. To minimize this function of a single variable, y, the first thing to try is to set its derivative to zero. Using the Leibnitz rule from calculus, it is not hard to show (see Exercise 12.20) that

$$\frac{d}{dy} \int_0^y h(y - z)f(z)\, dz = h \int_0^y f(z)\, dz$$

The derivative of the other integral used in $L(y)$ is not quite as easy to calculate because one of its limits of integration is infinity (so the Leibnitz rule does not immediately apply). However, with a little more analysis one can show (see Exercise 12.20) that

$$\frac{d}{dy} \int_y^\infty r(z - y)f(z)\, dz = -r \int_y^\infty f(z)\, dz$$

Thus,

$$\begin{aligned} \frac{d}{dy} [cy + L(y)] &= c + h \int_0^y f(z)\, dz - r \int_y^\infty f(z)\, dz \\ &= c + hF(y) - r[1 - F(y)] \end{aligned}$$

where the last equality follows using the definition of the distribution function F, namely:

> Given a nonnegative random variable D with density function f, the distribution function F is defined by
>
> $$F(y) = \text{Prob}(D \leq y) = \int_0^y f(z)\, dz$$

The distribution function F is a nondecreasing function that approaches 1 as y approaches $+\infty$. Figure 12.13 contains a graph of a typical distribution function for a nonnegative continuous random variable. Setting

$$\frac{d}{dy} [cy + L(y)] = 0$$

gives

$$hF(y) - r + rF(y) + c = 0$$

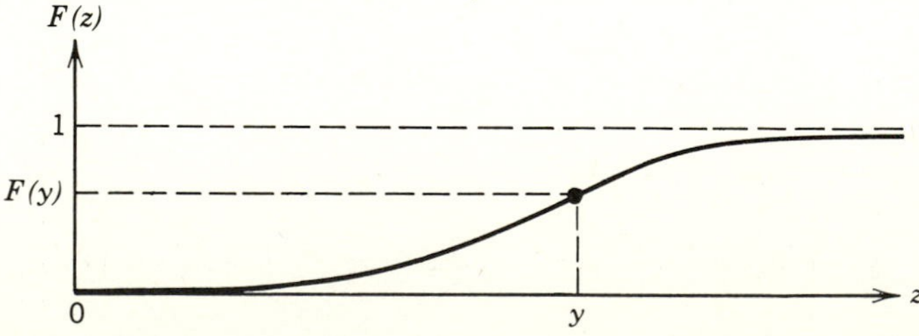

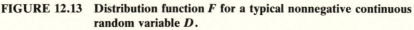

FIGURE 12.13 Distribution function F for a typical nonnegative continuous random variable D.

which implies that

$$F(y) = \frac{r - c}{r + h}$$

Note that

$$\begin{aligned}\frac{d^2}{dy^2}[cy + L(y)] &= \frac{d}{dy}[hF(y) - r + rF(y) + c] \\ &= (h + r)f(y)\end{aligned}$$

using the fact that the derivative of $F(y)$ equals $f(y)$. Note that the second derivative is positive provided $f(y) > 0$ for each y, which implies that $cy + L(y)$ is a strictly convex function. By definition, the density function is nonnegative for each y, so $cy + L(y)$ is a convex function. We shall assume that, over the range of the random variable, the density function f is positive (never zero), so we can be assured that $cy + L(y)$ is strictly convex. This assumption also assures that the distribution function F is strictly increasing.

Therefore, there is a unique minimizing point for $cy + L(y)$, namely, the point y^* satisfying

$$F(y^*) = \frac{r - c}{r + h}$$

so that

$$y^* = F^{-1}\left(\frac{r - c}{r + h}\right)$$

Because F is strictly increasing, the inverse function, F^{-1}, is defined, and the value of y^* is easily found from the distribution function as indicated in Figure 12.14. Note that $(r - c)/(r + h)$ is between 0 and 1 in practical situations. Also, recall that we wanted to minimize $cy + L(y)$ over all $y \geq 0$, and that y^* is nonnegative by construction.

It is important to appreciate the simplicity of this result. Given that the demand is known to be a random variable with distribution function F, to find the optimal amount y^* to order at the beginning of the period, we need only determine the parameters $r = p + s$, c, and h. To determine the parameters,

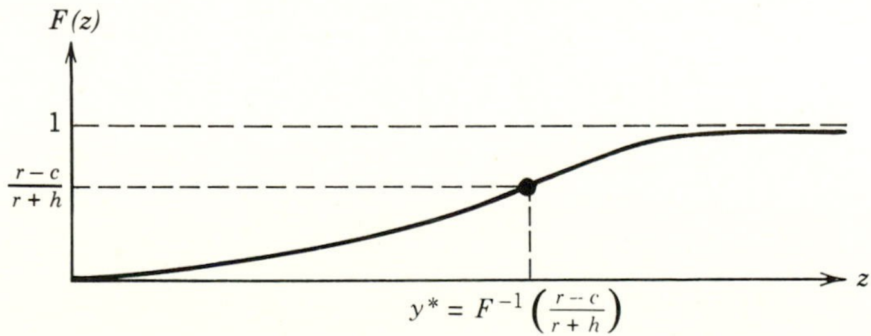

FIGURE 12.14 Finding the optimal amount y^* to make available given the parameters r, c, and h.

it is very important to keep in mind the expression for the profit, $P(D)$,

$$P(D) = \begin{cases} pD - cy - h(y - D) & \text{when } y > D \\ py - cy - s(D - y) & \text{when } y \le D \end{cases}$$

In the roadside stand example about selling gallons of cider during October, the parameters are

$$\begin{aligned} \text{selling price } p &= \$6.50/\text{gal} \\ \text{order price } c &= \$3.00/\text{gal} \\ \text{shortage cost } s &= \$0.50/\text{gal} \\ \text{holding cost } h &= -\$1.00/\text{gal} \end{aligned}$$

Note that, in this example, the holding cost is negative because each gallon of cider left over brings in \$1.00 of revenue. This additional revenue has to be accounted for in the term $-h(y - D)$ and so $h = -\$1.00$.

You might be tempted to reason that $h = \$2.00$ because each gallon left over cost \$3.00 and only sold for \$1.00 and so \$2.00 was lost. However, this would be an incorrect value for h because the original cost of the cider is already subtracted in the term cy.

With $p = 6.5$, $c = 3$, $s = 0.5$, and $h = -1$, we know from our derivation that the optimal amount, y^*, to order satisfies

$$F(y^*) = \frac{r - c}{r + h}$$

that is

$$y^* = F^{-1}\left(\frac{r - c}{r + h}\right) = F^{-1}\left(\frac{7 - 3}{7 + (-1)}\right) = F^{-1}\left(\frac{2}{3}\right)$$

where F is the distribution function for a random variable uniformly distributed between 100 and 200. Figure 12.15 contains a graph of this distribution function. Thus the value $y^* = 166\frac{2}{3}$ minimizes the strictly convex function $cy + L(y)$, so the actual integer amount we should order is either 166 or 167, depending on which value gives the smaller value for $cy + L(y)$.

Recall that

$$cy + L(y) = cy + \int_y^\infty r(z - y)f(z)\,dz + \int_0^y h(y - z)f(z)\,dz$$

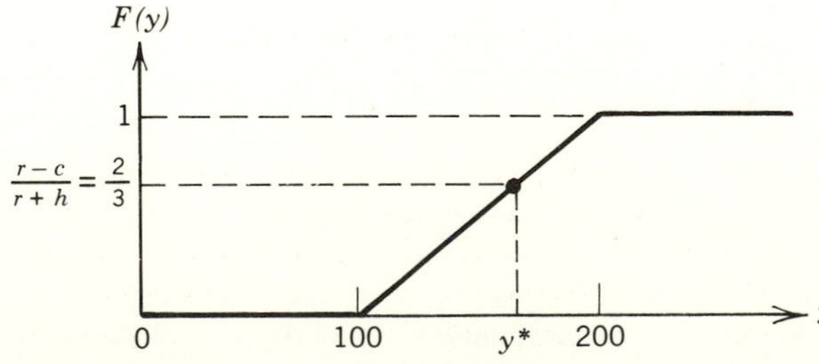

FIGURE 12.15 $F(y^*) = \frac{2}{3}$ **so** $y^* = F^{-1}(\frac{2}{3}) = 166\frac{2}{3}$.

and the density function for a random variable that is uniformly distributed between 100 and 200 is given by

$$f(z) = \begin{cases} \frac{1}{100} & \text{if } 100 \le z \le 200 \\ 0 & \text{otherwise} \end{cases}$$

Therefore, the preceding expression for $cy + L(y)$ becomes

$$\begin{aligned} cy + L(y) &= cy + \int_y^{200} r(z - y)\frac{1}{100}\,dz + \int_{100}^{y} h(y - z)\frac{1}{100}\,dz \\ &= 3y + \int_y^{200} 7(z - y)\frac{1}{100}\,dz + \int_{100}^{y} (-1)(y - z)\frac{1}{100}\,dz \\ &= \frac{3y^2}{100} - 10y + 1350 = \begin{cases} 516.68 & \text{when } y = 166 \\ 516.67 & \text{when } y = 167 \end{cases} \end{aligned}$$

Therefore, the optimal integer amount to be ordered is 166, and the maximum expected profit is

$$\begin{aligned} E[P(D)] &= pE(D) - 516.67 \\ &= (6.5)(150) - 516.67 = 458.33 \end{aligned}$$

Actually, given this explicit expression for $cy + L(y)$, y^* can be found by setting its derivative to zero, namely,

$$\frac{d}{dy}\left(\frac{3y^2}{100} - 10y + 1350\right) = 0 \quad \text{implies } 6y = 1000$$

so $y^* = 166\frac{2}{3}$ as found earlier.

A Single-Period Model with an Initial Inventory

In the case when there is an initial inventory, x, then the amount ordered is $y - x$ if y is the amount made available at the beginning of the period. The only change in the development is that the ordering cost becomes $c(y - x)$ instead of cy. The integrals for expected loss of revenue, expected shortage costs, and expected holding costs used in defining $L(y)$ do not change because they depend on the amount y made available at the beginning of the period (even if some of this amount y is from an initial inventory). Thus, to find the optimal value for y, we solve the problem

$$\begin{aligned} &\text{minimize} \quad c(y - x) + L(y) \\ &\text{subject to} \quad y \ge x \end{aligned}$$

Because x is a constant, minimizing $cy - cx + L(y)$ is equivalent to minimizing $cy + L(y)$. Thus, we need to solve the equivalent problem

$$\begin{aligned} &\text{minimize} \quad cy + L(y) \\ &\text{subject to} \quad y \ge x \end{aligned}$$

However, this problem is easy to solve because $cy + L(y)$ is a strictly convex

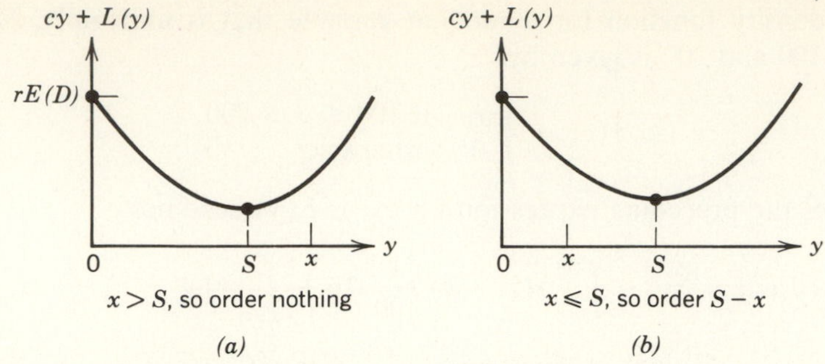

FIGURE 12.16 The optimal amount to order depends on the initial inventory, x.

function. From the graphs in Figure 12.16, we see that the minimum of $cy + L(y)$ over all $y \geq x$ can be found as follows:

Determine the minimizing point, S, for $cy + L(y)$ over all values of y, namely, the point S such that

$$F(S) = \frac{r - c}{r + h}$$

The optimal amount y^* to make available is given by

$$y^* = \begin{cases} S & \text{if } x \leq S \quad \text{so order } S - x \\ x & \text{if } x > S \quad \text{so order } 0 \end{cases}$$

In Figure 12.16*a* we see that adding to the initial inventory level only increases $cy + L(y)$ so nothing more should be ordered. In Figure 12.16*b* increasing the inventory level up to S (by ordering $S - x$) decreases $cy + L(y)$ as much as is possible.

Note in Figure 12.16 that when $y = 0$, $cy + L(y) = rE(D)$ as can be verified by checking the integrals in $L(y)$ when $y = 0$. Also, as y approaches infinity, $cy + L(y)$ also approaches infinity.

Single-Period Model with an Initial Inventory and a Fixed Cost of Ordering

Suppose that in addition to an initial inventory, x, there is a fixed cost, K, for placing an order. If y is the amount made available at the beginning of the period, the integrals used in defining $L(y)$ are exactly as in the earlier cases. If no additional inventory is ordered, so that x is the amount available at the beginning of the period, then the expected loss of revenue plus the expected shortage and holding costs are given by $L(x)$. If $y - x > 0$ is ordered, then the total expected cost must include the fixed cost K of placing an order as well as the ordering cost, $c(y - x)$, and the expected loss of revenue, shortage, and holding costs, $L(y)$. Thus,

$$\text{expected cost} = \begin{cases} K + c(y - x) + L(y) & \text{if } y > x \\ L(x) & \text{if } y = x \end{cases}$$

Given a value of $y > x$, we need to determine whether or not

$$L(x) \leq K + c(y - x) + L(y)$$

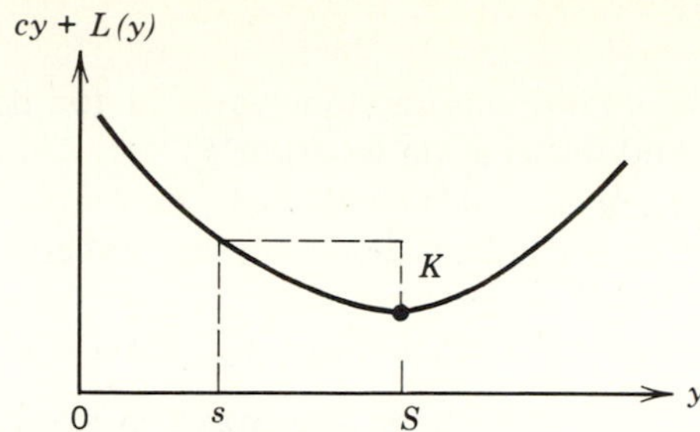

FIGURE 12.17 Determining an optimal (s, S) policy.

This inequality holds if and only if

$$cx + L(x) \leq K + cy + L(y)$$

Using this last inequality and the fact that $cy + L(y)$ is a strictly convex function, the optimal amount to make available at the beginning of the period is determined as follows:

Determine the minimizing point, S, for $cy + L(y)$ over all values of y, namely, the point S such that

$$F(S) = \frac{r - c}{r + h}$$

Let s be the smallest value such that

$$cs + L(s) = K + cS + L(S)$$

The optimal amount y^* to make available is given by

$$y^* = \begin{cases} x & \text{if } x > S \quad \text{so order } 0 \\ x & \text{if } s \leq x \leq S \quad \text{so order } 0 \\ S & \text{if } x < s \quad \text{so order } S - x \end{cases}$$

To see why this is true, consider the graph in Figure 12.17. From Figure 12.17, we make the following observations based on the fact that $cy + L(y)$ is a strictly convex function:

If $x > S$, then $cx + L(x) < cy + L(y)$ for all $y > x$
so $cx + L(x) < K + cy + L(y)$ for all $y > x$
and nothing should be ordered.

If $s \leq x \leq S$, then $cx + L(x) < K + cy + L(y)$ for all $y > x$
so nothing should be ordered.

if $x < s$, then $cx + L(x) > K + cy + L(y)$ for $x < y \leq S$,
and $K + cy + L(y)$ is minimized when $y = S$,
so order $S - x$.

A Two-Period Stochastic Inventory Model

The results for the single-period model can be extended to cases with two or more periods. To illustrate how this is done, we consider the case where there are two periods having demands given by the same random variable. In each period we make the same basic assumptions used in the single-period model, namely,

D = nonnegative continuous random variable for demand with density function f and distribution function F
p = selling price per unit of inventory
s = shortage price per unit of demand not satisfied
c = ordering cost per unit of inventory
h = holding cost per unit of inventory left at the end of the period

In addition, we assume that there is no fixed cost of ordering and we let

r_n = inventory level at the start of period $n = 1, 2$

The problem is to find the amount y_n that should be made available at the beginning of period n so as to minimize total expected cost.

A natural approach for solving this sequential decision problem is to use dynamic programming. Each period corresponds to a stage, and the state at any stage is the amount of inventory left from the previous period. To formulate the backward recursive relations, we consider the last period, period 2, first with r_2 as the initial inventory left over from period 1. Using the same dynamic programming notation used in Section 12.2 (and Chapter 10), we let

$f_2(r_2, y_2)$ = expected cost of the best ordering policy for period 2, given r_2 units on hand, and given that we make y_2 units available at the beginning of period 2

and let

$$\begin{aligned} f_2(r_2) &= \text{expected cost of the best ordering policy for period 2, given } r_2 \text{ units on hand} \\ &= \text{minimum}\{f_2(r_2, y_2) \mid y_2 \geq r_2\} \end{aligned}$$

Given r_2 units on hand at the beginning of period 2 with D denoting the random demand, we know from the single-period model that the optimal ordering policy is

select S_2^* that minimizes $cy + L(y)$ for period 2,
and then the optimal amount y_2^* to make available
at the beginning of period 2 is

$$y_2^* = \begin{cases} S_2^* & \text{if } r_2 < S_2^* \quad \text{so order } S_2^* - r_2 \\ r_2 & \text{if } r_2 \geq S_2^* \quad \text{so order } 0 \end{cases}$$

Therefore, for any r_2, we can calculate $f_2(r_2)$ as

$$f_2(r_2) = \begin{cases} c(S_2^* - r_2) + L(S_2^*) & \text{if } r_2 < S_2^* \\ L(r_2) & \text{if } r_2 \geq S_2^* \end{cases}$$

Now suppose we consider both periods and let

$f_1(r_1, y_1)$ = expected cost of the best ordering policy for periods 1 and 2, given r_1 units on hand, and given that we make y_1 units available at the beginning of period 1

and

$f_1(r_1)$ = expected cost of the best ordering policy for periods 1 and 2, given r_1 units on hand

If we make y_1 units available at the beginning of period 1, then $y_1 - r_1$ is the amount ordered, and the expected total costs for the two periods are

$c(y_1 - r_1)$ = ordering cost in period 1
$L(y_1)$ = expected loss of revenue plus the expected shortage and holding costs for period 1,
$E[f_2(r_2)]$ = expected cost incurred in period 2 given r_2 units on hand at the beginning of the period

It is important to realize that r_2, the amount on hand at the beginning of period 2, satisfies $r_2 = y_1 - D$, so we need to consider the expected value of $f_2(r_2)$ in calculating the expected costs over both periods. Thus, the recursive relation used to determine $f_1(r_1)$ is

$$f_1(r_1, y_1) = c(y_1 - r_1) + L(y_1) + E[f_2(r_2)]$$

Using the fact that $r_2 = y_1 - D$ and recalling that

$$f_2(r_2) = \begin{cases} c(S_2^* - r_2) + L(S_2^*) & \text{if } r_2 < S_2^* \\ L(r_2) & \text{if } r_2 \geq S_2^* \end{cases}$$

we see that

$$f_2(y_1 - D) = \begin{cases} c(S_2^* - y_1 + D) + L(S_2^*) & \text{if } y_1 - D < S_2^* \\ L(y_1 - D) & \text{if } y_1 - D \geq S_2^* \end{cases}$$

Thus, we calculate $E[f_2(r_2)] = E[f_2(y_1 - D)]$ as follows:

$$E[f_2(y_1 - D)] = \int_{y_1 - S_2^*}^{\infty} [c(S_2^* - y_1 + z) + L(S_2^*)]f(z)\, dz + \int_0^{y_1 - S_2^*} L(y_1 - z)f(z)\, dz$$

Note that the integrals define a function of a single variable, y_1. Therefore, we can calculate

$$f_1(r_1) = \text{minimum}\{f_1(r_1, y_1) \mid y_1 \geq r_1\}$$

by using a numerical method for minimizing a function of a single variable (as discussed in Chapter 9). It can be shown that there is a unique minimizing point y_1^* for this problem.

Thus, given y_1^* and y_2^*, the optimal ordering policy for this two-period model can be stated as follows:

At the start of period 1

if $r_1 < y_1^*$ then order $y_1^* - r_1$,
if $r_1 \geq y_1^*$ then order 0.

At the start of period 2

if $r_2 < y_2^*$ then order $y_2^* - r_2$,
if $r_2 \geq y_2^*$ then order 0.

This policy would remain the same even if r_2 were negative, that is, if shortages in period 1 were **backlogged** and could be filled in period 2.

This type of model can be extended to N periods, but we shall not show how this is done. The main result for the N-period model is that critical values

$y_1^*, \ldots, y_n^*$ can be found numerically so that the optimal ordering policy has the following form:

At the beginning of period i, $i = 1, 2, \ldots, N$

if $r_i < y_i^*$ then order $y_i^* - r_i$

if $r_i \geq y_i^*$ then order 0,

$y_1^* \geq y_2^* \geq \cdots \geq y_N^*$.

SELECTED REFERENCES

Buffa, E. S., *Modern Production/Operations Management* (6th ed.), Wiley, New York, 1980.

Hadley, G. and Whitin, T., *Analysis of Inventory Systems,* Prentice-Hall, Englewood Cliffs, NJ, 1963.

Hillier, F. S., and Lieberman, G. J., *Introduction to Operations Research* (4th ed.), Holden-Day, Oakland, CA, 1986.

Peterson, R., and Silver, E. A., *Decision Systems for Inventory Management and Production Planning,* Wiley, New York, 1979.

Veinott, A. F., Jr., The Status of Mathematical Inventory Theory, *Management Science* **12:**745–777 (1966).

Wagner, H. M., *Principles of Operations Research,* Prentice-Hall, Englewood Cliffs, NJ, 1969.

EXERCISES

12.1 A college bookstore buys pads of paper for 60 cents each and sells them to students for 80 cents each. Students use 200 pads per week, and it is reasonable to assume that the pads are continuously withdrawn. Each time an order is placed, there is a trucking cost of $50 (independent of the size of the order). The cost of maintaining the display racks where the pads are sold is estimated to be 3 cents per pad per week.

(a) Assuming that shortages are not allowed and that the same quantity is ordered each time an order is placed, what should the size of the order be and how often should it be placed to minimize total cost per week? What is the minimum total cost?

(b) If shortages are allowed at an estimated cost of 5 cents per pad per week, what is the optimal order size and how often should it be placed to minimize total cost per week?

(c) Suppose that the cost to the bookstore increases by 10 cents. How does the solution to part (a) change?

12.2 In the basic economic order-quantity model with no shortages, the total cost for the period $[0, Q/a]$ was calculated to be

$$K + cQ + (hQ^2/2a)$$

If Q is selected to minimize this total cost for the period instead of total cost per unit time, what would be the optimal value of Q?

12.3 A large trucking company continuously uses motor oil at the rate of 300 quarts per month. The oil costs 50 cents per quart, and each order that is placed incurs a fixed shipping cost of \$100. Prior to its use, oil must be stored in a special building at a cost of 20 cents per quart per month.
(a) Assuming that no shortages are allowed, determine the optimal amount to order and the resulting minimum total cost per month.
(b) When a shortage of oil occurs, the company must run its trucks on bad oil, and the cost of this is estimated to be \$10 per quart per month. If shortages are allowed, determine the optimal amount to order and determine the length of the shortage period.

12.4 In Exercise 12.3, suppose that the company has to pay 50 cents per quart if it orders less than 800 quarts. If it orders more than 800 but less than 2000 quarts, it pays 35 cents per quart. However, if more than 2000 quarts are ordered, then the price drops to 30 cents per quart. Assume no shortages are allowed.

Determine the optimal amount to order and the minimum total cost per month.

12.5 A large city hospital needs 1000 new nurses per year to replace nurses who leave or retire. Every group of new nurses has to go through a rather extensive orientation program that costs the hospital \$5000 each time it is given. Nurses must be available when they are needed so the hospital hires nurses before they are actually needed and puts them through the orientation program so that a pool of nurses is always available. Nurses in the pool are paid at the rate of \$7500 per yr even though they are just on call and not working.

What should be the size of each new group of nurses and how often should the orientation program be given, to minimize total costs per year?

12.6 A floor tile outlet store has a warehouse with 4000 ft^2 of space for storing tile. The outlet orders tile from two different manufacturers, one who makes classical tile and one who makes modern tile. Each manufacturer ships pallets of tile to the outlet store at a fixed cost of \$1000 per shipment from either manufacturer. Each pallet of classical tile takes up 10 ft^2 of warehouse space and each pallet of modern tile uses 20 ft^2. Warehouse storage costs for either tile are \$20 per pallet per year. The company pays \$200 per pallet of classical tile and \$400 per pallet of modern tile. Assume that the demand for classical tile is 800 pallets per year and the demand for modern tile is 1200 pallets per year. Assume also that these inventories are depleted in approximately a continuous manner.

Use the Lagrange multiplier method discussed in Section 12.1 to determine how many pallets of each type of tile the outlet store should order to minimize total costs per year.

12.7 For the floor tile problem in Exercise 12.6, show how to formulate (but do not solve) a nonlinear programming problem with a single variable whose solution will yield a solution to the original problem.

12.8 Demand for a product over the next 5 weeks is estimated to be

$$D_1 = 3 \qquad D_2 = 4 \qquad D_3 = 2 \qquad D_4 = 5 \qquad D_5 = 4$$

and these demands must be satisfied. There is a fixed cost of \$20 for each production run and the cost of producing x units in any period is \$$5x$. The holding cost charged on units left in inventory at the end of each period is 50 cents per unit.

(a) Use the dynamic programming approach for problems with concave costs to determine how much should be produced at the beginning of each period in order to minimize total production and holding costs.
(b) If there is no holding cost charged on items left in inventory at the end of a period, what is the optimal production policy?

12.9 Consider a three-period inventory problem with demands $D_1 = 4$, $D_2 = 1$, and $D_3 = 3$. Suppose that the fixed cost of ordering in the three periods is \$3, \$5, and \$5, respectively. The holding cost is \$3 per unit left in stock at the end of each period. For each period the ordering cost is \$10 for each of the first two units and \$5 per unit for each unit ordered above two.
(a) If this problem is solved as a dynamic programming problem ignoring the fact that the ordering and holding costs are concave, list the possible states s and the possible decisions x_n that need to be considered at each stage n, $n = 1, 2, 3$.
(b) Using the fact that the ordering and holding costs are concave, find the optimal ordering policy using the dynamic programming approach for such problems.

12.10 For the three-period inventory problem considered in Exercise 12.9, suppose that the ordering costs change as follows. For each period there is no fixed cost of ordering. Also, for each period, the ordering cost is \$5 for each of the first two units ordered and \$10 per unit for each unit above two that is ordered. The demands and holding costs remain unchanged.

Solve this problem using the algorithm discussed in Section 12.2 for periodic review models with convex costs.

12.11 (a) Solve Exercise 12.10 by dynamic programming using the *incorrect* assumption that ordering only occurs when the inventory level is zero. Does this yield an optimal solution?
(b) Solve Exercise 12.10 by dynamic programming without using the incorrect assumption that ordering only occurs when the inventory level is zero.

12.12 The demand for Christmas trees sold by a local store during the Christmas season is uniformly distributed between 0 and 20. These trees cost the owner of the store \$2 each and are sold for \$2.50 each. If a customer arrives at the store to buy a tree and one is not available, the owner has a policy of giving the customer 50 cents in cash as a goodwill gesture. All trees left over after Christmas must be hauled away at a cost to the owner of \$1 each.
(a) Write down integrals for the expected shortage cost, expected holding costs, and expected loss of revenue.
(b) How many trees should the store make available to maximize expected profit?
(c) Suppose that the store owner already has an initial inventory of four trees and that to order more trees there is a fixed cost of 90 cents for placing the order. Under these circumstances, should the owner order any additional trees? Explain.

12.13 The manager of Bag and Baggage, Inc. must order new purses to be sold during the summer season. The demand for purses during this season is a random variable uniformly distributed between 1000 and 2000 purses. Purses can be bought wholesale for \$5 each and sold for \$8 each by Bag and Baggage. No initial inventory is available. Any purses left over at the end of the summer season can be sold to a discount store for \$3 each. If the demand for purses is greater than the supply made available, the manager must special order purses at \$10 each but still must sell them for \$8 each.

(a) Let y denote the number of purses that the manager makes available at the beginning of the summer season. Write down an expression for the profit as a function of y and the random demand D.
(b) Write down, but do not evaluate, integrals giving the expected profit. Be sure to give the limits of integration.
(c) How many purses should the manager order to maximize expected profit?

12.14 Suppose that the only change in the Bag and Baggage problem in Exercise 12.13 is that the demand D is exponentially distributed with parameter $\lambda = \frac{1}{5000}$ so that its density function $f(x)$ is given by

$$f(x) = \begin{cases} \frac{1}{5000}e^{-x/5000} & \text{if } x \geq 0 \\ 0 & \text{otherwise} \end{cases}$$

and its distribution function $F(x)$ is given by

$$F(x) = 1 - e^{-x/5000} \quad \text{for } x \geq 0$$

(a) Write down, but do not evaluate, integrals giving the expected profit.
(b) With this distribution on the random demand D, how many purses should the manager order to maximize expected profit?

12.15 A local newspaper stand buys new magazines every month. The magazines cost the stand \$1.50 per copy and are sold for \$2.00 per copy. Demand for these magazines is a random variable uniformly distributed between 100 and 150 copies. At the end of the month, leftover magazines are returned for a full refund of \$1.50 but returning the magazines costs 10 cents per copy for postage. If demand is greater than supply, the magazines are special ordered for \$1.75 and still sell for \$2.00.
(a) Write down and evaluate the integrals giving the expected profit as a function of the amount y made available at the beginning of the month.
(b) Set the derivative of the expression obtained in part (a) to zero and solve to find the optimal number y^* of magazines to order.
(c) Check that the value for y^* obtained in part (b) agrees with that given by the formula

$$y^* = F^{-1}\left(\frac{r - c}{r + h}\right)$$

derived in Section 12.3.

12.16 The demand for cookies served at the department seminar series is uniformly distributed between 10 and 20. Cookies cost 5 cents each and are given out free. If a guest arrives and finds no cookies left, he writes a nasty note on the wall that cost 8 cents to remove. Cookies left over can be sold to passing students for 2 cents each.
(a) How many cookies should be ordered to minimize expected costs?
(b) Suppose there is a fixed cost of 30 cents for placing an order and that the number of cookies already on hand is denoted by x. Determine a (s, S) ordering policy for this problem.

12.17 If the demand for cookies in Exercise 12.16 is exponentially distributed with parameter $\lambda = 1/25$ and there is no initial inventory, how many cookies should be ordered to minimize expected costs?

12.18 Suppose that a function h is the sum of two other functions f and g so that

$$h(x) = f(x) + g(x)$$

(a) If f and g are convex functions, show that h is a convex function.
(b) Use part (a) to show that if f and g are concave functions, then h is a concave function.

12.19 Given a function f and a function g, suppose that h is the composite function defined by

$$h(x) = f(g(x))$$

(a) If f is a convex function and g is a linear function, show that h is a convex function.

(b) If f is a concave function and g is a linear function, show that h is a concave function.

12.20 The following is a statement of the Leibnitz rule for differentiating an integral.

If $g(z, y)$ is a continuous function of z and y, and
$\partial g/\partial y$ is continuous, and
$p(y)$ and $q(y)$ have derivatives everywhere, then,

$$\frac{d}{dy}\int_{p(y)}^{q(y)} g(z, y)\, dz = \int_{p(y)}^{q(y)} \frac{\partial g(z, y)}{\partial y}\, dz + g[q(y), y]q'(y) - g[p(y), y]p'(y)$$

where $q'(y)$ and $p'(y)$ are the derivatives of $q(y)$ and $p(y)$ with respect to y.

(a) Use the Leibnitz rule to derive the following result, which was used in Section 12.4:

$$\frac{d}{dy}\int_0^y h(y - z)f(z)\, dz = h\int_0^y f(z)\, dz$$

where f is a continuous density function and h is a constant.

(b) Use the Leibnitz rule to show that when M is a large positive constant,

$$\frac{d}{dy}\int_y^M r(z - y)f(z)\, dz = -r\int_y^M f(z)\, dz$$

where f is a continuous density function and r is a constant.

(c) How can this result be used to give a plausible argument (not a proof) that if M is replaced by ∞ then the result in (b) still holds?

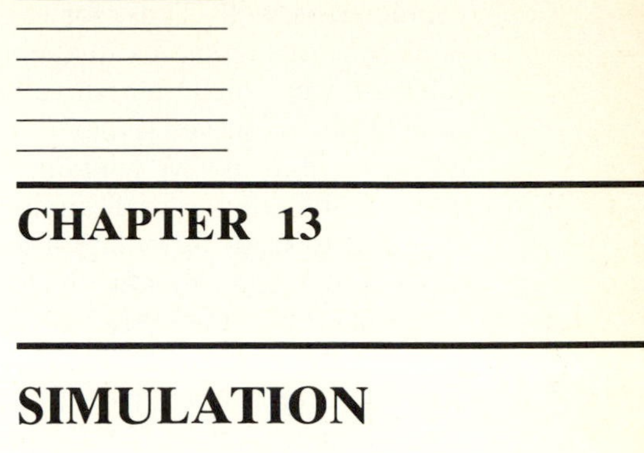

CHAPTER 13

SIMULATION

Chapters 2 through 12 all concern **mathematical models** of operations research. The idea of mathematical modeling is to answer questions about the real world, such as "how many chairs should we make" or "how long do the customers wait," by solving a simplified mathematical problem that represents the salient features of the real situation. Mathematical programs, the equations describing queueing and inventory systems, and other mathematical models yield precise and unambiguous results, and the insight they provide into the nature of real systems is frequently unobtainable in any other way. Formulating a good mathematical model requires considerable skill and effort, but may yield analytical results that are unsurpassed in conciseness and clarity.

Unfortunately, there are many problems of practical importance for which the simplest adequate mathematical models are analytically intractable. One conspicuous example is the problem of determining the transient behavior of a queueing system that is realistically complex. Unless a queueing system is extremely simple, the lag-differential equations that describe its behavior from moment to moment cannot be integrated analytically. Analytical results may be obtainable for the steady state, but some queueing systems go from one transient state to another without ever reaching a steady state. Any practical investigation of such a system's behavior must be based either on direct observation, which may be expensive or impossible, or on numerical computations of some kind.

One computational approach for dealing with a mathematical model that is analytically intractable is to use traditional numerical methods to solve the model itself. This is the approach that usually must be taken in solving mathematical programming problems, and it can be applied to queueing and inventory problems as well. Thus, for example, one can numerically integrate the state equations of a complex queueing system to obtain the transient response as a table of, say, queue length values at various times. This approach has several drawbacks, however. Numerical results are typically verbose and difficult to interpret, and always contain errors whose magnitudes are hard to estimate. These ambiguities are a disappointing reward for investing the considerable effort required to both formulate a good mathematical model and then apply a numerical algorithm to solve it. Moreover, there are some problems for which it has proved extremely difficult to formulate adequate mathematical models at all, and there are certainly mathematical models for which known numerical algorithms are ineffective.

Another computational approach that is sometimes possible is to forgo a mathematical model entirely and merely observe some mechanical imitation of the system under study. Such an imitation is called a **simulation model.** Simulation provides a natural approach to the quantitative study of queueing and inventory systems, and is much simpler in practice than solving intractable mathematical models numerically. In fact, simulation is conceptually simpler than the use of mathematical models for queueing and inventory systems even in cases when the models can be solved analytically. Simulation techniques also have some limited applicability for problems that lend themselves to formulation as mathematical programs.

Simulation suffers from many of the drawbacks of other numerical computations. In particular, simulation produces voluminous output data that must be summarized for interpretation, rather than explicit functional relationships, and the results are narrowly applicable to the particular system being modeled. This is in contrast to an analytical result, which might be a simple formula that is applicable to many different systems depending on the values of some parameters. Simulation is sometimes computationally expensive compared to obtaining results by numerical solution of a mathematical model. The results of a simulation study are statistical in nature and therefore always equivocal, but the nature of the simulation process also permits the use of statistical techniques for the determination of confidence intervals on the results. This is in contrast to results obtained by numerical solution of a mathematical model, for which estimates of uncertainty are usually quite difficult to compute.

Surveys of industrial operations research departments indicate that simulation is widely used, being roughly comparable to linear programming in popularity as a practical tool. Most curricula in operations research offer a separate course in simulation, so the introduction provided by this chapter is brief.

13.1 NEXT-EVENT SIMULATION

In Chapters 11 and 12 we considered some analytical techniques for modeling systems in which **events** such as customer arrivals, completions of service, and withdrawals from inventory occur at discrete instants in time. The basic idea of a simulation model is to construct a detailed time history of a typical sequence of events in such a system. This time history is called an **event list.** Then a statistical analysis of the event list is performed to estimate quantities such as average queue length or the probability of a stockout. In other words we treat the event list generated by the simulation as if it were obtained by direct observation of the system under study.

Observing a Queueing System

Consider a single-server queueing system in which the interarrival times are uniformly distributed between 1 and 5 min, and in which the service times are uniformly distributed between 2 and 3 min (Figure 13.1). If the system in question were actually in operation somewhere, the most naive way one could conceivably study it would be to stand nearby and watch what happened.

Customers would arrive, be served, and depart at random times. If the observer wrote down the times of arrival, service commencement, and departure for each customer passing through the system during the course of a day, it would be easy to use that record of events to calculate the average waiting time customers experienced during that day, the average queue length, and so forth. Of course, these numbers would only describe the particular day on which the observations were made, but they would also probably be reasonable estimates for the long-run average behavior of the queueing system because many arrivals and services would be represented in the observations for the one day. A simulation imitates the behavior of a real or hypothetical system, so that observations can be made of typical behavior without the need to be physically present at the site of the system or to actually spend a long time collecting observations.

Constructing an Event List

To see how a typical time history of events can be constructed, suppose an arrival occurs at time $t = 0$ in the example queueing system and that there are no customers already in the system when the arrival occurs. The customer will immediately go into service, and will finish service and depart the system at some time between $t = 2$ and $t = 3$ min. This sequence of events does not depend on subsequent arrivals, so the time of departure of the first customer could be determined at the time of arrival if it were known precisely how long the service would take. Also, the second arrival will surely occur at some time between $t = 1$ and $t = 5$ because interarrival times are distributed in that range. Therefore, it would be possible to determine the time of the second arrival as

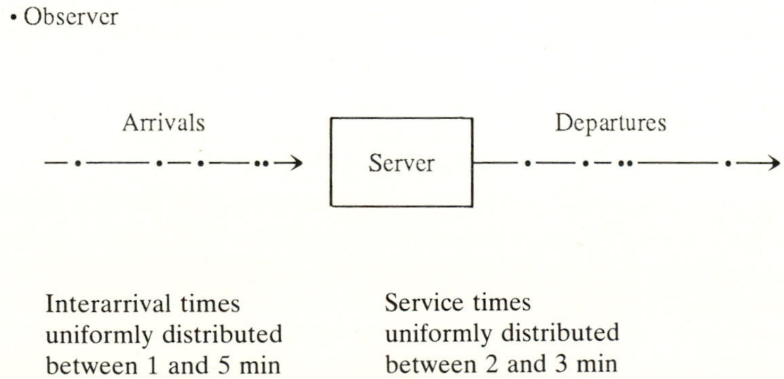

FIGURE 13.1 Observing a queueing system.

soon as the first one occurs, if it were known precisely what that particular interarrival time would be.

The description of our example queueing system includes the fact that, over the long run, any interarrival time between 1 and 5 min is equally likely to occur. Similarly, over the long run, any service time between 2 and 3 min is equally likely to occur. It is therefore reasonable to make the following key assumption:

> The long-run statistical properties of the real system will be the same as the long-run statistical properties of a hypothetical sequence of events constructed by selecting interevent times at random from their given distributions.

Because it is the long-run statistical properties of the system that we are ultimately interested in, rather than the precise details of what happens, this assumption permits us to construct a typical time history for the system without knowing, except in a probabilistic way, what each particular interarrival or service time actually would be in the real system.

Using this assumption, we can construct a typical time history of events for our example system by selecting interarrival times randomly from the interval [1, 5] and service times randomly from the interval [2, 3]. Thus, at the time of the first arrival (that is, at $t = 0$), we can schedule the second arrival at a time randomly generated between 1 and 5, and we can schedule the departure of the first customer at a time randomly generated between 2 and 3. Suppose the randomly generated interarrival time is 3.9 min, and the randomly generated service time is 2.2 min. Then our list of hypothetical observations would begin as follows:

	Time t	*Event*
$>$	0.0	customer 1 arrives and enters service
	2.2	customer 1 departs, server becomes idle
	3.9	customer 2 arrives and enters service

The pointer $>$ marks simulated time, and indicates that the list includes all the events that can be scheduled once the event at the indicated time and all earlier events have occurred. The next event that occurs in the system is the departure of customer 1 at $t = 2.2$, so there is no need to consider values of t in the interval $0.0 < t < 2.2$; the approach we are using is therefore called **next-event simulation.** We now think of advancing simulated time to the next event, at $t = 2.2$, permitting that event to occur, and asking whether the occurrence of that event permits any future events to be scheduled. Since no customers are waiting for service when customer 1 departs, nothing happens in the queueing system until the arrival of customer 2, and it is not possible to schedule any future events upon the departure of customer 1.

Advancing simulated time to $t = 3.9$, we next permit the event to occur that customer 2 arrives. Now it is possible to schedule the next arrival, by selecting an interarrival time at random from the interval [1, 5]. Suppose doing so yields a value of 1.7; then the arrival of customer 3 is scheduled to occur at $t = 3.9 + 1.7 = 5.6$. As soon as customer 2 arrives, it is also possible to schedule the departure of that customer, by randomly selecting a service time from the interval [2, 3]. Suppose doing so yields a value of 2.9; then the departure of customer 3 is scheduled for $t = 3.9 + 2.9 = 6.8$, and the event history looks like this:

	Time t	*Event*
	0.0	customer 1 arrives and enters service
	2.2	customer 1 departs, server becomes idle
>	3.9	customer 2 arrives and enters service
	5.6	customer 3 arrives and waits
	6.8	customer 2 departs
	6.8	customer 3 enters service

Note that the arrival of customer 3 before the departure of customer 2 means that customer 3 will wait 0.8 min for service, entering service only upon the departure of customer 2. Thus the arrival of customer 3 at $t = 5.6$ does not permit the scheduling of customer 3's departure. However, it does permit the scheduling of the arrival of customer 4. Suppose the random interarrival time generated is 1.0, so customer 4 arrives at $t = 6.6$; then the list looks like this:

	Time t	*Event*
	0.0	customer 1 arrives and enters service
	2.2	customer 1 departs, server becomes idle
	3.9	customer 2 arrives and enters service
>	5.6	customer 3 arrives and waits
	6.6	customer 4 arrives and waits
	6.8	customer 2 departs
	6.8	customer 3 enters service

Advancing simulated time to the next event, we permit customer 4 to arrive, generate another interarrival time of, say, 1.8 min, and schedule the arrival of customer 5 at $t = 6.6 + 1.8 = 8.4$ to get the following list:

	Time t	*Event*
	0.0	customer 1 arrives and enters service
	2.2	customer 1 departs, server becomes idle
	3.9	customer 2 arrives and enters service
	5.6	customer 3 arrives and waits
>	6.6	customer 4 arrives and waits
	6.8	customer 2 departs
	6.8	customer 3 enters service
	8.4	customer 5 arrives

At this point we do not know whether customer 5 will have to wait or not because we have not yet scheduled the departures of customers 3 and 4. With the departure of customer 2 at $t = 6.8$, customer 3 enters service, and we can schedule the departure of customer 3 by generating a random service time. If that value turns out to be 2.5 min, the departure will occur at $t = 6.8 + 2.5 = 9.3$ and the list grows to the following.

	Time t	*Event*
	0.0	customer 1 arrives and enters service
	2.2	customer 1 departs, server becomes idle
	3.9	customer 2 arrives and enters service
	5.6	customer 3 arrives and waits
	6.6	customer 4 arrives and waits
	6.8	customer 2 departs
>	6.8	customer 3 enters service
	8.4	customer 5 arrives and waits
	9.3	customer 3 departs
	9.3	customer 4 enters service

Now we can tell that customer 5 will find the server busy upon arriving, and will have to wait. Of course, while the event list is being constructed, there is no need to keep track of which customers have to wait or of when the server is idle because those things are consequences of the arrivals and service completions and can therefore be figured out after the list is complete. We have included them above only to help the reader imagine what is happening in the queueing system as the event list grows.

The event generation process is summarized schematically in Figure 13.2. The process of scheduling future events, advancing simulated time to the next scheduled event, permitting the scheduled event to occur, scheduling future events, and so on, could obviously be continued indefinitely to result in a time history of any desired length.

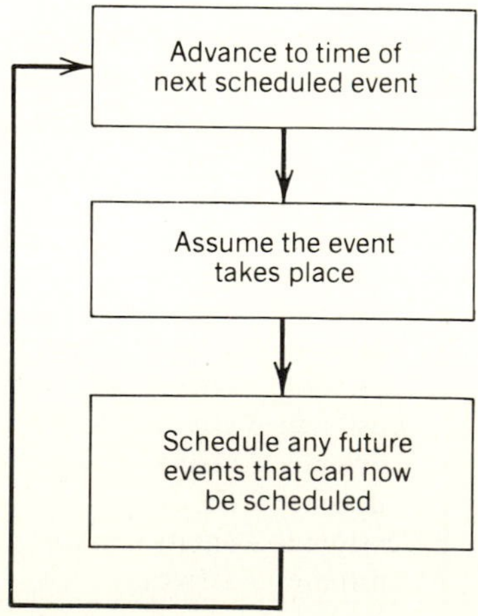

FIGURE 13.2 The event generation process.

13.2 STATISTICAL ANALYSIS OF SIMULATION RESULTS

An event list can be used, in the same way one would use a list of observations made on a real system, to estimate various long-run statistical properties of the system under observation.

Estimating L and L_q

For the preceding example the number of customers in the system as a function of time is shown in Figure 13.3. The average number of customers in the system is by definition

$$L = \frac{1}{T}\int_0^T n(t)\,dt$$

where $n(t)$ is the number at time t. Using Figure 13.3, we can estimate this quantity from our simulated event list as

$$\begin{aligned} L &= \frac{\text{area under curve}}{T} \\ &= \frac{\left[\begin{array}{lll} 1(2.2 - 0) & + 1(5.6 - 3.9) & + 2(6.6 - 5.6) \\ +\ 3(6.8 - 6.6) & + 2(8.4 - 6.8) & + 3(9.3 - 8.4) \end{array}\right]}{9.3} \\ &= \frac{12.4}{9.3} = \frac{4}{3} \text{ customers} \end{aligned}$$

Similarly, we calculate average queue length as

$$\begin{aligned} L_q &= \frac{\text{area under curve above } n(t) = 1}{T} \\ &= \frac{1(6.6 - 5.6) + 2(6.8 - 6.6) + 1(8.4 - 6.8) + 2(9.3 - 8.4)}{9.3} \\ &= \frac{4.8}{9.3} \approx 0.516 \text{ customers} \end{aligned}$$

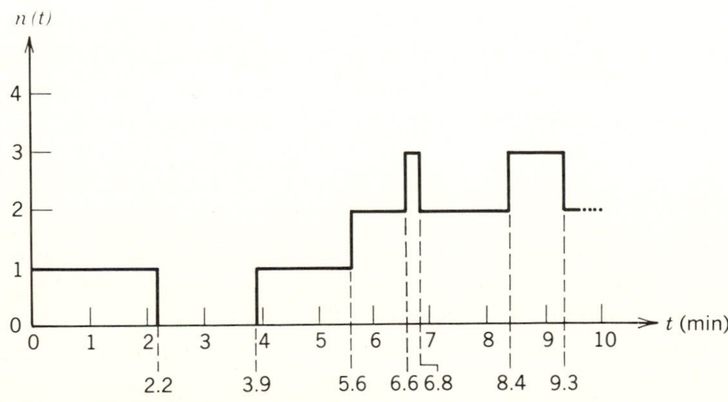

FIGURE 13.3 Customers in the system.

Estimating W and W_q

To estimate the average time each customer spends in the system, we compute the time measured for each customer from the event list and average the numbers, as follows:

Customer	*Time in System*	
1	$2.2 - 0.0 = 2.2$	
2	$6.8 - 3.9 = 2.9$	$W = (2.2 + 2.9 + 3.7)/3$
3	$9.3 - 5.6 = 3.7$	≈ 2.93 min
4	?	
5	?	

Note that at $t = 9.3$ customer 4 has just entered service and customer 5 is still waiting, so the total times they will ultimately spend in the system cannot be determined unless we continue the simulation. Simulation models are frequently used to study processes that do not have a natural ending time, and in those cases it is usual that some unfinished business must be omitted from the statistics no matter when they are calculated.

Average waiting time W_q is calculated in a similar way, as follows:

Customer	*Time Waiting*	
1	$0.0 - 0.0 = 0.0$	
2	$3.9 - 3.9 = 0.0$	$W_q = (1.2 + 2.7)/4$
3	$6.8 - 5.6 = 1.2$	$= 0.975$ min
4	$9.3 - 6.6 = 2.7$	
5	?	

Customer 4 enters service just as the simulation ends, so that waiting time can be found and included in this calculation.

Estimating Server Utilization

The fraction of time f that the server is busy is also easy to estimate from the observations, as

$$f = \frac{\text{time server is busy}}{T} = \frac{(2.2 - 0) + (9.3 - 3.9)}{9.3} \approx 0.817$$

Multiple servers could each be handled in this way because the event list would permit the determination of the time intervals during which each server was busy. This is in contrast to the complications that are introduced by considering multiple servers in a mathematical model for a queueing system.

Estimating Probability Distributions

In addition to the **summary statistics** already mentioned, it is also possible to construct a **histogram,** which serves as an approximate probability density function, for any random variable whose values can be calculated from the event list. In principle this permits the determination of any statistical property of the random variable in question.

For example, we might be interested in the probability distribution of the

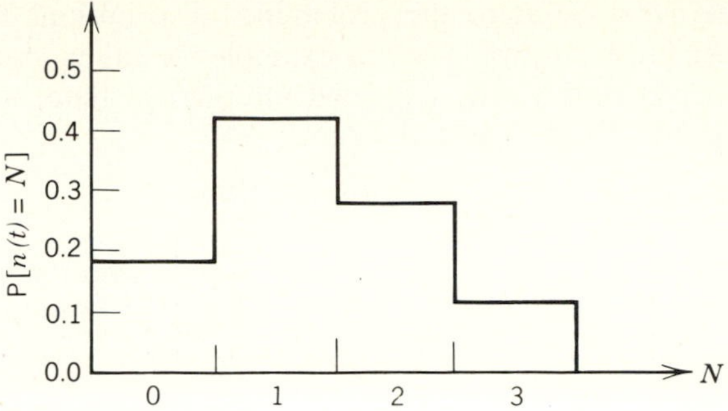

FIGURE 13.4 Histogram for $n(t)$.

total number of customers in the system. From Figure 13.3, there is exactly 1 customer in the system from $t = 0$ to $t = 2.2$ and from $t = 3.9$ to $t = 5.6$, for a total of $2.2 + (5.6 - 3.9) = 3.9$ min. This is a fraction 3.9/9.3 or about 41.8% of the total simulated time. Thus, an estimate for the probability of finding exactly 1 customer in the system is .418. Using the same method to find the probabilities that there are 0, 2, or 3 customers in the system results in the following table:

N	*Periods When N Customers in System*		$\Delta t(N)$ = *Total Time with N Customers in System*	$P[n(t) = N] = \Delta t(N)/T$
0	$(3.9 - 2.2)$	=	1.7	.183
1	$(2.2 - 0.0) + (5.6 - 3.9)$	=	3.9	.419
2	$(6.6 - 5.6) + (8.4 - 6.8)$	=	2.6	.280
3	$(6.8 - 6.6) + (9.3 - 8.4)$	=	1.1	.118

In this table and subsequently, $P[\cdot]$ denotes the probability that $\cdot$ occurs. The probability data from the table are plotted in the form of a histogram in Figure 13.4. This probability density function is consistent with our earlier determination that $L = \frac{4}{3}$. In fact, we can recalculate that result as the expected value of the probability distribution for which the histogram is the density function. Using the definition of the expected value of a discrete random variable,

$$L = \sum_{N=0}^{3} N \cdot P[n(t) = N]$$

$$= 0(.183) + 1(.419) + 2(.280) + 3(.118) = 1.333$$

Figure 13.4 contains additional information as well, however, including the suggestion that the true distribution might be bell shaped and the information required to find the variance and higher moments of $n(t)$.

Checking Random Interevent Times

The quality of the statistics yielded by a simulation depends on the number of events simulated and on the resemblance between the randomly generated interevent times and those that might actually be observed in the real system. A crude check on both requirements can be made by comparing the distributions of observed interevent times to the probability distributions from which the times are supposed to be drawn. For our example we can compare the average observed interarrival time to the expected interarrival time, and the average observed service time to its expected value:

Interarrival	*Time*		*Service*	*Time*	
1–2	3.9	average 2.1 min	1	2.2	average 2.53 min
2–3	1.7		2	2.9	
3–4	1.0		3	2.5	
4–5	1.8		4	?	
			5	?	

With service times uniformly distributed between 2 and 3, the expected value is 2.5, quite close to the measured value of 2.53. However, with interarrival times uniformly distributed between 1 and 5, the expected value is 3, far from the measured value of 2.1. The number of simulated events in our list is obviously far too small for any of our statistical estimates to be very trustworthy, and in practice we would need to use hundreds or thousands of simulated events.

13.3 GENERATING PSEUDORANDOM NUMBERS

The event generation process described in Section 13.1 requires that we supply interevent times drawn at random from the appropriate probability distributions. To formalize the intuitive idea of a random sequence, we use the following definition:

> A sequence of numbers $u_1, u_2, \ldots$ is **random** if a knowledge of $u_1, u_2, \ldots, u_k$ does not determine the value of u_{k+1} with certainty.

Note that it is not meaningful to refer to a single number as "a random number." Thus, for example, there is nothing random about the number 27. It only makes sense to refer to the randomness of a sequence of numbers.

Truly random sequences can be obtained by measuring natural random processes such as the spontaneous decay of a radioactive material or the noise in an electrical circuit, or by using mechanical devices such as cards or dice. If simulations were conducted by hand, we might obtain the required values by directly using just such means. In practice, simulations are conducted using computer programs, so random sequences must somehow be made available in a form that can be used by computers. In the early days of simulation this was done by recording, on magnetic tape, random sequences that were obtained by observing natural random processes. The simulation program could then read the numbers from the tape one by one as they were needed. This approach is expensive because of the need for physical measurements to be made before the tape can be written, it is slow because reading a tape is slow compared to doing calculations, and the random sequences produced by natural processes often come from probability distributions that differ in inconvenient ways from those desired for use in simulations.

The length of the random sequence available to a simulation program was also limited to the number of events recorded on the tape. At first it might seem that one could simply rewind the tape and start over again to obtain a longer sequence, but then the randomness property given above would not be satisfied. If a tape contains N numbers, repeating the sequence means

$$u_{N+k} = u_k$$

so u_{N+1} and all subsequent values are determined with certainty once u_1 through u_N are known.

Because of the various problems associated with stored sequences, modern simulation programs generate number sequences as they are needed, by using numerical algorithms.

Random and Pseudorandom Number Sequences

In view of the foregoing discussion of randomness, the idea of generating random sequences by a numerical algorithm should immediately strike the reader as silly. Algorithms of the sort one might implement by a computer program are by definition completely deterministic, and can therefore never produce a random effect of any kind. If a computer is working properly, it delivers precisely the same result each time it performs the same operation on the same data. For example, it produces the same result whenever it multiplies two particular numbers. Any sequence of numbers generated by an algorithm will therefore have the property that u_{k+1} can be determined with certainty from a knowledge of $u_1, u_2, \ldots, u_k$, and thus will not be random in the sense defined above. How could such perfectly predictable behavior ever be used to produce numbers that are usable in a simulation?

The answer to this seeming contradiction lies in the key assumption stated in Section 13.1, namely, that in simulation we are interested only in long-run statistics such as average waiting time and average queue length. Under the right conditions a sequence of numbers generated by an algorithm can have the same statistical properties as a sequence generated by a natural random process. That is, similar values will occur with similar frequencies in the two sequences, successive values will have similar correlations in both sequences, and so forth. Precisely what degree of similarity is required in which statistical measures in order for us to regard two sequences as effectively indistinguishable, or interchangeable for our purposes, depends on the details of the simulation to be conducted, so we adopt the following convention:

> Two number sequences are **interchangeable** with respect to a particular simulation if they yield simulation results that are statistically indistinguishable for the purposes of that simulation.

For an example of interchangeability, suppose the only purpose of a simulation is to determine the average time customers spend in the system. The simulation is conducted twice, using two different number sequences. In each run of the simulation, the time spent in the system is measured for numerous customers. A statistical test is then performed to determine whether the resulting average

times in the system measured in the two simulation runs are significantly different. If the test shows that the two results are not significantly different, the two number sequences are interchangeable for the purposes of the simulation.

It is easy to see that an algorithm can generate a sequence of numbers that has precisely the same statistics as a sequence of numbers generated by a natural random process, because in principle an algorithm can be made to produce any given number sequence of finite length. This sequence could be identical to the numbers on the tape previously mentioned, so it would obviously be statistically indistinguishable from that sequence, which was generated by a natural random process. Of course, it is not necessary to replicate the exact numbers appearing on the tape, because the statistical properties can be the same even if the sequences of numbers are not, and the sequences can therefore be interchangeable in the sense defined above.

A sequence of numbers that is generated by an algorithm but is interchangeable with a truly random sequence is called pseudorandom.

> A sequence of numbers is **pseudorandom** if every sufficiently short subsequence is interchangeable with a comparable sequence that is truly random.

The only unavoidable difference between a truly random sequence and a pseudorandom sequence is that the sequence generated by an algorithm might repeat, and this accounts for the phrase "sufficiently short" in the definition. Just as rewinding a tape that contains a truly random sequence and starting over yields a longer sequence that is no longer random, permitting the pseudorandom sequence generated by an algorithm to repeat might result in a longer sequence that is no longer interchangeable with a truly random sequence. To ensure this does not happen, the period of repetition of the algorithm must be longer than the sequence of numbers required by the simulation.

Random-Number Generators

After carefully distinguishing between the random sequences produced by natural random processes and the pseudorandom sequences produced by algorithms, it is conventional to use terminology that ignores the distinction. This is because in practice one never uses truly random sequences; it is always pseudorandom sequences that are under discussion, so no confusion can arise. Thus, an algorithm for generating a pseudorandom sequence of numbers is commonly called simply a **random-number generator.** It is also common to speak of **random numbers** with the understanding that a pseudorandom sequence of numbers is actually being referred to, and to speak of a single number in a pseudorandom sequence as *a* random number even though randomness is a concept that really applies only to the sequence as a whole.

There are many algorithms for generating random numbers, differing from one another in the extent to which they exhibit the following desirable properties:

- Long period of repetition

As already mentioned, it is desirable to be able to complete a simulation without having the random numbers begin to repeat. In practice, if the sequence is long

enough, then even if it does repeat that might have a small enough effect so the simulation results remain valid.

- Apparent statistical independence of successive numbers

This is necessary to ensure that the pseudorandom sequence is interchangeable with a truly random one. Knowing any subsequence should not help to determine the number that comes next.

- Uniform distribution

Sometimes a simulation requires random interevent times that are drawn from a uniform probability distribution, that is, a distribution whose values are equally probable within a given range. Even when that is not true, however, it is convenient to have a random-number generator produce uniformly distributed values. This is because, as we shall see, values from a uniform distribution are easy to transform into values from any arbitrary probability distribution that might be desired.

- Speed

Real simulations typically require a great many random numbers, so the speed with which an algorithm can be executed might have a big influence on the computer time consumed. Usually, simplicity is the key to high speed.

- Repeatability

Although it is important for the sequence to be pseudorandom in the sense already defined, it should be possible to use the *same* pseudorandom sequence from one run of a simulation to the next. The main reason for this is so that it is possible to debug the computer program that performs the simulation. Debugging is always necessary in practice and is extremely difficult unless the same number stream can be used repeatedly.

It is easy to design a random-number generator so that its output is repeatable. However, it is always necessary to make trade-offs between period of repetition, independence of successive numbers, uniformity of distribution, and speed. In order to produce pseudorandom sequences having good statistical properties, it is inevitably necessary to sacrifice speed. Thus, the selection of a suitable random-number generator for a given simulation can be a nontrivial task.

The Multiplicative Congruential Algorithm

The simplest random-number generator is the multiplicative congruential algorithm.

MULTIPLICATIVE CONGRUENTIAL ALGORITHM

0. Fix parameters p and m
1. u_0 = starting number, nonzero and odd
   $k \leftarrow 0$
2. $u_{k+1} = (mu_k) \bmod p$
3. $k \leftarrow k + 1$
   Go to 2

All the numbers used in the algorithm are nonnegative integers. The parameter m is called the **multiplier,** the parameter p is called the **modulus,** and the starting number u_0 is called the **seed.** The heart of the algorithm is the **recursion formula**

$$u_{k+1} = (mu_k) \bmod p$$

telling how each number in the sequence is generated from the previous one. The notation "mod p" in this formula refers to the result of an integer division.

$$x \bmod p = \text{remainder left over after dividing } x \text{ by } p$$

If no u_k is zero, the integers that can be generated range from 1 through $p - 1$, so $(u_k - 1)$ ranges from 0 through $p - 2$ and corresponding real numbers r_k in the range [0, 1] can be obtained as

$$r_k = \frac{u_k - 1}{p - 2}$$

To see how the multiplicative congruential algorithm works, consider the following example:

$$p = 17$$
$$m = 5$$
$$u_0 = 7$$
$$u_1 = (5 \cdot 7) \bmod 17 = \text{remainder of integer division } \tfrac{35}{17} = 1$$

$$\begin{array}{r} 2 \\ 17\overline{)35} \\ \underline{34} \\ 1 \end{array} = 35 \bmod 17$$

$$u_2 = (5 \cdot 1) \bmod 17 = 5$$
$$u_3 = (5 \cdot 5) \bmod 17 = 8$$

Continuing the process yields the sequence

7, 1, 5, 8, 6, 13, 14, 2, 10, 16, 12, 9, 11, 4, 3, 15, 7, 1, 5, 8, . . .
└──── period of repetition ────┘

The repetition is obvious in this example, with a period of 16 numbers. The period is always less than the parameter p, so p should be chosen as the largest value that can conveniently be used. Usually, because of efficiency considerations, this choice is influenced by the length of the words in the computer on which the algorithm is to be run. In particular, if the computer word length is w bits and we chose $p = 2^w$, the multiplication in the recursion formula yields a two-word result, the low-order word of which is the product modulo 2^w. The choice of $p = 2^{w-1}$ permits the desired value to be found almost as easily, and it is therefore also frequently used. In either case the fact that no division is required to perform the modulus operation provides a considerable speed advantage over algorithms using other values of p.

It is possible for the period of repetition to be much less than p if the multiplier m is inauspiciously chosen, and to get the longest possible period, $p - 1$, m should be chosen in such a way that $m - 1$ is a multiple of every prime number that divides p without a remainder.

It is also essential that no integer multiple of m be equal to p, as shown by the following example:

$$p = 10$$
$$m = 5$$
$$u_0 = 5$$
$$u_1 = (5 \cdot 5) \bmod 10 = 5$$
$$u_2 = (5 \cdot 5) \bmod 10 = 5$$
$$u_3 = (5 \cdot 5) \bmod 10 = 5$$

In such cases the period of repetition is only one number long.

Finally, for the sequence generated to have good statistical properties, m should be close to the square root of p, and the sequence used in a simulation should be no longer than the square root of the period of repetition.

Even if these rules are followed in selecting p and m, it is quite possible for the sequence generated by the multiplicative congruential algorithm to have significant **serial correlations** (that is, poor apparent independence of the successive numbers) or to depart significantly from a uniform distribution. In practice, therefore, the choice of values for p and m must be guided by statistical analysis of the sequences actually produced. The most commonly used multiplicative congruential generator, which is available on many computer systems by the subroutine name RANDU, has $p = 2^{31} = 2147483648$ and $m = 2^{16} + 3 = 65539$. These choices yield an algorithm that has a period of 2^{29} numbers and runs very fast on computers having a word length of 32 bits. However, the sequence it produces has the property that each number is related to the two previous ones by a simple formula, and this strong serial correlation makes the sequence generated by RANDU unsuitable for many simulations.

Some other multiplicative congruential generators in common use are reported to have statistical properties much better than those of RANDU. Two that are available in widely used commercial scientific subroutine packages have the following parameters:

$$p = 2147483648 = 2^{31} \qquad\qquad p = 2147483647 = 2^{31} - 1$$
$$\text{and}$$
$$m = 302875106592253 = 13^{13} \qquad\qquad m = 16807 = 7^5$$

The generator on the right runs slower than RANDU because the modulus operation is much easier to perform using $p = 2^{31}$ than with $p = 2^{31} - 1$.

Other Uniform Random-Number Generators

Improved statistical properties can be obtained by adding a constant term to the multiplicative congruential recursion formula, as follows:

$$u_{k+1} = (a + mu_k) \bmod p$$

The number a is called the **increment,** and an algorithm that uses this recursion formula is called a **mixed** or **linear congruential** generator. As in the case of the multiplicative congruential generator, p should be the largest value that can conveniently be used. The increment a should obviously be nonzero, but the other considerations involved in picking good values for a and m are rather complicated. One published algorithm uses the following parameter values:

$$p = 2147483648 = 2^{31}$$

$$m = 843314861 \approx p(\pi/8) + 5$$

$$a = 453816693 \approx p(3 - \sqrt{3})/6$$

This generator has excellent statistical properties, and although it runs slower than RANDU because of the extra addition in the recursion formula, it is faster on most machines with 32-bit words than multiplicative congruential generators having $p > 2^{32}$ or p not a power of 2.

Another commonly used random-number generator having good statistical properties is the **GPSS algorithm.** The idea of this algorithm can best be illustrated by an example using decimal integers, although it is actually carried out using binary numbers on a computer. Suppose for the sake of the example that it is desired to generate random numbers r_k from .0000 through .9999. We begin with a multiplier m, which is fixed, and a seed u_0.

$$m = 5167$$

$$u_0 = 3729$$

These are multiplied together to produce a product having eight digits.

$$mu_0 = 19267743$$

The middle four digits of this product are used to form the first random number r_1, and the rightmost four digits are used as u_1. Thus,

$$r_1 = .2677$$

$$u_1 = 7743$$

Next the product mu_1 is formed, its middle four digits are used as the digits of r_2, its rightmost four digits become u_2, and so forth. The GPSS generator gets its name from General Purpose System Simulator, a next-event simulation programming language discussed further in Section 13.5. The GPSS language is widely used, so many simulations generate random numbers this way.

The final category of random-number generators in wide use is the **generalized shift register** algorithms. These algorithms were developed by computational physicists for use in Monte Carlo simulations of the kind discussed in Section 13.4. In many such applications successive numbers are used as random coordinates in spaces of high dimension, and the sequences produced by otherwise acceptable multiplicative congruential generators were found to exhibit undesirable patterns when used in that way. Also, the periods of repetition of multiplicative congruential generators were found to be far too short for many Monte Carlo simulations. The generalized shift register approach avoids both shortcomings by using recursions in which u_k is determined from several previous values, extending back to u_{k-q}, rather than from only u_{k-1}. Results from number theory are used to select recursions having the maximum possible period of repetition, which is $2^q - 1$. One popular variant, called the R250 algorithm, has $q = 250$ and uses the recursion

$$u_k = u_{k-147} \oplus u_{k-250} \qquad k = 251 \ldots 250 + K$$

to generate K random numbers.

The symbol $\oplus$ in the recursion formula denotes the **exclusive-or** logical operation, applied to the bits of u_{k-147} and u_{k-250} when they are written as binary numbers. The exclusive-or operation is defined by the following table in which the marginal entries are the possible operand bits and the entries inside the table are the result bits obtained for the various input combinations:

		y = 0	y = 1
x	0	0	1
	1	1	0

$x \oplus y$ exclusive-or operation

The exclusive-or returns 1 if the input bits are different and 0 if they are the same. To see how the exclusive-or is used in a typical iteration of the R250 algorithm, suppose that at step k of the algorithm we have $u_{k-147} = 369$ and $u_{k-250} = 811$. To find u_k according to the recursion formula, we (or rather the computer) would proceed as follows:

$$\begin{aligned} u_{k-147} &= 369_{10} = 000101110001_2 \\ &\qquad\qquad\qquad \oplus \\ u_{k-250} &= 811_{10} = \underline{001100101011_2} \\ &\qquad\qquad\quad 001001011010_2 = 602_{10} = u_k \end{aligned}$$

Usually, of course, computers have words longer than the 12 bits used in this simplified example.

The 250 most recently generated values must be retained at each step of the R250 algorithm, so that subsequent steps will always have the values they require. This vector of the 250 previous u_k's can be thought of as a shift register whose contents are shifted left one element per iteration of the algorithm, and this accounts for the name. The first 250 values in the shift register may be thought of as seeds, and must be generated as part of initializing R250 by using another method such as a multiplicative congruential algorithm. The R250 algorithm achieves its maximum possible period of repetition of $2^{250} - 1$ ($\approx 2 \cdot 10^{75}$), and has been found empirically to have good statistical properties. The shifting of the shift register takes some computer time not required by the other algorithms we have discussed, and this algorithm also uses more storage, but the recursion itself runs very fast and the extremely long period makes R250 the algorithm of choice for some applications.

Nonuniform Distributions

Often it is necessary to provide interevent times that are randomly drawn from some probability distribution other than the uniform distribution. Various ad hoc techniques can be used to obtain random numbers from nonuniform distributions, but only one method is both universally applicable and widely used. That is the **method of inverse transformation.** The idea is to generate a random number from a uniform distribution and then calculate, from a formula (or a graph or table), a corresponding value that appears to be drawn from the desired distribution. The process is shown graphically in Figure 13.5.

To use the method of inverse transformation, we generate values u of a random variable U that is uniformly distributed on the interval [0, 1]. For each value u, we read across on the graph of Figure 13.5 until we meet the curve, and then read down to obtain a corresponding value x for the random variable X, which has the distribution we want. In other words, if the curve is a graph of the function $U = F(X)$, we calculate x as $x = F^{-1}(u)$.

For a hypothetical distribution function F, if $x = F^{-1}(u)$, then

$$P[X \leq x] = P[F^{-1}(U) \leq x] = P[U \leq F(x)]$$

The last equality above holds because the two probabilities are both represented by the shaded area in Figure 13.5. Next we observe that, by the definition of a uniformly distributed random variable,

$$P[U \leq F(x)] = F(x)$$

Thus $P[X \leq x] = F(x)$, and we see that the distribution function of X is in fact equal to $F(x)$. That is,

> if U is a random variable uniformly distributed on [0, 1], then $X = F^{-1}(U)$ is a random variable having the distribution function $F(x)$.

As an example of the method of inverse transformation, consider the problem of transforming uniformly distributed random numbers into random numbers drawn from an exponential distribution with parameter c. This comes up all the time in the simulation of queueing systems because exponentially distributed interevent times are often observed (or at least assumed). If X is an exponentially distributed random variable, its distribution function is

$$F(x) = 1 - e^{-cx}$$

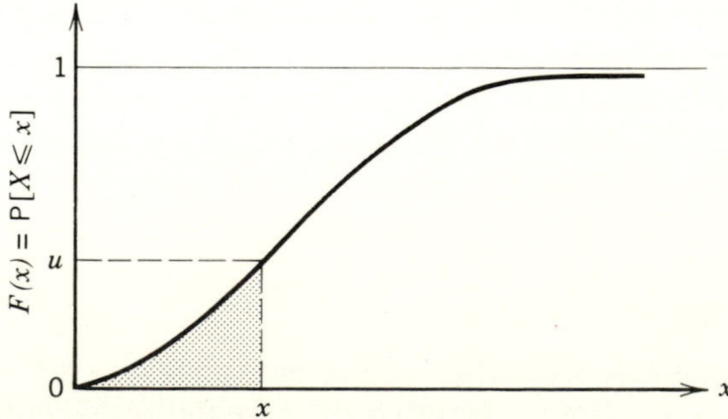

FIGURE 13.5 The method of inverse transformation.

Given a value u drawn at random from a distribution that is uniformly distributed on [0, 1], we can find the corresponding value of x by solving the equation $u = F(x)$ for x. If u is drawn from a distribution that is uniform on [0, 1], however, then we could just as well use $1 - u$ instead because it is also drawn from a distribution that is uniform on [0, 1]. Then it is easy to solve for x as follows:

$$\begin{aligned} 1 - u &= 1 - e^{-cx} \\ e^{-cx} &= u \\ -cx &= \ln(u) \\ x &= \frac{-\ln(u)}{c} = F^{-1}(1 - u) \end{aligned}$$

This result says that if U is distributed uniformly on [0, 1], then $X = -\ln(U)/c$ is distributed exponentially with parameter c. It also means that we can generate uniformly distributed random numbers and simply use the formula to get random numbers that are exponentially distributed with parameter c.

The method of inverse transformation can be used to transform uniformly distributed random numbers into random numbers from any distribution, so long as the inverse, F^{-1}, of the desired distribution can be found analytically or approximated numerically.

13.4 MONTE CARLO SIMULATIONS

In addition to next-event simulation, described in Section 13.2, there is another important technique, called **Monte Carlo simulation** or **static simulation,** in which random numbers and the laws of probability are used to approximate the solutions of problems that have nothing to do with queueing-type systems. Although this technique is referred to as simulation, it does not necessarily have any interpretation as the mechanical imitation of a real process. The name Monte Carlo derives from the fact that a city of that name is famous for its gambling casinos, and reflects the importance of random numbers in the method. Monte Carlo simulation, like dynamic programming, is a general approach rather than a specific algorithm. To give the flavor of this approach we shall discuss two of its standard applications, integration and optimization.

Evaluation of an Integral

Consider the problem of calculating the numerical value of the following definite integral:

$$I(a, b) = \int_a^b f(x)\,dx = \int_a^b \frac{\ln(x)}{(3 - x)}\,dx$$

This integral cannot be evaluated in closed form; that is, it cannot be expressed in terms of a finite number of elementary functions. One way of calculating $I(a, b)$ for given values of a and b would be to use a deterministic numerical procedure such as Simpson's rule. For $a = 1$ and $b = 2$, this yields $I(1, 2) \approx 0.294441$. Figure 13.6 shows a graph of the integrand function $f(x)$, with the area corresponding to $I(1, 2)$ shaded.

Another way of approximating the integral is to estimate what fraction of the rectangular box in Figure 13.6 is occupied by the shaded area, and then multiply that fraction by the known area of the box. The box can be chosen arbitrarily, so long as it contains the entire area corresponding to the value of the integral. For convenience the box in Figure 13.6 has for its base the interval on the x axis between the lower and upper limits of integration, and a height of 1, which makes the area of the box equal to 1.

To estimate the fraction of the box that is occupied by the shaded area, we can generate points (x, y) randomly located within the box, check each point to see whether it is above or below the graph of the function, and estimate the fraction of the box that is occupied by the shaded area from

$$\text{area fraction} \approx \frac{\text{number of random points } (x,\ y) \text{ having } y < f(x)}{\text{total number of random points tried}}$$

Each random point requires two random numbers, x and y. The x value needs to be uniformly distributed between 1 and 2, and the y value needs to be uniformly distributed between 0 and 1. We can generate a single stream of random numbers u that are uniformly distributed on [0, 1] and find the coordinates of the kth random point (x, y) from the following formulas:

$$x_k = 1 + u_{2k-1}$$

$$y_k = u_{2k}$$

Figure 13.6 shows the first 100 points obtained by using a multiplicative congruential algorithm to generate random numbers and the preceding formulas to find the point corresponding to each pair of random numbers. There are 32

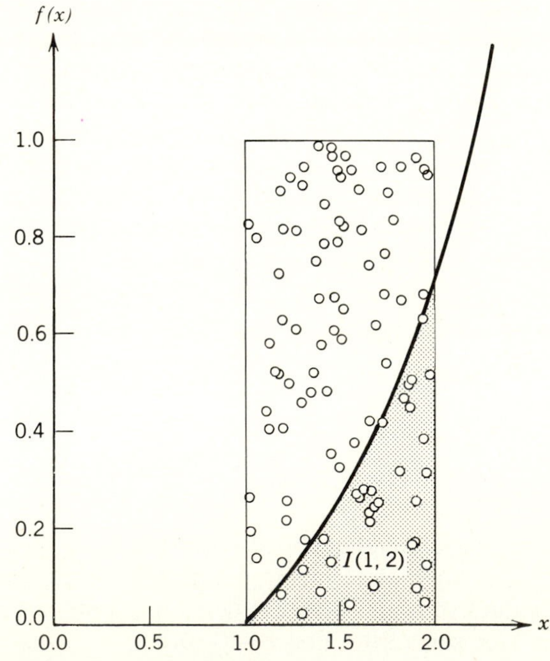

FIGURE 13.6 Monte Carlo integration.

points below the curve, so a crude estimate of the area fraction occupied by the shaded area is

$$\text{area fraction} = \tfrac{32}{100} = .32$$

Then the value of the integral is approximately

$$I(1, 2) = (\text{area fraction})\cdot(\text{area of box}) \approx (.32)\cdot(1) = .32$$

Using more random points yields the progressively better estimates given in the following table:

Total Number of Random Points (x, y)		Resulting Estimate of $I(1, 2)$
100		.320000
1000		.278000
10000		.299400
100000		.296040
1000000		.294538
True value	$\approx$	.294441

The estimate approaches (very slowly) the true value of the integral as the number of trial points increases.

One would never actually use Monte Carlo simulation to solve a simple problem like this example because a deterministic numerical method such as Simpson's rule gives a much more accurate result with much less computational work. For integrals in high dimensional spaces, however, the situation is often reversed. The integral value is then the volume of a region whose boundaries might be described by complicated formulas, so that it is difficult to determine what limits to use in a direct numerical integration. The need to integrate over many dimensions also greatly increases the computational work, so that a Monte Carlo estimate is often easier. Monte Carlo simulation for the evaluation of definite integrals (and for some closely related problems) is thus very important in fields such as nuclear physics and accounts for a large proportion of the supercomputer cycles devoted to scientific and engineering calculations today.

The Metropolis Algorithm

In 1953, N. Metropolis introduced a simple algorithm for simulating a collection of atoms in thermal equilibrium at a given temperature. Starting from some nonequilibrium initial arrangement of the atoms, the algorithm considers one atom at a time and computes the change in the energy of the system that would result from a small random displacement of that one atom. If the energy would decrease or stay the same, the displacement is accepted and the configuration with the displaced atom is used as the starting point for the next step. If the energy would increase, the displacement might still be accepted, with a probability that depends on the energy change proposed. The probability of accepting a higher-energy configuration decreases with decreasing temperature. For physical reasons Metropolis used the following probability formula:

$$P(\text{accept an energy increase of } \Delta E) = e^{-\Delta E/bT}$$

In the formula, T is temperature and b is a parameter that determines the **annealing schedule,** that is, how the probability changes with temperature T.

To simulate a collection of atoms, one fixes T and performs many iterations of the algorithm described above. Eventually the energy of the system stops changing systematically, and only random variations about the equilibrium energy are observed from one iteration to the next. If the temperature is then decreased, a new equilibrium arrangement is eventually reached. Decreasing T gradually to zero results in a final configuration that is a minimum energy state for the collection of atoms. This models the process of annealing in which a desirable arrangement of atoms is obtained in a metal by gradual cooling from a high temperature.

After Metropolis published his algorithm, S. Kirkpatrick observed that finding which arrangement of atoms has the lowest total energy is analogous to finding what values the decision variables of an optimiziation problem should have in order to minimize the objective function. He applied the Metropolis approach to the traveling salesman problem, which we discussed in Section 8.5, in the following way.

Represent the tour of the traveling salesman by a list of the numbers of the cities in the order that they are visited. For example, the tour in Figure 13.7*a* would be represented by the following list:

$$1, 2, 3, 4, 5, 6, 7, 8$$

It is understood that the tour is closed by returning from the city that is last on the list to the one that is first. A **perturbed tour** is formed by interchanging two city indices that are adjacent in the list. Thus, for example, we might switch 4 with 5, to form the following new list:

$$1, 2, 3, 5, 4, 6, 7, 8$$

This new list describes the tour in Figure 13.7*b*. The figure shows that the process of switching adjacent indices in the list corresponds to changing the direction in which one of the links in the tour is traversed. Thus, in Figure 13.7*b*, instead of going from city 4 to city 5, the salesman goes from city 5 to city 4. The list must be thought of as circular, so as to include the possibility of exchanging the beginning city with the ending one.

Some perturbations of the tour will yield a decrease in the total length, while others will yield an increase. To apply the Metropolis algorithm, we start with some suboptimal tour and generate a perturbed tour by randomly picking a pair of adjacent indices to exchange. For a list of N cities, this might be done

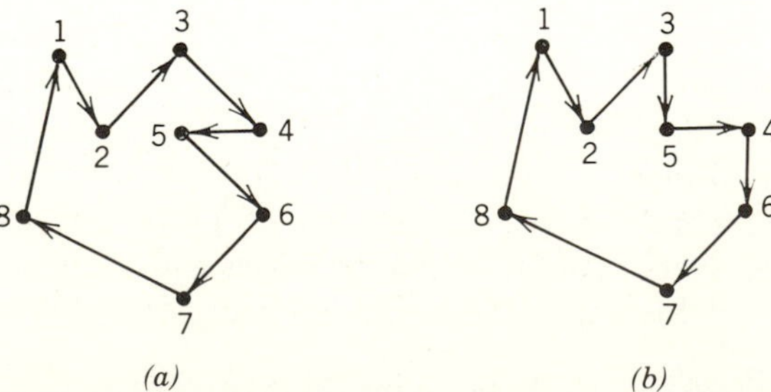

FIGURE 13.7 Traveling salesman tours.

by generating a random number u in [0, 1] and exchanging indices i and $(i + 1) \bmod N$ where

$$\frac{i-1}{N} < u \leq \frac{i}{N}$$

If the length of the perturbed tour is less than that of the starting tour, it is accepted. If the perturbed tour is longer, it is accepted only with a probability that depends on the amount by which the tour would be lengthened, as follows:

$$P(\text{accept a length increase of } \Delta L) = e^{-\Delta L/bT}$$

As in the original Metropolis algorithm, the parameter b determines a sort of annealing schedule. T can be thought of as a sort of temperature in that it measures the degree of disorder attributable to changes in the tour from one iteration to the next.

To find the tour of minimum length, we begin with a high temperature and run the algorithm until a steady state is reached. When the tour length no longer systematically decreases but only varies randomly from one iteration to the next, we decrease the temperature slightly and do some more iterations. Eventually a new equilibrium length is obtained that is lower than the previous value. Continuing this process until the temperature has been reduced to $T = 0$ yields a final configuration with a very low total tour length, perhaps even the optimal tour. This process is called **optimization by simulated annealing** because of the many similarities it has to the physical annealing that Metropolis originally simulated.

Simulated annealing can find approximate answers to very large traveling salesman problems much more quickly than deterministic algorithms. In fact, it has been used to find near-optimal solutions to problems with several thousand cities, whereas the best exact deterministic algorithms are at present impractical for problems with more than a few hundred cities.

Simulated annealing also is not limited to traveling salesman problems, nor even to integer programs. Some research has been conducted into continuous-variable nonlinear programming algorithms that use the Metropolis approach in searching for the global minimum of a nonconvex problem. The simplest algorithm of this kind selects a random direction from among those that are feasible, and moves a random distance in that direction. The resulting new point is accepted if its objective value is less than the current one, and is accepted with a certain probability if its objective value is greater. The temperature of the process is gradually reduced just as in the traveling salesman algorithm, so that the excursions become smaller as the minimum is approached. It is hoped that **random algorithms** of this kind will be able to avoid falling into local minima early in the solution process, and will therefore find global minima most of the time. If these algorithms are successful, they will constitute the first effective general way of solving nonconvex nonlinear programs.

Finally, the use of Monte Carlo methods in optimization is not limited to simulated annealing. Simple random sampling has been used to approximate the solutions of integer programs, and the random selection of pivot columns is sometimes used to prevent cycling in the simplex algorithm for linear programming. Random numbers also play a prominent role in the generation of test problems for use in the development and evaluation of mathematical programming algorithms and other numerical methods.

13.5 PRACTICAL CONSIDERATIONS IN SIMULATION MODELING

Both next-event simulation and Monte Carlo simulation are intuitively appealing and conceptually simple techniques, and it might seem from the preceding sections that the most difficult aspect of applying these methods is the choice or design of a random-number generator. Actually, although the generation of random numbers having the necessary statistical properties is very important to the success of a simulation model, the most serious difficulties encountered in practice have to do with other aspects of applying the technique. This section briefly discusses some of the most important problems that arise and what can be done about them.

Computer Programs for Simulation

The first and most obvious problem, already alluded to in earlier sections, is that simulation is not practical to do by hand because statistically valid results require that too many events be simulated. This problem can obviously be solved by writing a computer program to conduct the simulation. Simulations are sometimes programmed in FORTRAN, Pascal, and other procedural computer languages, but writing a complicated next-event simulation in a low-level general-purpose language is difficult, time consuming, and error prone. Fortunately, special-purpose high-level **simulation languages** have been developed in which simulations can be described, reliably and with comparatively little effort, in terms that are directly meaningful for the simulation model rather than in terms of basic computer operations. Thus, for example, simulation languages provide constructs for the representation of system entities such as customers and servers, and automatically handle such tasks as random-number generation and event scheduling. Among the many simulation languages now in wide use, some of the most popular are GPSS, Simscript, Simula, and SLAM.

The oldest and most commonly used simulation language is GPSS, mentioned in Section 13.3 in connection with the algorithm it uses for generating random numbers. In GPSS customers are represented by **transactions,** which are thought of as moving through the model and being acted upon at stationary **facilities,** which represent servers. This **worldview** is natural for many queueing-type systems in which customers (or parts, or messages, etc.) actually do move, in some phsyical sense, from one server (or machine, or radio station, etc.) to another.

A simple example will give the flavor of GPSS programming. Suppose we want to model a grocery checkout line with one cashier. Customers arrive at random, with interarrival times uniformly distributed between 120 and 240 sec. The cashier's service time is also uniformly distributed, between 120 and 200 sec. Note that elementary queueing theory results do not apply to this situation because the interevent times are not exponentially distributed. In the GPSS model, transactions will represent customers, and they will be acted upon at a facility representing the cashier. This can be shown by the **GPSS block diagram** in Figure 13.8*a*.

In addition to providing for the arrival, possible waiting, service, and departure of customers, it is also necessary to arrange for the collection of statistics. The left-hand **model segment** in Figure 13.8*a* therefore contains blocks that have

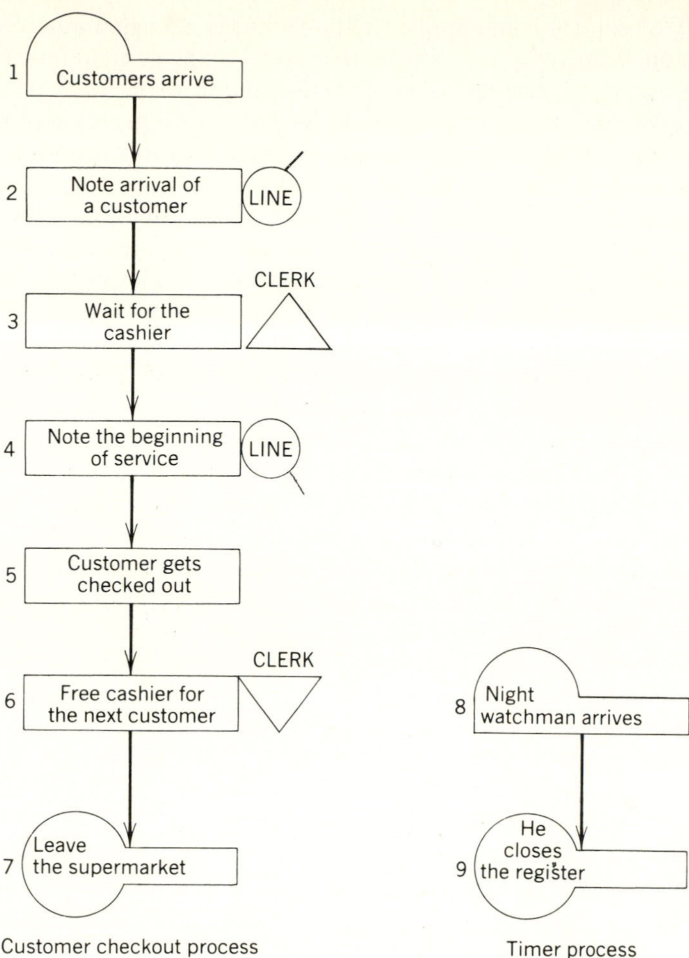

FIGURE 13.8*a* GPSS block diagram for the grocery checkout line.

the effect of causing events of interest to be recorded on the event list for the simulation as customers enter those blocks. The function of the right-hand model segment is to stop the simulation and order an analysis of the event list that has been constructed. This **timer segment** can be visualized as a night watchman who arrives at the end of the day to close the register and calculate the desired statistics about the operation of the checkout counter during the day. Each model segment can be thought of as modeling a different **process** taking place within the system.

Several other things should also be observed about Figure 13.8*a*. First, the generation of random interevent times, the scheduling of events, and other simulation management tasks are conducted automatically. These things are not indicated on the block diagram. Second, the block diagram specifies the flow of transactions through the model, *not* the flow of control from one instruction to another in the underlying computer program that actually performs the simulation. In this respect a GPSS block diagram is completely different from a block diagram that might be drawn to represent the flow of control in a computer program that is written in a procedural language like FORTRAN. In a simulation model, flow of control is not explicitly shown but is managed automatically by the simulation language processor, in this case GPSS, as required to perform the simulation. Third, transactions can be moving in many parts of the model simultaneously, and can accumulate (just like people in a queue) at blocks where they have to wait. Thus, for example, a queue of transactions might accumulate in the block marked "wait for the cashier." Of course, if the cashier is idle when a transaction arrives at that block, the wait will last zero seconds. If transactions are waiting for the cashier, then when the one being served is finished, its movement out of the store will be simultaneous with the movement of the next customer up to the register. In general a GPSS model can have many segments, and events can take place in all of them simultaneously. For an example of this, think of a model with additional segments to represent additional checkout lines, each with its own cashier.

For processing by a computer, a GPSS model must be encoded in a sequence of instructions, as shown in Figure 13.8*b*. The lines of Figure 13.8*b* numbered on the left correspond to the blocks in the block diagram of Figure 13.8*a*. There are also comments (lowercase remarks on the right, and any line beginning with an *), which are for human use only and are ignored by the GPSS processor. The remaining lines are control statements to tell the processor how to manage the simulation. The details of syntax used in the GPSS program are unimportant for the purposes of this introduction, but the pertinent features of the checkout line can be recognized in the comments and in the block operands. The first GENERATE block specifies uniformly distributed interarrival times of 180 ± 60 secs, and the ADVANCE calls for uniformly distributed service times of 160 ± 40 secs. The second GENERATE causes the night watchman to arrive exactly 28800 secs (= 8 hr) after the simulation begins.

When the night watchman ends the simulation, the GPSS processor analyzes the event history that has been collected so far and produces certain statistics automatically, as shown in Figure 13.8*c*. The output shows that the program

```
              SIMULATE
      *
      *       model segment to simulate a checkout line
1             GENERATE    180,60   customers arrive
2             QUEUE       LINE     note the arrival of a customer
3             SEIZE       CLERK    wait for the cashier
4             DEPART      LINE     note the beginning of service
5             ADVANCE     160,40   customer gets checked out
6             RELEASE     CLERK    free cashier for next customer
7             TERMINATE            leave the supermarket
      *
      *       timer segment to end the simulation
8             GENERATE    28800    night watchman arrives
9             TERMINATE   1        he closes the register
      *
      *       simulation control statements
              START       1        start the run
              END                  there are no more instructions
```

FIGURE 13.8*b* GPSS program for the grocery checkout line.

```
RELATIVE CLOCK: 28800          ABSOLUTE CLOCK: 28800

BLOCK CURRENT      TOTAL
1                    158
2                    158
3                    158
4                    158
5         1          158
6                    157
7                    157
8                      1
9                      1

              --AVG-UTIL-DURING--
FACILITY  TOTAL  AVAIL  UNAVL  ENTRIES   AVERAGE   CURRENT  PERCENT  SEIZING  PREEMPTING
          TIME   TIME    TIME            TIME/XACT  STATUS   AVAIL   XACT     XACT
  CLERK   .886                     158   161.670     AVAIL   100.0   159

  QUEUE  MAXIMUM   AVERAGE   TOTAL    ZERO     PERCENT  AVERAGE    $AVERAGE   QTABLE  CURRENT
         CONTENTS  CONTENTS  ENTRIES  ENTRIES  ZEROS    TIME/UNIT  TIME/UNIT  NUMBER  CONTENTS
   LINE         2     0.257  158      59       37.3     46.879        74.818               0

TOTAL BLOCK EXECUTIONS: 1106
```

FIGURE 13.8*c* GPSS output for the checkout line model.

stopped after simulating 28800 sec of real time (the computer time used amounted to only a few seconds, however). The block counts tell how many transactions moved through the various parts of the model, by the block numbers listed in Figure 13.8*b*, and the current count of one transaction in block 5 indicates that a customer was being checked out when the night watchman arrived and locked the cash register. The cashier, named CLERK in the program, was busy about 88.6% of the time, and on the average took about 162 sec to check a customer out of the store. The average service time we specified was 160 sec, so this last number seems reasonable. On the average there were 0.257 people waiting in line for the cashier, and of the 158 who went through the line during the day, 59 did not have to wait at all. The average waiting time was about 47 sec if all the customers are counted, or about 75 sec counting only the 99 who had to wait. The total of 158 arrivals during the course of the simulation is consistent with the average interarrival time of 180 sec that we specified, which would allow for 28800/180 = 160 arrivals. When the simulation ended, the queue was empty because the one person in the system was being served.

When we performed the next-event simulation example in Section 13.2 by hand, we kept an event list for the entire simulation and analyzed it at the end to obtain the statistics we wanted. GPSS uses an event list as well, but rather than recording the history of the whole simulation it retains only events that are scheduled to occur or are in the process of occurring. Statistics collection is therefore done by GPSS on the fly rather than retrospectively. In every other way the process followed by GPSS in conducting the simulation resembles what we would do by hand.

The foregoing simple example of GPSS programming suggests what it is like to specify a simulation in terms of high-level constructs such as transactions, facilities, and processes. Other next-event simulation languages provide this ability as well, and the simulations they provide are typically based on event lists. However, they differ considerably from GPSS and from one another in syntax, organization, and philosophy. Some simulation langauges adopt a fundamentally different worldview from the one used by GPSS, the most common alternative being to consider customers as static entities that are visited by active servers.

Logical and Statistical Validity

Numerous approximations are involved in the design of any simulation, and they can give rise to problems of two distinct kinds. The first sort of problem stems from approximations about the structure of the system being simulated. For example, in the supermarket checkout line it would be rather rude of the night watchman to lock the cash register while a customer is in the process of being checked out. It is sensible to ask whether the results of the simulation will accurately reflect reality if the model does not include the fact that the last customer is actually allowed to pay up and leave before the store is closed for the night. The second sort of problem has to do with the fact that the results obtained from a simulation model are statistical in nature. In the checkout line model we were interested in steady-state quantities such as the average number of people waiting for the cashier. When the simulation began, however, the server was obviously idle, so the first few customers probably did not have to wait as long as some of those who arrived later in the day after the store was full of people. Are the results obtained from the model affected by including the start-up period before steady-state conditions have been reached? For that matter, how can we even tell when steady state has been reached? Even worse, how long should the simulation be run so as to provide confidence intervals of a given size on the statistical estimates obtained?

Problems of the logical and statistical kinds, which have different causes and different cures, typically interact and imitate one another in such a way as to confuse the process of refining simulation models and frustrate their validation. At first this may seem like a perverse twist of fate, but in a certain sense it is only to be expected. After all, simulation is just a mechanical way of conducting an experimental investigation, and experimenters have always struggled with these same difficulties. Does the experimental setup faithfully reflect those aspects of reality that are under study? How precise are the measurements we make?

In addition to these problems, which are unavoidable hazards in using any experimental approach to problem solving, there are, of course, ordinary mistakes. Simulations are conducted using computer programs and are therefore subject to being wrong on account of errors that have nothing to do with the modeling process.

Various techniques have been developed for ensuring the correctness of experimental results and computer programs, and fortunately many of them are applicable to the validation of simulation models. The most prominent example of a technique that is applicable to both laboratory experiments and simulation experiments is statistical analysis, but any technique of laboratory science for which a reasonable analogy can be found in simulation modeling is fair game for adoption. In particular, every simulation modeling project should adhere to the following rules.

- Compare the predictions of the simulation model to observations of the real system whenever that is possible.

Often a simulation is used to find out what would happen if an existing system were changed in some way. In that case it makes sense to start with a simulation model whose results agree with the observed behavior of the real system, and only then modify the simulation in the way that it is proposed to modify the real system. Another reason for using a simulation is because the real system is difficult or expensive to observe. In that case a small number of observations of the real system should be made and used to verify that the simulation gives sensible results. In some cases, of course, it will be impossible to calibrate a simulation against observations of the real system, but whenever it is possible, even to a limited extent, such a comparison should be made.

- Check the simulation model for self-consistency.

It is not difficult to construct a model that contains logical inconsistencies, so that it does not make sense no matter what it is supposed to represent. The most obvious instances of this are provided by simulation programs that contain **bugs**, or accidental errors such as typing mistakes. For example, trying to generate random numbers from a distribution with negative variance would probably not make sense under any circumstances. Another way that logical inconsistencies can arise is that they are intentionally written into the simulation by an unwitting programmer. For example, the progress of transactions through one model segment might depend on the progress of transactions through another model segment and conversely, so that the transactions in each are waiting for those in the other. The program does exactly what the analyst thought he or she wanted, but nothing ever happens. Bugs and logic errors both can often be discovered by examining block counts or the equivalent, and by inspecting in detail the progress of particular transactions through the model. Simulation programs require debugging just as other computer programs do, but, if a simulation language is used, the debugging process is carried out at the level of transactions and facilities rather than at the level of machine instructions.

Troubles with random-number generation can also give rise to behavior that is not self-consistent in the sense that the simulation program does not reflect the intentions of the analyst. For example, the interevent times may have been assumed to come from a given distribution, but because of some defect in the algorithm used for generating the random numbers their actual distribution is something else. The reasonableness of the random numbers can be checked by comparing their actual distribution to that which was requested. We did simple checks of this kind in Section 13.2 and the first part of this section by comparing the observed average interevent times with those that were requested, but a more elaborate test could be performed by plotting histograms of measured interevent times and comparing them to the desired probability density functions.

Of course, even a simulation that does not have any of the problems mentioned above can contain errors that make the results incorrect. Another way of checking the reasonableness of a simulation model is to see if the results change in the right ways as the system parameters are varied. For example, in the supermarket checkout line we would expect that increasing the average service time would increase the average waiting time and the average length of the queue. Sometimes the most telling way to use this kind of a test is to consider **limiting cases** such as zero service time, which for our example should result in zero average waiting time and zero average queue length.

- Use experimental design and statistical analysis.

The design of a simulation model obviously depends greatly on the precise questions that it must answer because the pertinent features of the real system must be modeled so that the necessary measurements can be made. It is also important to design the statistical experiments that are to be performed upon the model in such a way that the results have statistical validity, including the required precision and freedom from bias. Assuming the random numbers are properly generated and the model is logically sound, the problem of statistical validity comes down to two questions. How can we assure that the measurements we make are of the behavior that is of interest? How can we assure that enough measurements have been made to provide acceptable confidence intervals on the results calculated from them?

The first question arises from the presence of transient effects, both in real systems and in simulation models. For example, when the supermarket checkout line opens in the morning, there is an interval during which no queue has yet developed and the time it takes to get checked out is less than usual. If the steady-state behavior of the system is of interest, the collection of measurements in the simulation should be delayed until after the transient effects have died out and steady state has been reached. One way of deciding when steady state has been reached is to run the simulation for a while and calculate an average that is desired, continue the simulation for an equal length of time and recompute the same average, and perform a statistical test of whether the two averages are significantly different. The process is repeated until two consecutive time intervals yield results that are not significantly different according to the test, and earlier measurements are discarded. The length of the intervals must be determined on the basis of the size and statistical power desired for the test.

If the transient effect is of interest, on the other hand, it is necessary to repeat the simulation of that part of the system's behavior many times, using a different stream of random numbers for each repetition. In this case it is also necessary to decide when the transient effect is over, so that measurements of the steady-state behavior can be excluded.

The second question, what **sample size** will provide given confidence intervals, is familiar from other experimental settings, and the same statistical techniques used there can be employed in the design of simulation experiments. A consideration for experimental design that is of special importance in simulation is that it might be quite expensive to conduct repeated trials by rerunning a simulation because of the time needed to get past the start-up transient. To avoid this problem a technique called **regeneration** is sometimes used. The basis for regeneration is that two subsequences of events in a steady-state simulation may be interchangeable if they have the same initial conditions. This will be true if the behavior of the system depends only on its present state and not on its state at any previous times. Thus, for example, suppose that at some time in the simulation the queue at the supermarket checkout counter is empty and one person has just entered service. If the same conditions prevail at some later time, the sequence of events between the two times can be taken as one sample and the sequence of events beginning at the second time can be taken as a second sample.

Much current research work is focused on the statistical aspects of simulation modeling, and courses in simulation typically consider questions of validation and error control in detail.

SELECTED REFERENCES

Birtwistle, G., Dahl, O.-J., Myhrhaug, B., and Nygaard, K., *SIMULA BEGIN,* Auerbach, Philadelphia, 1973.

Bonomi, E., and Lutton, J.-L., The *N*-City Travelling Salesman Problem: Statistical Mechanics and the Metropolis Algorithm, *SIAM Review* **26**:551–568 (1984).

Conley, W., *Computer Optimization Techniques,* Petrocelli, New York, 1980.

Fishman, G. S., *Principles of Discrete Event Simulation,* Wiley, New York, 1978.

Forsythe, G. E., Malcolm, M. A., and Moler, C. B., *Computer Methods for Mathematical Computations,* Prentice-Hall, Englewood Cliffs, NJ, 1977, Chapter 10.

Kirkpatrick, S., Gelatt, C. D., Jr., and Vecchi, M. P., Optimization by Simulated Annealing, *Science* **220**:671–680 (1983).

Kiviat, P. J., Villanueva, R., and Markowitz, H. M., *SIMSCRIPT II.5 Programming Language,* CACI, Los Angeles, 1975.

Knuth, D. E., *The Art of Computer Programming* (2nd ed.), Vol. 2, *Seminumerical Algorithms,* Addison-Wesley, Reading, MA, 1981, Chapter 3.

Law, A. M., and Kelton, W. D., Confidence Intervals for Steady-State Simulations: I. A Survey of Fixed Sample Size Procedures, *Operations Research* **32**:1221–1239 (1984).

Law, A. M., and Kelton, W. D., Confidence Intervals for Steady-State Simulations: II. A Survey of Sequential Procedures, *Management Science* **28**:550–562 (1982).

Metropolis, N., Rosenbluth, A., Rosenbluth, M., Teller, A., and Teller, E., Equation of State Calculations by Fast Computing Machines, *Journal of Chemical Physics* **21**:1087–1092 (1953).

Pritsker, A. A., *Introduction to Simulation and SLAM-II* (2nd ed.), Wiley, New York, 1984.

Rubinstein, R. Y., *Simulation and the Monte Carlo Method,* Wiley, New York, 1981.

Schreiber, T. J., *Simulation Using GPSS,* Wiley, New York, 1974.

Sowey, E. R., A Chronological and Classified Bibliography on Random Number Generation and Testing, *International Statistics Review* **40**: 335–371 (1972).

Tausworthe, R. C., Random Numbers Generated by Linear Recurrence Modulo Two, *Mathematics of Computation* **19**:201–209 (1965).

EXERCISES

13.1 List several advantages and drawbacks of simulation as a modeling technique.

13.2 Consider the example discussed in Sections 13.1 and 13.2.

(a) Using the probability distribution plotted in Figure 13.4, calculate the variance of $n(t)$,

$$s^2 = \sum_{N=0}^{3} N^2 \cdot \mathrm{P}[n(t) = N] - L^2$$

(b) Continue the event generation process of Section 13.1 to extend the event list by 8 additional event times. The first 10 randomly generated interarrival and service times for this example are:

Interarrival:	3.9	1.7	1.0	1.8	4.2	2.2	1.5	1.1	5.0	2.6
Service:	2.2	2.9	2.5	2.1	2.5	2.0	2.7	2.6	3.0	2.3

The values to the left of the line were used in Section 13.1. Use the subsequent values, in the order they are listed, to find the new event times.

(c) Repeat the statistical analysis of Section 13.2 and part (a) of this question, using the enlarged event list. How do the results differ from those obtained using only the original event list?

(d) Suggest a naive way of deciding, as a simulation progresses, whether enough events have been generated so the statistical results are trustworthy.

13.3 A next-event simulation is to be conducted to study the behavior of a single-server queueing system in which interarrival times are uniformly distributed between 20 and 40 min and service times are uniformly distributed between 10 and 50 min. The following two sequences of interevent times can be considered to have been drawn at random from the appropriate probability distributions.

Interarrival:	22	36	30	38	26
Service:	18	38	34	26	50

(a) Assuming customer 1 arrives and enters service at time zero, construct an event list including events 1 through 7 using simulated interarrival times from the first sequence and simulated service times from the second.

(b) Use the event list constructed in part (a) to calculate the average number of customers in the system, the average time a customer spends in the system, and the fraction of the time the server is busy.

13.4 Suppose that in a sequence of numbers $u_1, u_2, \ldots$, the value of each number after the second one is determined as follows:

$$u_k = \begin{cases} .5(u_{k-1} + u_{k-2}) \text{ with probability } .99 \\ .5u_{k-1} + 2u_{k-2} \text{ with probability } .01 \end{cases}$$

(a) Is the sequence random? Why or why not?
(b) If some truly random process could be used to determine which formula to use at each iteration, would the sequence be a good source of pseudorandom numbers for a simulation? Explain.

13.5 Consider the following multiplicative congruential generator:

$$u_{k+1} = 14u_k \bmod 23$$

(a) Compute the first five values that are generated when the seed is $u_0 = 7$.
(b) It is proposed to check the period of repetition of the above generator by generating values until the repetition is observed. What is the maximum number of values that must be generated in order to carry out this plan?
(c) Determine the period of repetition of the generator.

13.6 If a zero is produced by a multiplicative congruential pseudorandom number generator, all of the subsequent numbers generated are zero and the period of repetition shrinks to one number in length. This will obviously occur immediately if the seed is zero.
(a) Can a zero ever be produced if the seed is not zero? Consider poorly designed generators in addition to good ones, but assume that the multiplier and modulus are positive.
(b) Find the period of repetition of the following generator when the seed $u_0 = 5$ and when $u_0 = 1$.

$$u_{k+1} = (2u_k) \bmod 17$$

(c) Assuming the seed is not zero, does its value affect the period of repetition of a multiplicative congruential generator? Consider all positive values as possible seeds, not just odd numbers as required by the statement of the algorithm. Explain your answer.

13.7 The RANDU algorithm described in Section 13.3 has $u_{k+1} = (2^{16} + 3)u_k \bmod 2^{31}$. Show that this results in a sequence that is serially correlated according to the following formula:

$$u_{k+2} = 6u_{k+1} - 9u_k$$

13.8 Consider the following linear congruential generator:

$$u_{k+1} = (5 + 14u_k) \bmod 23$$

(a) Compute the first five values that are generated when the seed is $u_0 = 7$.
(b) What happens to the sequence produced by a linear congruential generator if a zero occurs? Explain.
(c) It is proposed to check the period of repetition of the generator by generating values until the repetition is observed. What is the maximum number of values that must be generated in order to carry out this plan?
(d) Determine the period of repetition of the generator.

13.9 Continue the iterations of the GPSS generator described in Section 13.3, to generate r_2 through r_5.

13.10 Suppose it is desired to generate random numbers x_k that are uniformly distributed on the interval $[a, b]$.

(a) Use the method of inverse transformation to show that if u_k is drawn at random from a distribution uniform on [0, 1], corresponding values of x_k can be found from the formula

$$x_k = a + u_k(b - a)$$

(b) The following sequence of numbers can be considered to have been drawn from a probability distribution that is uniform on [0, 1]:

.1, .8, .5, .9, .3

Find the corresponding values of x_k if x_k is uniformly distributed between 10 and 50.

13.11 Find a transformation suitable for generating random numbers that are drawn from a logistic distribution, which has the following probability density function where a and b are positive constants:

$$f(x) = \frac{e^{-(x-a)/b}}{b[1 + e^{-(x-a)/b}]^2}$$

13.12 For the Monte Carlo integration example of Section 13.4, use the following random number sequence to estimate the value of $I(1, 2)$ using 10 points.

.186, .129, .395, .067, .679, .242, .186, .888, .725, .938
.582, .268, .490, .931, .196, .312, .166, .520, .659, .232

13.13 Describe a Monte Carlo integration that estimates π by using the formula for the area of a circle.

13.14 Describe a Monte Carlo simulation for estimating the expected value of a random variable X whose probability density function is

$$f(x) = \begin{cases} 2e^{-2x} & x \geq 0 \\ 0 & x < 0 \end{cases}$$

APPENDIX:

NOTATIONAL CONVENTIONS FOR MATRIX ALGEBRA

A.1 MATRICES

A **matrix A** is a rectangular array of real numbers denoted by

$$\mathbf{A} = \begin{bmatrix} a_{11} & a_{12} & a_{13} & \cdots & a_{1n} \\ a_{21} & a_{22} & a_{23} & \cdots & a_{2n} \\ \vdots & \vdots & \vdots & & \vdots \\ a_{m1} & a_{m2} & a_{m3} & \cdots & a_{mn} \end{bmatrix}$$

The ith **row** of **A** is denoted by

$$[a_{i1} \quad a_{i2} \quad a_{i3} \ldots a_{in}]$$

and the jth **column** of **A** is the **vector** denoted by

$$\begin{bmatrix} a_{1j} \\ a_{2j} \\ \vdots \\ a_{mj} \end{bmatrix}$$

The preceding matrix **A** has m rows and n columns and is called an $m \times n$ matrix. The (i, j)th **element** of the matrix **A** is

$$a_{ij} = \text{the entry in the } i\text{th row and } j\text{th column}$$

For example, in the 3×5 matrix

$$\mathbf{A} = \begin{bmatrix} 2 & 0 & 0.33 & 2 & -7 \\ -1 & 9 & 5 & -5 & 0 \\ 0.53 & 3 & -4 & 6 & 27 \end{bmatrix}$$

$$[-1 \quad 9 \quad 5 \quad -5 \quad 0] = \text{the 2nd row}$$

$$\begin{bmatrix} 0.33 \\ 5 \\ -4 \end{bmatrix} = \text{the 3rd column}$$

and

$$a_{23} = 5$$

Equality of Two Matrices

Two $m \times n$ matrices **A** and **B** are said to be **equal** if their corresponding entries are equal, that is,

$$\mathbf{A} = \mathbf{B} \quad \text{if } a_{ij} = b_{ij} \text{ for each pair } (i, j)$$

It does not make sense to talk about equality of two matrices that do not have the same number of rows and the same number of columns.

Multiplying Two Matrices

When the number of columns in a matrix **A** equals the number of rows in a matrix **B**, the product **AB** of the two matrices is the matrix whose (i, j)th element is

$$(\mathbf{AB})_{ij} = a_{i1}b_{1j} + a_{i2}b_{2j} + \cdots + a_{im}b_{mj}$$

If **A** is an $m \times n$ matrix and **B** is an $n \times p$ matrix, then the product, **AB**, is an $m \times p$ matrix. If the number of columns in **A** does not equal the number of rows in **B**, the product **AB** is not defined.

The right-hand side of the expression for $(\mathbf{AB})_{ij}$ is called the **dot product** of the ith row of **A** with the jth column of **B**.

We illustrate matrix multiplication by finding the product of the following two matrices **A** and **B**:

$$\mathbf{A} = \begin{bmatrix} 1 & 3 & 1 & 0 \\ -1 & 2 & 0 & -1 \\ 3 & 5 & -2 & 4 \end{bmatrix} \qquad \mathbf{B} = \begin{bmatrix} 9 & 5 \\ 2 & 3 \\ 0 & -5 \\ 6 & 1 \end{bmatrix}$$

The entry, $(\mathbf{AB})_{11}$, in the 1st row and 1st column of the product is the dot product of the 1st row of **A** with the 1st column of **B**. This dot product is defined because there are four columns in **A** and four rows in **B**. Thus

$$(\mathbf{AB})_{11} = 1 \cdot 9 + 3 \cdot 2 + 1 \cdot 0 + 0 \cdot 6 = 15$$

Also,

$$(\mathbf{AB})_{12} = 1 \cdot 5 + 3 \cdot 3 + 1 \cdot (-5) + 0 \cdot 1 = 9$$

and the first row of the product has only these two entries. It is helpful to consider the following schematics, which illustrate these two dot products:

$$\begin{bmatrix} \boxed{1 \quad 3 \quad 1 \quad 0} \\ * \quad * \quad * \quad * \\ * \quad * \quad * \quad * \end{bmatrix} \begin{bmatrix} \boxed{\begin{matrix} 9 \\ 2 \\ 0 \\ 6 \end{matrix}} & \begin{matrix} * \\ * \\ * \\ * \end{matrix} \end{bmatrix} = \begin{bmatrix} \boxed{15} & * \\ * & * \\ * & * \end{bmatrix}$$

$$\mathbf{A} \qquad\qquad \mathbf{B} \qquad\qquad \mathbf{AB}$$

$$\begin{bmatrix} \boxed{1 \quad 3 \quad 1 \quad 0} \\ * \quad * \quad * \quad * \\ * \quad * \quad * \quad * \end{bmatrix} \begin{bmatrix} \begin{matrix} * \\ * \\ * \\ * \end{matrix} & \boxed{\begin{matrix} 5 \\ 3 \\ -5 \\ 1 \end{matrix}} \end{bmatrix} = \begin{bmatrix} * & \boxed{9} \\ * & * \\ * & * \end{bmatrix}$$

$$\mathbf{A} \qquad\qquad \mathbf{B} \qquad\qquad \mathbf{AB}$$

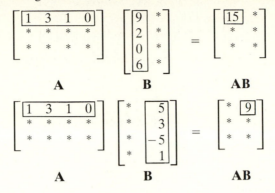

The dot product of the 2nd row of **A** with the successive columns of **B** gives the second row of the product, and the final row of the product is found by taking the dot product of the final row of **A** with the successive columns of **B**. Thus

$$\begin{bmatrix} 1 & 3 & 1 & 0 \\ -1 & 2 & 0 & -1 \\ 3 & 5 & -2 & 4 \end{bmatrix} \begin{bmatrix} 9 & 5 \\ 2 & 3 \\ 0 & -5 \\ 6 & 1 \end{bmatrix} = \begin{bmatrix} 15 & 9 \\ -11 & 0 \\ 61 & 44 \end{bmatrix}$$

$$\mathbf{A} \qquad\qquad \mathbf{B} \qquad = \qquad \mathbf{AB}$$

The product **BA** might not be defined even though the product **AB** is defined. For example, the product **BA** is not defined for the preceding two matrices because the number of columns in **B** does not equal the number of rows in **A**.

A.2 SYSTEMS OF LINEAR EQUATIONS

Matrix notation provides a very compact way of describing a system of linear equations. For example, consider the following system of three linear equations with four variables, x_1, x_2, x_3, and x_4:

$$\begin{aligned} 2x_1 + 8x_2 - 9x_3 + 7x_4 &= 14 \\ - 7x_2 + x_3 + 2x_4 &= 9 \\ 9x_1 - 5x_2 + 11x_3 - 3x_4 &= 17 \end{aligned}$$

Let **A**, **x**, and **b** denote the following matrices:

$$\mathbf{A} = \begin{bmatrix} 2 & 8 & -9 & 7 \\ 0 & -7 & 1 & 2 \\ 9 & -5 & 11 & -3 \end{bmatrix} \qquad \mathbf{x} = \begin{bmatrix} x_1 \\ x_2 \\ x_3 \\ x_4 \end{bmatrix} \qquad \mathbf{b} = \begin{bmatrix} 14 \\ 9 \\ 17 \end{bmatrix}$$

The matrices **x** and **b** only have one column and are therefore vectors. The matrix product **Ax** is defined because the number of columns in **A** equals the number of rows in **x**. **Ax** is the 3×1 matrix

$$\mathbf{Ax} = \begin{bmatrix} 2x_1 + 8x_2 - 9x_3 + 7x_4 \\ 0x_1 - 7x_2 + 1x_3 + 2x_4 \\ 9x_1 - 5x_2 + 11x_3 - 3x_4 \end{bmatrix}$$

Thus, the original system of linear equations can be represented by the single **matrix equation**

$$\mathbf{Ax} = \mathbf{b}$$

This equation states that the entries of **Ax** are equal to the corresponding entries of **b**, which is exactly what the linear system requires.

A.3 THE SUM OF TWO MATRICES

The sum of two $m \times n$ matrices **A** and **B** is also an $m \times n$ matrix denoted by $\mathbf{A} + \mathbf{B}$, and the (i, j)th entry, $(\mathbf{A} + \mathbf{B})_{ij}$, of the sum is

$$(\mathbf{A} + \mathbf{B})_{ij} = a_{ij} + b_{ij}$$

If two matrices do not have the same number of rows and the same number of columns, then their sum is not defined. The following is an example of matrix addition:

$$\begin{bmatrix} 3 & -4 & 5 & 8 \\ 2 & 3 & -8 & 2 \\ 1 & 1 & 4 & 1 \end{bmatrix} + \begin{bmatrix} 9 & 7 & 1 & 2 \\ -1 & -3 & 4 & 6 \\ 0 & 5 & 0 & 9 \end{bmatrix} = \begin{bmatrix} 12 & 3 & 6 & 10 \\ 1 & 0 & -4 & 8 \\ 1 & 6 & 4 & 10 \end{bmatrix}$$

A.4 MULTIPLYING A MATRIX BY A REAL NUMBER

To multiply a matrix **A** by a real number r, we simply multiply every entry of **A** by r and denote the product by $r\mathbf{A}$. For example,

$$\text{if } r = 3 \text{ and } \mathbf{A} = \begin{bmatrix} 2 & 1 & 6 \\ 3 & -2 & 4 \\ 1 & 1 & -1 \end{bmatrix} \quad \text{then} \quad r\mathbf{A} = \begin{bmatrix} 6 & 3 & 18 \\ 9 & -6 & 12 \\ 3 & 3 & -3 \end{bmatrix}$$

A.5 THE TRANSPOSE OF A MATRIX

The transpose of an $m \times n$ matrix **A** is an $n \times m$ matrix denoted by $\mathbf{A}^T$ whose jth column is the jth row of **A**. For example,

$$\text{if } \mathbf{A} = \begin{bmatrix} 2 & 9 \\ 3 & -7 \\ -5 & 6 \end{bmatrix} \quad \text{then} \quad \mathbf{A}^T = \begin{bmatrix} 2 & 3 & -5 \\ 9 & -7 & 6 \end{bmatrix}$$

A.6 SOME SIMPLE MATRIX IDENTITIES

In the following identities r is a real number, and it is assumed that the matrices **A**, **B**, and **C** have the appropriate number of rows and columns so that the indicated operations are defined.

$\mathbf{A} + \mathbf{B} = \mathbf{B} + \mathbf{A}$	Addition is commutative.
$\mathbf{AB} \neq \mathbf{BA}$, usually	Not all multiplications are commutative.
$\mathbf{A}(\mathbf{B} + \mathbf{C}) = \mathbf{AB} + \mathbf{AC}$	Multiplication is distributive.
$(\mathbf{A} + \mathbf{B})\mathbf{C} = \mathbf{AC} + \mathbf{BC}$	
$\mathbf{A} + (\mathbf{B} + \mathbf{C}) = (\mathbf{A} + \mathbf{B}) + \mathbf{C}$	Addition is associative.
$\mathbf{A}(\mathbf{BC}) = (\mathbf{AB})\mathbf{C}$	Multiplication is associative.
$\mathbf{A}(r\mathbf{B}) = r(\mathbf{AB})$	
$(\mathbf{AB})^T = \mathbf{B}^T\mathbf{A}^T$	The transpose of a product is the product of the transposes in reverse order.
$(\mathbf{A} + \mathbf{B})^T = \mathbf{A}^T + \mathbf{B}^T$	The transpose of a sum is the sum of the transposes.
$(r\mathbf{A})^T = r(\mathbf{A}^T)$	
$(\mathbf{A}^T)^T = \mathbf{A}$	

All of these identities can easily be established using only the definitions provided in Sections A.1 through A.5. Each book in the selected references gives a proof of these identities in its first few pages.

SELECTED REFERENCES

Kolman, B., *Introductory Linear Algebra with Applications,* Macmillan, New York, 1976.

Noble, B., and Daniel, J. W., *Applied Linear Algebra,* Prentice-Hall, Englewood Cliffs, NJ, 1977.

ANSWERS TO SELECTED EXERCISES

CHAPTER 2

2.1 (a) If the price falls below \$10, the new optimal vector is $x^* = (3.125, 6.25)^T$.

(b) No, for any price $p > 10$, $20x_1 + px_2$ is maximized at the point $(2.25, 8)^T$.

(c) No.

2.4 (a) The optimal solution is $x^* = (0, 4, 0)^T$.

(b) With the new objective function, all points on the line segment from $(0, 0, 4)^T$ to $(4, 0, 0)^T$ are optimal.

2.11 (a) The graph of the feasible set shows that all points on the line where $x_1 = x_2$ are feasible for all $x_1 \geq 0$ and $x_2 \geq 0$. Along this line the objective function can be made as small as desired by increasing x_1 and x_2.

(b) With the new constraint the optimal solution is $(2, 0)^T$.

2.14 Let x_i be the amount of product i produced.

(a) maximize $x_1 + x_2$

subject to $1.1x_1 + 2.0x_2 \leq 1000$ (lathe)

$3.0x_1 + 4.5x_2 \leq 2000$ (sander)

$$\begin{aligned} &2.5x_1 + 1.3x_2 \le 1500 \quad \text{(polisher)} \\ &x_1 \ge \tfrac{1}{3}(x_1 + x_2) \quad \text{(proportionality)} \\ &x_1 \ge 0, \quad x_2 \ge 0 \end{aligned}$$

(b) maximize $\sum_{i=1}^{200} x_i$

subject to $\sum_{i=1}^{200} r_{ij} x_i \le c_j, \quad j = 1, \ldots, 300$

$$x_1 \ge \frac{1}{3}\left(\sum_{i=1}^{200} x_i\right)$$

$$x_i \ge 0, \quad i = 1, \ldots, 200$$

2.17 (a) The optimal solution is $x^* = (271\tfrac{3}{7}, 0)$.

(b) The redundant constraints are

$$7x_1 + 9x_2 \ge 500$$
$$5x_1 + 3x_2 \le 1500$$
$$2x_1 + 4x_2 \le 1000$$

CHAPTER 3

3.1 (a) minimize $-2x_1 - 3x_2$

subject to
$$\begin{aligned} x_1 + x_2 + x_3 \quad &= 4 \\ -x_1 + 2x_2 \quad - x_4 &= -1 \\ x_i \ge 0, \quad i = 1, \ldots, 4 \end{aligned}$$

(c) Optimal vector $x^* = (0, 4, 0, 9)^T$.

3.5 (a) Optimal vector $x^* = (0, 2.5, 0)^T$. (b) Yes

3.10 (a) Third row; (b) second row; (c) x_2 column, first constraint row; (d) write the x_7 column before the x_1 column.

3.16 (a) $x^* = (0, 3, 0, 4, 1)^T$.

(b) $x^* = (0, 5, 15, 0, 25, 0, 0, 0)^T$.

3.17 Optimal $x^* = (0, -6, -5)^T$.

3.23 (a) False; (d) true; (h) false; (i) false.

3.29 (a)

$$\begin{bmatrix} 1 & 1 & 0 & 0 \\ 0 & \frac{1}{2} & 0 & 0 \\ 0 & \frac{1}{2} & 1 & 0 \\ 0 & -2 & 0 & 1 \end{bmatrix}$$

CHAPTER 4

4.2 (a) There are two optimal extreme points: $(15, 0, 0, 10, 5, 0)^T$ and $(0, 15, 0, 10, 20, 0)^T$.

(b) $v = (1, 0, 0, 1, 0, 1)^T$

4.3 (a) Three

(d) The third best extreme point is $(2, 0, 0, 0, 5)^T$.

4.6 The set of optimal points has the form

$$\left\{ x \,\middle|\, x = \begin{bmatrix} 0 \\ 1 \\ 0 \\ \frac{1}{2} \end{bmatrix} + t \begin{bmatrix} 0 \\ 2 \\ 1 \\ \frac{3}{2} \end{bmatrix} \quad \text{for } t \geq 0 \right\}$$

4.10 (a) There are three optimal basic feasible solutions: $(5, 0, 0, 0, 2)^T$, $(1, 0, 0, 2, 0)^T$, and $(0, 1, 0, 4, 0)^T$.

(b) No

CHAPTER 5

5.1 (a)

$$\begin{aligned} &\text{minimize} \quad 8y_1 + 0y_2 + 12.5y_3 \\ &\text{subject to} \quad \begin{bmatrix} 0 & 2 & 2 \\ 1 & -1 & 1 \end{bmatrix} \begin{bmatrix} y_1 \\ y_2 \\ y_3 \end{bmatrix} \geq \begin{bmatrix} 20 \\ 15 \end{bmatrix} \\ &\qquad\qquad y \geq 0 \end{aligned}$$

(d) \$10

5.3 (b) The optimal dual vector is $y = 5$.

5.7 (a) The minimum price is 2.

(b) Up to 40 units

5.8 (c)

$$\begin{aligned} &\text{maximize} \quad b^T x \\ &\text{subject to} \quad A^T x \geq c \\ &\qquad\qquad B^T x = a \\ &\qquad\qquad x \text{ is free (not constrained to be nonnegative)} \end{aligned}$$

5.12 The optimal vector is $x^* = (\frac{10}{3}, \frac{14}{3}, 0, \frac{1}{3}, \frac{7}{3}, 0, 0)^T$.

5.16 (a) False; (d) true

CHAPTER 6

6.1 (a) The optimal vector is $(13, 3, 26, 0, 0, 27, 2)^T$.

(d) The optimal vector is $(12, 0, 21, 1, 0, 31, 3)^T$.

6.3 (c) The minimum price to charge is 24.

(f) The new optimal vector is $x^* = (5, 0, 25, 0)^T$.

(h) The new optimal vector is $x^* = (5, 15, 0, 0)^T$.

(j) The smallest selling price is 22.

(l) The new optimal vector is $x^* = (10, 5, 0, 0)^T$.

6.4 (e) If the price is not less than 5, the optimal basic sequence does not change. The price can increase indefinitely without changing the optimal vector.

6.5 (b) The new optimal production program is $x^* = (55, 10, 15, 0)^T$.

6.7 (a) No, the new optimal vector is $x^* = (0, 5, 0, 10)^T$.

CHAPTER 7

7.1 (b) Letting x_{ij} be the amount shipped from warehouse i to store j, the optimal solution has $x_{11} = 10$, $x_{22} = 15$, $x_{23} = 5$, $x_{31} = 10$, $x_{33} = 10$, and $x_{ij} = 0$ otherwise.

7.3 (b) The optimal solution has $x_{12} = 10$, $x_{21} = 15$, $x_{32} = 5$, $x_{33} = 5$, and $x_{ij} = 0$ otherwise. Optimal cost $\alpha(x) = 65$.

7.7 (a) False; (c) false

7.16 (b) The optimal solution has $x_{13} = 10$, $x_{23} = 40$, $x_{34} = 25$, $x_{35} = 25$ with flow equal to zero on all other arcs.

CHAPTER 8

8.5 (a) $\mathbf{x}^* = [3, 12]^T$, $z^* = -27$

8.7 No

8.9 (c) $\mathbf{x}^* = [0, 1, 1, 0]^T$, $z^* = -1$

8.11 (b) $z^* = 2$

8.15 (b) 40

8.18 (a) $y_1 + y_2 + \cdots + y_m \leq m - k$

CHAPTER 9

9.5 (a) $[1, 1]^T, [-1, -1]^T$

9.11 (c) $u_1 = -3, u_2 = 1$

9.13 (c) $\mathbf{x}^* = [-1/\sqrt{2}, 1/\sqrt{2}]^T$, $u^* = 1/\sqrt{2}$

9.23 (a) Yes

9.28 (a) $\mathbf{x}^* = [-1/\sqrt{3}, 1/\sqrt{3}, -1/\sqrt{3}]^T$, $u^* = \sqrt{3}/2$

9.34 (b) $\mathbf{x}^* = [\sqrt{2}, \sqrt{2}]^T$

9.38 (a) $\mathbf{x}^* = [1, 0]^T$

9.45 $\mathbf{x}^1 = [\frac{21}{16}, \frac{3}{8}]^T$

9.50 $\mathbf{x}^1 = [\frac{1}{3}, 0]^T$

9.55 $\mathbf{x}^1 = [-\frac{13}{3}, 0]^T$

9.59 (b) $\boldsymbol{\delta}^* = [\frac{1}{2}, \frac{1}{2}, \frac{1}{2}, \frac{3}{4}]^T$

CHAPTER 10

10.1 (a) There are three optimal routes from state 3 to 13, namely,

3 to 5 to 9 to 11 to 13
3 to 7 to 8 to 10 to 13
3 to 7 to 9 to 11 to 13

with an optimal cost of 11.

(b) There are three optimal routes from state 1 to state 8, namely,

1 to 2 to 5 to 8
1 to 2 to 6 to 8
1 to 3 to 7 to 8

with an optimal cost of 7.

10.4 There are two optimal paths from node 1 to node 13, namely,

1 to 2 to 6 to 11 to 13
1 to 4 to 8 to 11 to 13

with a minimum cost of 8.

10.7 The optimal assignment is to assign

2 extra teams to site 1
1 extra team to site 2
0 extra teams to site 3

with a minimum probability of failure of .07.

10.10 The optimal policy is to produce

15 cars in month 1
40 cars in month 2
35 cars in month 3
0 cars in month 4

with a minimum cost of $112,400.

10.17 The minimizing point is $x_1^* = \frac{6}{5}$ and $x_2^* = \frac{18}{5}$ with a minimum value of $-\frac{156}{5}$.

CHAPTER 11

11.1 (a) $1 - e^{-2}$; (b) $e^{-2} - e^{-3}$; (c) $1 - e^{-1}$; (d) $e^{-1} + (e^{-2})/2 + (e^{-3})/6$; (e) 1

11.5 (a) Rate diagram: states (0) (1) (2) (3) ... with arrival rate 3 on each forward transition and service rate 5 on each backward transition.

(b) $P_0 = \frac{2}{5}$, $P_n = (\frac{3}{5})^n P_0$, $n \geq 1$.

(c) $L = \frac{3}{2}$ (d) $W = \frac{1}{2}$ hr, $W_q = \frac{3}{10}$ hr, $L_q = \frac{9}{10}$

11.8 (b) $L_q = \frac{27}{14}$ (c) $\frac{24}{105}$ hr (d) $\frac{9}{14}$

11.10 (a) $P_1 + 2(P_2 + P_3)$ (b) P_3 (c) $1 - P_3$

11.15 (b) $L = 2.006$, $L_q = 0.63$, $W = 0.73$ hr, $W_q = 0.23$ hr

11.19 (a) $\frac{9}{17}$ (b) $\frac{8}{13}$ (c) \$2.36 per hr for Ralph, \$2.31 per hr for Jack. (d) Hire Jack to minimize total cost per hour.

CHAPTER 12

12.1 (a) Optimal order is $Q^* = 816.5$ pads every 4.08 weeks for a minimum total cost of \$144.50 per week.

(b) Optimal order is 1033 pads every 5.16 weeks.

(c) The optimal order size or period does not change. The minimum total cost per week increases by \$20.

12.3 (a) $Q^* = 547.72$ quarts for a minimum cost of \$259.54.

(b) Optimal order quantity is 553.17 quarts and the length of the shortage period is 0.036 months.

12.6 Order $Q_1^* = 140.4$ pallets of classical tile every 0.175 yr (or about 64 days) and order $Q_2^* = 129.8$ pallets every 0.108 yr (or about 39 days).

12.9 (a) For $n = 3$, states $s = 0, 1, 2, 3$ and decisions $x_3 = 3, 2, 1, 0$. For $n = 2$, states $s = 0, 1, 2, 3, 4$ and $x_2 = 4, 3, 2, 1, 0$. For $n = 1$, $s = 0$ and $x_1 = 4, 5, 6, 7, 8$.

(b) Optimal $x_1^* = 5$, $x_2^* = 0$, $x_3^* = 3$ with $f_1(0) = 71$ as the optimal cost.

12.13 (a)

$$P(D) = \begin{cases} 5D - 2y & \text{if } y > D \\ -2D + 5y & \text{if } y \leq D \end{cases}$$

(b)

$$\int_{1000}^{y} \frac{5z - 2y}{1000}\, dz + \int_{y}^{2000} \left(\frac{-2z + 5y}{1000}\right) dz$$

(c) $y^* = 1714$

12.16 (a) 15 cookies (b) $(s, S) = (5, 15)$ so if $x < 5$, then order $15 - x$, otherwise order 0.

CHAPTER 13

13.3 (b) $L = \frac{98}{94}$

13.5 (b) 22

13.9 $r_5 = 0.2135$

13.11 $x = a - b \ln\left(\dfrac{1 - u}{u}\right)$

13.12 $I(1, 2) \approx 0.4$

INDEX